BETWEEN SCIENCE AND INDUSTRY

Institutions in the History of Materials Research

A World Scientific Encyclopedia of the Development and History of Materials Science

---

Published

Between Science and Industry: Institutions in the History of Materials Research
Editor-in-Chief: Arne Hessenbruch
Edited by: Robert P Crease
ISBN: 978-981-12-8433-5

Between Nature and Society: Biographies of Materials
Editor-in-Chief: Arne Hessenbruch
Edited by: Bernadette Bensaude-Vincent
ISBN: 978-981-12-5174-0

Between Making and Knowing: Tools in the History of Materials Research
Editor-in-Chief: Arne Hessenbruch
Edited by: Joseph D Martin, Cyrus C M Mody
ISBN: 978-981-12-0762-4

A World Scientific Encyclopedia of the Development and History of Materials Science – Vol. 3
Editor-in-Chief: Arne Hessenbruch

BETWEEN SCIENCE AND INDUSTRY

Institutions in the History of Materials Research

Editor

Robert P Crease
Stony Brook University, USA

World Scientific

NEW JERSEY • LONDON • SINGAPORE • BEIJING • SHANGHAI • HONG KONG • TAIPEI • CHENNAI • TOKYO

Published by

World Scientific Publishing Co. Pte. Ltd.
5 Toh Tuck Link, Singapore 596224
USA office: 27 Warren Street, Suite 401-402, Hackensack, NJ 07601
UK office: 57 Shelton Street, Covent Garden, London WC2H 9HE

Library of Congress Control Number: 2024931690

British Library Cataloguing-in-Publication Data
A catalogue record for this book is available from the British Library.

BETWEEN SCIENCE AND INDUSTRY
Institutions in the History of Materials Research

ISBN 978-981-12-8433-5 (hardcover)
ISBN 978-981-12-8434-2 (ebook for institutions)
ISBN 978-981-12-8435-9 (ebook for individuals)

For any available supplementary material, please visit
https://www.worldscientific.com/worldscibooks/10.1142/13625#t=suppl

Desk Editor: Joseph Sebastian Ang

Typeset by Stallion Press
Email: enquiries@stallionpress.com

About the Contributors

Iwnetim (Tim) Abate is an Assistant Professor in the Department of Materials Science and Engineering at the Massachusetts Institute of Technology. Previously, he was Miller and Presidential Post-Doc Fellow at UC Berkeley and received his PhD in Materials Science and Engineering from Stanford University. He is also a Co-founder and President of a non-profit organization, www.scifro.org, working on empowering the African youth to solve local problems through scientific research and innovation.

Cristoforo Benvenuti graduated in physics from Milano University and entered CERN in 1966, where he worked until retirement in 2005. Active in technologies such as cryogenics, vacuum and superconductivity, he is the author of many patents and innovations, some of which are described in this book. His work has been rewarded with prizes, including the European Achievement Prize (1998) and the Gaede–Langmuir Award (2002). In 2005, he created a company, SRB Energy Research, to develop and market a high-performance thermal solar panel he had patented in 2003.

Luca Bottura has trained as a Nuclear Engineer at the Engineering Faculty of the University of Bologna (Italy) and was awarded a PhD from the University College of Swansea (Wales, UK) for the physical modeling, scaling and numerical analysis of quench in large-scale, force-flow cooled superconducting coils. He has experience in the design and testing of superconducting cables and magnets for fusion (NET and ITER), as well as accelerator magnet design, manufacturing and testing (CERN LHC, High Field Magnets). He is the author of computational methods used for magnet design and analysis, and from 1995 to 2004, he created the Field Description for the LHC (FiDeL), now an embedded system of the LHC control of field quality and corrections. He presently contributes to magnet R&D at CERN, especially for muon colliders, and the development of accelerators for particle therapy.

Panos Charitos is currently the Scientific Information Manager at CERN's Experimental Physics Department. He is the Editor-in-Chief of CERN's EP Newsletter and Co-editor of Accelerating News. Panos acted as spokesperson for the Future Circular Collider (FCC) collaboration during the Conceptual Design Report Phase (2015–2019), coordinating the communication activities for the project and the supporting H2020 projects. Presently, he co-chairs the communication group overseeing the communication activities and the dissemination of the scientific results of the FCC Feasibility Study. Panos joined CERN in 2011 as a Member of the ALICE experiment while previously he held various positions in the publishing and media industry. He holds a bachelor's and a master's degree in Astrophysics from Imperial College London and an MSc in Media and Communications from the London School of Economics.

Samuel Chigome is a Senior Researcher in the Nanomaterials Division of the Botswana Institute for Technology, Research and Innovation (BITRI). In 2017, he set up the Secretariat Office of the African Materials Research Society (African MRS) in Botswana and hosted at BITRI. Subsequently, he took up the successive roles of Vice President and President of the African MRS over the period 2017–2022. In 2021, he set up the first ever African Regional Office of the International Union of Materials Research Societies (IUMRS), which is hosted in Botswana at BITRI.

Katharina C. Cramer is a historian and social scientist, specializing in the history, politics and organization of Big Science and Research

Infrastructures. She is the author of *A Political History of Big Science: The Other Europe* (Palgrave) and Co-editor of *Big Science and Research Infrastructures in Europe* (Edward Elgar). Katharina is currently a Fellow at the universities of Bonn and Lund.

Robert P. Crease is a former Chair of and Professor in the Department of Philosophy at Stony Brook University, where he teaches history and philosophy of science. He works on the history of physics institutions, especially Brookhaven National Laboratory; his books include *Making Physics: A Biography of Brookhaven National Laboratory, 1946–1972* (Chicago: University of Chicago Press, 1999), and *The Leak: Politics, Activism, and Loss of Trust at Brookhaven National Laboratory* (with Peter D. Bond, MIT Press, 2022). In 2021, he was awarded the William Thomson, Lord Kelvin Medal and Prize (Institute of Physics) for "describing key humanities concepts for scientists, and explaining the significance of key scientific ideas for humanists." For the past 24 years, he has written a monthly column, "Critical Point," for *Physics World.*

Olof Hallonsten is a sociologist of science at the Lund University School of Economics and Management in Sweden. His work mainly concerns the institutional and historical development of the science–society interface, and he has previously written several books and articles on the organization and funding of synchrotron radiation and neutron facilities in Europe and the United States.

Armando Fernández Guillermet is a PhD and Docent from the Royal Institute of Technology (Stockholm, Sweden) and Senior CONICET Researcher at Centro Atómico Bariloche (CAB-CNEA, Argentina). He was awarded the Professor Emeritus title at the National Universities of Cuyo (UNCUYO) and of Comahue (UNCO) in Argentina. He has academic interests in the fields of materials science, education, humanities and social sciences.

Thomas Heinze is a Professor of Organisational Sociology at the Bergische Universität Wuppertal in Germany. He has published several studies on the funding system for scientific research and the institutional conditions for creativity and breakthroughs in research, including comparative studies of Germany and the United States.

Klaus Hentschel studied physics, philosophy and history of science at the University of Hamburg. Since 2005, he is Full Professor and Head of the Section for History of Science & Technology at the University of Stuttgart. He has published 10 books (on the theory of relativity, the Einstein Tower, spectroscopy, materials science, visual science cultures, physics in the Nazi period and on the mentality of German physicists after 1945), 24 anthologies and editions, and more than 150 papers. For further information and downloads, see https://www.researchgate.net/profile/Klaus-Hentschel.

Linn Walker Hobbs OBE is Professor Emeritus in the Departments of Materials Science & Engineering and Nuclear Science & Engineering at the Massachusetts Institute of Technology and was the inaugural holder of the John F. Elliott Chair in Materials. He began his education and research career in the early years of the Department of Materials Science & Engineering at Northwestern University, under Professors Morris Fine, Johannes Weertman and Lyle Schwartz, and pursued his doctoral studies in electron microscopy of alkali halides at Oxford University's Department of Metallurgy (later Department of Materials) under Professor Sir Peter Hirsch FRS, as a Marshall Scholar. He remained in the UK for a decade, successively as NSF Postdoctoral Fellow, Research Fellow of Wolfson College, Oxford under the inaugural presidency of Sir Isaiah Berlin CH — serving also as college wine steward — and a section leader at the Atomic Energy Research Establishment at Harwell. Returning to the US, he was Associate Professor of Ceramic Science at Case Western Reserve University prior to joining the MIT faculty, where he served as Associate Chair of the Institute faculty. He has been a director of three materials societies: the *Materials Research Society*, the *American Ceramic Society*, and the *Microscopy Society of America*, is a fellow of the last two societies, and served as President of the last. His research and teaching interests are in the areas of radiation effects in non-metals; nuclear waste remediation; hightemperature corrosion of metals and alloys; cryogenic low-dose transmission electron microscopy;

amorphization and atomistic structures of non-crystalline solids; mechanisms of bone formation; bone-tendon junctions and biomaterials for orthopaedic prosthesis; materials archaeology of historic cementitious materials and natural polymer fibers; histories of materials technologies in the human experience; oenology; amateur radio; forte pianos; and 17th–19th century English horology. He was made an Officer of the Order of the British Empire by Queen Elizabeth II in 2001.

Thomas Kaiserfeld is a Professor of Technology and Social Change at the Department of Thematic Studies, Linköping University, Sweden. His research focuses primarily on historical and institutional aspects of science and technology. He is the author of *Beyond Innovation: Technology, Institution and Change as Categories for Social Analysis* (Basingstoke, 2015).

Hiroto Kono is a curator of the history of science and technology at the National Museum of Nature and Science, Tokyo, and a PhD candidate at the Tokyo Institute of Technology. His research focuses on the history of the sciences of matter, such as statistical and condensed matter physics and materials science in the twentieth century. His study on the Japanese theoretical physicist Ryogo Kubo (1920–1995) examined the trajectory of the physicist's learning and research in wartime by drawing from his unpublished manuscripts. His PhD project focuses on the development of the sciences of matter in Japan, especially on the history of *Busseiron*, which emerged in wartime and later developed into a major sub-discipline of physics.

Paul Lecoq received his diploma as Engineer in Physics Instrumentation at the École Polytechnique de Grenoble in 1972 under the leadership of Nobel Laureate Louis Néel. After two years of work at the Nuclear Physics laboratory of the University of Montreal, Canada, he got his PhD in Nuclear Physics in 1974. Since then he has been working at CERN in five major international experiments on particle physics, one of them led by Nobel Laureate Samuel Ting. His action on detector instrumentation, and particularly on heavy inorganic scintillator materials, received strong support from the Nobel Laureate, Georges Charpak. He has been the Technical Coordinator of the electromagnetic calorimeter of the CMS experiment at CERN, which played an important role in the discovery of the Higgs boson. He is the Founder of the CERN-based International Crystal Clear Collaboration regrouping 31 institutes and companies worldwide contributing to the development of scintillator science. He also created the SCINT conference series in 1991, which gathers every second year the international community working on fundamental aspects, production technologies and applications of scintillators. Member of a number of advisory committees and of international societies, he was the promoter of the CERIMED initiative (European Center for Research in Medical Imaging, installed in Marseille) for networking physics and medicine in the field of medical imaging in 2002.

Liu Lin is the Head of the Accelerator Division at the Brazilian Synchrotron Light Laboratory (LNLS). She completed undergraduate studies in physics at the Federal University of Paraná and master's studies in X-ray optics at the University of São Paulo. Her PhD at the University of São Paulo investigated accelerator physics issues applied to the first synchrotron light source built in Brazil, UVX. She has worked at LNLS since 1986 as an accelerator physicist and participated in the teams that designed, constructed, commissioned, and operated the two Brazilian synchrotron light sources. Her current interests focus on optimizing the performance of the fourth-generation light source, SIRIUS.

Joseph D. Martin is an Associate Professor in the Department of History, Durham University and the Co-editor-in-chief of *Physics in Perspective.* His research focuses on the history of solid state and condensed matter physics and materials science in the 20th century. He is the author of *Solid State Insurrection: How the Science of Substance Made American Physics Matter* (Pittsburgh, 2018) and co-editor, with Cyrus Mody, of *Between Making and Knowing: Tools in the History of Materials Research* (World Scientific, 2020).

W. Patrick McCray is a Professor in the History Department at the University of California, Santa Barbara. He is the author of five books including, most recently, *Making Art Work: How Cold War Engineers and Artists Forged a New Creative Culture* (The MIT Press, 2020).

Michael Moll is working on the characterization and development of radiation hard silicon detectors. He obtained his PhD from the University of Hamburg in 1999 and then went to CERN to work on the operation of radiation facilities, development of radiation sensors and the study of radiation damage in silicon detectors. Presently, he is a senior staff in the Detector Technologies Group and since 20 years Co-spokesperson of the CERN RD50 Collaboration "Radiation hard semiconductor devices for very high luminosity colliders." He has extensive expertise in radiation monitoring and radiation damage to semiconductor devices with more than 140 publications on the subject.

Wei Qian is a Professor and the Director of the Institute of Cultural Heritage and History of Science & Technology, University of Science and Technology, Beijing. He focuses on research on the history of metallurgy and material sciences, and industrial heritage, and also takes active part in programs on science, technology and society. He received his doctoral degree in History of Science and Technology from the same university in 2001 and spent his time in Princeton for a Fulbright Scholarship Program from August 2014 to July 2015.

Terry Quinn graduated in physics from Southampton University in the UK in 1959 and then moved to Oxford for his PhD in what was then the Metallurgy Department. His thesis supervisor was Professor William Hume-Rothery FRS. In 1962, he joined the National Physical Laboratory in Teddington where he worked on high-temperature standards. He later made an acoustic determination of the Gas Constant. In 1967–1968, he spent a year at the then National Bureau of Standards in Washington. He moved to the BIPM in Paris in 1977 as the Deputy Director and became Director later in 1988. He was Lady Margaret Distinguished Visiting Fellow at the Cavendish Laboratory Cambridge in 1983–1984. In 2002, he was elected a Fellow of the Royal Society and on his retirement at the end of that year was made a CBE by the Queen. He has continued to be active in international affairs in metrology. He has published more than 150 papers in scientific journals and two books, *Temperature* (Academic Press London, 1983, and 2nd edition in 1990) and *From Artefacts to Atoms: The BIPM and The Search for Ultimate Measurement Standards* (Oxford University Press, New York, 2011).

Brit Shields is a Senior Lecturer and Director of the Engineering Ethics Initiative in the School of Engineering and Applied Science at the University of Pennsylvania. Shields is a historian and sociologist of science, mathematics and technology. She researches and writes about the cultural history of scientific communities, institutions and ideas.

Wole Soboyejo is the President of SUNY Polytechnic Institute. Prior to joining SUNY Polytechnic Institute in October 2023, he served as the Interim President, Provost and Bernard M Gordon Dean of Engineering and Engineering Leadership at Worcester Polytechnic Institute 2016–2023. The Senior Vice President and Provost at the Worcester Polytechnic Institute (WPI). Prior to joining WPI as the Bernard M. Gordon Dean of Engineering and Engineering Leadership in 2016, he was a Professor at the Princeton Institute of Science and Technology of Materials (PRISM) and the Department of Mechanical and Aerospace Engineering at Princeton University (1999–2016), where he established the NSF-funded US/Africa Materials Institute (USAMI). He is also one of the founders of the African Materials Research Society (AMRS), the Nelson Mandela Institutions in Abuja and Arusha, and the Mathematics and Science for Sub-Saharan Africa (MS4SSA) program.

Guillermo Solórzano is a Professor of Materials Science and Engineering at the PUC-Rio de Janeiro. He is currently an Associate at the School of Engineering and Applied Science, Harvard University. He is the Founding President of the Brazil MRS, the Former President of the Brazilian Society of Microscopy and Microanalysis, and also the Inter-American Society of Electron Microscopy. He was formerly a Board Member of the International Federation of Microscopy Societies, a Member of the Senate International Union of Materials Research Societies (IUMRS), a Visiting Scientist at the Max Plank Institute, Stuttgart, and at the Institute Nationale Politechnique de Grenoble (1990–1991), a Visiting Faculty at MIT (1991–1994), a Thinker/Visiting Professor at Stanford University (2008), the Director of the International Relations Committee of the Materials Research Society (MRS) (2004–2006). In 2017, he was elected a Fellow of the Microscopy Society of America (MSA).

Daniel Ucko is the Head of Ethics and Research Integrity for the American Physical Society journals. Previously, he has worked as part of the editorial staff of *Physical Review Letters* and as a research fellow at the University of Birmingham, UK. He holds a PhD in Condensed Matter Physics from University College London, UK, and a PhD in Philosophy of Science from Stony Brook University. He is the author of the forthcoming book, *The Subjectivities and Objectivities of Peer Review*, which will be published by Brill in 2023.

Catherine Westfall is a retired Michigan State University Professor. With co-authors, she has published two books, *Critical Assembly: A Technical History of Los Alamos During The Oppenheimer Years, 1943–1945* and *Fermilab, Physics, the Frontier, and Megascience.* She has also published more than two dozen articles that focus on the development of US laboratories and the national laboratory system. Most recently, she has worked on the rise of the New Big Science with the advent of large materials science accelerators.

Introduction: Materials Science Institutions

Robert P. Crease

Institutions are often overlooked in the history of science. They are treated as backdrops to the experiments and discoveries comprising the history of science itself. But institutions are indispensable and fundamental features of research; if you picture scientific experiments as performances staged to further research, then institutions provide the theaters, the staging, props, and trained performers necessary to produce such performances. The word "institution" in this volume is therefore conceived in a broad sense: as a basic physical, organizational, educational, or regulatory structure needed for the operation of research.

Materials science institutions are different from other kinds of science institutions in several respects. The first reason is their origin stories. To paraphrase Dan Ucko in his article about materials science journals, some were born that way, some became that way, and some had it thrust upon them. The reason is that materials science as a unified discipline is of recent vintage and that it appeared around the time of the birth of Big Science. Many large scientific institutions, such as networks of national laboratories, had already been created, and charged with missions, in a variety of fields. The federal government's rapidly growing interest in materials research therefore meant that the research had to be rapidly integrated into this already existing network — which had to take place sometimes from scratch, sometimes by evolution or cooperative engagement, and sometimes by fiat. The "Rule of Three," as one might call it, therefore shows up in the origin stories of many types of materials science institutions.

The second reason why materials science institutions are different is related to the diversity of their patrons, clients, dynamics, and products, as well as to the fact that they must be responsive to things, such as national policy, industrial requirements, and commercial desires. National laboratories engaged in materials research tend to be mission-driven and respond to government priorities. Metrological laboratories have (until recently) focused on making standards and testing materials. Industrial labs are product-driven and can target single materials (such as Corning) or a range of them (IBM). Military labs, meanwhile, study defence-related materials and devices.

The third reason for the difference is that materials science institutions are linked in especially strong and complex networks together with other institutions, including funding agencies, professional societies, educational institutions, and journals. Materials science institutions therefore come in various "nodes" in these networks, including the following: facilities, or the large-scale, historically evolving establishments specially suited to materials science research; laboratories, or the physical "staging grounds" of research; university centers and funding agencies that make possible the education and training of researchers; professional societies and publications that provide regulative and networking support and disseminate information. But these types overlap, as reflected in many of the articles in this volume. Moreover, while many materials science institutions are internationally networked and have developed and are structured similarly, others evolved initially in isolation and have a deep independent history.

This volume is organized into seven parts. The first covers the historical evolution and disciplinary formation of materials science institutions. The second and third parts treat examples of this evolution of three different types — neutron, synchrotron, and spallation sources — and of three different classes — industrial, national, and individual labs — of facilities. The fourth and fifth parts deal with different kinds of institutional patrons and promoters: university centers, funding agencies, professional societies, publications, and conferences. The sixth and seventh parts are about networks of materials science institutions and about institutions, whose principal role is to coordinate other materials science institutions.

This volume is not an encyclopedia, for it would be impossible to cover every one of the numerous institutions in each category. The chapters therefore often present paradigmatic examples of institutions in each part, which reveal characteristic paths, features, or behaviors of the institutions covered. In Part I, for instance, the development of European synchrotron radiation sources is used to illustrate the developmental and disciplinary theses of the other two. Other parts use specific facilities, nations, or networks as examples. This volume therefore seeks a comprehensive integrity by including articles on the templates of each category, and exemplary cases in each, so as to illustrate the formative features and developments of institutions in them. For instance, this volume includes articles illustrating the process as it took place in Japan, China, and Africa. Certain international institutions, most notably the International Bureau of Weights and Measures (BIPM), moreover, were created to coordinate national institutions and thus might be considered institutions of institutions.

Scientific institutions strive to be both stable and flexible. On the one hand, they seek to allow research to be performed without interference, and thus they need to be supportive and remain in the background as much as possible. On the other hand, they must be highly active and anticipate and adapt to the latest needs of research and researchers; to technical, industrial, and funding resources; and to the surrounding social and political climate. This balancing act is particularly challenging for materials science institutions because of the diversity of their patrons, clients, dynamics, and products. Furthermore, networks of these institutions also rapidly evolve, with existing networks sometimes changing focus or disappearing and new networks sometimes springing up, such as in nanotechnology.

Most of the articles in this volume focus on institutions in the second half of the 20th century after materials science began to emerge as a discipline and therefore as warranting specialized institutions. As the articles by Ucko, McCray, and others reveal, in the 21st century the development of things such as biomaterials, nanomaterials, and 2D and quantum materials, as well as the blending of the border between 'hard' and 'soft' materials, with polymers the transition, may eventually cause materials science institutions to lose their distinctive character, therefore providing this volume with yet another kind of unity and comprehensiveness.

Contents

Part I
Historical Evolution and Disciplinary Formation

Part I consists of three chapters addressing different facets of the evolution of materials science institutions and the disciplinary formation that made them possible. Until recently, materials such as ceramics, glasses, semiconductors and metals were produced with significant amounts of practitioner know-how, and their production was studied separately with different instruments and theories. Only in the past few decades did those fields draw together into a more or less single and coherent field of science, though with varying scope and given different names, such as solid-state physics, condensed matter physics, and materials science. How and why were institutions implicated in this disciplinary consolidation, and how did it justify, in turn, creating a distinct kind of institution to foster it?

The first chapter, "Institutions and Technology in Materials Research" by Thomas Kaiserfeld, takes up the chief factors shaping materials science institutions: the technologies that provided access to knowledge about materials, the social and political pressures on the research, the changing availability of funding, and the social concerns about missions and safety. Institutions, Kaiserfeld argues, not only provided the sites for integrating those factors into materials science research but also developed norms and traditions necessary to foster its continuance. "If technology is the hardware, institutions are the software," writes Kaiserfeld. He concludes the chapter by picking out several institutional features that characterize materials science institutions, including their historically layered research tools and theories, their cross-disciplinary application-driven efforts, and their core cluster of subjects.

If Kaiserfeld describes the development of horizontal structures of materials science institutions — their relation to technology, materials, demand, appeal, and dependence on social consent — Klaus Hentschel focuses on vertical structures, on the developments that took materials science from "scattered research on diverse materials" to "a concerted quasi-discipline of materials science." While his chapter, "From Materials Research to Materials Science: Disciplinary and Institutional Developments," focuses mainly on Germany and the United States, Hentschel argues that the institutions in those countries played a leading role in shaping the eventual discipline and modeled developments elsewhere. In the US, materials science achieved disciplinary status after pressures exerted by institutions such as the Department of Defense and the Advanced Research Project Agency (ARPA), as well as by initiatives such as the American Presidential Science Advisory COSMAT (Committee on the Survey of Materials Science and Engineering) Report. Hentschel also examines and evaluates the spectrum of different positions taken on the disciplinary status of materials science. His appraisal is that the final stage to that status was signaled when institutions started dropping the final "E" from MSE — Materials Science and Engineering. Thanks to the role of institutions in coordinating various kinds of knowledge to various kinds of demands, materials science has "achieved disciplinary status in a historically rare and enduring process of discipline formation through 'emergence by integration' rather than by differentiation."

The third chapter in this part, "Institutional Development of the Organizational Field of Synchrotron Radiation Facilities in Europe," by Olof Hallonsten and Thomas Heinze, uses the development of synchrotron radiation sources to illustrate the concepts of institutional analysis and organization theory as these apply to materials science institutions. Synchrotron sources nicely exhibit what can be called the "Rule of Three" in the history of materials science institutions: some were created from scratch (from the second generation on), some were converted (from obsolete high-energy equipment), while in some cases they were the result of still-viable high-energy accelerators being given a new mission. Hallonsten and Heinze discuss how most synchrotron radiation facilities were established by and within existing institutions and how they quickly grew out of those institutions to form a new organizational field requiring a new kind of institution. This development then helped consolidate technical designs and user interfaces, transformed organizational designs and practices, and ultimately caused the facilities to mirror each other technologically and structurally.

Chapter 1.1

Institutions and Technology in Materials Research

Thomas Kaiserfeld

The relation between the material or physical world and the immaterial one is an old philosophical problem engaging thinkers such as Descartes, who in the 17th century claimed that the essence of matter was its spatial extension. After Newton had formulated his laws of gravitation and motion a few decades later, material ontologies started to gain popularity, the perhaps most common exponent of this shift being the watchmaker analogy for God. Since then, a number of social and cultural processes, mechanization, secularization and industrialization have contributed to a slowly growing obsession with material phenomena. Madonna has become mainstream when she claims "that we are living in a material world."[1]

Leaving philosophy and popular culture behind, material and immaterial objects, entities or phenomena, although intimately linked to each other, still constitute a dichotomy. In the social sciences, a similar conceptual pair can be found in the notions of technology and institutions where the former is often used to denote material aspects and the latter immaterial. A standard definition of institutions is given by Douglass North as "the rules of the game in a society, or more formally, [...] the humanely devised constraints

[1] Brown and Rans (1984).

that shape human interaction."[2] If technology is the hardware, institutions are the software. In materials research, it is clear that both technology and institutions need to be taken into account when understanding the development of the field.

In order to separate out material from matter, the former has been defined as "a substance which is useful or of value for human purpose."[3] This notion of materials complicates things since both usefulness and value are relative to context. Historically, categories of materials judged the most useful or most highly valued have certainly varied. Lewis Mumford has used this insight to divide the history of the Western world into three phases, each depending on the dominating materials.[4] In the eotechnic era (1000–1800), wood and organic fibers dominated making agriculture, and sloyd common modes of production, with the family as the most important entity. In the paleotechnic age (1700–1900), metals dominated making mass production in factories the characteristic form of production according to Mumford. Finally, in the neotechnic phase (from 1900), the mixing of materials and methods led to composites relying on scientific analyses of different material properties and making the laboratory central to production.

Mumford's perspective on the neotechnic phase seems influenced by the early 20th-century evolution of materials science from metallurgy, which had been methodologically enriched by the introduction of X-ray diffraction used to visualize and thus better understand the structure of different metals on the atomic level.[5] The result was the formation of a new research field, X-ray crystallography, only one of the many examples of how technology supplied the tools needed to acquire knowledge about materials. Another important vein for the nascent materials science was chemistry and the efforts to make plastics from synthetic components, something the Belgian chemist Leo Baekeland achieved in 1907 by controlling

[2] North (1990, p. 3).

[3] Bensaude-Vincent (2001, pp. 223–248).

[4] Mumford (1934).

[5] Bensaude-Vincent (2001, p. 225).

temperature and pressure when studying reactions of different proportions of phenol and formaldehyde. Throughout the 20th century, technology became ever more crucial for materials science.

During WWII and the Cold War, materials research developed through the promotion of some of the most powerful of institutions: the military–industrial complex, complemented by academic research and education. It is no coincidence that the term "materials science" was coined in the early 1960s, just as the President's Science Advisory Committee had identified materials as a research priority and the US Department of Defense contracted a handful of American universities to study materials for weapons and aerospace applications.[6] If anything, this is an indication of the importance of supporting institutions for the formation of new research fields, especially those depending on advanced research tools, specialized knowledge and practical skills as is the case for materials science.[7]

The crucial aspect here is the accessibility to technologies in the form of advanced research tools. Although other components of materials science surely demand funding, some of the experimental technologies in demand, for instance, high-intensive neutron beams, need to rely on resources on a much higher scale. Here, support from strong institutions is of the essence. Only by creating norms, for instance, expressed through the formation of traditions or organizational support, is it possible to gather the resources necessary to develop and realize the experimental technologies identified as promising.

Institutional work with this aim often needs to be performed on a multitude of levels. Since much research relies on public funding, political lobbying is often crucial. Political consent is also required for changes of regulation that may hinder certain research efforts such as human embryonic stem cell research or technologies for gene editing. Engaging resource-intensive sectors of society such as the military or industry, where there is a demand for the potential, but not yet realized benefits, is the second form of institutional work.

---

[6] Leslie (1993).

[7] McCray (2005, pp. 177–203).

In addition, it has been proven effective to launch public campaigns, whether general regarding visionary aspects of a technology such as controlled energy production by fusion or more specific regarding a certain problem such as Kennedy's promise in May 1961 to land a human on the moon and have them return safely to the earth before the decade was out. The fourth form of institutional work can be performed in the legal system, for example, to solve patent disputes or other legal matters that may influence the possibility of pursuing specific research problems. In practice, these forms are hard to separate out and often influence each other so that public campaigns act on policymaking, which in turn influences sectors such as the military, or the other way around.

The tight connections between technology and institutions and how they act in tandem are evident from the rise of nanotechnology as a concept in the 1990s. As Cyrus Mody has shown, the technological development in the 1960s and 1970s of scanning tunneling microscopy combined with a community of users in the 1980s distributed over academic departments and the industry contributed to the institutionalization of nanotechnology as a relevant field of knowledge and practice.[8] Materials science and its derivatives are never the consequence of technology or institutions, but always of both.

As in the case of materials science in the early 1960s, the resulting National Nanotechnology Initiative announced by President Clinton in 2000 and the subsequent tsunami of research funds secured the establishment of the label as well as the diverse research methods and problems it represented.[9] In this context, the rhetorical importance of a narrative that mirrored the development from materials science and its practitioners' ability to create new materials and substances to the "molecule-by-molecule control" popularly defining nanotechnology should not be underestimated. Neither should the end of the Cold War and a more general shift in global

[8] Mody (2011).

[9] McCray (2005, p. 179).

research policy agendas from a focus on military interests to commercial ones.[10]

Historically, the presentation of popular visions including new technologies has proved to be an efficient way to secure research funding. The examples abound from the civilian use of nuclear power in the "Atoms for Peace" program of the 1950s to the present buzzword "NewSpace" for the growing privatization of spaceflight. More specifically, as Arne Hessenbruch has claimed using the example of nanotechnology, the harder the practical difficulties are to realize an emerging technology, the more promising the long-term benefits need to be in order for the efforts to seem worthwhile.[11]

As has been argued, external relevance is of the essence to all knowledge-seeking processes that require resources beyond individual curiosity. Furthermore, some research problems call for more support than others. Clearly, research into materials has depended heavily on the relevance to institutions external to the community of scientists, irrespective of whether the means are political decision-making or public campaigning and the ends war or money.

Having said this, it is equally important to note that science itself may be called an institution with its own norms and traditions. Being part of this institution of science, the internal conditions specific to materials science should not be neglected. Following are also some institutional features that can be said to characterize materials science.

First of all, the general story of materials science can be said to include a number of historically layered research tools and theories.[12] The development of X-ray diffraction and subsequently X-ray crystallography in the early 20th century has already been mentioned. Another important contribution to our understanding of materials layered on to this methodology was the band theory of metals in the 1930s, which in combination with quantum theory

---

[10] Bensaude-Vincent (2016, pp. 45–60).

[11] Hessenbruch (2004, pp. 135–144).

[12] Regarding the concept of layering in the framework of new institutionalism, see Wolfgang Streeck and Kathleen Thelen (2005, pp. 3–39).

could later be used for the creation of yet another layer of knowledge about semiconductors and ceramics. Solid-state physics and chemistry have thus been important foundations for knowledge about materials. What the future holds is a different matter although further layering of theories of biomaterials and nanotechnical aspects is a bet as good as any.[13]

Secondly, materials science as well as nanotechnology have been labeled multidisciplinary (meaning researchers from different disciplines working together to solve problems that demand knowledge from a combination of research fields), thus constituting a common ground for cross-disciplinary efforts, at least in American contexts. Here, the creation and funding of interdisciplinary labs (where knowledge and methods from different disciplines were supposed to be integrated to form syntheses of approaches) in the early 1960s promoted the crossing of disciplinary boundaries at universities. An important consequence was the creation of interdisciplinary courses. Teaching and training were essential for the development of common concepts used when studying different types of materials. This model, however, did not conquer the world. In Europe, for instance, materials research was more commonly conducted within the framework of traditional disciplinary science forming sub-disciplines, which in some cases also invigorated traditional research orientations.[14]

Implicit in the notion of multidisciplinarity (and also interdisciplinarity) is that research conducted in this vein, including materials science, is more problem-driven than even disciplinary organized research efforts. Behind the rationale is the idea that multidisciplinary research has a larger palette of perspectives and methods to draw from and is thus better suited to finding solutions to problems not generated within a disciplinary tradition or frame of mind. In the case of nanotechnology, the ambitions and hopes for interdisciplinarity were raised even higher involving engineering and to some extent also social sciences. Concepts mirroring these developments

[13] Bensaude-Vincent and Hessenbruch (2004, pp. 345–347).

[14] Bensaude-Vincent (2016).

such as "Converging Technologies" were launched.[15] Whether this reasoning holds water, or not, when confronted with empirical evidence is still an unknown, both in the general case and more specifically regarding materials science.

Thirdly, as an alternative to the historical layerdness and the presently connoted multidisciplinarity or interdisciplinarity of materials science, a characterization has been offered, in which its core is called a cluster of subjects (rather than problems) taken from disciplines, such as physics and chemistry.[16] With such a perspective, it seems close at hand to view the emergence of the field as some sort of labeling process in order to achieve raised research resources for a certain type of research tools and problems. If so, the method has indeed proved to be successful in promoting research funding for those activities that have gone under the heading of materials research in the 1960s and nanotechnology 40 years later.

Much of the accomplishments should be attributed to the flexibility and dynamic character of the labeling process. Over time, materials science has proved to be able to assimilate a number of different research problems and traditions as well to incorporate several methods and technologies. This flexibility and dynamics have resulted in a heterogeneous field of knowledge depending not only on geopolitical or professional contexts and traditions but also on an ability to expand by importing relevant novelties in terms of potential applications irrespective of their disciplinary origin. This guiding principle has been formulated so as to "catch the latest hype wave."[17] Admittedly tendentious, such a portrayal still hits the nail on the head. Given the course of global research policies over the past decades, this does not need to be read as altogether negative for materials science practitioners but rather the reverse, at least from an institutional point of view.

---

[15] Ibid., pp. 51–52.

[16] Bensaude-Vincent and Hessenbruch (2004, p. 347).

[17] Hessenbruch (2020).

References

Bensaude-Vincent, B. (2001). The construction of a discipline: Materials science in the United States. *Historical Studies in the Physical and Biological Sciences* 31(2), 223–248.

Bensaude-Vincent, B. (2016). Building multidisciplinary research fields: The cases of materials science, nanotechnology and synthetic biology. in: M. Merz and P. Sormani (eds.), *The Local Configuration of New Research Fields: On Regional and National Diversity*, Sociology of the Sciences Yearbook 29 (Basel: Springer International Publishing), pp. 45–60.

Brown, P. and Rans, R. (1984). "Material Girl," performed by Madonna on the musical album *Like a Virgin*.

Hessenbruch, A. (2004). Nanotechnology and the negotiation of novelty. In: D. Baird, A. Nordmann and J. Schummer (eds.), *Discovering the Nanoscale* (Amsterdam: IOS Press), pp. 135–144.

Hessenbruch, A. (Spring 2020). Overview of the History of Materials Research. *American Physical Society, Forum on the History and Philosophy of Physics, Newsletter*. https://engage.aps.org/fhpp/resources/newsletters/spring-2020 (accessed 15 July 2021).

Leslie, S. W. (1993). *The Cold War and American Science: The Military-Industrial-Academic Complex at MIT and Stanford* (New York: Columbia University Press).

McCray, W. P. (2005). Will small be beautiful? Making policies for our nanotech future. *History and Technology* 21(2), 177–203.

Mody, C. C. M. (2011). *Instrumental Community: Probe Microscopy and the Path to Nanotechnology* (Cambridge, Mass: The MIT Press).

Mumford, L. (1934). *Technics and Civilization* (New York: Harcourt, Brace & Co.).

North, Douglass (1990). *Institutions, Institutional Change and Economic Performance* (Cambridge: Cambridge University Press), p. 3.

Streeck, W. and Thelen, K. (2005). Introduction: Institutional change in advanced political economies. In: Wolfgang Streeck and Kathleen Thelen (eds.), *Beyond Continuity: Institutional Change in Advanced Political Economies* (Oxford, Oxford University Press), pp. 3–39.

Chapter 1.2

From Materials Research to Materials Science: Disciplinary and Institutional Developments*

Klaus Hentschel

Introduction

This chapter pursues a dual goal: (1) to provide a broad overview of the literature on the history of materials research (with the focus placed on the German-speaking realm) and (2) to present the disciplinary evolution of this field of research and its progressive institutionalization in the 20th century. The metallurgists Cyril Stanley Smith (1903–1992) and Robert W. Cahn (1924–2007) have conducted groundbreaking work on this history.[1] Central to this story are the following major core fields from which important branches of research on materials ensued: the theoretical conceptions, experimental procedures and instrumental techniques of analysis in metallurgy, as well as research technologies in solid-state physics or physical chemistry.

*This chapter is Ann M. Hentschel's translation of an updated version of my German paper in *NTM* 19 (2011), pp. 5–40.

[1] Smith (1961, 1969, 1977, 1986a) or Cahn (2001a). For documentations of important stages compiled by material scientists, see Smith (1965, 1968), Tylecote (1976, 1992), Psaras and Langford (1987), Weaire and Windsor (1987), Sass (1998), and Wagner (2000). There are general overviews in Cunliffe (1994), Dresselhaus (1992), Strahm (1995), Maier (2000), Wagner (2000), and Braun and Herla (1994).

Disciplinary attributions that were still uncontroversial around 1900 have become increasingly problematic, considering the hybrid status of materials science at modern universities, polytechnics, independent research institutions and industrial laboratories. Joseph Martin (2018) has recently shown how long solid-state physics had to struggle to gain full acceptance and sub-disciplinary status in physics proper. It is conspicuous that since 1975, growing numbers of technical laboratories and courses of study involving materials decided to change their names to the more uniform designation: "materials science and engineering," to obscure their roots in chemistry, materials technology, metallurgy and physics, just to name a few of these origins. Has "materials science" meanwhile become a discipline of its own, perhaps even a kind of "superdiscipline," as some of their actors appear to believe? Or is its emergence (resembling nanotechnology very recently (Schummer, 2011) rather a sign of disciplinary coherence gradually falling apart in favor of interdisciplinary, increasingly informal research collaborations? This point is still controversial even among materials scientists. In 2006, Martin Stratmann, the then Director of the Max Planck Institute for Iron Research and now President of the Max–Planck Society, argued in favor of viewing it as a kind of cross-disciplinary field, whereas the British metallurgist Robert W. Cahn (1924–2007) advocated the view that since the 1990s, materials science has established a more coherent canon of research methods out of a rather loose assortment, leading to the development of a convergent discipline of its own. The conclusion that Bernadette Bensaude-Vincent drew from her historical consideration of science in 2001 was that such a disciplinary maturation of this field of research has not occurred. She studied the case of the failed French interdisciplinary program on materials research (*Programme interdisciplinaire de recherche sur les matériaux*, PIRMAT) and described it rather as a loose agglomerate (Bertrand and Bensaude-Vincent, 2011).

This chapter compares the positions taken by influential actors over time and sorts them against the backdrop of their different disciplinary origins, interests and ranges of problems. Familiar models and criteria for the genesis of disciplines from the history and sociology of science as well as from epistemology will be consulted in making this evaluation. I shall use eight structural and generally accepted specific indicators as the basis of my systematic analysis of the concept of "discipline" to check whether materials science may claim this status.[2] They are as follows: a coherent set of paradigms and methods; textbook literature; academic teaching units and courses of study; professorships and departments; specific journals; professional associations; professional and personal exchanges; and group identity as a defining and discriminatory attribute.

After presenting a largely chronological review of the traditions in materials research, this chapter runs through the phases of its institutionalization. Materials testing stations, industrial materials science laboratories, and interdisciplinary laboratories that were specifically oriented toward materials research at American universities are characteristic markers along the way. The COSMAT Report by the American National Academy of Sciences from 1974 and its subsequent reports constituted another milestone in our collection of the scientific topics also covered in public advertising about the interests and methods of materials research. We close by examining the present state of this complex field of research against the above indicators and modify Bensaude-Vincent's assessment with the conclusion that by 2000 materials science has indeed developed into a veritable discipline. In particular, agreement has now been reached about suitable generic descriptors (materials catholic) for characterizing many classes of materials.[3] Thus, the formation phase lasting

---

[2] See Chubin (1976), Lemaine *et al.* (1976), Stichweh (1984), Thiel and Peckhaus (1999), Laitko (2002).

[3] Rolf Hummel combines a historical and systematic approach according to the class of materials in his textbook from 1998. Brief introductions to the most important classes of materials also in manufacturing, combining other historical and systematic aspects, are offered by Hall (2014) and Weitze and Berger (2013). For wood as a natural workable material, see Radkau (2007); for biomaterials, see Geison and Creager (1999); for ceramics, see Mämpel (1985), Kingery (1986), Aldinger (1999); for composite materials, see Haka (2020); for polymers, see footnote 45.

practically an entire century for this unusual process of discipline formation has come to completion as an "emergence by integration."[4]

From Manufacturing Know-How to Diverse Materials-Related Sciences

For a long time, knowledge about materials was the know-how of practitioners. It was mostly conveyed orally from master to apprentice in the form of recipes. Often, it was purposefully kept secret (see, e.g., Scheel, 1989 on Egyptian metal working). The smelting processes of iron were held safe in the Hittite empire at least since the 17th century B.C., for instance. The decline of this empire in the early 12th century A.D. allowed them to escape and gradually spread from the Near East into the Mediterranean region (Raymond, 1986, Chapter 3; Hummel, 1998, 2004, Chapter 7). Toward the end of the middle ages, more texts were written by practitioners (such as booklets for testers, miners and stonemasons, books about machines, and treatises by gunsmiths or shipbuilders), preserving quite pertinent knowledge about materials and procedures, albeit often in a form that — for us today at least — is very hard to follow (Kranzberg and Smith, 1979, p. 16f). This knowledge was nevertheless largely left out of the teaching canon. In early modern times, these strains of practical knowledge were also gradually picked up by systematic scholars in quest of a theory. Agricola observed lifting technology used in mines and described it vividly in *De Re Metallica* from 1556, Kunckel published his *ars vitraria experimentalis* in 1679, and Galileo gathered inspiration in the Venetian arsenals for his new science about the strength of materials, as he described it at the beginning of his *Discorsi* from 1638. Otto von Guericke (1602–1686), finally, worked together with pump-makers and coppersmiths, who made his vacuum vessels and assisted him with their sealing.

---

[4] Cahn (2001b, p. 307) introduces it explicitly as the countermodel to traditional discipline formation of the 19th century, which occurred through differentiation (cf. here footnote 43). Other candidates for "discipline formation by integration" are geology and molecular biology as well as computer science: Ceruzzi (1989). Smith (1986b, p. 3) proposes an interesting analogy to "aggregation of electrons into atoms and matter."

Other early zones of systematic materials research apart from the mechanics of strength and elasticity theory (Leon, 1912; Timoshenko, 1953; Kahlow, 1991; Johnson, 2009) formed in alchemy/chemistry, metallurgy, and specialized areas of glass and porcelain production. The glass-making treatises by Johannes Kunckel (ca. 1630–1703) come to mind. Around 1660, the curator of the London Royal Society, Robert Hooke (1635–1703), inspected various materials under the microscope and drew a connection between their strength and torsion properties and their structure. Owing to this application of microscopy, his *Micrographia* from 1665 has even occasionally been judged a preliminary stage of later metallurgy (Smith, 1969; Mehl, 1948, p. 12). The "physical metallurgist," Cahn (1995, p. 1506f), on the other hand, interpreted these early microscopic investigations as a "subsidiary science," defining the neologism "subsidy" as a "subsidiary scientific domain" and hence below the level of disciplines. In the 18th century, René Antoine Ferchault de Réaumur (1683–1757) tried to use such findings, supplemented by chemical insights about the influence of salts, sulfur and carbon, to produce refined steels.[5] Michael Faraday (1791–1867) worked for years at the Royal Institution in London on analyzing and optimizing steels and highly refractive lead glass.[6] These examples may already represent systematic forms of materials research or technology, but they remain isolated, highly specialized areas of analysis practiced by entirely differently titled groups, such as artisan mechanics (*Maschinenkünstler*) and "millwheel doctors" (*Mühlenärzte*), artist engineers, mining assessors or gentleman scientists. This is by no means an early form of materials technology or materials science (in singular). Their origins rather lie at the turn of the 20th century. Two reasons for this are set forth in the following: the interdependence between materials research and research technologies and the explosive variety of materials triggered by the two world wars.

---

[5] Réaumur (1722), Leon (1912, pp. 13–16), Smith (1969, pp. 483–487), Tylecote (1976, 1992, p. 177). For the pivotal changes in categories of materials in chemistry under other experimental sciences during the 18th century, see Hacking (2002) Klein and Lefevre (2007) and Bensaude-Vincent (2011).

[6] Day (1995) goes so far as to interpret Faraday retrospectively as a "materials scientist" *avant la lettre*.

The Interdependence between Materials Research and Research Technologies

Great attention to systematic testing and measurement procedures — as opposed to intuitive handling skills — is one distinguishing marker between the scientific approach to materials and artisanal practice. This is why the development of materials technology or materials science was closely associated with the history of "research technologies," which Terry Shinn has shown to have the following structural characteristics (Shinn and Joerges, 2001): (1) generic applicability to a great variety of disciplines; (2) an "interstitiality" among its actors, who can skip back and forth between different disciplines, often between a university and industry and non-academic research institutions; (3) a unique metrology.[7]

Most research technologies of the 20th century exerted at times considerable influence on materials testing. X-rays, discovered in 1896, made it feasible to look at the interior of materials partially permeable to the radiation without causing damage. X-ray diffraction (cf. Bragg, 1945; Ewald, 1962; Eckert, 1993, pp. 45–48; Heuck and Macherauch, 1995; Lemmerich, 1996; Luxbacher, 2018) revealed the crystal structure by generating interference patterns by reflection off the crystal's periodic lattice levels. Transmission electron microscopy allowed wafer-thin material samples to be examined for microscopic cracks or other inhomogeneities.[8] Scanning tunneling microscopy granted insight into structure and partly also into alterations in a material on an atomic scale.[9]

---

[7] Consider, for instance, the mirror extensometers by August Wöhler, Adolf Martens and Ernst Preuß to determine the elongation of materials under strain or the universal testing machine attributed to Ludwig Werder (1808–1885), described, e.g., in Baumann (1912), Domke (1964, 1999, p. 290f), Nüßlein (1965), Opitz and Dude (1964, 1972, p. 53f), Kloos and Berger (2007, pp. 312–314).

[8] Ramsauer (1943), Qing (1995), Rasmussen (1997), Newbury and Williams (2000).

[9] Binnig and Rohrer (1985, 1987), Wickramasinghe (2000), Hennig (2004, 2011), Mody (2006), Granek and Hon (2008); other examples are provided in Ashby and Jones (1980), Cockayne (2002), Mody (2006).

A few research technologies were introduced as early as the 19th century, such as mineralogical analysis, chemical spectral analysis around 1860 (Hentschel, 2002) or electric resistance measurement in 1827, with Georg Simon Ohm (1789–1854) elucidating the underlying concept. This was of great practical relevance for the transatlantic submarine cabling from the 1850s on, for instance, to test the electric conductivity not only of different copper alloys but also of the insulating sleeve out of natural rubber (*gutta percha*), which was decisive (Smith, 1977, pp. 22–28; Hunt, 1997). However, implementation of research technologies for the purposes of materials research along a broad front only occurred after X-rays, electrons and other forms of radiation connected with radioactivity had been discovered around 1900. It then became possible to conduct examinations at the atomic or molecular level. These investigative options intensified with the development of methods from nuclear physics and solid-state physics. Many of these research technologies originated from physics but were soon applied beyond it in other entirely different fields. Nuclear magnetic resonance, for example, became established in medicine. The first of Shinn's three criteria, namely genericity, was thus satisfied.[10] Cahn (1995, p. 1550) also regards the investigative methods that physics supplied as its most important contribution to materials science which otherwise was rather defined by chemistry and engineering.

Another technology — metallography — came from mineralogy and crystallography. The finest structural properties of samples could be made visible by chemically etching a polished section for observation under the microscope or for photographing. Henry Clifton Sorby (1826–1908) from Sheffield pioneered this procedure,[11] inspired by Widmanstätten's similar meteorite preparations. This method was developed further in Germany by the railway engineer and Professor at the Charlottenburg polytechnic

---

[10] This transgression of the original disciplinary boundary and the related transformations in research style is discussed by Reinhardt (2006, 2011).

[11] Sorby (1897), Mehl (1948, pp. 11–13), Higham (1963), Smith ed. (1965, 1969, p. 491f), Hammond (1989), Tylecote (1992, p. 180), Edyvean and Hammond (1997).

in Berlin, Adolf Martens (1850–1914), who was also the Director of the materials testing office in Dahlem near Berlin founded in 1904. His collaborator was the ferrous metallurgist Emil Heyn (1867–1927).[12] Metallography then began to be applied on a large scale during the early 20th century for examining microstructure, for example, to study cracks and fissures in materials, grain and phase boundaries, and deformations formed by pressure, torsion, temperature fluctuations or other external influences, such as oxidation and erosion.[13] Aspiring practitioners poured over large plate volumes with hundreds of such metallographs to study what transparent cuts and etchings of the various metallic materials normally looked like and how to recognize the subtlest sign of weak spots.[14]

Industrialization and the massive expansion of railway systems since the mid-19th century led to constantly heightening demands on the strength properties of metallic construction materials, especially steel, which had to be tested, classified and appraised (Ude, 1935). The first materials testing series for the Krupp Steel Works in the 1860s were conducted in the private establishment of David Kirkaldy (1820–1897) in London (Robertson, 2013), the supplier of Krupp's first testing devices (cf. Fig. 2).

At the instigation of a technical committee of the German Association for Railway Administrators, memoirs were published establishing the first minimum values for tensile strength and constriction at breakage (Wöhler, 1877) and calling for the founding of

---

[12] See, e.g., Martens (1880, 1898), Piersig (2007, 2008) and Werner (1964). Another milestone in the establishment of metallography was the textbook Preuß (1927), which first appeared in 1913. Ernst Preuß was a pupil of Martens and from 1907 materials testing assistant at the Darmstädter mechanisch-technischen Materialprüfungsanstalt.

[13] Bach and Baumann (1914, 1921) containing almost 1,000 black-and-white illustrations of different material defects, and *Atlas Metallographicus* by Hanemann and Schrader (1933, 1952); cf. Baumann (1920), Pusch (1979).

[14] Historic steam-boiler explosions and railway accidents were caused by material defects or fatigue: Sonnenberg (1968) and Krankenhagen and Laube (1983).

Fig. 1. David Kirkaldy's "Museum of Destruction" in which he showcased samples of breakage caused by deficiencies in material which during the 19th century often led to catastrophic events like train accidents or steam vessel explosions. From Kirkaldy (1891, p. x); on the "Museum," cf. Robertson (2013, p. 130).

state-run testing stations for these inspections (Bauschinger, 1879).[15] Further standardization of testing procedures and equipment was resolved at the first conferences on establishing uniform testing methods in Munich 1884 and Dresden 1886, out of which emerged the German Federation for Materials Testing in Engineering in 1896.[16] Novel visual methods in materials analysis were developed. The reduced grain boundaries induced by annealing are clearly visible in the material on the right-hand side of Fig. 2. (The scale of microscopic magnification is indicated at the bottom right.)

Martens's microscope with its strongly skewed lighting of metallic samples produced good contrast and sharp shadows of depressions but rather low resolution, which is why Martens (and

[15] *Verein deutscher Eisenbahnverwaltungen*. *Cf.* Schulz (1949), Kloos and Berger (2007, p. 311). Maier (2009, 2019) emphasizes the importance of expert boards (*Fachausschüsse*) within larger societies in the consolidation of materials research.

[16] *Verband für die Materialprüfungen der Technik* (cf. DVM, 1997). On the steep rise since 1895 in the number of testing machines and processes, see Bauman (1912, pp. 154, 177–184).

incidentally Sorby too) also sketched most of his metallographically examined samples instead of photographing them. No tool, he contended, schools the observer's eye better than sketching at the microscope (Martens, 1880). Intense training in these techniques of preparation of microscopic samples and skilled visual analysis of images such as Fig. 1 led to specific visual regimes within visual cultures (For the emergence and stabilization of these visual regimes within broader visual cultures in science and technology, see Hentschel (2014)).

This influx of novel visual research technologies into metallography largely altered physical metallurgy as a discipline. In 1896, Adolf Martens founded the national materials testing federation, *Deutscher Verband für die Materialprüfungen der Technik*, later renamed *Deutscher Verband für Materialforschung und -prüfung e.V.*, which still exists (DVM, 1997). The history of the institutions founded specifically for the purpose of materials testing also starts in the 1870s: the *Mechanisch-Technisches Laboratorium* in Munich and the *Eidgenössische Materialprüfungs- und Forschungsanstalt* in Zurich, Switzerland, were founded in 1879, as well as analogous materials testing stations in Stuttgart or Darmstadt (abbreviated as MPAs) were founded in 1884 and 1907, respectively, each with certification seal for state and

Fig. 2. A metallographic comparison between cold-drawn brass (left) and annealed brass (right) (Baumann, 1920, Fig. 28).

private clients.[17] The measurement and testing procedures were standardized further if only to assure fair treatment among all their clients. The subsequent meter designs, textbooks and manuals document the progressive refinements obtained as a result of other applied research technologies.[18] Early pioneers of materials testing, including Franz Carl Achard (1753–1821), August Wöhler (1819–1914) and Johann Bauschinger (1834–1893), worked as physicists, chemists or engineers at polytechnics or materials testing stations.[19] In 1933, the Berlin MPA instituted a specially designated X-ray Department (Röntgenstelle) directed by Rudolf Berthold (1898–1960) who at the same time also became the Head of the Laboratory for X-ray materials testing at the Siemens & Halske Company (also in Berlin; cf. Richter (1999, p. 15ff)). In 1933, Berthold joined forces with engineers Hermann Hellmich (Head of the Technical Supervision Societies), Ernst Lupberger (1893–1971), Head of the union of steam vessel owners, and the President of the Berlin MPA, Robert Otzen (1872–1934) — they jointly founded the *Deutsche Gesellschaft für Zerstörungsfreie Prüfung (DGZfP)* in order to consolidate the various new methods of analysis which allowed the destruction-free examination of materials (cf. Richter, 1999, p. 15ff; Luxbacher, 2018). Similar societies were soon founded in other countries, for

---

[17]For German and Swiss materials testing institutions, see, e.g., Föppl *et al.* (1959), Amstutz (1955), Zweckbronner (1985), Blind and Werner (1984), Ditchen (2009), and Kloos and Berger (2007). Other non-Prussian materials testing stations are introduced in Anonymous (1930) and Sonnemann and Friedrich (1978). For the Berlin *Königliches Materialprüfungsamt*, see Martens and Guth (1904), Ruske (1971), and Ruske *et al.* (1996). For the American Institute of Mining and Metallurgical Engineers (founded in 1871), see Mehl (1948). For the general boost in materials testing in the 19th century, see Schulz (1949), Sigwart (1986), and Kahlow (1991).

[18]I mention as a suitable example the official textbook for engineering students of the former GDR by Opitz/Dude, which appeared in seven editions between 1964 and 1977; the various editions of Domke (1964) serve as a West-German correlate.

[19]For their new classifications of materials and testing instruments, see Föppl *et al.* (1959), Kahlow (1991, 299f), and Krankenhagen and Laube (1983, Chapters 4 and 5), and the appendix there with reprints of contributions to discussions about the classification of iron and steel underway at that time.

instance, the *American Society for Non-destructive Testing* (founded in 1941, then still named *American Industrial Radium and X-Ray Society*, and linked to the journal *Industrial Radiography and Non-Destructive Testing, later* renamed *Materials Evaluation.*[20]

The first German academic chair for materials technology (*Werkstoffkunde*) was established in 1927 at the Grand-Ducal Polytechnic in Darmstadt. Others followed, such as those in Munich, Berlin (1935), Hannover (1935), Braunschweig and Stuttgart, often operating in unison with the head positions at the regional MPAs — partly even sharing identical staff. In the major industry, there was a parallel development of private materials testing establishments for quality control in production and supply, on the one hand, and well-equipped materials-specific research laboratories, on the other. Well-studied cases besides the chemical industry (e.g., Hounshell and Smith (1988) on DuPont) include the research laboratory of General Electric, founded in 1900, Bell Laboratories, founded in 1925 (Birr, 1957; Wise, 1980; Reich, 1985; Clayton and Algar, 1989) and the Siemens research laboratory, opened in 1920 (Trendelenburg, 1975).[21]

Materials Research during World War II and the Cold War

The shortages during the First and Second World Wars triggered the quest for new materials: either for materials better than the familiar ones or for substitutes owing to deficiencies as a result of war blockades. These particularly included artificial rubber, various fuels and oils, as well as light metals or surrogate materials. While during World War I, such bottlenecks were often noticed too late for any effective countermeasures to be taken, during World War II, efforts to solve or compensate scarce resources were made on both sides of

[20] See https://www.asnt.org/ASNT_75th/Timeline_and_History.aspx (last accessed Sept. 7, 2023).

[21] According to Cohen (1979, pp. 35, 102, footnote 83), in 1940, only about 10% of all American industrial firms in the metals sector had a materials department, but 20 years later, over half of them did. For a comparison between (hesitant) universities and MSE-related industry, see the contribution by N.B. Hannay from Bell Labs vs. George Krauss's from Lehigh University, in Roy ed. (1970).

the front line. National Socialist foundings of such agencies as the Office for Economic Development (*Reichsamt für Wirtschaftsausbau*) and a Research Council (*Reichsforschungsrat*) were supposed to find redress just as much as the US National Research Council. Despite the polycratic power structures under National Socialism, shortfalls were in fact reduced in many sectors through the development or industrial supply of substitute materials such as aluminium–zinc–magnesium alloys, the so-called "German metals" — *Deutsche Metalle* (Maier, 2002, 2006; Luxbacher, 2010, 2011, 2020). Engineering associations, professional committees and institutions set standards and passed important resolutions (e.g., the German Society for Materials Science, *Deutsche Gesellschaft für Materialkunde*, with its influential special boards (Maier, 2019)). In any event, the centralized structures that were created could only develop full effectiveness later on in the subsequent phase of the Cold War.

In the 1950s, the US Presidential Science Advisory Committee recommended "materials as a priority" in research and development.[22] The Sputnik shock of 1957 accelerated efforts to make space flight a reality with its extreme demands on materials. The United States Department of Defense funded many R&D projects on armaments and space technology also in the Air Force and Navy. The Advanced Research Project Agency (ARPA) clinched contracts with a number of universities, for example, in 1961 with Harvard, MIT, Brown, Stanford and Chicago for $13 million.[23] In 1957, 35% of invested monies went into solid-state physics, 29% into metallurgy and 10% into ceramics research. Altogether, 12 "Interdisciplinary Materials Laboratories" were founded by ARPA between 1960 and 1962, following the already successful interdisciplinary model of research laboratories on radar, electronics and atomic science.

[22] For the context, see Harwood in Roy ed. (1970, p. 210); N.E. Promisel (member of the Materials Advisory Board, National Academy of Sciences) in Roy ed. (1970) and Bensaude-Vincent (2001).

[23] Excerpts from one of these contracts are provided by Julius J. Harwood in Roy ed. (1970, p. 8); the appendix to these published *Proceedings of the National Colloquy on the Field of Materials* from 1969 contains a survey of all these university programs and important industrial research centers in the area of MSE, including staffing data.

Together with the major industrial research laboratories, they formed another nucleus for the consolidating institutionalization of materials research (For the background, see Turnbull (1983), Schwartz (1987), Sproull (1987), and Bensaude-Vincent (2001)).

From Materials Research to Materials Science

In this context, from the 1950s onwards, "material/s science/s" became a frequently used term by some of the important actors in the field, often with the appendix "and engineering" — MSE. The renaming of "Interdisciplinary Materials Laboratories" into "Material Research Labs" (MRL), such as the ones at Cornell and the Universities of Pennsylvania and Northwestern (Schwartz 1987), along with similar rechristening of entire academic departments on American campuses around 1970, continued the march of institutionalization of the field.[24] Here are some pertinent graphs: Figure 3 displays the first wave of foundings triggered by ARPA's support from 1960 on and the multiplying renamings into MSE after 1981, which goes back to an initiative by the National Science Foundation for the promotion of "university industry cooperative research centers" [Groenewegen/Peters 2002: 121–123].

We see stagnation in research fields such as ceramics or polymer science and a steady decline in metallurgy in Fig. 4. Between 1970 and 1980, solid-state physics drops and stays stagnant until 1990, whereas MSE is the only field almost continually on the rise.

As will be shown in the following, the wave of renaming of societies, academic institutes and laboratories hit Europe at least a whole decade later. The trend is perceptible in industry as well, for

[24] See Nielsen in Roy ed. (1970, p. 17) for the names of 85 departments in the US and Canada; in 1969, 25% of them included "materials science," with more in the offing, cf. Schartz (1987), Groenewegen and Peters (2002), where the dependence of these phases of institutionalization on types of funding is underscored. For how "intellectually suspicious" terms like "interdisciplinary" were to various practitioners of materials research around 1960, see Sproull (1987, pp. 9–11). For Cornell, in particular, see Mody and Choi (2012); for the University of Pennsylvania, see Choi and Shields (2015).

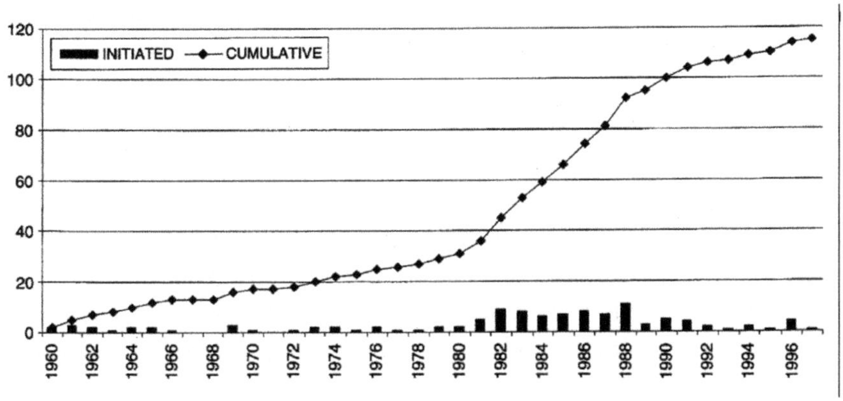

Fig. 3. Research centers for MSE in the USA 1960–1997 (Groenewegen and Peters, 2002, p. 129).

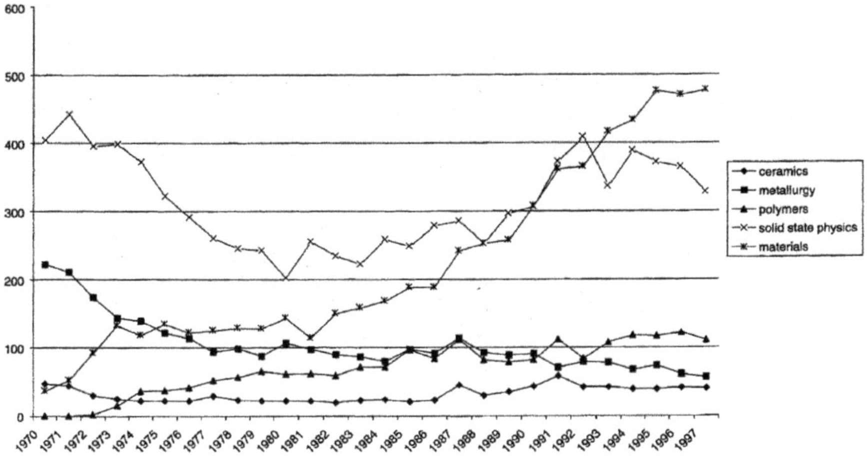

Fig. 4. Doctoral degrees at American universities, 1970–1998, grouped into disciplinary fields (Groenewegen and Peters, 2002, p. 130).

example, in the Bell Telephone Laboratories in New Jersey, where the transistor had been developed in 1947 (Hoddeson, 1977) and in the General Electric Corporate Laboratory in Schenectady. By 1987, there were already 17 state MRLs in the USA. If the materials researchers among their staff are divided up into their original academic disciplines, the following were the approximate distribution results: 35% were physicists, 25% were chemists, 19% were metallurgists, 16% were engineers, and 5% came from other fields of study.[25] The US Department of Energy also funded MRLs, for instance, conducting analyses on neutron scattering and material fatigue or participating in reactor programs and synchrotron radiation experiments in accelerators at Stanford or Brookhaven.

In 1958–1959, the first Department of Metallurgy at Northwestern University in Illinois was renamed as the Graduate Department of Materials Science. Others followed suit, including Texas in 1967, which was renamed as Metallurgy and Materials Science and MIT in 1974, which was renamed as Materials Science and Engineering. By 1969, 30% of all American departments of metallurgy had already decided to add the terms 'materials' or 'materials science' to their official designations. Meanwhile, hardly any genuine departments of metallurgy existed in English- or German-speaking countries anymore. France, however, has maintained the traditional tight bond between metallurgy and chemistry to the present day. At Northwestern University, the following justification for renaming their department was offered in 1959 (Fellman, 2005):

> "Traditionally, the field of material[s] science has developed along somewhat separate channels — solid state physics, metallurgy, polymer chemistry, inorganic chemistry, mineralogy, glass and ceramic technology. Advance in materials science and technology is hampered by this artificial division of the whole science into separate parts. [... We see the] advantages of bringing together a group of specialists in the various types of materials and allowing and encouraging their cooperation and free interchange of ideas."[26]

[25] A chart on how this distribution changes between 1970 and 1985, based on Flemmings (1999) and the yearbook of the American Society for Metals (ASM), is available in Bensaude-Vincent (2001, p. 236).

[26] The sometimes quite hefty renaming debates are discussed, e.g., by Cahn (2002, p. 3; 2003b, p. 468). Only some years after the metallurgist Robert Mehl had retired was the Pittsburgh Department of Metallurgy and Materials Science renamed Department of MSE, cf. Bever (1988) and Flemings (2001, pp. 854–857) on analogous developments at MIT.

Other reasons for this albeit long and heftily debated name changing included much greater interest in microstructure as a verifiable indication of properties along with the related analysis techniques from solid-state physics and its concepts like lattice and point defects.[27] These new ways to explain structure scientifically changed the way metallurgists viewed themselves. Formerly strongly oriented toward improving and controlling industrial production of metallic materials, their self-understanding shifted in the direction of structure analysis toward understanding solid-state properties and partly also toward purposeful alteration (design materials),[28] which had already started to happen at the close of the 19th century in chemistry and alloy science. Therefrom resulted a "transformation of metallurgy from an empirical art to a field of study based on principles of physics and chemistry." Or else, as the manager of the research laboratory for metals and ceramics at General Electric, J. Herbert Hollomon (1919–1985) formulated it in 1958 in the form of a chemical transformation: "out of metallurgy, by physics, comes materials science."[29] A solid-state physicist, William B. Shockley (1910–1989), headed the semiconductor group at Bell Laboratories, which was staffed just as interdisciplinarily. Shockley had successfully developed the bipolar transistor in 1947, for which he was awarded the Nobel Prize in Physics for 1956 together with John Bardeen and Walter Houser Brattain (Queisser, 1985, p. 104; Hoddeson, 1977; Hoddeson *et al.*, 1992, pp. 406–482; Riordan and Hoddeson, 1997).

---

[27] On the history of these concepts of *Verhakung* and dislocation as well as flaws or dopants and their diffusion in crystal lattices, see Mott (1980, pp. 125–177), Smith (1986b, pp. 8–10), Weairde and Windsor (1987), Cahn (1995, pp. 1520–1535), and Seeger (2009).

[28] For this process, see Smith (1977, p. 3): "professionalism in materials, until very recently, meant facilitating and cheapening the large-scale production"; (ibid., p. 27): "adoption in some measure of the physicist's view of the world without loss of an interest in practical diversity."

[29] Cited after Cahn (2002, p. 3; 2003b, p. 468). Hollomon considered 1958 to be the birth year of 'materials science.' (cf. Kranzberg and Smith (1979, p. 29)). Here, he advocates a view that is otherwise rather found among physicists. Holloman himself was an engineer trained in metallurgy at MIT and later founded the National Academy of Engineering in the US.

The hybrid mixture of natural and technical sciences put at the service of materials research was already functioning very well in industrial laboratories, long before a new name for it had been found.

Speaking of the early 1960s when materials science slowly began to cohere, physicist Robert L. Sproull (1918–2014), who was the Director of the *Defense Advanced Research Projects Agency* (DARPA, a successor of ARPA) between 1963 and 1965, writes in a still very telling account three decades later:

> "Physics was at last bringing something to the party, and metallurgists and chemists needed physicists, if only physicists would rise above their snobbery. Physicists needed chemists and metallurgists, since increasingly sophisticated experiments required detailed knowledge of chemical and physical imperfections and structures. Of even more consequence was the conviction that the design and creation of new materials, such as composites, high-temperature coatings, and catalysts, would require true collaboration among chemists, physicists, and engineers."[30]

In retrospect, Sproull considered the materials initiative a success since it mobilized many resources.

The COSMAT Report from 1974

A second, more intense phase of institutionalization took place in the 1970s. It was ushered in by the National Colloquy on the Field of Materials in 1969 (Roy ed., 1970) from which emerged the Committee on the Survey of Materials Science and Engineering (COSMAT) in 1970. Incidentally, it is notable that besides scientists and engineers, its members also included the metallurgist Cyril S. Smith (1903–1992) and the historian of technology Melvin Kranzberg (1917–1995), thus explicitly condoning a historical perspective. In 1974, the related report by the National Academy of Sciences

[30] Sproull (1987, p. 3); for the remaining disciplinary competition, ranking and in–out issues, see Reinhardt (2011).

appeared under the title: *Materials and Man's Needs. Materials Science and Engineering* (COSMAT, 1974) and became the start signal for similar endeavors in other countries. There are a number of reasons why this second institutionalization phase of MSE occurred precisely at this time: the oil crisis of 1973, growing environmental problems, the establishment of the White House Energy Policy Office, the passage of legislation about air quality control (the Clean Air Act), from 1970 onwards greater impetus in research on catalysts, as well as the founding in 1972 of the environmental program of the United Nations, more intense materials research in specialized areas (such as, for nuclear reactors, the automotive industry, air and space flight) as well as semiconductors for microelectronics. One novelty was the incorporation of material cycles and waste products, prompted by the rapidly mounting rubbish heaps of the consumer society.

The 1974 COSMAT report estimated that roughly 500,000 scientists and engineers were working on topics in materials research, of which at most 50,000 had academic degrees in which the word "material" explicitly occurred. All the others held conventional degrees in chemistry, physics or one of the diverse areas of engineering, such as metallurgy or materials processing. In 1974, there were 35 national associations, for instance, for materials technology and the materials sciences. The report counted in the English-speaking realm about 90 'materials-designated' undergraduate courses of study (BA) and approximately 50 graduate ones explicitly employing the word "material." In this report, the term "materials science" was defined as follows: "Materials science and engineering is concerned with the generation and application of knowledge relating the composition, structure, and processing of materials to their properties and uses."

The detailed exposition covered a wide field of tasks, ranging from chemical and structural analysis to manufacturing and synthesis in the laboratory as well as mass production at the industrial level. Thus, it incorporated classical tasks of chemistry, physics, as well as plant design and processing technology in addition to environmental impact control (at least somewhat). Just a year earlier, in 1973, the Materials Research Society was founded at Penn State University. At that time, the more restrictive, virtually disciplinary label "materials science" was consciously left aside in favor of the broader term "materials research." The university's program presented it broadly as "interdisciplinary goal-oriented basic research on materials of technological importance."[31] This definition made sure that no researcher working on materials could feel left out anymore, regardless of the discipline he or she was officially affiliated with.

The German expression *Materialwissenschaft* (materials science) was not yet in common usage. During the 1970s, one rather referred to *Werkstoffkunde* or *Werkstoffwissenschaft* — roughly translated as the field of material resources, manufacturing materials or materials technology. Examples include the textbook by Werner Schatt (1923–2009) from Dresden, originally published in 1972 until its ninth edition in 2003, (Schattm, 1972) or the didactical considerations written that same year by Klaus Urban (1972).[32] Interestingly enough, this term seems to have crossed through the Iron Curtain from East to West. Initially, it was used almost exclusively in the German Democratic Republic, then was hesitantly borrowed by Köster (1969, p. 666) in "scare quotes," before West-German publishers (such as Shaker-Verlag in Aachen) adopted it for their specialized series. Nowadays, it is practically synonymous with *Materialwissenschaft.* The more neutral, older terminology, such as *Werkstoffkunde* or *Werkstoffprüfung* (materials testing), along with other composites were used equally in both Germanies after World War II and continue to be a firm component of basic training in all the engineering sciences. Despite the increasingly conventional usage of 'materials science/s' (frequently still plural), respectively, *Werkstoffwissenschaft* (mostly singular), around 1974, the actors in East and West largely concurred that no convergence toward a new discipline existed yet:

---

[31] Quoted after Cahn (2003b, p. 469), whose wry remark about this choice was an "inspired form of words." For the history of this society, see also Jackson *et al.* (1993, p. 8f, 1993, pp. 69–87), where the distinction is made between a "nucleation stage," a "period of rapid growth" and a phase of maturation.

[32] For the style of thought and linguistic aspects, see Vogler (2008).

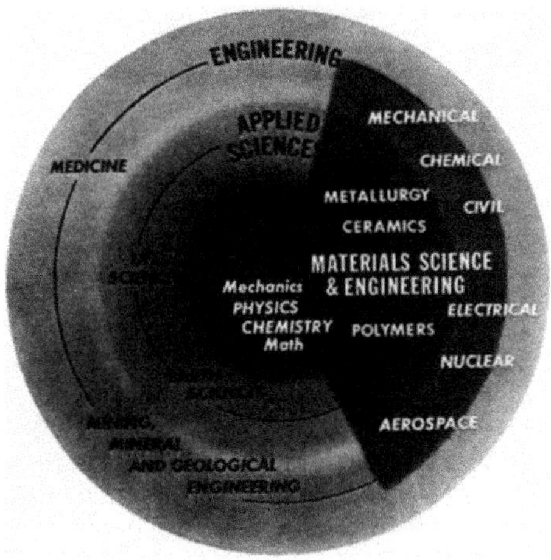

Fig. 5. Multi-disciplinary materials science and engineering (COSMAT, 1975, p. 26).

> "Materials science and engineering is not a single discipline, like physics or chemistry, but rather a multi-disciplinary area of research and engineering that has evolved gradually over the last twenty years into a coherent aggregation of related activities, but within each of the component disciplines retains its identity."[33]

In the subsequent report in 1975, the COSMAT Committee agreed on adopting the handy formula "materials science and engineering (MSE) as a multi-discipline" (see Chapter 3 of Kranzberg and Smith (1979, p. 28)). This rhetoric was supported by a straightforward pie chart that cuts up the large field into handy disciplinary slices (Fig. 5). Physics, chemistry and mathematics occupy core regions, with successive zones positioned radially outwards with

[33]Melvin Calvin to Dr. Handler, in December 1973, quoted in the Foreword to COSMAT (1974). The subsequent report emphatically defends the singular: "one should speak of materials science and engineering as an it rather than them" (COSMAT, 1975, Section 2.2). There is a parallel case: the renaming of "computer sciences" as "computer science," as Ceruzzi (1989, p. 266f), for instance, describes it, although even there the addition "science(s) and engineering" is still dragged along around 1959, only to later vanish.

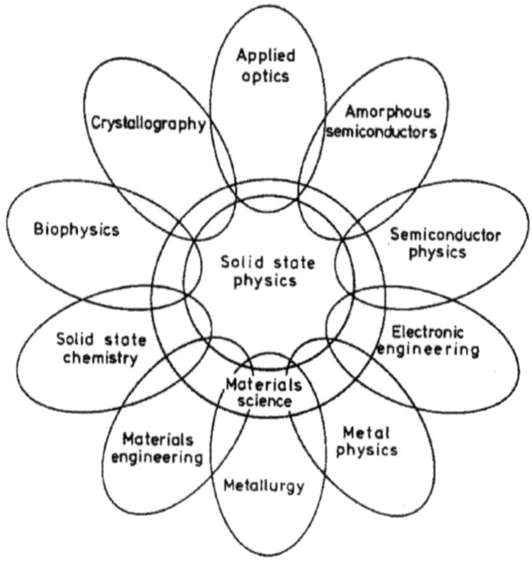

Fig. 6. A physicalistic vision of materials science/s as solid-state science (Weaire and Windsor, 1987, p. x).

increasing proximity to applications.[34] This chart is misleading among other things because in this "multi-discipline," MSE is not, in fact, a directed gradient from science to technology. Rather, both operate as hybrids in teaching and research without any hierarchical arrangement between them.

Another competing diagram (Fig. 6) more strongly biased toward the physical perspective set solid-state physics alone at the center, surrounded by materials science, around which all the other professional fields are grouped, tangentially overlapping each other like flower petals. Some physicists imagined materials science as a kind of elaborate solid-state physics[35] and therefore as a combination of, for example, applied quantum mechanics and quantum chemistry, crystallography, applied optics, electronics, physics of

[34] For the history and criticism of this linear model, see Grandin (2004).

[35] For the history of solid-state physics and a few pioneers of MSE, see Turnbull (1983, p. 2f), Weaire and Windsor (1987), Weart (1988), Hoddeson *et al.* (1992), Riordan and Hoddeson (1997).

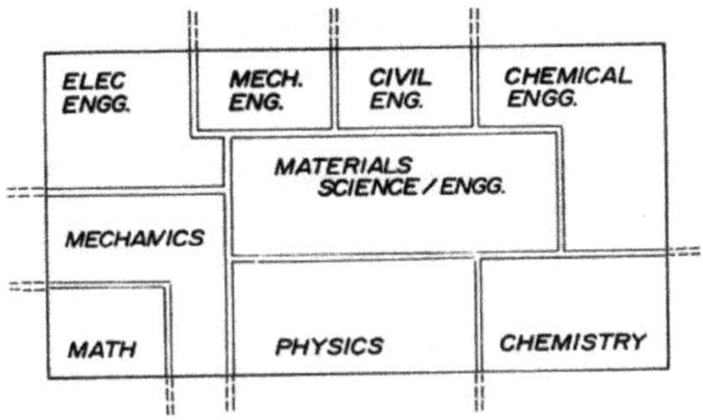

Fig. 7. MSE at the center of materials research, depicted here as a solidly framed contact zone of various elder disciplines (Roy, 1970, p. 161).

metals, statistical mechanics, etc., which surely yields the rest: "Solid State Physics is at the core of nearly all [these] fields, while Materials Science embraces the wider application of the basic studies." (Weaire and Windsor, 1987, p. x). Unfortunately, this vision by two physicists draws a false picture. Solid-state physics came (too) late in history, was too long too strongly focussed on retrodictively explaining the already observed properties of materials and concerned itself too little with predicting and constructing new materials.[36]

Rustum Roy (1924–2010), Founder of the Materials Science Laboratory at Pennsylvania State University, developed a more refined scheme for the volume he edited about the state of the art in MSE (Roy, 1970). It situates materials research as a kind of contact zone (preempting Galison's "trading zones" as it were) between its elder disciplinary neighbors from the natural and technical sciences. All these disciplines are granted the ability to conduct materials research just as much as the "neodiscipline" MSE. In Fig. 7, MSE

[36]The physicist Nevill F. Mott, for example, in his autobiography from 1986. To some extent, the Stuttgart Max Planck Institute for Solid State Physics has followed this option. There is also the Max Planck Institute for Metal Research (Köster, 1949, 1961; Maier, 2009), originally founded in 1921 in Neubabelsberg as the *Kaiser-Wilhelm-Institut für Metallforschung*, where a central office for scientific and technical analyses had already existed. In 1922, it was relocated to Dahlem, a suburb of Berlin, and in 1934 (the year after being shut down for lack of funding), it was resurrected again in Stuttgart.

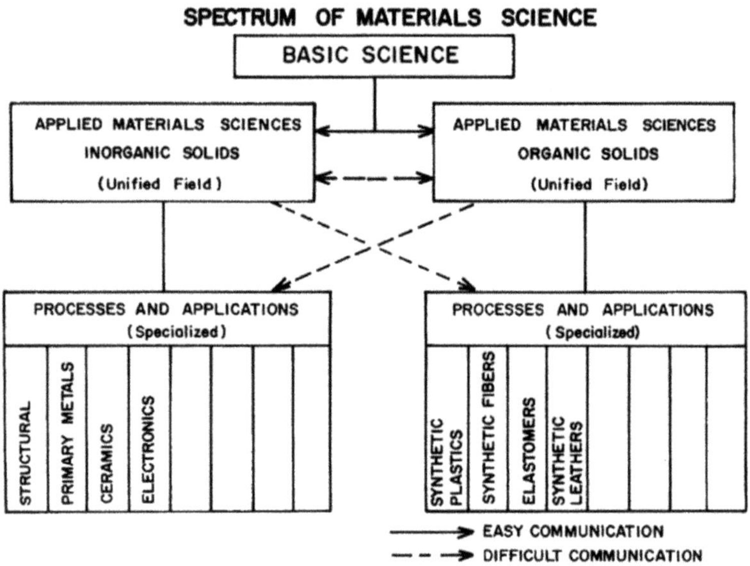

Fig. 8. Disciplinary relations, as depicted by George Krauss (1970, p. 41).

is situated at the center instead of solid-state physics, but its distinguishing factor is that materials research is its sole activity. This explains the dashed lines leading away from the solid rectangular core of materials research.

The communicational cross-links between these disciplinary neighbors is the topic of a chart designed by the metallurgist Georg Krauss (*1933) at Lehigh University in his contribution to the same volume (Figure 8). There the split between organic and inorganic materials is viewed as the most important gaping "internal front" in communications within MSE. The "linear model" is evidently presumed not to be a problem here either and is even rendered hierarchically, with the top-down communications between basic and applied research not judged to be questionable.

If one presents the scenarios about MSE offered by active experts as a function of time, a notable spread of possibilities results with an interesting trend:

- "a new synthesis of disciplines" (Harwood, 1970, p. 1),
- "coherent aggregation" of related activities (Calvin, 1973),

- "multi-discipline" (COSMAT, 1975, 1989; Flemings, 1999, p. 16–19),
- "the array of disciplines and subdisciplines that overlay the materials cycle" (Cohen, 1979, p. viii),
- "many disciplines involved [...] but multidisciplinary cohesion" (Kranzberg and Smith, 1979, p. 33),
- "superdiscipline" (Turnbull, 1983, p. 2),
- "materials is not to be thought of as a single discipline, but rather as a broad and vital field of knowledge and techniques" (Johnson, 1987),
- "ultramolecular" (Turnbull, 1983, p. 1) or "Supramolecular Science" (in *Neue Zeitschrift*, 1995),
- "a very loose alliance of recognized subdisciplines such as Metallurgy, Continuum Mechanics, Crystallography, Polymer Chemistry, and Solid State Physics. Or one might think of it as a confederation of innumerable research specialities" (Ziman, 1995, p. 2047),
- "is materials science a proper discipline or is it like a tangle of jungle creepers, hanging heavily between disciplinary trees? [...] just a multi-disciplinary chimera?" (Cahn, 1997, p. 160)
- "it is time to view the core of our field as a discipline itself" (Flemings and Cahn, 2000, p. 380),
- "a discipline (while simultaneously part of a larger discipline)" (Flemings and Cahn, 2000, p. 371),
- "an independent interdisciplinary science" (*European White Book*, Rühle *et al.*, 2002, p. 115),
- "cross-cutting area between scientific research and industrial development" (Stratmann, 2006).

It is conspicuous that the first clear vote in favor of viewing materials science as a new discipline in its own right comes quite late, around 2000, and it is equally remarkable that its main advocates, the materials scientists Merton C. Flemings (*1929), who was the first Director of the Materials Processing Center at MIT from 1980 to 1982, and the above-mentioned Robert W. Cahn, who filled the first British professorship for materials science from 1965 to 1981 at the University of Sussex, were still so unsure of themselves about this matter that they also offered an alternate framework elsewhere in the very same publication. Given this mixture of rhetoric and vision, exaggerated claims and dashed expectations, it is advisable to apply the systematic criteria acknowledged in scientific research in trying these claims to judge whether we have a discipline or not now. If we go through the indicators listed at the beginning of this chapter for the cases of materials science and technology, we quickly see that — with the exception of the first point (paradigms and coherent methods) — all exist there right now.

Let's begin with the indicator "specific professional journals." In the German-speaking world, one periodical dominates: the *Internationale Zeitschrift für Metallographie*, which had been founded in 1909 as the central publishing organ for the entire theory and practice of metals and alloys by William Guertler (1880–1959), at that time a private lecturer at the Charlottenburg polytechnic in Berlin, who joined the National Socialist Party in 1931 and got full tenureship as Professor of Physical Metallurgy (*Metallkunde*) in 1933. From this central organ emerged in 1919 the *Zeitschrift für Metallkunde*, renamed thus on the occasion of the founding of the German society that same year, *Deutsche Gesellschaft für Metallkunde* (DGM).[37] Up to the 1930s, science and technology were equally strongly represented, afterwards scientific contributions increased as the technical articles began to diminish. In 1992, this journal gained an English subtitle connected with the internationalization of the host professional society: *International Journal of Materials Research.* Starting with the 97th volume in 2006, it became the main title and by now all its articles appear in the English language. There are many other journals for materials scientists, mostly in English; to name a few important examples, *Materials Science and Engineering* (since 1966, now divided into a number of subseries), the *Journal of Materials Education*, the *International Journal of*

[37] Guertler is discussed in Anonymous (1934). For the *Zeitschrift für Metallkunde*, see the editorial of the commemorative volume of the renamed *International Journal of Materials Research*, 2006, p. 8. For the DGM and its journal, cf. DGM, ed. (1969, 1994).

Educational Modules for Materials Science and Engineering (edited since 1979 by members of the Materials Research Laboratory at Penn State), and the *Acta Metallurgica* (available since 1954). The latter, subtitled *Materialwissenschaft und Werkstofftechnik* since 1974, was renamed as *Acta Metallurgica et Materialia* around 1990 and then *Acta Materialia* six years later.[38] It also incorporates the journal *Nanostructured Materials* which was published independently from 1992 to 1999 but then merged with the main journal. *Scripta Materialia* was established in 1967 as a companion journal, publishing rapid communications and viewpoints. A curious case is the international journal, *Physica Status Solidi*, which was founded in 1963 by reputed solid-state physicists from both sides of the iron curtain in a conscious effort to bridge the East–West divide (Hoffmann, 2013).

At the levels of professional associations and academic institutes or professorial chairs, there are numerous examples of renaming during the two decades between 1970 and 1990, in which the word "metal" is exchanged with "materials." One typical instance that is also historically significant is the former *Institut für allgemeine Metallkunde* at Göttingen, which goes back to the metallurgist originally trained as a physical chemist, Gustav Tammann (1861–1938).[39] The associated academic department underwent a name change every time this chair changed hands during this period. Between 1970 and 1993, it first was known as the *Institut für Metallphysik und allgemeine Metallkunde*, then *Institut für Metallphysik*, and finally at the end of the 1990s, *Institut für Materialphysik*. The renaming of the

---

[38] For the intense cooperation between solid-state physicists and metallurgists since the appearance of *Acta Metallurgica*, see Cahn (1997, p. 163): "I still recall the tremendous intellectual impact created by the first volume." Cf. Cahn (1995, p. 1556) on the predecessor publication founded by Paul Rosbaud (1896–1963), *Progress in Metal Physics*, among other journals and book series.

[39] For Tammann's more scientific approach to processes of importance in metallurgy, such as oxidation and corrosion, recrystallization and phase transitions, with which he and his many pupils in the first half of the 20th century exerted great influence on metallurgy in Germany, see Mehl (1948, p. 23) as well as Cahn (1995, pp. 1540, 1554f).

Department of Mining and Metallurgy at MIT went through an analogous series of face changes, albeit somewhat earlier. The first designation had been held since 1888. In 1967, it became the Department of Metallurgy and Materials Sciences, and in 1974, it was redubbed the Department of Materials Science and Engineering.[40] In the UK, the Department of Materials Science and Metallurgy — in that order! — at the University of Cambridge united the examinations held at the Institute of Metals and the Institute of Materials, which between 1966 and 1976 were still separate academic entities (Charles and Greer, 2005). In 1989, the professional society *Deutsche Gesellschaft für Metallkunde* was renamed as *Deutsche Gesellschaft für Materialkunde* at the initiative of Werner Köster (1896–1989).[41] The French society was analogously named *Société Française de Métallurgie et de Matériaux*. The London Institute of Metals, founded in 1908, was first renamed Institute of Materials in 1976 and then Institute of Materials, Minerals and Mining in 2002. In the USA, the Materials Research Society was founded in 1973; and the European Materials Research Society was established — with typical tardiness of about a decade — in 1983, soon followed by the Federation of European Materials Societies in 1987 (FEMS) (Peter Paul Schepp in Maier (2019, pp. 577–598)) as well as the International Union of Materials Research Societies as the umbrella organization. In 2002, the first *White Book of the European Union on Fundamental Research in Materials Science* appeared; in 2009,

[40] For the history of MSE at MIT, see Bever (1988) and Flemings (1999, pp. 9–11). The term "*Metallurgie*" on its own has a much broader scope in German than does "metallurgy" in English, which refers to practical metal manufacture. I owe this point to Helmut Maier.

[41] Köster (1969), DGM ed. (1969), DGM ed. (1994), Maier (2019, pp. 253–264); Köster played a very influential role until 1945. He was among other things Head of the Research Laboratory of the steel consortium *Vereinigte Stahlwerke*, Editor of *Zeitschrift für Metallkunde* and Member of the *Reichsforschungsrat*, furthermore Professor of Physical Metallurgy at the Stuttgart polytechnic and Director of the local Kaiser Wilhelm Institute for Metals Research as well as being a member of the organization for metal manufacturers during the Third Reich; see Maier (2007, 2009), and the sources cited there. For Köster and the founding of the DGM, which emerged out of the metals board of the *Verein Deutscher Ingenieure*, see Köster (1969), DGM ed. (1969, 1994), and Maier (2019, Chapters 1–3).

another one was published specifically about "nanomaterials science" (Dosch and van der Voorde, 2009; Schummer, 2011).

Another indicator of the growing institutionalization of the field is the tendency during the 1980s and 1990s to replace the plural form of "materials sciences" into the singular "materials science." The loose assortment of individual activities evoked by the former (as visualized in Fig. 6) becomes a more consolidated, coherent discipline (as depicted in Fig. 7). The considerable expansion of university and industrial research, coupled with major building projects for thematically pertinent research clusters, accelerated the establishment of advanced courses for graduates, bachelors and masters students headed by increasing numbers of designated chairs. In Germany, the reassignments of academic institutes and curricula away from chemistry, physics and engineering began in 2000.

Here are some statistics to demonstrate this burgeoning of the field: By 2000, 41 pure MSE departments had been established in the US alone. Add to that 14 specialized departments in the neighboring sciences of physics, chemistry and engineering, along with 41 others at least partly covering such topics. Between 1990 and 1994, approximately 1,100 BA degrees in MSE were issued in the US each year. Consequently, by around 2000, about a third of all collaborators on related projects were graduates with specific expertise in the field. In 1997, about 625 faculty positions existed in MSE. In other countries, the situation is basically similar. In the UK, by 1998, 21 pure MSE departments had been established. Adding the departments only covering subfields, one arrives at a total of roughly 61. In Germany, the approximate total is 79, in Italy 72, in France 48, in Poland 50, in Switzerland 7, and in Sweden 4.[42] Much effort was expended on materials science in Japan too, where a National Institute of Materials Science was founded in 2001 (having emerged from the National

[42] These figures come from Flemings (1999, p. 57) and Bensaude-Vincent (2001). The graphs in Flemings (2001, p. 857f) indicate, from 1982 on, graduates ranging from 1,700 in number to just above 1,000 (2000) each year in a constantly declining trend. Until 1991, these figures run parallel to the sinking numbers of graduations in chemical engineering, after which the latter rise again as MSE continues to stagnate.

Research Institute for Metals in existence since 1956) alongside the National Institute of Advanced Industrial Science and Technology and the older Institute of Physical and Chemical Research (Riken) more closely oriented toward basic science, along with numerous smaller research units also performing research on materials.

Figure 9 provides an overview of the numerous research institutions active in materials science and engineering solely in Germany,

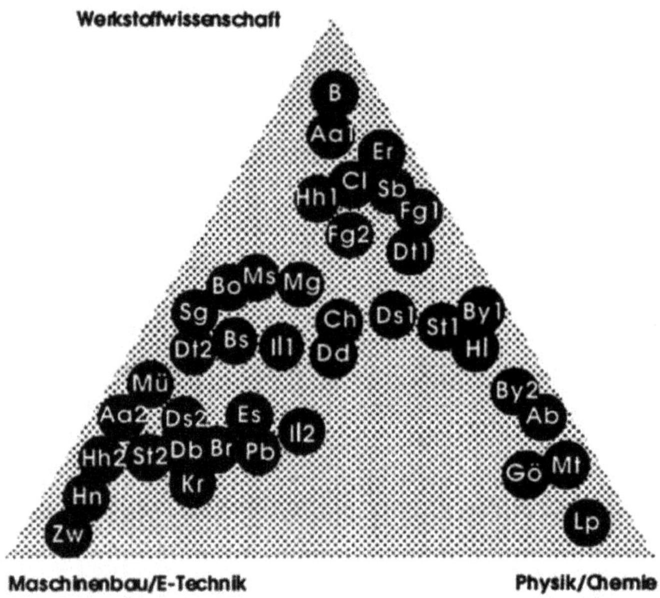

Aa1	Aachen (BHG)	Dt2	Dortmund (Mb)	Kr	Karlsruhe
Aa2	Aachen (Mb)	Dd	Dresden	Lp	Leipzig
Ab	Augsburg	Db	Duisburg	Mg	Magdeburg
By1	Bayereuth	Er	Erlangen-Nürnberg	Ms	Merseburg
By2	Bayreuth(Ch,Fh)	Fg1	Freiberg (WW)	Mü1	München (TU)
B	Berlin (TU)	Fg2	Freiberg (Met)	Mü2	München (BW)
Bo	Bochum	Gö	Göttingen	Mt	Münster
Br	Braunschweig	Hl	Halle-Wittenberg	Pb	Paderborn
Ch	Chemnitz	Hh1	Hamburg (TUHH)	Sb	Saarbrücken
Cl	Clausthal	Hh2	Hamburg (BW)	Sg	Siegen
Ds1	Darmstadt (MW)	Hn	Hannover	St1	Stuttgart (Ch)
Ds2	Darmstadt (Mb)	Il1	Ilmenau (ETIN)	St2	Stuttgart Mb)
Dt1	Dortmund (Ch)	Il2	Ilmenau (Mb)	Zw	Zwickau

Fig. 9. Distribution of the focal specialties of 40 German MSE-related research institutes (DGM, 1994, p. 8).

indicating their relative proximities to the three main poles of MSE: materials science (*Werkstoffwissenschaft*); physics and chemistry; and the engineering sciences, mechanical and electrical engineering. The remarkable thing about this diagram from 1994 for the *Deutsche Gesellschaft für Materialkunde* is that even these materials scientists themselves seldom viewed their institutes as equicentrally located between the poles (with the exception of Chemnitz or Dresden, for instance). The research profiles of the great majority of the institutes continued to be aligned very close to one of the areas. There was still no notion of a uniform Materials Science — in Germany, at least, no claim to disciplinary status was yet being raised.

Thus, if Robert W. Cahn is right about his contention that it was time to regard materials science as a new discipline in 2000, then this would be what he himself considered to be a rare but certainly not singular pattern of "emergence by integration." Whether materials science was in fact a full-fledged discipline was still under dispute, however. Cahn significantly asks himself in his textbook "Can a discipline also be interdisciplinary? Is materials science a real discipline?" (Cahn, 2001a, p. viii; 1997, p. 161). The most important indicator of such a convergence, he and Flemings identify, is a gradual integration of the two largest fields, which had been conducting research independently for so long (Flemings and Cahn, 2000, p. 380):

> "Metallurgy and ceramics are fast becoming absorbed in our thinking into the broader field of MSE, just as nonferrous metallurgy, ferrous metallurgy and extractive metallurgy became subsumed in metallurgy a generation or more ago [...] the rationale [behind viewing the core of our field itself as a discipline] is the intellectual cohesiveness the field has developed over the past 30 years, the industrial recognition it increasingly enjoys over what are becoming the narrower subdisciplines of metallurgy, ceramics, polymers, etc."[43]

[43]Verbatim in Flemings (1999, p. 17) and similarly in (2001, p. 857). The blossoming of metallurgy as the lead science in materials science and its subsequent decline are also discussed in Cahn (2003b, p. 468), citing further literature.

These coauthors also mentioned another indicator that is well known in the literature on discipline formation, namely, the beginnings of insider–outsider demarcation. Sociologists refer to this as "boundary work" (Gieryn, 1983, 1995; Shumway and Messer, 1991; Reinhardt, 2011):

> "We distinguish ourselves from our peers in chemical engineering (who have largely taken over extractive metallurgy) by our expertise in structure and structure–property relations as well as our breadth across materials classes. [...] We similarly distinguish ourselves from mechanical engineering by our focus on the internal structure of materials, specifically on atoms and molecules. [...] Our fundamentals are ... structure, processing, properties and performance. We emphasize relationships among these elements more than other academic departments." (Flemings and Cahn, 2000, p. 381)

The multiple occurrence of the third person plural signals group awareness. Indeed, "we distinguish ourselves from our peers" was clearly laying down a boundary to set themselves apart from their neighboring disciplines, which were considered competitors to MSE in its pursuit of students, resources and ideas.[44]

The "fundamentals" mentioned toward the end of the longer quotation above insinuate the so-called four elements of materials science (COSMAT, 1989), which have often been depicted since in the form of a tetrahedron in textbooks (cf. Kranzberg and Smith (1979, p. 31) and Flemings (1999, p. 2). Its faces show the four approaches of materials science pointing in entirely different directions, ranging from structure analysis to production, harmoniously united within a single regular Platonic shape (see Fig. 10). Depending on the background and priorities of the given author, sometimes

[44]First signs of such a conscious boundary-setting can already be found in Roy ed. (1970, p. viii, 11) (repeated "we"-usage), and (13): an us-*vs.*-them distinction between the "impatience of the metallurgist" and the "occupational disease of the physicist [...] that he typically jumps into a field with a highly simplified model to describe a complex question."

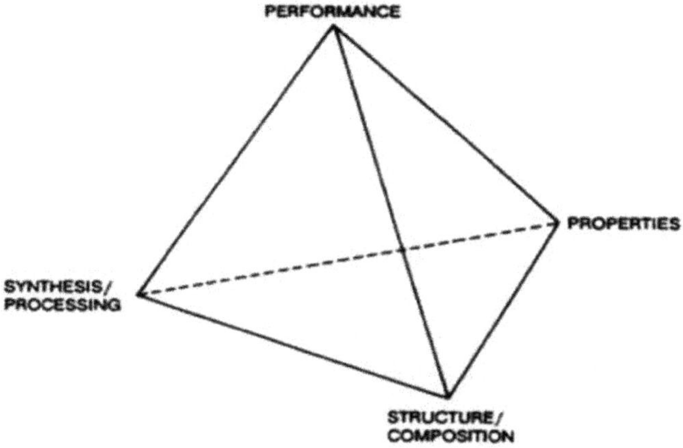

Fig. 10. The four elements of MSE (COSMAT, 1989, p. 29).

"structure/composition" is set at the peak of this regular pyramid. However, it is clear to all that in order for MSE to be successful, all of its four constituent parts must be well developed and fruitfully interactive. The unity in such a diagram was rather symbolic, but it was an important step toward finding a common paradigm.

Robert W. Cahn (2002) recalled in his moody introductory piece for the first issue of *Nature Materials* — itself another indicator of effectuating institutionalization — that, formerly, members not only of the fundamental sciences physics and chemistry but also of engineering would have had great reservations about having materials science as an independent discipline. One only has to think of the distain with which Arnold Sommerfeld's "greasy physics" (*Schmierölphysik*) were received (Eckert, 1993) or Wolfgang Pauli's horrified reaction, "One shouldn't wallow in dirt," to Rudolf Peierl's cautious suggestion to posit a theory of electric conductivity in solids (Cahn, 2002, p. 4). The rapid developments in semiconductor physics and the revolution in microelectronics that it made possible (Queisser, 1985; Riordan and Hoddeson, 1997) transformed this prejudice against impurities and deviations from ideal crystal lattices into mounting interest. The influence of tiny admixtures, henceforth subsumed under the term "doping," gained a positive connotation. Nowadays, their targeted control and influencing, in parts per trillion, have become one of the core tasks of materials science among other microstructural specialties. It has *become* a "science of dirt" (cf. Cahn, 2002; Schummer, 2011). Nevertheless, at least five indicators remain, which — at least in the view of Bensaude-Vincent (2001) — speak against convergence into a single discipline.

First of all, the considerable spectrum of topics and the diversity of methods remain confusing (depending on the special area of application, they range broadly from photonic and magnetic materials to ceramics and composite materials) (Canby, 1989; Sass, 1998, Chapter 14; Haka, 2011, 2020). Second, the core of knowledge transmitted in all textbooks about the field is minimal and is given specific tinctures, depending on the orientation of the local research. Third, there are few canonical subjects in teaching, among them many imports from crystallographic and solid-state physics, such as lattices, defects and surface structures. Fourth, polymers with complex chemistries, for instance, continue to figure rather marginally in MSE, and polymer chemistry is still quite definitely institutionalized in chemistry.[45] Finally, fifth, the self-understanding of the experts continues to weigh unfavorably, with them tending to adopt such conventional professional attributions as solid-state physicist, polymer chemist or engineer.[46]

This general impression of weak coherence prevailing in the substance of materials science led Bensaude-Vincent to pass negative judgment on the question of its homogeneity as a discipline:

> "The overall impression is one of an aggregation of fragments of knowledge taken from various disciplines around a core of atomic physics [...] no inner logic presides over the organization of the discipline forty years after its emergence. Rather, the organization

---

[45] For an internalistic point of view of polymer research and its history, see Morawetz (1985), Seymour (1981, 1989), Carraher (1981), Furukawa (1998), Bijker (1995), Lemstra (2002, pp. 48–51).

[46] Kranzberg and Smith (1979, p. 29) write as follows: "the sense of coherence is still only rudimentary [...] neither a metallurgist nor a polymer chemist nor a solid-state physicst working in the field of MSE tends to think of himself primarily as a materials scientist or engineer."

of most textbooks simply reflects the successive layers of the field's historical development [... there is a] clustering of specialists from various disciplines and an amalgamation of science and technology."[47]

Almost a decade later, upon reconsidering all these indicators of disciplinary development, I arrive at a different conclusion. Pertinent and influential international and national professional associations exist with this denomination along with professorial chairs and entire institutes; likewise, numerous periodicals that comprehend the entire breadth of materials science not just special subdivisions of it. Furthermore, integrative textbooks and courses of study are becoming increasingly acknowledged. An "invisible college" of persons at the core of this new field has long since come together, who see each other regularly at conferences and strategy meetings to coordinate research programs and define new focus areas and who actively perform rhetorical, social and institutional boundary work against neighboring fields and the mother disciplines.[48]

Even with regard to the formation of compulsory paradigms, which for a long while proved to be the most problematic indicator, this field of research has made great progress in the past decade. New course literature uses as its backbone the notion of "structural descriptors" to characterize generally valid properties of materials irrespective of class. The textbook coauthored by the two MIT faculty members, Samuel M. Allen (Professor of Physical Metallurgy at the local Stevens Institute of Technology) and Edwin L. Thomas (Head of the Department of Materials Science and Engineering),

---

[47] Bensaude-Vincent (2001, p. 242; 2011). "Aggregation" is the term that Calvin and others used in 1974 (cf. footnote 33). Thus, Bensaude-Vincent's assessment appears to be more firmly grounded in the state of MSE around 1975. For more recent debates, see also Dusastre (2002, 2003) and Bensaude-Vincent and Hessenbruch (2004).

[48] A draft listing of present-day protagonists in the core field could be drawn up by taking some 100 names in the tables of contents of the two *European White Books* — coordinated by the *Max–Planck-Institut für Metallforschung* at Stuttgart (Rühle, 2002; Dosch and van de Voorde, 2009).

for instance, chooses examples from among crystals, glass types, ceramics or liquid crystals to illustrate morphological defects such as vacancies, migrations, grain boundaries or pores, altered ordering parameters in phase transitions, as well as diffusion or other random processes.[49] The historically oriented textbook already mentioned before, written by the teacher at the College of Engineering of the University of Florida, Rolf E. Hummel (*1935), also demonstrates a balanced view of the different classes of materials and ably integrates this knowledge.

Correlated with this scientification and institutionalization, we observe a growing tendency to dismiss applied and practice-related techniques as mere "materials technology";[50] and the formerly omnipresent abbreviation "MSE" is increasingly abandoned. Other authors rather view the specific situation of materials science as somewhere between the natural and technical sciences on the meso-scale of matter — in the typical orders of magnitude of 10^{-8} m to 10^{-3} m used in their models, experiments and increasingly also in simulations. One of them was the specialist in high-tech ceramics at Stuttgart, Fritz Aldinger (*1941), who directed not only the research department dedicated to materials synthesis and grit texture design at the *Max–Planck-Institut für Metallphysik* but also the *Institut für nichtmetallische anorganische Materialien* at the University of Stuttgart (Aldinger and Baumard, 2002, p. 28; 1999, p. 34). Even smaller

[49] See Allen and Thomas (1998): vi for their (in my opinion legitimate) claim to generic structural description across the "materials catholic." This approach toward covering, unifying and simultaneously classifying all groups of materials is already explicitly recognized by Schatt (1978, p. 361), which he explains is the reason "why there is only one materials science" ("warum es nur eine Werkstoffwissenschaft gibt", original emphasis). His textbook from the early 1970s does not reflect this view, however, cf. furthermore Ashby and Jones (1980).

[50] There is no entry for the term "materials technology" in the *Oxford English Dictionary.* It is first applied as a strategic delimiting term in Turnbull (1983, p. 1). For more recent usages, see, e.g., Rühle (2002, p. 37), where it appears along with the equally unusual designation "solid-state chemistry." There MSE is situated midway between the two extremes of physico-chemical exploratory research, on the one hand, and in proximity to the final product of material technology, on the other.

scales are covered by the rhetorics introduced in the 1990s for "nanoscience and nanotechnology," which are likewise struggling to become institutionalized.

There is a bewildering variety of materials in all economically relevant areas of application, ranging from the automotive and aircraft industries to the building trade. Practically, all classes of materials are continually being optimized in intense research. (Materials for the aircraft industry, for instance, are suspended in a state somewhere between "evolution and revolution" (Hassinger, 2013).) Complex databases being made available to researchers, consumers, and developers interested in practical applications aim to assist in providing some overview. Another database based at my own section for the history of science and technology in Stuttgart presents the historical development of such materials.[51]

Obviously, in attempting to embrace the full range of metals, semiconductors, alloys and nonmetals, glass and ceramics, polymers and composite materials, as well as increasingly also biomaterials, this new field of materials science remains complex and multifarious. Even the editors of the first issue of *Nature Materials* in 2002 had to concede this. Referring to the existence of a "culture gap between scientists involved in research into 'soft' and 'hard' matter," they admitted as follows: "The diversity and 'frontier' spirit can sometimes be problematic." (Cahn, 2002, p. 1) The remaining skepticism about their discipline is attributed to its origins in the technical sciences. In the editorial to the second volume of *Nature Materials*, we read under the dramatic headline "A Matter of Survival" the following appeal, which urges materials scientists to evade the decline of conventional materials-intense industries in the areas of metallurgy and aerospace by diversifying the topics they cover, to include all the different classes of materials even as they stay united within the discipline (Dusastre, 2003, p. 1):

---

[51] For a comparative overview, see https://www.engineeringclicks.com/material-database/. The Stuttgart database *Historische Entwicklung von Werkstoffen* (HEW) is accessible at https://www.hi.uni-stuttgart.de/gnt/hew/.

> "Most materials science departments in the US and UK have their roots in nineteenth century metallurgy and/or mining departments. If they are to flourish and justify their existence in years to come they will need to shake off their traditional image as engineering departments and adapt to new industrial and societal needs. [... T]he need to educate students not only across all materials classes [...] but also across other traditional disciplines [...] is increasingly putting pressure on the materials science curriculum. One solution to this problem may be that, although the field should remain a multidiscipline, it should be thought and taught as a coherent and single discipline."

In his response, Cahn (2003a) pointed out that this is already happening. MSE has dropped its appendix "engineering" at many sites already to become "materials science." Thereby, the list of recognized indicators of the existence of a scientific discipline is complete, and it is indeed proven to be one. The pattern by which this occurred differed fundamentally from the way that disciplines from the 19th and 20th centuries are known to have typically emerged, however (Stichweh, 1984). The institutionalization pattern followed by materials science definitely does not correspond to the classical genesis of disciplines, such as physics and chemistry (around 1800) or their merger in physical chemistry (around 1900).[52] This does not speak against the disciplinary status for materials science though. The actors within MSE or materials technology do not claim "transdisciplinarity" — at best "multidisciplinarity" (such as in the last quotation), but especially since 2000 they often couple this claim with unique disciplinary identity. This self-localization is entirely understandable rationally: There already exist stable institutions,

[52]For the genesis of these disciplines, see, e.g., Meinel (1985, 1987), Stichweh (1984, 1992, 1993), and Servos (1990). On the contrast between the typical development of a discipline and that of MSE, see Kranzberg and Smith (1979, p. 30). Cahn (1997, p. 160f; 2001b) draws explicit comparisons with the genesis of physical chemistry, involving longstanding conflicts between physicists and chemists, as well as with geology. Alternative models are discussed by Ziman (1996) and Shinn (2002) with further literature cited there.

journals, canonical educational content and successful teaching programs that permit systemic reproduction of the material sciences. Since the 1990s, increasingly purposeful boundary-making is ascertainable along with rising rivalry with other disciplines for pecuniary and instrumental means as well as upcoming generations of scientists.[53]

This so awkward borderline between materials science and the major traditional disciplines — particularly between the natural and technical sciences — perhaps also explains why the history of this research terrain has been so little explored by historians of physics, chemistry and engineering, who themselves are often equally divided and focused on their traditional disciplinary terrains.

References

Aldinger, F. (1999). Werkstoffe, die die Welt verändern — Hochleistungskeramiken machen den etablierten Materialien Konkurrenz. *Physikalische Blätter* 55(11), 31–37.

Aldinger, F. and Baumard, J.-F. (2002). Advanced ceramic materials: Summary of possible applications. In: Manfred Rühle *et al.* (eds.), *European White Book on Fundamental Research in Materials Science* (Stuttgart: Max-P-Institut für Metallforschung), 26–31.

Allen, S. M. and Thomas, Edwin L. (1998). *The Structure of Materials* (New York: Wiley).

Amstutz, E. (1955). EMPA: Rückblick und Ausblick. *Schweizer Archiv für angewandte Wissenschaft & Technik* 21, 347–359.

Anonymous (1930). Entwicklung der nichtpreußischen Materialprüfungsanstalten. *Mitteilungen der deutschen Materialprüfungsanstalten,* (6/8), 81–98.

Anonymous (1934). Ein neues Ordinariat für Metallkunde. *Metallwirtschaft* 13, 25–26.

[53] Dusastre's editorial in *Nature Materials* 2003 is cogent: "[O]ther academics departments, notably physics, chemistry, and chemical and mechanical engineering, are also facing waning undergraduate enrolment. Because these departments increasingly tend to view materials science and engineering as a subset of their own field, they all end up competing for the same students, thereby threatening their own existence."

Ashby, M. F. and Jones, David R. H. (1980). *Engineering Materials* (Oxford: Pergamon Press); German trans. by P.P. Schepp: *Ingenieurwerkstoffe. Einführung in ihre Eigenschaften und Anwendungen* (Berlin: Springer, 1986).

Bach, C. and Baumann, R. (1914, 1921). *Festigkeitseigenschaften und Gefügebilder der Konstruktionsmaterialien* (2nd exp. edn.) (Berlin: Springer).

Baumann, R. (1912). Das Materialprüfungswesen und die Erweiterung der Erkenntnisse auf dem Gebiete der Elastizität und Festigkeit in Deutschland während der letzten 4 Jahrzehnte, *Beiträge zur Geschichte der Geschichte der Technik und Industrie* 4, 147–195.

Baumann, R. (1920). Metallographie. In: Otto Lueger (ed.), *Lexikon der gesamten Technik und ihrer Hilfswissenschaften,* supp. vol. 1 (Stuttgart; Leipzig: Dt. Verlags-Anstalt), pp. 449–454.

Bauschinger, J. (1879). Ueber Einrichtung und Ziele von Prüfunganstalten für Baumaterialien und die Classification der letzteren, insbesondere des Eisens und Stahls. *Zeitschrift des Vereines deutscher Ingenieure* 23, 49–66 and plates 4–6.

Bensaude-Vincent, B. (2001). The construction of a discipline: Materials science in the United States. *Historical Studies in the Physical Sciences* 31, 223–248.

Bensaude-Vincent, B. (2011). The concept of materials in historical perspective. *NTM. Zeitschrift für Geschichte der Wissenschaften, Technik und Medizin* 19(1), 107–123.

Bensaude-Vincent, B. and Hessenbruch, A. (2004). Materials science: A field about to explode? *Nature Materials* 3, 345–347.

Bertrand, E. and Bensaude-Vincent, B. (2011). Materials research in France: A short-lived national initiative (1982–1994). *Minerva* 49, 191–214.

Bever, M. B. (1988). *Metallurgy and Materials Science and Engineering at MIT: 1865–1988* (Cambridge, MA: MIT Press).

Bijker, W. E. (1995). *Of Bicycles, Bakelites, and Bulbs. Towards a Theory of Sociotechnical Change* (Cambridge, MA: MIT Press).

Binnig, G. and Rohrer, H. (1985). The scanning tunneling microscope. *Scientific American* 253, 50–56.

Binnig, G. and Rohrer, H. (1987). The scanning tunneling microscope — From birth to adolescence. *Review of Modern Physics* 59, 615–625.

Birr, K. (1957). *Pioneering in Industrial Research: The Story of the General Electric Research Laboratory* (Washington, D.C.: Public Affairs Press).

Blind, D. and Werner, G. (1984). 100 Jahre Materialprüfung in Stuttgart. *Wechselwirkungen*, pp. 65–75.

Bragg, Lawrence (1945). *The History of X-Ray-Analysis* (London: Longmans Green & Co.).

Braun, Hans-J. and Herla, A. (eds.) (2002). *Materials: Research, Development and Applications* (Turnbout: Brepols).

Cahn, R. W. (1995). Physics of materials. In: Laurie M. Brown, Abraham Pais and Brian Pippard (eds.), *Twentieth-Century Physics*, Vol. 3 (Bristol: Institute of Physics Publ.), pp. 1505–1564.

Cahn, R. W. (1997). History of physical metallurgy and of materials science. *Acta Metallurgica Sinica* 33, 2 (197), 157–164.

Cahn, R. W. (2001a). *The Coming of Materials Science* (Oxford: Pergamon).

Cahn, R. W. (2001b). Genesis by definition. *Nature* 410, 307.

Cahn, R. W. (2002). The science of dirt. *Nature Materials* 1, 3–4.

Cahn, R. W. (2003a). Materials departments are facing up to education and recruitment concerns. *Nature Materials* 2, 128–129.

Cahn, R. W. (2003b). Has the child of metallurgy walked out on its parent? *MRS Bulletin* 28, 468–469.

Canby, T. Y. (1989). Advanced materials — Reshaping lives. *National Geographic* 176, pp. 746–781.

Carraher, C. E. (1981). History of polymer education — USA. In: R. B. Seymour (ed.), *History of Polymer Science and Technology* (New York: Dekker), pp. 1237–1261.

Ceruzzi, P. (1989). Electronics, technology and computer science, 1940–1975. A coevolution. *Institute of Electrical and Electronics Engineers Annals of the History of Computing* 10(4), 257–275.

Charles, J. A. and Greer, A. L. (2005). *Light Blue Materials. The Department of Materials Science and Metallurgy, University of Cambridge — A History* (Cambridge: Cambridge University Press).

Choi, H. and Shields, B. (2015). A place for materials science: Laboratory buildings and interdisciplinary research at the University of Pennsylvania, *Minerva* 53, 21–42.

Cohen, M., (ed.) (1979). Materials science and engineering: Its evolution, practice and prospects. *Journal of Materials Education* 1, 1–102.

Chubin, D. E. (1976). The conceptualization of scientific specialties. *Sociological Quarterly* 17, 448–476.

Clayton, R. J. and Algar, J. (1989). *General Electric Company Research Laboratory 1919–84* (New York: IEE).

Cockayne, D. (2002). Electron microscopy. Better vision with electron lenses. *Nature Materials* 1, 7–8.

COSMAT (1974). *Materials and Man's Needs: Materials Science and Engineering*, Washington, D.C. Summary Report by the Committee on the Survey of Materials Science and Engineering. http://books.nap.edu/catalog.php?record_id=10435 (accessed 26 Setember 2010).

COSMAT (1975). *Materials and Man's Needs: Materials Science and Engineering.* Vol. 1: *The History, Scope and Nature of Materials Science and Engineering.* Vol. 2: *The Needs, Priorities, and Opportunities for Materials Research.* Vol. 3: *The Institutional Framework for Materials Science.* Vol. 4: *Aspects of Materials Technology Abroad. Supplementary Report of the Committee on the Survey of Materials Science and Engineering* (Washington, D.C.: National Academy of Sciences).

COSMAT (1989). *Materials Science and Engineering for the 1990's. Report by the Committee on Materials Science and Engineering National Research Council* (Washington, D.C.)

Cunliffe, B. (ed.) (1994). *The Oxford Illustrated Prehistory of Europe* (Oxford: Oxford University Press).

Day, P. (1995). Michael Faraday as a materials scientist, *Materials World* 3, 374–376.

DGM, ed. (1969). *50 Jahre Deutsche Gesellschaft für Metallkunde e.V. im Spiegel der Zeitschrift für Metallkunde* (Berlin: Riederer).

DGM, ed. (1994). *Zur Geschichte der DGM im Spiegel der Zeitschrift für Metallkunde* (Oberursel: Deutsche Gesellschaft für Materialkunde).

Ditchen, H. (2009). *Die Beteiligung Stuttgarter Ingenieure an der Planung und Realisierung der Reichsautobahn unter besonderer Berücksichtigung des Netzwerks von Fritz Leonhardt und Otto Graf* (Berlin: Logos).

Domke, W. (1964) 1999. *Werkstoffkunde und Werkstoffprüfung* (10th edn.) (Essen: Girardet).

Dosch, H. and Van de Voorde, M. H. (eds.). 2009. *GENNESYS White Paper. A New European Partnership between Nanomaterials Science & Nanotechnology and Synchrotron Radiation and Neutron Facilities* (Stuttgart: Max-Planck-Institut für Metallforschung).

Dresselhaus, M. S. (1992). Materials science and cultural change. *New Literary History*, 23, 839–844.

Dusastre, V. (2002). Editorial: Our changing nature. *Nature Materials* 1, 1.

Dusastre, V. (2003). Editorial: A matter of survival. *Nature Materials* 2, 1.

Dusastre, V. (2009). Editorial. *International Journal of Materials Research* 100, 1, 2–8.

DVM (1997). *Deutscher Verband für Materialforschung und -prüfung. Festschrift 1896–1996* (Berlin: DVM).

Eckert, M. (1993). *Die Atomphysiker. Eine Geschichte der theoretischen Physik am Beispiel der Sommerfeldschule* (Braunschweig: Vieweg).

Edyvean, R. G. J. and Hammond, C. (1997). The metallurgical work of Henry Clifton Sorby and an annotated catalogue of his extant metallurgical samples. *Historical Metallurgy* 31(2), 54–85.

Ewald, P. P. (1962). *Fifty Years of X-Ray Diffraction* (Utrecht: Oesthoek).

Fellman, M. (2005). World's first materials science and engineering department turns 50 years old. *Northwestern University Observer Online*, 20 October 2005. http://www.northwestern.edu/observer/issues/2005/10/20/materials.html (accessed 20 December 2010).

Flemings, M. C. (1999). What next for departments of materials science and engineering. *Annual Review of Materials Science* 29, 1–24.

Flemings, M. C. (2001). Why materials science and engineering is good for metallurgy. *Metallurgical and Materials Transactions* 32 A, 853–860.

Flemings, M. C. and Cahn, Robert W. (2000). Organization and trends in materials science and engineering education in the US and Europe. *Acta Materialia* 48, 371–383.

Föppl, L., Wintergerst, S. and Rüsch, H. (1959). Die Materialprüfung an der Technischen Hochschule München. *Materialprüfung* 1, 142–146.

Furukawa, Y. (1998). *Inventing Polymer Science: Staudinger, Carothers and the Emergence of Macromolecular Chemistry* (Philadelphia: University of Penn. Press).

Geison, G. L. and Creager, A. N. K. (1999). Introduction: Research materials and model organisms in the biological and biomedical sciences. *Studies in the History and Philosophy of the Biological and Biomedical Sciences* 30, 315–318.

Gieryn, T. F. (1983). Boundary-work and the demarcation of science from non-science. *American Sociological Review* 48, 781–795.

Gieryn, T. F. (1995). Boundaries of science. In: Jasanoff, Sheila (ed.), *Handbook of Social and Technology Studies* (London; New Delhi: Society for Social Studies of Science), pp. 398–403.

Grandin, K. (ed.) (2004). *The Science Industry-Nexus: History, Policy* (Sagamore Beach, MA: Science History Publications Implications). Nobel Symposium no. 123 (2002).

Granek, G. and Hon, G. (2008). Search for asses, finding a kingdom: The story of the invention of the scanning tunnelling microscope (STM). *Annals of Science* 65, 101–125.

Groenewegen, P. and Peters, L. (2002). The emergence and change of materials science and engineering in the United States. *Science, Technology and Human Values* 27, 111–133.

Hacking, I. (2002). *Historical Ontology* (Cambridge, MA: Harvard University Press); German trans.: *Historische Ontologie* (Zürich: Chronos, 2005).

Haka, A. (2011). Flügel aus „Schwarzem Gold". Zur Geschichte der Faserverbundwerkstoffe. *NTM. Zeitschrift für Geschichte der Wissenschaften, Technik und Medizin* 19, 69–105.

Haka, A. (2020). *Werkstofftechnische Entwicklungen und konstruktive Anwendungen von Hybridwerkstoffen im 19. und 20. Jahrhundert.* submitted Habilitation thesis.

Hammond, C. (1989). The contribution of Henry Clifton Sorby to the study of reflected light microscopy of iron and steel. *Historical Metallurgy* 23(1), 1–8.

Hall, C. (2014). *Materials. A Very Short Introduction* (Oxford: Oxford University Press).

Hanemann, H. and Schrader, A. (eds.). 1933–1952. *Atlas Metallographicus* (Berlin: Bornträger).

Hassinger, P. (2013). *Zwischen Evolution und Revolution. Der Werkstoffwandel im Flugzeugbau* (Karlsruhe: KIT Scientific Publ.).

Hennig, J. (2004). Changes in the design of scanning tunneling microscopic images from 1980 to 1990. *Techné* 8, 1–20.

Hennig, J. (2011). *Bildpraxis. Visuelle Strategien in der frühen Nanotechnologie* (Bielefeld: transkript).

Hentschel, K. (2002). *Mapping the Spectrum. Techniques of Visual Representation in Research and Teaching* (Oxford: Oxford University Press).

Hentschel, K. (2011). Von der Werkstoffforschung zur Materials Science. *NTM. Zeitschrift für Geschichte der Wissenschaften, Technik und Medizin* 19, 5–40.

Hentschel, K. (2014). *Visual Cultures in Science and Technology — A Comparative History* (Oxford: Oxford Universty Press).

Hentschel, K. and Webel, J. (eds.) (2016). *Geschichte und Praxis der Materialforschung* (Diepholz: GNT-Verlag).

Heuck, F. H. W. and Macherauch, Eckard (eds.) (1995). *Forschung mit Röntgenstrahlen: Bilanz eines Jahrhunderts (1895–1995),* (Berlin: Springer).

Higham, N. (1963). *A Very Scientific Gentleman: The Major Achievements of Henry Clifton Sorby* (Oxford: Pergamon Press).

Hoddeson, L. (1977). The roots of solid-state research at Bell Labs. *Physics Today* 30(3), 23–30.

Hoddeson, L., Braun, E., Teichmann, J. and Weart, S. (1992). *Out of the Crystal Maze. Chapters from the History of Solid-State Physics* (Oxford: Oxford University Press).

Hoffmann, D. (2013). Fifty Years of physica status solidi in historical perspective. In*: Physica Status Solidi* B 250(2013) 4, 871–887.

Hounshell, D. and Smith, J. K. (1988). *Science & Corporate Strategy, DuPont R & D, 1902–1980* (Cambridge, MA: Cambridge University Press).

Hummel, R. (1998, 2004). *Understanding Materials Science: History, Properties, Applications* (2nd exp. edn.) (Berlin: Springer).

Hunt, B. (1997). Doing Science in a global empire: Cable telegraphy and electrical physics in Victorian Britain. In: Bernard Lightman (ed.), *Victorian Science in Context* (Chicago: Chicago University Press), pp. 312–353.

Jackson, K., Roy, R.,Voorhoeve, R. J. H. and Picraux, Thomas (eds.) (1993). MRS 20th anniversary. *Materials Research Society Bulletin* 18, 69–87.

Johnson, A. (2009). Material experiments: Environment and engineering institutions in the early American republic. *Osiris* 24, 53–74.

Kahlow, A. (1991). Materials in 19th century Germany. *History and Technology* 7, 255–266.

Kingery, W. D. (1986). *High-Technology Ceramics: Past, Present and Future* (Westerville, OH: American Ceramics Society).

Kirkaldy, D. (1891). *Illustrations of David Kirkaldy's System of Mechanical Testing as Originated and Carried on by Him During a Quarter Century: Comprising a Large Selection of Tabulated Results, Showing the Strength and Other Properties of Materials Used in Construction, with Explanatory Text, and Historical Sketch* (London: Low, Marsden, Searle & Rivington).

Klein, U. and Lefèvre, W. (2007). *Materials in 18th-Century Science. A Historical Ontology* (Cambridge, MA: MIT Press).

Kloos, K. H. and Berger, C. (2007). 100 Jahre staatliche Materialprüfanstalt Darmstadt und 80 Jahre Teilgebiet und Institut für Werkstoffkunde an der TH Darmstadt. *Materialwissenschaft und Werkstofftechnik* 38, 310–346.

Köster, W. (1949). *25 Jahre Kaiser-Wilhelm-Institut für Metallforschung 1921–1946* (Stuttgart: Riederer).

Köster, W. (1961). Max-Planck-Institut für Metallforschung in Stuttgart. *Jahrbuch der Max- Planck-Gesellschaft*, part II, 600–626.

Köster, W. (1969). 50 Jahre Deutsche Gesellschaft für Metallkunde. *Metall* 23, 661–666.

Krankenhagen, G. and Laube, H. (1983). *Werkstoffprüfung. Von Explosionen, Brüchen und Prüfungen* (Reinbek bei Hamburg: Rowohlt).

Kranzberg, M. and Smith, C. S. (1979). Materials in history and society. *Materials Science and Engineering* 37, 1–39.

Krauss, G. (1970). Report of rapporteurs for session II — The value of the university. In Rustum Roy (ed.) (1970). Materials Science and Engineering in the United States (University Park, PA: Pennsylvania State University Press), pp. 37–43.

Kunckel, J. (1679, 1972). *Ars Vitraria Experimentalis, oder Vollkommene Glasmacherkunst* (reprinted edition) (Frankfurt; Günther Stein, Hildesheim: Ohms).

Laitko, H. (2002). Die Disziplin als Strukturprinzip und Entwicklungsform der Wissenschaft. In: Ekkehard Höxtermann and Uwe Hoßfeld (eds.), *Die Entstehung biologischer Disziplinen* (Berlin: Verlag für Wissen und Bildung), pp. 19–55.

Lemaine, G., MacLeod, R., Mulkay, M. and Weingart, P. (eds.) (1976). *Perspectives on the Emergence of Scientific Disciplines* (Mouton; Paris: Maison des Sciences de l'Homme).

Lemmerich, J. (1996). *100 Jahre Röntgenstrahlen* (exhibition catalogue), trans. by A.M. Hentschel: *Röntgen Rays Centennial* (Würzburg: Universty Würzburg).

Lemstra, P. J. (2002). Soft materials and polymers: The rise and decline of polymer science & technology in Europe. In: M. Rühle, H. Dosch, E. J. Mittemeijer and M. H. Van de Voorde (eds.), *European White Book on Fundamental Research in Materials Science* (Stuttgart: Max-Planck-Institut für Metallforschung), pp. 48–50.

Leon, A. (1912). *Die Entwicklung und die Bestrebungen der Materialprüfung* (Vienna: Verein des österreichischen Verbandes für die Materialprüfungen der Technik).

Luxbacher, G. (2010). Werkstoff, Sparstoff, Heimstoff. Metallforschung in Deutschland 1920–1970. In: K. Orth and W. Oberkrome (eds.), *Die Deutsche Forschungsgemeinschaft 1920–1970* (Stuttgart: Steiner), pp. 163–181; trans. by A.M. Hentschel: Thrift Materials, Domestic Materials, and Substitutes — Metals Research in Germany 1920–1970. In: Mark W., K. Orth, U. H. and Rüdiger vom B. (eds.), *The German Research Foundation 1920–1970. Funding Poised between Science and Politics* (Stuttgart: Steiner, 2013) pp. 151–168.

Luxbacher, G. (2011). „Für bestimmte Anwendungsgebiete best geeignete Werkstoffe ... finden". Zur Praxis der Forschung an Ersatzstoffen für

Metalle in den deutschen Autarkie-Phasen des 20. Jahrhunderts. *NTM. Zeitschrift für Geschichte der Wissenschaften, Technik und Medizin* 19, 41–68.

Luxbacher, G. (2018). *Durchleuchten und Durchschallen. Geschichte der Deutschen Gesellschaft für Zerstörungsfreie Prüfung von 1933 bis 2018* (München: Hanser).

Luxbacher, G. (2020). *Ersatzstoffe und Neue Werkstoffe — Metalle, Technik und Forschungspolitik in Deutschland im 20. Jahrhundert* (Stuttgart: Steiner).

Mämpel, U. (1985). *Keramik. Von der Handform zum Industrieguß* (Reinbek: Rowohlt).

Maier, H. (2000). Materials science. In: Arne Hessenbruch (ed.), *Reader's Guide to the History of Science* (London; Chicago: Fitzroy Dearborn), pp. 442–444.

Maier, H. (2002). Ideologie, Rüstung und Ressourcen. Das Kaiser-Wilhelm-Institut für Metallforschung und die „deutschen Metalle" 1933–1945. in: *Rüstungsforschung im Nationalsozialismus: Organisation, Mobilisierung und Entgrenzung der Technikwissenschaften* (Göttingen: Wallstein), pp. 357–388.

Maier, H. (2006). Leichter als Krupp-Stahl? Zur ‚material culture' des Aluminiums seit der Mitte des 19. Jahrhunderts. In: Torsten Meyer and Marcus Popplow (eds.), *Technik, Arbeit und Umwelt in der Geschichte* (New York: Waxmann), pp. 265–279.

Maier, H. (2007). *Forschung als Waffe. Rüstungsforschung in der Kaiser-Wilhelm-Gesellschaft und das Kaiser-Wilhelm-Institut für Metallforschung 1900 bis 1945/48* (Göttingen: Wallstein).

Maier, H. (2009). Ideology, armament and resources: The KWI for metal research and the German metals 1933–1945. In: S. Heim, C. Sachse and M. Walker (eds.), *The Kaiser-Wilhelm Society under National Socialism* (New York: Cambridge University Press), pp. 253–282.

Maier, H. (2019). *100 Jahre Deutsche Gesellschaft für Materialkunde.* Bochumer Studien zur Technik- u. Umweltgeschichte no. 11 (Bochum).

Martens, A. (1880). Das mikroskopische Gefüge und die Kristallisationen des Roheisens, *Zeitschrift des Vereines Deutscher Ingenieure* 24, 397–406 and plates XX–XXI.

Martens, A. (1898). *Handbuch der Materialkunde für den Maschinenbau* (Berlin: Springer).

Martens, A. and Guth, M. (1904). *Das königliche Materialprüfungsamt der Technischen Hochschule Berlin* (Berlin: Springer).

Martin, J. (2018). *Solid State Insurrection: How the Science of Substance Made American Physics Matter* (Pittsburgh: University of Pittsburgh Press).

Mehl, R. F. (1948). *A Brief History of the Science of Metals* (New York: American Institute of Mining and Metallurgical Engineers).

Meinel, C. (1985). Reine und angewandte Chemie — Die Entstehung einer neuen Wissenschaftskonzeption in der Chemie der Aufklärung. *Berichte zur Wissenschaftsgeschichte* 8, 25–45.

Meinel, C. (1987). Zur Sozialgeschichte des chemischen Hochschulfaches im 18. Jahrhundert, *Berichte zur Wissenschaftsgeschichte* 10, 147–168.

Mody, C. (2006). Corporations, universities and instrumental communities. Commercializing probe microscopy, 1981–1996. *Technology and Culture* 47, 56–80.

Mody, C. and Choi, H. (2012). From materials science to nanotechnology: Interdisciplinary center programs at Cornell University, 1960–2000. *Hist. Studies in the Natural Sciences* 43, 121–161.

Morawetz, H. (1985). *Polymers: The Origins and Growth of a Science* (New York: Wiley).

Mott, Nevill F. (ed.) (1980). The beginnings of solid state physics, *Proceedings of the Royal Society London*, A 371, 1–177.

Mott, Nevill F. (1986). *A Life in Science* (London: Taylor & Francis).

Newbury, D. E. and Williams, David B. (2000). The electron microscope: The materials characterization tool of the millenium. *Acta Materialia* 48, 323–346.

Nüßlein, R. (1965). Die Geschichte der ersten deutschen Universalprüfmaschine. *Materialprüfung* 7, 425–432.

Opitz, H. and Dude, We. (1964 1977). *Allgemeine Werkstoffprüfung für Ingenieurschulen* (7th edn.) (Leipzig: VEB Fachbuchverlag).

Piersig, W. (2007). *Adolf Martens — Erinnerungen an den Nestor — Materialprüfungen der Technik* (Norderstedt: GRIN-Verlag).

Piersig, W. (2008). *Emil Heyn. Nestor der Technikwissenschaften — Metallkunde und Metallographie* (Norderstedt: GRIN-Verlag).

Preuß, E. (1913, 1927). *Die praktische Nutzanwendung der Prüfung des Eisens durch Ätzverfahren mithilfe des Mikroskops* (3rd edn.) (Berlin: Springer).

Psaras, P. A. and Langford, H. Dale (eds.) (1987). *Advancing Materials Research* (Washington: National Academy Press).

Pusch, R. (1979). Die Geschichte der Metallographie mit besonderer Berücksichtigung der mikroskopischen Prüfverfahren. *Praktische Metallographie* 16, 26–35, 79–87, 294–300. Compiled In: Richard Pusch (ed.), *Die Geschichte der Metallographie, unter Berücksichtigung der mikroskopischen Prüfverfahren* (Düsseldorf: Verein Deutscher Eisenhüttenleute 1976) Fachausschussbericht Nr. 9005.

Queisser, H. (1985). *Kristallene Krisen. Mikroelektronik* (München: Piper).

Qing, L. (1995). *Zur Frühgeschichte des Elektronenmikroskops* (Stuttgart: GNT-Verlag).

Radkau, J. (2007). *Holz. Wie ein Naturstoff Geschichte schreibt* (München: Oekom).

Ramsauer, C. (1943). *Elektronenmikroskopie: Bericht über Arbeiten des AEG Forschungsinstituts 1930–1942* (3rd exp. edn.) (Berlin: Springer).

Rasmussen, N. (1997). *Picture Control: The Electron Microscope and the Transformation of Biology in America, 1940–1960* (Stanford: Stanford University Press).

Raymond, R. (1986). *Out of the Fiery Furnace — The Impact of Metal on the History of Mankind* (University Park, PA: Pennsylvania State University Press).

Réaumur, R.-A. F. de (1722). *L'art de convertir le fer forgé en acier* (Paris: Brunet).

Reich, L. (1985). *The Making of American Industrial Research. Science and Business at GE and Bell, 1876–1926* (New York: Cambridge University Press).

Reinhardt, C. (2006). *Shifting and Rearranging. Physical Methods and the Transformation of Modern Chemistry* (Sagamore Beach, MA: Science History Publ).

Reinhardt, C. (2011). Habitus, Hierarchien und Methoden: „Feine Unterschiede" zwischen Physik und Chemie. *NTM. Zeitschrift für Geschichte der Wissenschaften, Technik und Medizin* 19, 125–146.

Richter, H.-U. (1999). *Chronik der zerstörungsfreien Materialprüfung* (Berlin: Verlag für Schweißen und verwandte Verfahren DVS-Verlag GmbH).

Riordan, M. and Hoddeson, L. (1997). *Crystal Fire. The Birth of the Information Age* (New York: W.W. Norton).

Robertson, F. (2013). David Kirkaldy (1820–1897) and his museum of destruction: The visual dilemmas of an engineer as man of science. *Endeavour* 37, 125–132.

Roy, R. (1970). Materials research. In: Rustum Roy (ed.), *Materials Science and the Field of Materials.* pp. 115–123.

Roy, R. (ed.) (1970). *Materials Science and Engineering in the United States* (University Park, PA: Pennsylvania State University Press).

Rühle, M., Dosch, H., Mittemeijer, E. J. and Van de Voorde, M. H. (eds.) (2002). *European White Book on Fundamental Research in Materials Science* (Stuttgart: Max-Planck-Institut für Metallforschung).

Ruske, W. (1971). *100 Jahre Materialprüfung in Berlin* (Berlin: Bundesanstalt für Materialprüfung).

Ruske, W., Becker, G. W. and Czichos, H. (1996). *125 Jahre Forschung und Entwicklung, Prüfung, Analyse, Zulassung, Beratung und Information in Chemie- und Materialtechnik.* (Berlin: Bundesanstalt für Materialforschung und -prüfung).

Sass, S. L. (1998). *The Substance of Civilization. Materials and Human History from the Stone Age to the Age of Silicon* (New York: Arcade).

Schatt, W. (1972) 2003. *Einführung in die Werkstoffwissenschaft* (9th edn.) (Leipzig: Deutscher Verlag für Grundstoffindustrie, Weinheim: Wiley-VCH).

Schatt, W. (1978). Wesen und Bestimmung der Werkstoffwissenschaft. *Neue Hütte* 23, 360–362.

Scheel, B. (1989). *Egyptian Metalworking and Tools* (Aylesbury: Shire).

Schulz, E. H. (1949). 100 Jahre Werkstoffprüfung. *Zeitschrift des Vereins Deutscher Ingenieure* 91, 141–147.

Schummer, J. (2011). Nanotechnologie: Eine neue Dynamik an der Schnittstelle zwischen Wissenschaft und Öffentlichkeit. *NTM. Zeitschrift für Geschichte der Wissenschaften, Technik und Medizin* 19, 147–167.

Schwartz, L. H. 1987. Materials research laboratories: Reviewing the first twenty-five years. In: P. A. Psaras and H. D. Langford (eds.), *Advancing Materials Research* (Washington: National Academy Press), pp. 35–48.

Seeger, A. (2009). Verhakungen, dislocations, solitons, and kinks. *International Journal of Materials Research* 100, 24–36.

Servos, J. W. (1990). *Physical Chemistry from Ostwald to Pauling. The Making of a Science in America* (Princeton: Princeton University Press).

Seymour, R. B. (ed.) (1981). *History of Polymer Science and Technology* (New York: Dekker, special issue nos. 6/7 of *J. of Macromolecular Science* A15, 1980).

Seymour, R. B. (ed.) (1989). *Pioneers in Polymer Science* (Dordrecht: Kluwer).

Shinn, T. and Joerges, B. (2001). *Instrumentation between Science, State and Industry* (Dordrecht: Kluwer, Sociology of Science Yearbook no. 22).

Shinn, T. (2002). The triple helix and the new production of knowledge. Pre-packaged thinking on science and technology. *Social Studies of Science* 32, 599–614.

Shumway, D. and Messer-D., E. (1991). Disciplinarity: An introduction. *Poetics Today* 12(2), 201–225.

Sigwart, H. (1986). *Aus der Geschichte der Werkstoffprüfung und der Festigkeitslehre* (Berlin: Deutscher Verband für Materialprüfung).

Smith, C. S. (1960) 1988. *A History of Metallography: The Development of Ideas on the Structure of Metals before 1890* (Chicago: University of Chicago Press, reprinted Cambridge, MA: MIT Press).

Smith, C. S. (1961). Retrospective notes on a changing profession. *Archaeomaterials* 1, 3, 11.

Smith, C. S. (ed.) (1965). *The Sorby Centennial Symposium on the History of Metallurgy* (New York: Gordon & Breach).

Smith, C. S. (ed.) (1968). *Sources for the History of the Science of Steel, 1532–1786* (Cambridge, MA: MIT Press).

Smith, Cy. S. (1969). The development of ideas on the structure of metals. In: Marshall Clagett (ed.), *Critical Problems in the History of Science* (Madison, WI), pp. 467–498.

Smith, C. S. (1977). *Metallurgy as a Human Experience* (Materials Park, OH: American Society for Metals).

Smith, C. S. (1986a). On material structure and human history. *Annual Review of Materials Science* 16(1), 11.

Smith, C. S. (1986b). Retrospective notes on a changing profession. *Archaeomaterials* 1, 3–11.

Sonnemann, R. and Friedrich, E. (1978). Die Werkstoffwissenschaft im Spiegel der Geschichte der Technischen Universität Dresden. *Neue Hütte* 23, 357–360.

Sonnenberg, G. S. (1968). *100 Jahre Sicherheit. Beiträge zur technischen und administrativen Entwicklung des Dampfkesselwesens in Deutschland* (Düsseldorf: Verein Deutscher Ingenieure).

Sorby, H. C. (1897). *Fifty Years of Scientific Research. An Address Delivered before the Members of the Shefield Literary and Philosophical Society* (Sheffield: Independent Press).

Sproull, R. L. (1987). The early history of the materials research laboratories, *Annual Review of Materials Science* 17, 1–12.

Stichweh, R. (1984). *Zur Entstehung der wissenschaftlichen Disziplinen: Physik in Deutschland, 1740–1890* (Frankfurt a.M.: Suhrkamp).

Stichweh, R. (1992). The sociology of scientific disciplines. *Science in Context* 5, 3–15.

Stichweh, R. (1993). Wissenschaftliche Disziplinen — Bedingungen ihrer Stabilität im 19. und 20. Jahrhundert. In: J. Schriewer, E. Keiner and C. Charle (eds.) *Sozialer Raum und akademische Kulturen* (Frankfurt a.M.: Peter Lang), pp. 235–250.

Strahm, C. (1995). Die Anfänge der Metallurgie in Mitteleuropa. *Helvetia Archaeologica* 25(97), 2–39.

Stratmann, M. (2006). Materials science: A cross-cutting area between scientific research and industrial development. *Max Planck Forum* 7, 209–227.

Thiel, C. and Peckhaus, V. (eds.) 1999. *Disziplinen im Kontext* (München: Fink).

Timoshenko, S. P. (1953). *History of the Strength of Materials* (New York: McGraw Hill).

Trendelenburg, F. (1975). *Aus der Geschichte der Forschung im Hause Siemens* (Düsseldorf: Verein Deutscher Ingenieure, Technikgeschichte in Einzeldarstellungen, no. 31).

Turnbull, D. (1983). A commentary on the emergence and evolution of materials science. *Annual Review of Materials Science* 13, 1–7.

Tylecote, R. F. (1976, 1992). *A History of Metallurgy* (2nd edn.) (London: The Institute of Materials).

Ude, H. (1935). Zur Geschichte der Eisenbahnwerkstoffe. *Technikgeschichte* 24, 38–60.

Urban, K. (1972). *Die Ziel-Inhalt-Methode-Relation in der Lehre der Werkstoffwissenschaft für Ingenieure des Maschinenbaus an Hochschulen der DDR* (TU Dresden: Dissertation A).

Vogler, D. (2008) *Der technikwissenschaftliche Denkstil in seiner sprachlichen manifestation, dargestellt am Beispiel der Werkstoffwissenschaft* (Hamburg: Kovač).

Wagner, D. B. (2000). Metallurgy. In: A. Hessenbruch (ed.), *Reader's Guide to the History of Science* (London; Chicago: Fitzroy Dearborn), pp. 472–473.

Weaire, D. J. and Windsor, Colin G. (eds.) (1987). *Solid State Science. Past, Present and Predicted* (Bristol: Hilger).

Weart, S. (1988). The birth of the solid-state physics community. *Physics Today* 41(7), 38–45.

Werner, O. (1964). Adolf Martens — zur 50. Wiederkehr seines Todestages. *Materialprüfung* 6, 249–250.

Weitze, M.-D., and Berger, Christina (2013). *Werkstoffe. Unsichtbar aber unverzichtbar* (Berlin; Heidelberg: Springer).

Wise, G. (1980). A new role for professional scientists in industry: Industrial research at general electric 1909–16. *Technology and Culture* 21, 408–429.

Wickramasinghe, H. K. (2000). Progress in scanning probe microscopy. *Acta Materialia* 48, 347–358.

Wöhler, A. (1877). Denkschrift über die Einführung einer staatlich anerkannten Classification von Eisen und Stahl. *Zeitschrift des Vereines deutscher Ingenieure* 21, 518–523.

Ziman, J. (1995). Some reflections on physics as a social institution. In: Laurie M. Brown, Abraham Pais and Brian Pippard (eds.), *Twentieth-Century Physics,* Vol. 3 (Bristol: Institute of Physics Publications), pp. 2041–2059.

Ziman, J. (1996). Post-academic science: Constitution knowledge with networks and norms. *Nature* 382, 751–754.

Zweckbronner, G. (1985). Materialprüfung in Wechselwirkung zur technisch-industriellenEntwicklung, dargestellt am Beispiel Carl Bach. *Ferrum* 56, 39–42.

Chapter 1.3

Institutional Development of the Organizational Field of Synchrotron Radiation Facilities in Europe

Olof Hallonsten and Thomas Heinze

Introduction

The growth of the use of synchrotron radiation as a technical tool for experimental science in the past 50 years has been dramatic. Originally, a lab curiosity and a very unreliable tool, synchrotron radiation, has become a major experimental resource with tens of thousands of users worldwide in a range of scientific disciplines, including many branches of physics, chemistry, biology, materials science, and environmental science. The development has been enabled and driven not only by both technical and organizational improvements of infrastructure, user interfaces, and many crucial complementary services, such as sample preparation and scheduling of experiments, but also by an increasing demand in scientific communities for the high-intensity X-ray (and UV and IR) radiation that these experimental facilities produce.

This chapter chronicles the development, with specific focus on Europe and with the help of institutional analysis, using neoinstitutional organization theory and conceptual tools from the complementary historical-institutionalist tradition. It takes as the point of departure the realization, established in previous publications (Hallonsten and Heinze, 2013, 2015; Heinze, *et al.* 2015a, 2015b; Hallonsten, 2015, 2016), that synchrotron radiation facilities and the organizational fields they form are very promising objects of study from the perspective of multilevel and multidimensional continuity and change, and thus how organizational fields in science form, expand, and mature in gradual and cumulative processes. Synchrotron radiation facilities are highly complex but simultaneously very distinct examples of continuity and change in science: they operate large-scale scientific instrumentation (Big Science) for a wide range of ordinary small-scale research projects (Little Science) and therefore take active part in the crucial development of scientific instrumentation and its use, enmeshed in a complex web of alliances with various scientific communities, research funders, policy actors, and other organizations. The growth of the organizational field of synchrotron radiation facilities in Europe began in the 1960s, with the launch of small-scale experimental activities within pre-existing accelerator-based facilities for particle physics. Today, some 40 dedicated and purpose-built synchrotron radiation facilities are in operation worldwide, with close to 15 in operation in Europe, depending on the count (Hallonsten, 2016, pp. 258–260). This growth is in itself remarkable, but it is the delineation of its dynamics and subprocesses that catches the interest of historically interested sociologists of science and that hence is the topic of this chapter.

The chapter begins with an empirical section, which summarizes the emergence and growth of synchrotron radiation as an experimental resource and the concurrent growth of the number of labs in Europe. This historical account is, naturally, very condensed or even curtailed, and we therefore urge the interested reader to seek complementary sources that give more detailed accounts (for Europe, e.g., Cramer, 2017, 2020; Hallonsten and Heinze, 2013, 2015; Hallonsten, 2011, 2016; Heinze *et al.*, 2015a, 2015b, 2017; for the United States, e.g., Crease, 2008, 2009; Hallonsten, 2015; Westfall, 2008, 2012). In the third section, the theoretical frameworks are presented. Thereafter, in the fourth section, comes the analysis and discussion, followed by a brief section with conclusions.

The key findings of the chapter are the following: First, in line with neoinstitutional organization theory, we observe that most synchrotron radiation facilities were originally established within pre-existing research organizations, such as universities and national research institutes. Second, we note that the formation of the organizational field gained momentum particularly once dedicated single-mission synchrotron radiation facilities began to emerge outside pre-existing research organizations. Third, we conclude that field expansion occurred primarily because of the eventual consolidation of customary technical designs and reliable user interfaces. Fourth, in line with historical-institutionalist theory, we identify several gradual and cumulative changes on the organizational level as well as on the field level that characterized the whole process. Fifth, and finally, we conclude that once a period of expansion and consolidation ensued, the field began to exhibit isomorphic change, as facilities started to imitate each other technologically and organizationally.

History

The early exploration of synchrotron radiation as a tool in experimental science began in the early Cold War period when particle physics was in its most expansive period and new accelerator facilities were built at a fast pace on both sides of the Iron Curtain. Synchrotron radiation was originally an unwanted waste product of particle acceleration and collision, and so the first attempts to make use of this very intense radiation for advanced studies of materials happened in "parasitic" mode at particle physics facilities, including at the West German national lab the German Electron-Synchrotron (Deutsches Elektronen-Synchrotron, DESY) in Hamburg (Heinze *et al.*, 2015a). Accelerator development in the 1960s not only brought drastic improvements in the performance of collision experiments in particle physics but also for the exploitation of synchrotron radiation: When a new type of machine design, called storage rings, reached enough maturity to be turned into actual infrastructure, this meant a quantum leap in the quality of the synchrotron radiation because storage rings emit continuous beams of light whereas the predecessor, synchrotrons (from which synchrotron radiation owes its name), only deliver short (millisecond) flashes. DESY was one of the facilities that pioneered storage ring construction in the 1960s, and in 1974, its DORIS machine started operation and began to be used scientifically both in particle physics and for experiments with synchrotron radiation in spectroscopy, scattering, microscopy, and imaging. These and other activities elsewhere in the world (most notably at Stanford University in California) in the mid-1970s exceeded expectations on what synchrotron radiation could deliver and spurred efforts to make technical improvements in order to simplify user interfaces and increase reliability without interfering with the particle physics research program on whose generosity the activities still operated (Hallonsten and Heinze, 2013, p. 596). One key innovation was the so-called insertion devices, put in the straight sections of storage rings, that significantly enhance the quality and not least the intensity of the radiation and enable the fine-tuning of wavelength spans, as well as the emission of radiation in far more focused beams (Hallonsten, 2016, p. 245). This technology has since been optimized for various specialized purposes and is nowadays an infrastructural backbone of synchrotron radiation facilities. In the mid-to-late 1970s, results coming out of the parasitic synchrotron radiation labs at DESY and in other parts of the world (most notably the United States, Japan, and the Soviet Union) inspired initiatives to make synchrotron radiation break free from the unreliable and parasitic status at particle physics labs and design and construct dedicated, non-parasitic facilities. In Europe, the expansion of the intergovernmental particle physics facility CERN meant the monopolization of national particle physics budgets, with the effect that a number of particle accelerator facilities closed and their designers and operators became unemployed (Hallonsten, 2014, p. 36). Making use of this abandoned competence, new dedicated synchrotron radiation facilities were launched, including the Synchrotron Radiation Source (SRS) in Daresbury, UK, opened in 1981, and the

MAX laboratory in Lund, Sweden, opened in 1986 (Hallonsten, 2011). These were second-generation facilities, succeeding the first-generation, parasitic synchrotron radiation labs.

During the 1980s, the third generation emerged: Synchrotron radiation facilities designed and built with storage rings and all other equipment optimized for the use of insertion devices. The pan-European, collaborative European Synchrotron Radiation Facility (ESRF) in Grenoble (opening to scientific use in 1994), the ELETTRA facility in Italy (1991), the DELTA in Germany (1995), the MAX II in Lund, Sweden (1997), and the BESSY II in Germany (1998) were all third-generation facilities, built in response to the growing demand for high-quality synchrotron radiation in European scientific communities and taking a major role in the continued expansion of the user base throughout the 1990s. Most of them were optimized for somewhat longer wavelengths (UV and the so-called soft X-rays), but the ESRF stood out in this regard, as it was built specifically to deliver shorter wavelength radiation (the so-called hard X-rays), which especially the growing applications in biology and some parts of condensed matter physics demanded. But at the time, the construction of such hard X-ray synchrotron radiation facilities required very large (several hundred meters circumference) accelerator rings, which typically was beyond the reach of the science budgets of individual European countries — hence, the ESRF was a collaboration between 12 countries and one of only three such large, hard X-ray, third-generation facilities in the world, with the other two built in Japan and the United States in the early-to-mid-1990s (Hallonsten and Heinze, 2015: 843–846). These facilities allowed a significant expansion of the user base for synchrotron radiation and not least the broadening of the range of applications to structural biology, where synchrotron radiation is today used as a key tool to map and analyze the structure of macromolecules, such as proteins. In addition, the stable operations enabled by third-generation facilities also made synchrotron radiation more accessible and user interfaces more inviting, not to mention the giant leaps in reliability and throughput. The success was monumental: From 1990 to 2010, the use of synchrotron radiation in structural biology multiplied several times (Hallonsten, 2016, pp. 251–252), and since 1997, a number of Nobel Prizes in chemistry have been awarded to discoveries that rely heavily on the use of synchrotron radiation (Hallonsten and Heinze, 2015, p. 846).

The technical improvements accomplished in the 1990s concerned almost all components of synchrotron radiation facilities and both improved performance and lowered costs. In the late 1990s, innovation on the side of accelerator technology enabled the design of what came to be called third-generation *intermediary energy* synchrotron radiation facilities, which combine the performance of the hard X-ray sources and the soft X-ray/UV sources in smaller and less costly designs. The Swiss Light Source (SLS) at the Paul Scherrer Institute (PSI) in Villigen, Switzerland, was the first such facility, opening for scientific use in 2001 and followed by the SOLEIL in France (2006), DIAMOND in the UK (2007), ALBA in Spain (2010), MAX IV in Sweden (2016), and SOLARIS in Poland (2016), and a number of similar facilities elsewhere in the world (Hallonsten, 2016, pp. 259–260). With this growing number of facilities in an organized and structured field came the proliferation of a number of best practices for accelerator and instrument operation, scientific user support, user access (including peer review of proposals for experiments), sample preparation and handling, data management, parallel accommodation of vastly different scientific and technical requests from users, and the continuous development and refinement of technologies as well as organizations. Indeed, whereas in earlier times, the pace of development of storage ring and instrument technologies made competitive edge hinge on unique designs to enable exceptional capabilities in niched areas, there is significantly more copying today. An extreme example is the SOLARIS facility in Poland, which uses an exact copy of the storage ring design of MAX IV in Lund, Sweden.

Theory

Institutional theory of various shades has reached an enormously popular position in the social sciences in the past decades, even to the extent that some social scientists have claimed that "we are all institutionalists now" (Pierson and Skocpol 2002, p. 706). Especially neoinstitutional organization theory, which operates on the analytical level of organizational fields and which theorizes imitation of practices and structures (isomorphism) as a key feature of such fields, has won great acclaim and become mainstream not only in organization studies but also in the social sciences generally. Among other things, neoinstitutional theory postulates that new organizational fields emerge from pre-existing organizational settings and that once a field reaches some level of maturity, it will exhibit strong isomorphic developments, meaning that the organizations within the field imitate each other in order to build and maintain legitimacy and status (DiMaggio and Powell, 1983; Greenwood *et al.*, 2008). These postulations are used in the analysis of this chapter (see the following section), along with tools from another branch of institutional theory that developed mostly within political science, but has been used in recent studies of evolution and transformation of institutional setups of national science systems (e.g. Heinze and Münch, 2016; Heinecke, 2016; Hallonsten, 2018), namely, historical institutionalism.

Importantly, historical institutionalism challenges previous assumptions in institutional theory that institutions are characterized by continuity and stability in the long term and that their formation or change can only be accomplished by exogenous shocks or ruptures at the so-called critical junctures. In contrast with this view, historical institutionalism identifies that change is gradual, incremental, and cumulative, and operates on several levels and in several dimensions. Change processes conceptualized within this tradition include *layering*, *conversion*, *displacement*, *dismantling*, and *drift*, and an application of these processes as conceptual tools has been shown to be useful in analyses of long-term change that is neither exogenous nor disruptive, but instead conducive to long-term, macro-level institutional persistence through adaptation and renewal (Mahoney, 2000; Thelen, 2004; Streeck and Thelen, 2005; Hallonsten and Heinze, 2012; Heinze and Münch, 2016). In this chapter, we make use of the concepts of layering, conversion, and displacement to analyze the development and transformation of the organizational field of synchrotron radiation facilities in Europe in the past few decades. The concepts are defined as follows: Layering is a process whereby new arrangements are added on top of pre-existing structures, which enables the accommodation of new structures without excessive compromise of the logic of the pre-existing ones. Conversion is a process whereby capacities for one purpose are redirected to other purposes, and neither new additions are made or existing capacities terminated. Displacement is similar, but means the succession of capacities by the dismantling of existing ones, and the creation of new ones. Examples, in the current setting, include the following. The launch of entirely new facilities is a case of layering, as is the addition of new instrument capacities at an existing lab. The redirection of a facility away from one area of use (in this context, particle physics) to a new area of use (in this context, synchrotron radiation) is a case of conversion, as is the redirection of specific instruments from one area of use to another (such as from a subfield of materials science to a subfield of structural biology, within the same synchrotron radiation facility). The dismantling of a facility and the creation of a new one in its place would be a case of displacement.

Analysis

The initiation period of the organizational field of synchrotron radiation occurred in the mid-to-late 1970s when the results of early pioneering experiments proliferated and the first dedicated facilities were planned. The parasitic use of particle physics facilities in the 1960s, most notably at DESY, is defined as a *pre-initiation phase* that paved the way, but nonetheless occurred in the context of the predecessor organizational field of particle physics facilities. While results proved the usefulness of synchrotron radiation and provided the motivation for continued investments, significant obstacles remained, most of all the limitations of parasitic operations, and these had to be overcome in order for the field to be established in its own right. The initial efforts to establish synchrotron radiation facilities as parasitic operations within particle physics facilities thus constituted *intra-organizational layering*, as it was accomplished by adding new technical equipment and scientific/technical staff to existing particle physics infrastructure within existing organizations. The prestigious nature of these particle physics organizations (derived from the combination of scientific breakthroughs and science policy support during the Cold War) inhibited developments of synchrotron radiation in the short run but enabled it in the long run: The invention of insertion devices to overcome the limitations of existing storage-ring-based radiation sources would prove momentous to the longer-term development, especially later on, when insertion devices became the infrastructural backbone of the successful third-generation facilities. In other words, crucial field-level elements of what allowed the formation and expansion of the field in the 1980s and especially the 1990s had already been initiated at this early parasitic stage (Hallonsten and Heinze, 2015, p. 844).

The impressive scientific results that emerged from parasitic synchrotron radiation activities in the mid-1970s eventually led research funders to begin efforts of securing the supply of synchrotron radiation for a growing and demanding user community, and the alliances thus forged between scientific communities, instrument and infrastructure design and construction communities, and science funders and thus be named pivotal to the initiation and first expansion of the organizational field, well in line with expectations and with what neoinstitutional theory would predict. But the initiation period, i.e., the launch of the second-generation sources, mostly happened through *intra-organizational layering*, as new facilities were built within existing labs and universities (MAX at Lund University, ASTRID at Aarhus University). The result is, however, not unambiguous: exceptions exist in the shape of SRS at Daresbury and BESSY in Berlin, which were both greenfield projects and thus constituted *field-level layering* (Hallonsten and Heinze 2015, p. 845).

The step from formation to expansion of the field occurred when second-generation facilities proved their worth, and insertion device technology simultaneously became reliable and mature enough to build entire facility designs on. Considerable investment in new facilities brought the organizational field into the phase of expansion, with two distinct types of facilities — hard X-rays and UV/soft X-rays — through *intra-organizational layering* (MAX II in Lund), *intra-organizational displacement* (BESSY II replacing BESSY I), and *field-level layering* (ELETTRA, DELTA, and the ESRF). Though the ESRF was not the first facility of the third generation (designed and built purposefully for the use of insertion devices), it demonstrated decisiveness on behalf of policymakers and funders, expanded capacity significantly by its potential to serve thousands of users annually, and became arguably the most influential facility in the institutionalization of best practices at the field level, through its purposeful optimization of technologies and organizational practices, including user support. Field expansion continued into the 2000s, with *field-level layering* (SOLEIL, ALBA), *intra-organizational layering* (SLS, ANKA, ASTRID II), and *field-level displacement* (DIAMOND). Meanwhile, the emergence of several intermediary energy facilities in the first decade of the new millennium also indicates that the field had entered a period of consolidation, most of all shown by the proliferation of best practices and the eventual imitation of designs, which constitutes isomorphism as conceptualized in neoinstitutional organization theory (Hallonsten and Heinze, 2015, p. 846).

Conclusions

The analysis in the previous section showed how a combination of concepts from neoinstitutional organization theory and historical institutionalism can be of great use in the sociology and history of science, and science policy and management studies, by contributing a broadening of their conceptual and empirical reach. In this endeavor, this chapter follows the examples of Heinecke (2016), Hallonsten and Heinze (2012, 2013, 2015), and Hallonsten (2016, 2018), and presents a snapshot of what institutional analysis of the development and expansion of new organizational fields in science can provide in terms of new insights into the evolutionary dynamics of science systems. Arguably, the field-level perspective and the longitudinal analysis are instrumental for deepening the understanding of how contemporary publicly funded science is organized and how it renews itself on the basis of scientific advance and technological innovation in connection with broader shifts in society, economy, and its priorities. Specifically, the analysis in this chapter makes the following contributions.

The organizational field of synchrotron radiation facilities emerged after the waste product of prestigious particle physics laboratories proved useful for experimental work in its own right, with its own infrastructure and organizations. Technical limitations at the early, pre-initiation stage sparked technological innovation that would later form the infrastructural backbone of the entire organizational field, namely, insertion device technology. The pre-initiation phase was characterized by small-scale internal organizational modifications (*intra-organizational layering*) that would eventually prove very influential. The real formation of the organizational field however occurred once dedicated single-mission synchrotron radiation facilities were established on greenfield sites on the basis of an acknowledgment of the potential of such standalone exploitation of the radiation and the ability of funders, scientific communities, and instrumentation design and construction communities to jointly provide the means for realizing this potential. Field expansion was enabled by the consolidation of the insertion device-based storage ring technology, in whose tracks followed the development of reliable infrastructural setups operated by streamlined organizations and routinely used by scientists from a wide range of disciplines.

Verifying the postulations of historical institutionalism, the chapter has demonstrated that several gradual but cumulative processes of institutional change occurred at different phases of the overall development, both within single organizations and on the field level: *intra-organizational layering*, meaning the adding of major new experimental facilities within pre-existing research organizations, which happened over the entire period covered, and *field-level layering*, meaning the launch of new, purpose-built synchrotron radiation facilities on greenfield sites, which was the single most important process in the expansion phase. There are also signs of consolidation in the later phase of expansion, paired with the evident tendencies of isomorphic change on the technological and organizational side, and culminating in the intermediary energy third-generation facilities that are very much alike in size, technological setup, and organizational structures.

Through this analysis and these conclusions, the particular case of the emergence and growth of synchrotron radiation as an experimental resource in science, and the formation and expansion of the new organizational field of synchrotron radiation facilities, are better understood. Change is essential to science, and the analysis of change is essential to history and sociology alike and highly relevant to science policy studies, as it not only adds new perspectives and angles but also provides crucial insight into the long-term growth and proliferation of new techniques, new experimental tools, and new ways of organizing the utilization of these tools in experimental science. Specifically, the game of priorities in science policy would be well served by the deeper analysis of institutional change with the use of historical institutionalism, given the keys that it provides for understanding how layering, conversion and displacement operate in the launch of new infrastructures parallel to the reconfiguration or downsizing of old ones. The analysis in this chapter shows that the ability of science funders and policymakers to invest in seemingly marginal and esoteric efforts — or, more specifically, parasitic synchrotron radiation research at particle physics labs — was what provided the growth environment where synchrotron radiation as an experimental tool could eventually flourish. The lesson is clear: also seemingly peripheral activities, and activities that challenge dominant structures, deserve attention and consideration for support. In the longer term, however, the formation and expansion of the organizational field of synchrotron radiation facilities were only possible by their liberation from the organizational and infrastructural exile in particle physics facilities. Therefore, it can be concluded that a growth environment alone does not suffice, and successful experimental science requires standalone organizations and institutional independence.

References

Cramer, K. C. (2017). Lightening Europe: Establishing the European Synchrotron Radiation Facility (ESRF). *History and Technology* 33(4), 396–427.

Cramer, K. C. (2020). *A political history of Big Science: The Other Europe* (London: Palgrave Macmillan).

Crease, R. (2008). The national synchrotron light source, Part I: Bright idea. *Physics in Perspective* 10, 438–467.

Crease, R. (2009). The national synchrotron light source, Part II: The bake-out. *Physics in Perspective* 11, 15–45.

DiMaggio, P. and Powell, W. W. (1983). The iron cage revisited: Institutional isomorphism and collective rationality in organizational fields. *American Sociological Review* 48, 147–160.

Greenwood, R., Oliver, C., Sahlin, K. and Suddaby, R. (eds.) (2008). *The Sage Handbook of Organizational Institutionalism* (London: Sage).

Hallonsten, O. (2011). Growing big science in a small country: MAX-lab and the Swedish research policy system. *Historical Studies in the Natural Sciences* 41, 179–215.

Hallonsten, O. (2014). The politics of European collaboration in Big Science. In: M. Mayer, M. Carpes, and R. Knoblich (eds.), *The Global Politics of Science and Technology*, Vol. 2. (New York: Springer).

Hallonsten, O. (2015). The parasites: Synchrotron radiation at SLAC, 1972–1992. *Historical Studies in the Natural Sciences* 45, 217–272.

Hallonsten, O. (2016). *Big Science Transformed. Science, Politics and Organization in Europe and the United States.* (London: Palgrave Macmillan).

Hallonsten, O. (2018) Development and transformation of the third sector of R&D in Sweden, 1942–2017. *Science and Public Policy* 45(5), 634–644.

Hallonsten, O. and Heinze, T. (2012). Institutional persistence through gradual adaptation: analysis of national laboratories in the USA and Germany. *Science and Public Policy* 39(4), 450–463.

Hallonsten, O. and Heinze, T. (2013). From particle physics to photon science: Multidimensional and multilevel renewal at DESY and SLAC. *Science and Public Policy* 40(5), 591–603.

Hallonsten, O. and Heinze, T. (2015). Formation and expansion of a new organizational field in experimental science. *Science and Public Policy* 42(6), 841–854.

Heinecke, S. (2016). The gradual transformation of the Polish Public science system. *PLoS ONE* 11(4), e0153260.

Heinze, T. and Münch, R. (eds.) (2016). *Innovation in Science and Organizational Renewal. Historical and Sociological Perspectives* (London: Palgrave Macmillan).

Heinze, T., Hallonsten, O. and Heinecke, S. (2015a). From periphery to center: Synchrotron radiation at DESY, Part I: 1962–1977. *Historical Studies in the Natural Sciences* 45(3), 447–492.

Heinze, T., Hallonsten, O. and Heinecke, S. (2015b). From periphery to center: Synchrotron radiation at DESY, Part II: 1977–1993. *Historical Studies in the Natural Sciences* 45(4), 513–548.

Heinze, T., Hallonsten, O. and Heinecke, S. (2017). Turning the Ship: The transformation of DESY, 1993–2009. *Physics in Perspective* 19, 424–451.

Mahoney, J. (2000). Path dependence in historical sociology. *Theory and Society* 29, 507–548.

Pierson, P. and Skocpol, T. (2002). Historical institutionalism in contemporary political science. In: I Katznelson and H. V. Miller (eds.), *Political Science — State of the Discipline* (New York: Norton).

Streeck, W. and Thelen, K. (2005). Introduction: Institutional change in advanced political economies. In: W. Streeck and K. Thelen (eds.), *Beyond Continuity. Institutional Change in Advanced Political Economies* (New York: Oxford University Press).

Thelen, K. (2004). *How Institutions Evolve* (Cambridge, MA: Cambridge University Press).

Westfall, C. (2008). Retooling for the future: Launching the advanced light source at Lawrence's laboratory, 1980–1986. *Historical Studies in the Natural Sciences* 38, 569–609.

Westfall, C. (2012). Institutional persistence and the material transformation of the US National Labs: The curious story of the advent of the advanced photon source. *Science and Public Policy* 39, 439–449.

Part II

How Sources Evolve: Neutron, Synchrotron, and Spallation

The chapters in Part II, "Facilities," are about the networks of institutions involved in three different kinds of large-scale materials science facilities: reactors in the US and Europe, the ESS spallation source in Sweden, and Brazilian synchrotron sources. These articles show how the processes outlined in the preceding section play out in practice.

Thomas Kaiserfeld's chapter, "Three Generations of Institutions of Neutron Sources," challenges the periodization of neutron sources and therefore the institutions built to support them. What is the dividing line of the first two generations — enriched radioactive isotopes followed by fission reactors, or pre- and post-Atoms for Peace/Cold War reactors? Does the third generation consist of pulsed reactors, spallation sources, or inertial fusion devices? Various periodizations move sources from one generation to another. Generations of neutron sources are defined and redefined, Kaiserfeld writes, in the service of the promotion of potential future sources, indicating the "relative character of the concept of generations of neutron sources," and how "far-fetched" it is to group them into "the holy numeral of three." A more convincing ordering, Kaiserfeld writes, "would likely depart from knowledge, skills, engineering experiences, political and financial opportunities, research demands and instrumentation possibilities, resulting in a number of generations characterized by different measures of both inbreeding and crossbreeding."

As reactors became costlier to build, and less popular politically and socially, more attention began to be paid to spallation — accelerator-based — neutron sources. Hallonsten's chapter, "The Interplay of Actors and Institutions in the Campaign to Localize the European Spallation Source (ESS) to Lund, Sweden," shows how the establishment of this yet more powerful kind of neutron source, likely to replace reactors as the largest tools for neutron scattering, required the efforts of a new kind of multinational institution, the European Research Infrastructure Consortium (ERIC), comprising 15 internationally distributed partner members. The ESS story shows the continuing and likely future evolution of the template for collaborative forefront Big Science illustrated in the previous chapter.

The third chapter in this part, Liu Lin's "Brief History of Synchrotron Radiation Sources in Brazil," shows a specific case of the process described by Hallonsten and Heinze — a specific institution, a synchrotron radiation facility, established within an existing institution, quickly growing large enough to require a new kind of institution, which in turn inspired the building of yet more ambitious synchrotron facilities. The synchrotron source was the Brazilian Synchrotron Light Laboratory, created in 1986 by the Ministry of Science and Technology, leading to the establishment of the first modern national laboratory in Brazil. Lin's story also illustrates the increasing role of industry and international collaboration in such large-scale projects.

Chapter 2.1

Three Generations of Institutions of Neutron Sources

Thomas Kaiserfeld

Introduction

Before World War II, access to neutrons had primarily been achieved using small-scale devices where neutrons were generated through mixing radioactive isotopes that emit α-particles with isotopes of light elements. An alternative was spontaneous fission by decay in radioactive atomic nuclei. Neutrons were thus accessed through nuclear decay occurring either naturally or by manipulation. After World War II, however, research reactors developed within the American Manhattan Project supplied new means to obtain neutrons on a much larger scale through nuclear fission in cooled and moderated chain reactions.

The understanding of neutrons was crucial for developing nuclear technologies, including both nuclear weapons and nuclear reactors for electric power production, simply because the neutrons are the atomic nuclear particles (also called nucleons) that create the chain reactions so important for the release of energy in these devices. This line of inquiry was thus important for both military and civilian purposes in the wake of World War II. One vital step for developing civil uses of atomic power production was the establishment of the US Atomic Energy Commission in 1946, an organization that took over and managed the different American laboratories and production sites originally created within the Manhattan Project.[1]

In the 1950s, another initiative was taken by the American administration to promote what could be perceived as civilian uses and applications of nuclear physics now in the international arena. It resulted in the famous "Atoms for Peace" conference, organized in Geneva in 1955 under the auspices of the United Nations for scientists and policymakers from both sides of the iron curtain to meet and discuss the spread of nuclear technologies for civilian uses as well as for acquiring goodwill for American achievements in nuclear science.[2] An additional purpose was to induce a positive stance toward nuclear technologies among the general public.

During the 1950s and 1960s, the building of sources for neutron physics continued to be intimately connected to basic nuclear research as well as the application of nuclear technologies, at first military, but from the mid-1950s to a growing extent also civilian. Consequently, neutron physics developed as "a parasitic activity at nuclear reactors."[3] In reality, of course, matters were more complicated. At some labs, reactor research was tied to military interests while at others, civilian uses dominated. Especially important developments in the 1960s at Brookhaven included a research reactor, the High Flux Beam Reactor, developed for neutron research purposes.[4] When it started operation in 1965, it broke the record for thermal flux of neutrons achieving about 10 times the flux of the former record holder.[5]

The Development of European Neutron Sources

The combination of civilian and military interests connected to the uses of neutron science was mirrored in the differentiation of

[1] Westwick (2003).
[2] Krige (2006), pp. 161–81.
[3] Pynn and Fender (1985), p. 47.
[4] Crease (1999).
[5] Hanna (1997), p. 172.

reactor technology in the sense that nuclear reactors in the early 1950s were more often custom-built for different specific purposes: research, production of fissile material, or electric power production. Of special interest here is research reactors being developed to achieve high neutron flux, a necessity for a better understanding of the atomic nucleus. In addition, research reactors could be used to study the effects of long-term radiation of materials and important knowledge when developing new and more efficient reactors for the production of fissile materials or power. Such research tasks also fed back on the development of neutron physics and engineering since clients at research reactors, apart from nuclear physicists, also included material engineers, metallurgists etc.[6] During the 1950s and 1960s, research reactors were developed with a neutron flux of about a hundred times the flux of the reactors from the 1940s, an achievement further improved in the 1960s when new reactors would have a flux of about a thousand times that of the reactors from the 1940s.[7]

The sharp increase in the number of research reactors in the late 1950s and early 1960s (Fig. 1) was primarily a result of the 1955 "Atoms for Peace" conference on peaceful uses of atomic energy, facilitating declassification and dissemination of relevant information. But the field of neutron research also expanded rapidly during the cold war due to the improved opportunities offered by the increasing number of research reactors to use low-energy neutrons to probe condensed matter where there are great variations in atomic and molecular composition and structure arrangement.[8] The background was that different types of uranium reactors produced different intensities of neutrons at different energies, the so-called cold, thermal and hot neutrons, which in turn could be used for different types of experiments. Instruments could also be constructed to compensate for the lower neutron flux given, for instance, by low enriched uranium.

[6] Johnston (2012), p. 114–18.

[7] Richter and Springer (1998), p. 13.

[8] The National Academies (1977).

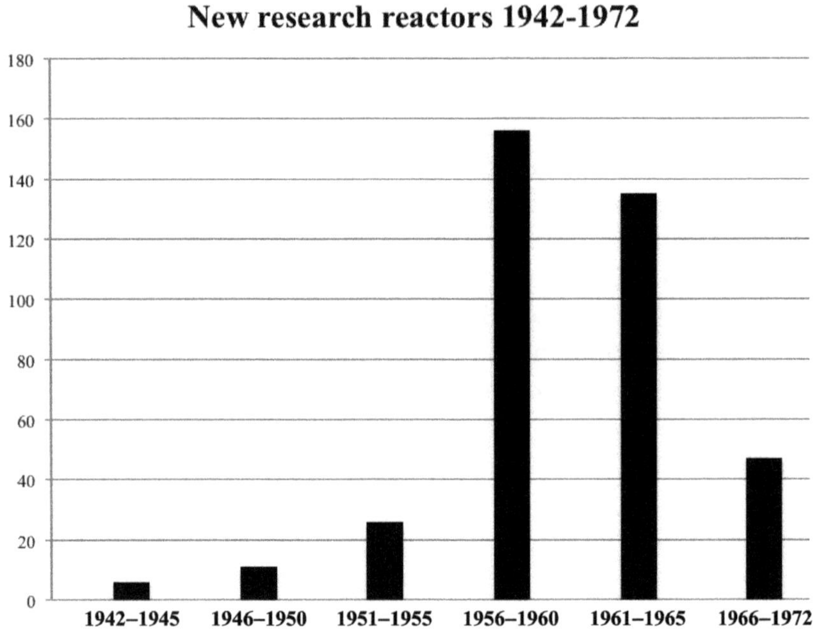

Fig. 1. Number of new research reactors built between 1942 and 1972 with five-year interval. During the total period, approximately 380 research reactors were built in 44 countries of which 160 were built in the US. Among these, 88 had been shut down after various periods of operation (graph compiled from information in Simnad (1981)).

In Europe, one important organizational consequence of the "Atoms for Peace" conference, leading to an early initiative to coordinate research efforts in a specific branch, was the forming of the European Atomic Energy Community (EURATOM) in 1958 to promote science and development of nuclear technologies on the continent. Its context was to some extent the European research infrastructure CERN established in 1952, in the name Conseil Européenne pour la Recherche Nucléaire devoted to nuclear research although in practice focused on high-energy nuclear physics, soon identified as particle physics. More important, however, were two other politically negotiated organizations for European economic cooperation, the European Steel and Coal Community and EEC.[9] Together, they constituted the three European communities of the late 1950s, which after the end of the cold war were absorbed into the European Union. Notable in this context is that the European Steel and Coal Community and EURATOM, both developed from the idea that the control of energy sources was an important factor behind the political stability necessary to prevent future conflicts in Europe. Although of very different kinds, coal being traditional and atomic power the future energy sources of the 1950s and 1960s demanding quite extensive research, both were seen as platforms for European cooperation, not the least in research and development.[10]

EURATOM soon supported an effort to construct a periodically pulsed fast reactor in northern Italy. The project, named "fast source" in Italian, *Sorgente Rapida* (SORA), was conceptualized in 1962 with further planning managed by EURATOM. Important inspiration was supplied by the successful construction of a Soviet pulsed research reactor in Dubna close to Moscow after an idea in late 1955 to fix a part of a reactor on a rim of a rotating disk so that at each revolution, this part passed close enough to the stationary part of the reactor to create a supercritical mass for a short time. This idea was realized through the construction of the pulsed Dubna reactor, which became operational in 1960. Its advantage was a much more economic use of neutrons since it did not mechanically chop up the normally continuous flow of neutrons from a reactor.

Another advantage, in comparison to continuous neutron sources such as common reactors, was that the velocity of neutrons produced in pulsed sources could be calculated by knowing the time of travel from target to instrument as well as the distance between them. If the pulses were short as well as intense enough, it could even be possible to study fast chemical reactions and other processes impossible to probe by a steady flow of neutrons from conventional high-flux sources, such as those built from the mid-1960s.

---

[9] Mersits (1987), Nieburg (1963), Weilemann (1983), Skogmar (2004), Kohlrausch and Trischler (2014), pp. 205–241, and Reitbauer (2015).

[10] Benner (2018), p. 118.

In 1965, a mock-up reactor design of SORA as well as a rotary pulsation device was prepared for tests at Oak Ridge. Simultaneously, the new high flux reactor at Brookhaven made it easier to argue for the realization of SORA with reference to the importance to construct a corresponding European neutron source. If this was not achieved, a EURATOM report stated, the effects for "the development of neutron physics in Europe does not have to be explained in further detail."[11]

The tests at Oak Ridge were carried out in 1967 with what appears to have been positive results making it possible to move forward with a construction plan aiming for operation in 1970.[12] The EURATOM's assessment was as follows:

> "In many fields of neutron physics, and especially inelastic scattering, neutron fluxes of greater intensity are required. The only technically feasible ways of meeting this need in the immediate or near future are pulsed reactors and in the more distant future pulsed reactors and spallation machines."[13]

Despite these conclusions, SORA was never to be realized and was finally given up in the early 1970s.

By then, other new and improved reactors had been built in Europe with the High Flux Reactor at the Institut Laue-Langevin (ILL) in Grenoble, France, as a climax. By a reactor design running on high-grade fuel, it proved able to outcompete even the Brookhaven reactor in neutron flux. Operational in 1971, it was constructed by France and West Germany with the UK joining as an equal partner two years later when plans for a national British research reactor had been thwarted. Later, additional member states joined.

Parallel to the efforts to develop research reactors, other types of neutron sources were developed. An important site for this work had been built up immediately after the war at a Royal Air Force

---

[11] Ritter (1966), p. 5.
[12] Kley (1968).
[13] EURATOM (1967), p. 99.

airfield about 10 km south of Oxford in the UK, called Harwell. In 1957, a new National Institute for Research in Nuclear Science was formed here, in the vicinity of the already established military Atomic Energy Research Establishment, to focus resources on larger and more expensive instruments that otherwise had to be financed by different universities. Although this generated complaints from some of the university researchers, the new national institute was nevertheless formed and soon work started on a proton synchrotron while the military Atomic Energy Research Establishment also supplied equipment, such as a proton linear accelerator.[14]

In the late 1950s and early 1960s, the military Atomic Energy Research Establishment constructed an electron linear accelerator, the Harwell Neutron Project, to produce X-rays to be absorbed by uranium to produce neutrons primarily to be used for reactor technology research. A similar device was set up at Oak Ridge National Laboratory in 1969 called Oak Ridge Electron Linear Accelerator. In the mid-1970s, work began to build a new similar, but much more powerful neutron source at Harwell named HELIOS, which started operation in 1979 as a state-of-the-art instrument producing about twice the neutron flux available at Oak Ridge Electron Linear Accelerator, which by now was 10 years old.[15]

Before the opening of HELIOS, work had already begun in 1977 on a new so-called spallation neutron source at the non-military laboratory at Harwell, now named the Rutherford Laboratory and soon to become the Rutherford Appleton Laboratory. There were many background prerequisites to this decision, for instance, the construction and operation of a British polarised beam detector at ILL in 1975.[16] A more important reason was the dismantling of the proton synchrotron that had been in operation at Harwell since 1964. A proposal suggested that a new spallation source should be built in the space of the former proton

---

[14] Litt (1979)

[15] HELIOS (1979), Carpenter and Yelon (1986), p. 99–100. HELIOS was primarily built for the study of nuclear power technologies, see Lynn (1980).

[16] Meadows (2009), p. 33.

synchrotron scavenging on some of its used equipment to make it as cheap as possible and saving up to two-thirds of the costs had it been built from scratch.[17] There was to be sure opposition to overcome, primarily from those who favored a more powerful reactor neutron source with the suspicion that the spallation source was a clever way of keeping the particle physicists busy. But much of this criticism seems to have evaporated once the spallation source was up and running from late 1984.

ISIS, as the new British spallation source was named after a local term for the river Thames, was inaugurated in 1985 and soon judged to be the best spallation source in the world although its peak flux was only somewhat higher than the continuous flux of the ILL reactor. The success of ISIS is clearly mirrored in the number of visiting researchers. In 1990, there were 500 UK users alone with a 100 new users per year so that the number in 2005 had reached 1,600 visiting users.[18] Another clear indication is that ISIS already by 1988 was the main source of income for the Rutherford Appleton Laboratory.[19]

European Advantages in Instrumentation

Together, ISIS and ILL gave the global centre of gravity of neutron science a secure foothold in Europe in the late 1980s. In addition, they were supplemented with different national research reactors constructed in the 1960s with the French in Saclay and German in Jülich being the best known. But although both ISIS and ILL constituted rather firm steps toward increased neutron intensity, they achieved only about four times the thermal neutron intensity the reactors used in the pioneering neutron scattering experiments of the early 1950s had managed.[20] This was even more frustrating since intensity is what really makes a difference in scattering experiments.[21]

[17] Findlay *et al.* (2006), Meadows (2009), pp. 33–34.

[18] Meadows (2009), pp. 34 & 37.

[19] Meadows (2009), p. 110.

[20] Gutberlet and Mezei (2002).

[21] Finney and Lengeler (1994).

The decisive advantage in comparison to American facilities was instead the instruments used, which according to American reports opened up new applications in the field of neutron scattering. Especially ILL was admired by American neutron researchers who claimed it alone served 25% more users than all American facilities taken together, noting however that its annual budget of 25 million dollars in the mid-1980s corresponded to the total American expenditures on neutron science.[22] In total, the annual Western European budget for neutrons scattering experiments exceeded the American by more than three times. More specifically, two pulsed spallation sources that had been developed at Argonne and Los Alamos cost 10 million dollars per year corresponding to about 40 percent of the total American neutron scattering budget of 25 million dollars per year. Between 1970 and the early 1990s, American estimates were that European investments in neutron facilities were about five times the American ones.[23] Consequently, confidence grew among European neutron scientists.

To be sure, the drift of neutron science from USA to Europe was slow and in the second half of the 1970s, Americans could claim, "it may still fairly be said that the most sophisticated neutron scattering in the world has consistently been done at the U.S. reactor centers."[24] Simultaneously, however, the funding of neutron research in USA remained at a relatively flat rate while a feeling of self-complacency nevertheless spread among American neutron researchers.[25] The result was clearly visible 15 years later when the American appreciation of European neutron research had grown considerably.[26]

In addition to the new sources with a higher flux of thermal and cold neutrons, i.e., neutrons with energies corresponding to or below room temperature and with the most likely speed of a few kilometers per hour, better instruments were developed too.[27]

---

[22] The National Academies (1984).

[23] US Department of Energy (1993), p. 57.

[24] The National Academies (1977), p. 146.

[25] Rush (2015), p. 141.

[26] US Department of Energy (1993), p. 1.

[27] The National Academies (1977), pp. 144–6.

When nuclear reactors of different constructions had been built after World War II, instruments had lagged behind the higher neutron flux now available. But during the 1950s, instrument makers had caught up, making it possible to use neutrons for measurements in new areas, such as alloy structures or atomic structures in liquids.[28] In order to use neutrons for yet new fields, a higher flux of neutrons was needed and to a growing extent achieved in Europe from the early 1970s where instrumentation was simultaneously taken to new heights.[29] In this way, possibilities to use neutrons for new research areas coevolved with neutron sources and instrumentation.

Twenty years later, an American report admitted that the superior European neutron facilities

> "did not result only from the design of the reactor, which produces about the same flux as existing U.S. high-flux reactors. The technical innovation was in instrumentation, cold neutron moderators to enhance the cold neutron production, and extensive use of neutron guides to increase the number of instruments that could be employed. These technical developments enabled the growth of a large user community."[30]

Taken together, the construction of new neutron sources as well as their instrumentation made the European community of neutron researchers both bigger and better equipped than their American colleagues from the 1970s. In the early years of the decade, research reactors in Germany as well as France and, above all, the ILL reactor in Grenoble, together with the development of better instrumentation, lay the foundation for the European success in neutron scattering. In addition, a new research reactor for materials research named Orphée had opened in Saclay in 1980. By the mid-1980s, there were 15 research reactors available as neutron sources for

---

[28] Brugger (1968), p. 26.
[29] The National Academies (1984).
[30] US Department of Energy (1993), p. 49.

research purposes in Western Europe, with the most recent being ILL and Orphée in France.[31]

To some extent, the higher flux of thermal and cold neutrons produced at European facilities and their better instrumentation in comparison to corresponding American facilities was the result of European countries pooling resources and by this managing to exceed the American neutron research budgets. As mentioned, while American funding of neutron research had remained rather constant, European investments increased considerably after the establishment of EURATOM and the build-up of research reactors in the 1970s.[32] However, the resurgence of European neutron facilities occurred before the European Community had managed to coordinate efforts to promote pan-European research infrastructures or increase the research budget above a few percent of the total revenue. Although most of the new European facilities at least partly relied on EURATOM funding, initiatives were national or the result of cooperation between a limited number of European countries.

The consequences were nevertheless important for the access to neutron-research facilities. In contrast to the American national laboratories, more or less permanently staffed with researchers, some of the European facilities were open to proposals from research groups competing for beamtime to carry out their experiments. As a result, these facilities with more distributed access to beamtime attracted neutron users representing a wider range of different areas. As stated by contemporary practitioners, "Neutron scattering in Europe appears set to flourish for the rest of the century."[33]

An Alternative to Reactors: Spallation as a New Technology to Achieve Higher Neutron Flux

On the neutron supply side, ideas for new ways of producing neutrons were also developed. They primarily relied on spin-offs from

---

[31] Commission of the European Communities (1991), pp. 12–13.

[32] Barry and Walters (2003).

[33] Pynn and Fender (1985), p. 48.

tests of producing neutrons by letting high-energy protons or other charged nucleons hit a target consisting of some heavy element rich in neutrons. Early trials along these lines had been pursued at the Livermore Naval Air Station in California and at the Chalk River Nuclear Laboratories, Ontario, in eastern Canada in the early and mid-1950s to find methods to produce fissile materials for atomic weapons and more efficient power production in nuclear reactors, respectively.

These facilities were early forerunners of the mentioned spallation neutron sources developed in the 1970s. Here, accelerated particles were grouped together in coordinated bunches that hit a target consisting of an element rich in neutrons. It accordingly emitted pulses of neutrons that were then guided to different experiments. These spallation technologies thus created a new form of pulsed neutron sources providing much higher neutron flux in the peaks of the pulses. An additional advantage was that less heat was released for each fast neutron produced, typically about a sixth in comparison to fast neutrons from reactors.

The downside was that beams in spallation sources could be discontinued more or less frequently due to malfunctioning equipment making experiments impossible while reactors were a far more secure neutron source. While the problem at spallation sources was to secure a continuous beam of neutrons, reactors demanded the opposite, i.e., assured methods of discontinuing the chain reaction in the core and halting the flow of neutrons. Guarantees of neutrons were of course important when experiments were pursued by visiting research groups using neutron beams as one of many different methods to gather information, for instance, about a material or a substance. In reactors, though, the handling of radioactive uranium often made the facilities fall under security regulations applying for any nuclear power plant, something that often complicated matters. This was a disadvantage that became more important in the 1970s and later.

The idea to produce neutrons by letting accelerated charged particles hit different types of targets was quite straightforward. Also, lighter particles like electrons could be accelerated to hit a target of, for instance, beryllium or mercury to emit electromagnetic radiation to be absorbed by uranium, a process that was used for producing fast neutrons, for example, to measure nuclear cross-sections and other characteristics of interest to the nuclear power industry.[34] From the late 1950s, several electron linear accelerator neutron sources were operated in different parts of the world, such as the UK, the USA, the Soviet Union, Japan and Canada. In the US, electron linear accelerator neutron sources were running at Rensselaer Polytechnic Institute from December 1961 and at the end of the decade also at Oak Ridge National Laboratory.[35] At Los Alamos National Laboratory, proton accelerators were used to study neutrons from 1972.

The earliest examples of the new more powerful sources relying on protons or hydrogen ions stripped of electrons were erected in the USA, both at Argonne National Laboratory in the vicinity of Chicago and at Los Alamos National Laboratory in New Mexico, as well as in Japan. The Argonne case, analyzed by historian of science, Catherine Westfall, has shown how this development partly relied on scavenging on older particle accelerators and the rebuilding of existing facilities for new purposes.[36] These efforts of the mid-1970s resulted in operable facilities in Japan and USA in 1980 and 1981, respectively. The trend was further strengthened by the construction of the first European neutron spallation source and simultaneously the world's most powerful, the already mentioned ISIS at the Rutherford Appleton Laboratory.[37]

Early on, users of neutron spallation sources mainly consisted of material scientists, but later the target group of potential users were to be expanded. These initiatives to develop new techniques for achieving higher thermal and cold neutron flux should be seen as development work in order to circumvent the limitations of the existing fission reactors, primarily their relatively low efficiency in

[34] Cockroft (1949).
[35] Danon *et al.* (2005), Kohler (2005).
[36] Westfall (2010).
[37] Meadows (2009).

producing low-energy neutrons (which are better suited for probing materials) and their less controlled neutron flux.

Plans for Spallation Sources in Europe

In the early 1980s, there were also plans for a spallation source at the Nuclear Research Site (Kernforschungsanlage) of Jülich in West Germany close to the Dutch and Belgian borders, from 1990 renamed as Forschungszentrum Jülich. This site had housed two research reactors from the early 1960s and one of them was to be shut down in the mid-1980s. This spurred development work together with Kernforschungszentrum Karlsruhe on a spallation source called Spallation Neutron Source (Spallation Neutron Quelle (SNQ)) from the late 1970s and during the first half of the 1980s, a powerful spallation source with a proton beam power of about 5 MW designed to produce long pulses with typically a thousand times longer duration than short pulses of neutrons at a cost of about 680 million German Marks.[38]

The SNQ was primarily designed for use in solid-state research and nuclear physics and had originally been supported by interest in research in condensed materials. The arguments supporting SNQ also included research on neutrons as well as other particles. To these arguments were added national sentiments claiming that the spallation neutron source would give West Germany a world-leading position in research areas connected to high neutron flux. One could perhaps hope that German neutron research could rely on the successful research reactor ILL in Grenoble with France, West Germany and the United Kingdom as partners. In a report of 1981, however, the proponents of SNQ pointed out that ILL was not enough to accommodate all the different experiments needed to be carried out and that new neutron sources were planned and built both in France, as a reactor, and in the UK, as a spallation source later to become ISIS.[39]

---

[38] Bauer *et al.* (1981).

[39] Ibid., pp. 23–24.

The Jülich spallation neutron source was indeed a promising project and was reviewed as "perhaps of greatest potential impact on the long-term development of pulsed neutron sources."[40] Despite these positive judgements, however, the German federal minister of science and technology eventually stopped SNQ in June 1985. The main reason for this was the exploding project costs, first doubling to 1.2 billion and then rising even steeper to 2.9 billion German Marks, almost five times the original estimates.[41] Some of the design elements, such as the ambitiously high beam power of 5 MW and the target constructed as a rotating slab cooled with water or helium, were later to be picked up in other projects, such as the pan-European project, the European Spallation Source (ESS).[42]

From a broader perspective, different factors such as raised criticism against nuclear power in many European countries, especially after the Chernobyl disaster in April 1986, non-proliferation agreements and the rules and regulations surrounding any research reactor under construction and operation, all quenched the development of new research reactors. Such circumstances made it seem as if there would be less resistance and thus more realistic to promote the construction of a spallation source in comparison to a research reactor, especially a reactor producing a higher flux of neutrons since these demanded higher grades of fissile material such as plutonium. Another disadvantage of research reactors was the relatively large amounts of heat generated per neutron in fission processes. As a consequence of these considerations, there were no new research reactors taken into operation in Europe between 1984 and 2004, when the FRM-II in Munich (Forschungsreaktor München II) was opened after a process that had started in the mid-1970s.[43] In this context, spallation sources were an interesting alternative although immensely more complicated to construct. However, a challenge is

---

[40] The National Academies (1984), p. 32.

[41] Gläser *et al.* (1994), p. 299. Joint Publication Research Service (1986), pp. 44 & 47.

[42] Bauer (1982, 1999, 2010).

[43] Gläser *et al.* (1994).

not always regarded as a drawback among advanced design engineers and instrument developers.

Resistance to research reactors was also the reason behind the construction of the Swiss Spallation Neutron Source (SINQ) at the Paul Scherrer Institute. Planning of SINQ had started already in the mid-1980s, at about the time the SNQ of Jülich was stopped.[44] It would take a decade of continuous design and construction work as an organizational merger in 1988 of two Swiss research institutes to realize SINQ, which started operation in 1996 as the world's first continuous spallation source with a neutron flux about a 10th of the 25-year older reactor at ILL. Although ISIS could achieve higher flux in pulses, SINQ had the most powerful beam on target, protons with energies of about 1 MW.[45]

To sum up, there were a number of national initiatives to construct neutron spallation sources in Europe in the 1980s and 1990s. Many of them were however hard to realize due to construction costs and other uncertainties. A way to come to terms with the challenges would be to create a spallation source as an international European collaborative effort, and in 1990, ideas formed to launch a European project to pool national resources to construct a larger and better spallation source than earlier been attempted: the ESS.

The Problem of Neutron Science: "We Need More Neutrons!"

Neutron scientists in Europe grew more and more self-confident during the 1990s claiming that "European science is fortunate in having available to it the world's most intense reactor and pulsed spallation neutron sources, namely the Institut Laue-Langevin in Grenoble and the ISIS Facility in Oxfordshire, UK."[46] Instead of arguing for new and more powerful neutron sources by referring to a need to catch up with the Americans, it was the physical limits of the contemporary neutron sources together with the observation that many of the present European neutron sources were getting

---

[44] Atchison *et al.* (1986).

[45] Bauer (1997).

[46] Finney and Lengeler (1997), p. 129.

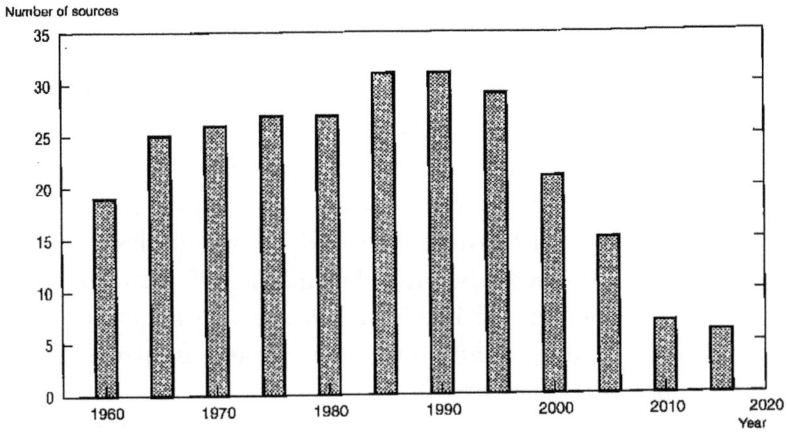

Fig. 2. Graph with the number of neutron sources in OECD countries until the mid-1990s and beyond, projecting a decline of OECD neutron sources due to the closing of research reactors and other sources reaching calculated lifetime. In reality, the projected decline proved correct until the mid-2000s, but after that, the predicted decrease was less radical due to the building of new sources and the upgrading of existing ones (Riste (1993), p. 68).

old, approaching the end of their technical and economic viability creating a threatening lack of neutron sources, a "neutron gap" or "neutron drought," that provided the drive for new and ambitious designs for the future materialized in the ESS.[47]

The idea was developed through projections of the number of European neutron sources compared with projections of the number of users made by the Norwegian solid-state physicist, Tormod Riste. Graphs were compellingly displayed in an analysis of material available and presentations made at the 1993 meeting of the OECD Megascience Forum's expert group for neutron sources as well as the report of the US Department of Energy on future neutron sources.[48] By bringing together information about different existing neutron sources and their calculated time of operation, Riste constructed a graph (Fig. 2) suggesting a dramatic decrease in the

[47] Commission of the European Community (1991), Riste (1993), and European Science Foundation (2000), p. 7.

[48] Riste (1993).

number of neutron sources in less than 10 years time. This graph was then used and reused throughout the 1990s in order to gather support for the realization of ESS and other neutron sources.

The danger of a neutron gap does not seem to have been diminished by the Swiss continuous spallation source SINQ having started operation in 1996 complementing the four existing pulsed spallation sources, an observation confirmed by an analysis made by OECD and the European Science Foundation.[49] Neither did the plans to build a new reactor in Munich nor a new spallation source in Austria diminish the exclamations of European neutron researchers, "We need more neutrons!"[50]

Riste's calculations proved to hold throughout the 1990s so that by the end of the decade, there were 21 neutron sources in operation in the OECD countries of 1993.[51] The argument was still reiterated in the spring of 2002 when it was claimed that out of the 26 neutron sources in operation, only five were expected to still be running in 2015.[52] In the longer run, however, time has proven Riste's calculations to be wrong. In 2012, there were 19 neutron research facilities in operation in OECD countries as compared to the seven or eight that Riste had projected back in 1993.[53] Not all of these had existed in the early 1990s when Riste first drew attention to the risk of a decreasing number of neutron sources, but many of those in operation then are still running after having been upgraded. In addition, there are still today new neutron research facilities under construction. Thus, it is not surprising to find that the argument departing from a decreasing number of neutron sources launched in support of building new ones in order to counter a projected "neutron gap" died out during the first decade of the 21st century.

---

[49] Ibid., OECD Megascience Forum (1998), Richter and Springer (1998).

[50] Clery and Lawler (1995), p. 952.

[51] Richter and Springer (1998), pp. 17–18.

[52] ESS Project (2002), p. 7.

[53] National Institute of Standards and Technology, Center for Neutron Research (2012).

Concerns about a diminishing number of neutron sources were combined with calculations pointing to an increase in the number of neutron beam users in Europe making the situation even more urgent. At the OECD meeting in 1993, the community of about 4,000 neutron beam users in the OECD countries — virtually all of Europe including Switzerland as well as the US, Canada, Japan and Australia — was projected to grow to about 7,000 in 2000.[54] Studies of European neutron users in the mid-1990s showed a similar growth counting 3,500 users in 1995 and 4,400 users in 1997.[55] A year later, the total number of users of neutron scattering experiments in the world was estimated at 6,000.[56] This could be compared to the number of neutron researchers in the US, which had grown from about 750 to 1,500 from 1983 to 1991.[57]

The projections made about the growing number of neutron beam users turned out to be quite correct since in 2000, the community of European neutron beam users was estimated to hold about 5,000 members or two-thirds of the world's total number of neutron beam users.[58] Despite this growth, however, the estimated six years later were still about 5,000 European users of neutron spectroscopy and, in 2007, the supplied figures were down to 4,000 European users out of a total of 6,000 in the whole world.[59] According to a more recent survey, though, made by the European Neutron Scattering Association (ENSA), the figure was again up to 6,000 and there were estimates pointing to 10,000 European neutron users "during the 21st century."[60] Clearly, estimates and surveys made by the ENSA pointed toward higher numbers of European neutron users than other studies.

---

[54] Riste (1993), pp. 80–81.

[55] ENSA (1998), p. 34.

[56] OECD Megascience Forum (June 1998), p. 3.

[57] US Department of Energy (1993), p. 32.

[58] European Science Foundation (2000), p. 7.

[59] European Strategy Forum for Research Infrastructures (2006), p. 24, Editorial (2007).

[60] European Spallation Source (2011), p. 4.

The European Spallation Source Fills the "Neutron Gap"

The rhetoric of the anticipated problem of a "neutron gap" also included a solution to the problem departing from the notion that fission reactors were reaching their practical limits as neutron sources in terms of possible neutron flux. Even if they could be improved and their efficiency could have risen with respect to generated neutron flux, it was not possible to increase the reactor output significantly. Instead, OECD and others claimed that more powerful spallation sources would lead "to a quantum leap in neutron science."[61]

The slow technological and geographical transformations as well as the expanded potential uses of neutrons paved the way for discussions of a European spallation source. This was the situation in 1990 when the Commission of the European Community (CEC) started to promote an activity named Access to Large-Scale Facilities, which was part of an institutional framework where larger European resources were pooled in framework programs for research and development.[62] Access to Large-Scale Facilities not only meant that a handful of national European neutron sources signed access contracts opening national facilities to non-national users.[63] In addition, five study panels were set up in 1990 to facilitate the use of large-scale facilities in Europe by making priorities and report back which facilities to support, the very first of these study panels was organized in the area of neutron sources.

This study panel met three times during 1990 with members from countries of the European Community housing research neutron sources, i.e., Germany, France, the UK, the Netherlands, Belgium, Denmark Greece, Italy and Spain, while Ireland and Portugal made no nominations, in the Portuguese case despite the existence of a national research reactor. The panel concluded that there was indeed demand for a pan-European neutron source and in mid-1993, a council was formed for the creation of ESS with

[61] Richter (2002), p. 818.

[62] Commission of the European Community (1994), Guzzetti (1995).

[63] Commission of the European Community (2001).

representatives from the eight countries already involved as well as observers from France and Spain. In May 1994, after the establishment of the ESS Council, European neutron users were also engaged in efforts to create ESS through the founding of a support organization, the European Neutron Scattering Association, formed as an umbrella organization for different national user associations, especially after initiatives from the German and Swiss groups.

This was the situation in the 1990s when the first ESS design was developed. The proposal has been described as an inflated version of the already operating spallation source ISIS in the UK. Two short-pulse target stations were to be fed with a proton beam each: one of 5 MeV and one of 1 MeV. Construction costs were calculated to be less than one billion euros.[64]

While European collaboration resulted in a feasibility report of ESS, the OECD Megascience Forum expert group for neutron sources transformed into a Neutron Sources Working Group with the task to estimate future demand for neutron research facilities and to identify technical problems connected to upgrading existing neutron sources. In their final report, it was stated that the "most important class of next-generation sources consists of the accelerator-based spallation facilities, whose increased power will lead to improvements in the quantity and quality of research, and enable expansion into new scientific areas."[65] An ambitious plan was laid out to build three different new-generation accelerator-based neutron spallation sources in Europe, America and Japan, respectively.[66]

The rationale behind such an aspiring proposal to build three new powerful neutron sources on three different continents was not only the growing number of neutron scientists and the risk of a diminishing number of neutron sources. An additional reason was the fear that the social and industrial impact of a facility of this kind tended to be more regional than global, especially considering the

[64] Lengeler (1994).
[65] OECD Megascience Forum (1998), p. 5.
[66] Tindemans (2010).

diversity of the user community and the effects of cross-disciplinary interactions. The ambitious plan to construct new neutron sources on three different continents was endorsed in May 1999 by the OECD ministerial conference in Paris as a global strategy for neutrons together with support for the maintenance and refurbishment of a number of smaller national neutron sources and more powerful regional sources.

The support from OECD created momentum for the ESS and a more advanced, as well as more expensive, proposal was presented at a big meeting in Bonn in 2002 where a number of researchers, policymakers and journalists had been invited. It consisted of a far bolder design to equip ESS with two very different target stations, one short-pulse target station as in the earlier proposal, but now accompanied by a long-pulse target station, each to be fed with a beam of 5 MeV. Consequently, projected construction costs had increased by about 40%.[67]

The Bonn meeting introduced five different proposals for the location of ESS. Two of them came from the UK (Oxford and Yorkshire), two from Germany (Jülich and Leipzig) and one from Scandinavia (Lund). But after the Bonn meeting and just as the selection process for ESS was to be initiated, the Germans withdrew from the project after a negative evaluation from the German Council of Science and Humanities.[68] This proved a catastrophe for ESS and the plans were on the brink of termination.

In the spring of 2003, the situation for ESS thus looked grim.[69] Sensing the danger, the ESS Council regrouped and shrank both budget and office. But through the formation of the European Strategy Forum for Research Infrastructures (ESFRI) in early 2002, help was closer at hand than expected. ESFRI was a body established by the European Union for working out guidelines for infrastructure support. Soon, it formed a Working Group of Neutron Facilities, which immediately started to discuss different scenarios for neutron

[67] ESS Project (2002).

[68] Wissensachaftsrat (2002).

[69] Tindemans and Clausen (2003).

research in Europe deciding to analyze both the building of ESS and the upgrading of ILL and ISIS.[70] The expectations were raised even further by neutron research being the first case for the ESFRI roadmap.

Again, ESS was redesigned to fit the new political situation and to get priority on the roadmap, ESS was downsized to only one target station for long pulses with a beam power of 5 MeV at a projected cost of somewhat less than 1.1 billion euros. The operation was crowned with success when ESS got the highest priority on the first European roadmap for research infrastructures in 2005.[71]

Without wasting any time, a call for site proposals was once again published, this time attracting three bids: from Hungary, Spain and Sweden. After a long and quite careful evaluation round interfered with national and European policymaking reaching its climax in the spring of 2009, Lund in southern Sweden was to host ESS. Despite negotiations on the ministerial level, funding was still only partial however, and it took another five years to get the agreements needed to start construction.

The Third-Generation Neutron Sources

Parallel to the filling of the neutron gap by the development of spallation sources as an alternative neutron source to reactors, a discourse emerged of a "third-generation neutron source." It departed from the view that there was a first generation consisting of enriched radioactive isotopes followed by a second generation consisting of fission reactors. The more powerful spallation sources were expected to constitute the third generation.

When brought together in graphs with their respective neutron flux as a function of time, the image was suggestive in the sense that the future of neutron sources seemed more or less inevitably devoted to pulsed (from the 1970s, spallation) sources. Although advanced research reactors such as ILL and Brookhaven High Flux

[70] ESS Project (2003), p. 8.

[71] ESFRI (2005).

Beam Reactor were still recognized as world-leading in the 1990s, the call for third-generation neutron sources lumped them together with much older and less efficient reactors thought of as constituting a second generation with limited possible flux visible in its envelope becoming parallel to the time axis. This can be compared to the imagined third generation of pulsed spallation sources, which were characterized by an envelope pointing upward along the scale of flux rather than parallel to the time axis.

Such diagrams of three generations of neutron sources had occurred as early as 1968 when there were no running spallation sources. At that time, the third generation of neutron sources instead consisted of another type of source, pulsed reactors such as SORA planned to be constructed in northern Italy by EURATOM. Later on, when spallation sources had become a reality, it seemed reasonable to substitute the pulsed reactors for pulsed spallation sources as the main component of imagined third-generation neutron sources.

This idea of three successive generations of neutron sources as presented in the graphs in Fig. 3 constituted a strong argument for the necessity, almost inevitability, to fund and build new facilities that could be plotted further out along the horizontal time axis on the trajectory of ever more powerful spallation sources. By presenting generations of neutron sources in diagrams like those in Fig. 3, more powerful spallation sources seemed more or less predestined and it became more a matter of where rather than when they were to be realized. Europe was clearly a potential location.

The image of third-generation neutron sources has lived on. In fact, more recently, pulsed spallation sources have been substituted for a new imagined future neutron source, one relying on inertial fusion, as shown in Fig. 4. And in order to keep the holy numeral of three in the number of imagined generations, the first generation existing in earlier images of generation upon generation of neutron sources has been given up so that reactors that earlier constituted the second-generation of sources have now been degraded to the first and different spallation sources that earlier constituted the third generation have now lost its position

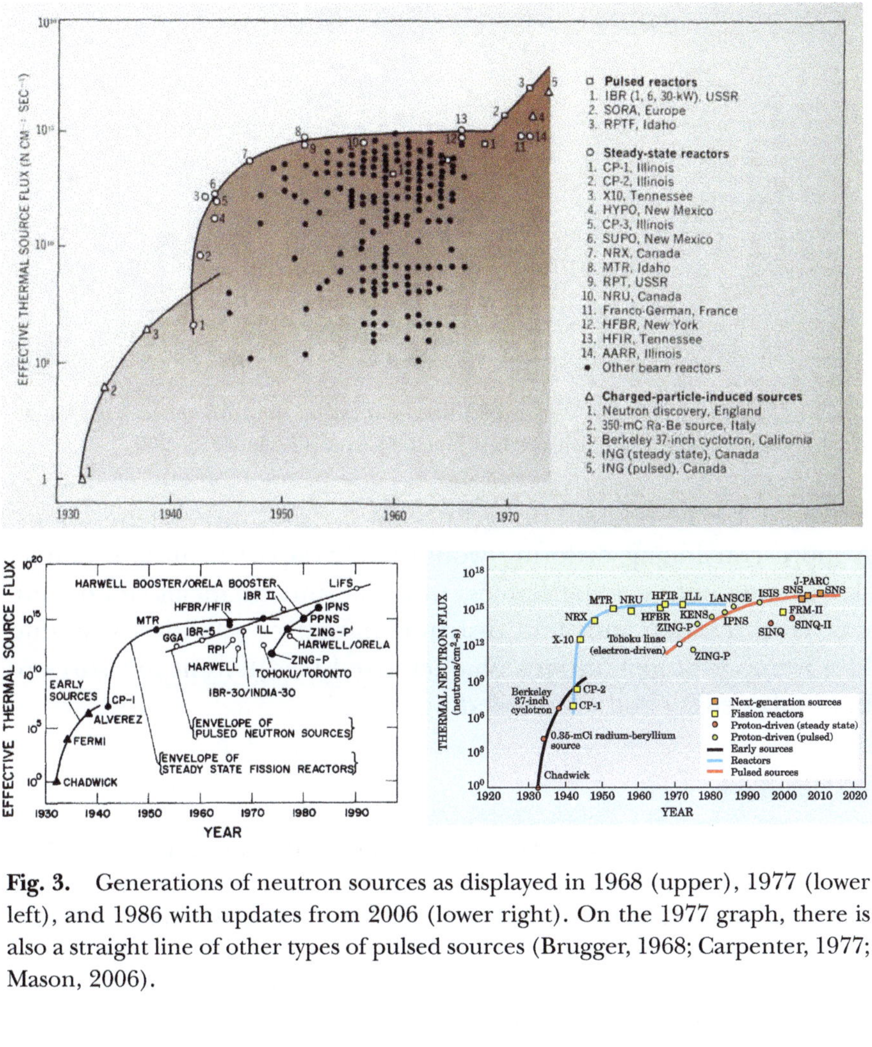

Fig. 3. Generations of neutron sources as displayed in 1968 (upper), 1977 (lower left), and 1986 with updates from 2006 (lower right). On the 1977 graph, there is also a straight line of other types of pulsed sources (Brugger, 1968; Carpenter, 1977; Mason, 2006).

as the carrier of hopes for the future and instead been categorized in a second generation.

That generations of neutron sources can be defined and redefined according to contemporary views of potential sources in the future that vary over time indicate the relative character of the concept of generations of neutron sources. Although this historical overview of neutron sources certainly has revealed different construction principles resulting in qualitatively diverse neutron beams,

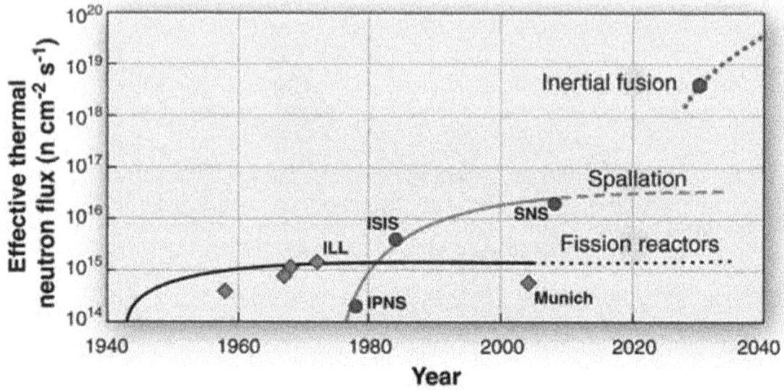

Fig. 4. The more recently imagined third-generation neutron sources no longer depart from spallation techniques but inertial fusion (Taylor *et al.*, 2007).

it would be far-fetched to group them into three generations. A more convincing ordering would likely depart from knowledge, skills, engineering experiences, political and financial opportunities, research demands and instrumentation possibilities, resulting in a number of generations characterized by different measures of both inbreeding and crossbreeding.

References

Atchison, F., *et al.* (1986). The spallation-neutron-source project SINQ. *Physica B* 136, 97–99.

Barry, A. and Walters, W. (2003). From euratom to 'complex systems': Technology and European government. *Alternatives* 28(3), 305–329.

Bauer, G. S. (1982). General concept for a neutron spallation source in the Federal Republic of Germany. *Atomkernenergie, Kerntechnik* 41(4), 234–242.

Bauer, G. S. (1997). Commissioning of the 1 MW spallation neutron source SINQ. In: *Proceedings of the 1997 Particle Accelerator Conference, May 12–16, 1997.* http://ieeexplore.ieee.org/stamp/stamp.jsp?tp=&arnumber=753416 (accessed 8 March 2013).

Bauer, G. S. (1999). Concept of a fault tolerant linear accelerator as proposed in the SNQ project. In: *Utilisation and reliability of high power proton accelerators, Workshop proceedings of OECD, Nuclear energy Agency, October 13–15, 1998*, pp. 211–225.

Bauer, G. S. (2010). Overview on spallation target design concepts and related material issues. *Journal of Nuclear Materials* 398, 19–27.

Bauer, G. S., Sebening, H., Vetter, J.-E., and Willax, H. (June 1981). Realisierungsstudie zur Spallations-Neutronenquelle. Kernforschungsanlage Jülich & Kernforschungszentrum Karlsruhe Arbeitsgemeinschaft Spallations-Neutronenquelle, Jül-Spez 113, KfK 3175, p. 19. http://bibliothek.fzk.de/zb/kfk-berichte/KFK3175.pdf (accessed 5 February 2013).

Benner, M. (2018). *The New Global Politics of Science: Knowledge, Markets and the State* (Cheltenham: Edward Elgar).

Brugger, R. M. (December 1968). We need more intense thermal-neutron beams. *Physics Today* 21(12), 23–30.

Carpenter, J. M. (1977). Pulsed spallation neutron sources for slow neutron scattering. *Nuclear Instruments and Methods* 145, 91–113.

Carpenter, J. M. and Yelon, William B. (1986). Neutron sources. *Methods of Experimental Physics* 23(A), pp. 99–196.

Clery, D. and Lawler, A. (February 17, 1995). The looming neutron gap. *Science* 267 (5200), 952–954.

Cockroft, J. D. (June 4, 1949). High-energy electron accelerators as pulsed neutron sources. *Nature* 163(4153), 869.

Commission of the European Communities (1991). Neutron beam sources. Access to Large-Scale Facilities, Study panel report 1. http://bookshop.europa.eu/en/neutron-beam-sources-pbCG6495001/ (accessed 24 October 2016).

Commission of the European Community (1994). The large installations plan 1989–1992. Access to Large-Scale Facilities. http://en.youscribe.com/catalogue/reports-and-theses/access-to-large-scale-facilities-1114245, (accessed 24 October 2016).

Commission of the European Community (2001). Research Infrastructures: Neutron sources: Project Report, a cura di Campbell Warden (Luxembourg: Office for Official Publications of the European Communities).

Crease, R. P. (1999). *Making Physics: A Biography of Brookhaven National Laboratory, 1946–1972* (Chicago: The University of Chicago Press).

Danon, Y., *et al.* (2005). The gaerttner LINAC laboratory review and current activity. In: *Proceedings of the Oak Ridge Electron Linear Accelerator (ORELA, 2006), Workshop, ORNL/TM-2005/272,* Oakridge, TN, July 14–15, 2005, pp. 41–46.

Editorial (2007). The big picture. *Nature* 450(7170), 586.

ENSA (August 1998). Survey of the neutron scattering community and facilities in Europe. The European Neutron Scattering Association, Unpublished Report by The European Neutron Scattering Association. https://www-llb.cea.fr/rapport/ensa_survey-1998.pdf (accessed 25 September 2023). p. 34.

ESFRI (March 2005). Towards new research infrastructures for Europe: The ESFRI 'List of Opportunities'. European Commission, EUR 2167 22 EN. ftp://ftp.cordis.europa.eu/pub/era/docs/esfri_list_opportunities_290305.pdf (accessed 22 December 2011).

ESS Project (May, 2002). *European Source of Science*, Vol. I.

ESS Project (September 2003). *Foundation for the Next European Spallation Source.*

EURATOM (April 1967). Reactor and neutron physics. European Atomic Energy Community EURATOM Documentation attached to the tenth General Report on the Activities of the Community, (March 1966–February 1967), pp. 99–103.

European Science Foundation (December 2000). The scientific-strategic case for a next-generation European spallation neutron source for science and research (ESS Project). ESF Studies on Large Research Facilities in Europe, Interim Report (European Science Foundation).

European Spallation Source (2011). *ESS Activity Report 2010–2011.* (Lund: ESS AB).

European Strategy Forum for Research Infrastructures (2006). *European Roadmap for Research Infrastructures: Report 2006* (Luxemburg: Office for official publications of the European Communities).

Findlay, D. J. S., *et al.* (2006) ISIS upgrades — A status report. in: *Proceedings of European Particle Accelerator Conference*, Edinburgh, UK, June 26–30, 2006, pp. 935–939.

Finney, J. L and Lengeler, H. (July/August 1997). The next generation European spallation source. *Europhysics* 28(4), 129–132.

Finney, J. L. and Lengeler, H. (1994). Proposed European spallation source takes shape. *Neutron News* 5(3), 25–28.

Gläser, J., *et al.* (1994). If people become afraid of your research methods: Conflicts over research reactors in Berlin and Munich. in: Uwe Schimank and Andreas Stucke (eds.), *Coping with Trouble: How Science Reacts to Political Disturbances of Research Conditions* (Frankfurt: Campus Verlag), pp. 293–332.

Gutberlet, T. and Mezei, F. (2002). The European spallation source ESS: A quantum leap in exploring matter with neutrons. *Synchrotron Radiation News* 15(4), 32–35.

Guzzetti, L. (1995). *A Brief History of European Union Research Policy* (Luxembourg, European Communities).

Hanna, C. G. (1997). Accelerators and fusion. in: *Canada Enters the Nuclear Age: A technical history of Atomic Energy of Canada Limited as seen from its research laboratories* (Quebec: Atomic Energy of Canada Limited), pp. 169–190.

HELIOS (July 19, 1979). Harwell's electrons knock out faster neutrons. *New Scientist* 83(1164), 194.

Johnston, S. F. (2012). *The Neutron's Children: Nuclear Engineers and the Shaping of Identity* (Oxford: Oxford University Press).

Joint Publication Research Service (March 17 1986). West Europe report: Science and technology. JPRS-WST-86-010. http://www.dtic.mil/cgi-bin/GetTRDoc?Location=U2&doc=GetTRDoc.pdf&AD=ADA348057 (accessed 5 February 2013).

Kley, W. (1968) The use of the SORA-reactor for neutron physics experiment. European Atomic Energy Commission (EURATOM), Report EUR 4082e (Joint Nuclear Research Center, Ispra Establishment).

Kohler, P. E. (2005). Oak ridge national laboratory neutron sciences overview. In: *Proceedings of the Oak Ridge Electron Linear Accelerator (ORELA, 2006) Workshop, ORNL/TM-2005/272,* July 14–15, 2005, pp. 19–23.

Kohlrausch, M. and Trischler, H. (2014). *Building Europe on Expertise: Innovators, Organizers, Networkers* (Basingstoke: Palgrave Macmillan).

Krige, J. (2006). Atoms for peace, scientific internationalism, and scientific intelligence. in: *American Hegemony and the Postwar Reconstruction of Science in Europe* (Boston, MA: The MIT Press).

Lengeler, H. (1994). Proposals for spallation sources in Europe. in: *Fourth European Particle Accelerator Conference, June 27-July 1, 1994,* pp. 249–253. http://epaper.kek.jp/e94/PDF/EPAC1994_0249.PDF (accessed 16 December 2011).

Litt, J. (ed.) (1979). *Nimrod: The 7GeV proton synchrotron* (Chilton: Rutherford Laboratory). http://epubs.cclrc.ac.uk/bitstream/1945/NIMROD%20 The%207GeV%20proton%20syncatron.pdf (accessed 28 February 2013).

Lynn, J. E. (1980). HELIOS: The new Harwell linear accelerator, and its scientific programme. *Contemporary Physics* 21(5), 483–500.

Mason, T. E. (May 2006). Pulsed neutron scattering for the 21st century. *Physics Today* 59(5), 44–49.

Meadows, J. (2009). *Big Science: Fifty Years of the Rutherford Appleton Laboratory* (Rayleigh, N.C: Lulu Press).

Mersits, U. (1987). From cosmic-ray and nuclear physics to high-energy physics. in: Armin Hermann, John Krige, Ulrike Mersits and Dominique Pestre (eds.), *History of CERN*, Vol. I (Amsterdam: North-Holland), pp. 3–60.

National Institute of Standards and Technology, Center for Neutron Research (2012). http://www.ncnr.nist.gov/nsources.html (accessed 16 February 2012).

Nieburg, H. L. (1963). EURATOM: A study in coalition politics. *World Politics* 15(4), 597–622.

OECD Megascience Forum (June 1998). Report of the Neutron sources working group. Submitted to the 13th meeting of the Megascience Forum.

Pynn, R. and Fender, B. E. F. (January 1985). Neutron scattering in Europe. *Physics Today* 38(1), 46–53.

Reitbauer, M. (2015). The origins, formation and development of Euratom. *Zeitgeschichte* 42(5), 299–306.

Richter, D. (2002). The European spallation source. *Applied Physics A* 74(Suppl.), 818–822.

Richter, D. and Springer, T. (November 1998). A twenty years forward look at scattering facilities in the OECD countries and Russia. European Science Foundation & OECD Megascience Forum. Technical Report.

Riste, T. (1993). Analytical report. in: *Neutron Beams and Synchrotron Radiation Sources, Megascience: The OECD Forum* (Paris: OECD), pp. 63–94.

Ritter, G. (1966) The SORA project ready for its realization. European Atomic Energy Commission (EURATOM), Report EUR 3057e (Joint Nuclear Research Center, Ispra Establishment).

Rush, J. J. (2015). US neutron facility development in the last half-century: A cautionary tale. *Physics in Perspective* 17(2), 135–155.

Simnad, M. T. (1981). Perspective on research reactors and their fuel elements. *Journal of Nuclear Materials* 100, 78–92.

Skogmar, G. (2004). *The United States and the Nuclear Dimension of European Integration* (Basingstoke: Palgrave Macmillan).

Taylor, A., *et al.* (February 23, 2007). A route to the brightest possible neutron source? *Science* 315, 1092–1095.

The National Academies (1977). Neutron research on condensed matter: A study of the facilities and scientific opportunities in the United States. http://books.google.se/books?id=DTsrAAAAYAAJ&pg=PA15&lpg=PA15&dq=Neutron+research+on+condensed+matter+1977&source=bl&ots=i0S1d60x57&sig=0bhPwDOANwH9j2xb7qdr4GypmhM&hl=sv&sa=X&ei=r63RUObxMKqO4gSSx4DABg&redir_esc=y#v=onepage&q=Neutron%20research%20on%20condensed%20matter%201977&f=false (accessed 24 October 2016).

The National Academies (1984). Current Status of Neutron-Scattering Research and Facilities in the United States. Panel to assess the current status of facilities and research in neutron scattering (Washington D.C: National Academies Press).

Tindemans, P. (2010). Politics of major facilities. *Neutron News* 21(1), 51–53.

Tindemans, P. and Clausen, K. (2003). ESS on hold: Europe urgently needs a strategy. *Neutron News* 14(3), 2.

US Department of Energy (January 1993). Neutron sources for American's future. Office of Science; http://science.energy.gov/~/media/bes/besac/pdf/neutron_source_america_future.pdf (accessed 24 October 2016).

Warden, C. (2001). *Research Infrastructures: Neutron sources: Project Report.* Commission of the European Community, (Luxembourg: Office for Official Publications of the European Communities).

Weilemann, P. (1983). *Die Anfänge der Europäische Atomgemeinschaft: Zur Gründung von EURATOM 1955–1957* (Baden-Baden: Nomos Verlagsgesellschaft).

Westfall, C. (2010). Surviving to tell the tale: Argonne's intense pulsed neutron source from an ecosystem perspective. *Historical Studies in the Natural Sciences* 40(3), 350–398.

Westwick, P. J. (2003). *The National Labs: Science in an American system, 1947–1974* (Cambridge, MA: Harvard University Press).

Wissensachaftsrat (July 12 2002). *Stellungsnahme der neun GrossgeräteStellungnahme zu neun Großgeräten der naturwissenschaftlichen Grundlagenforschung und zur Weiterentwicklung der Investitionsplanung von Großgeräten*, Drs. 5363/02.

Chapter 2.2

Interplay of Actors and Institutions in the Campaign to Localize ESS to Lund

Olof Hallonsten

Introduction

The European Spallation Source (ESS) is a very large neutron scattering facility under construction in Lund in Southern Sweden, estimated to open to scientific use in 2023. The ESS will use a linear accelerator and a so-called target station to produce highly intensive bursts of neutrons that will be used for a variety of experiments at the 15 instruments currently under construction. The ESS is owned and organized by 15 European partner countries through the European Spallation Source, a *European Research Infrastructure Consortium* (ERIC), created specifically for the purpose of constructing and operating the facility as a joint European undertaking. The project itself — a state-of-the-art, pan-European neutron facility — was conceived in the late 1980s and first became a concrete proposal in 1993, when a number of European scientific organizations formed the ESS Council. In 1997, the first design was presented, and in 2003, the scientists' intentions were to bring the project to the attention of politicians and get a decision in place for its funding and realization. However, it took another 12 years before representatives of European governments put ink to paper and construction of the ESS could start. This happened in Lund, a rather small village in southern Sweden, on the periphery of the European knowledge-based economy and in a country with very limited experience of the use of neutrons for experimental science and no prior experience of hosting joint European Big Science facilities.

The story of how this could happen is, therefore, intriguing. What made European neutron strongholds, France, Germany, and the United Kingdom, bow out in the competition for hosting the ESS? How could Sweden, with its limited experience and comparably minor role in European politics and science, bring this major trophy home? Why did it take almost a quarter of a century for the ESS project to go from idea to start of construction?

This chapter analyzes these questions from an institutional perspective and contrasts individual decision-making and action with the power, resilience and inertia of institutions, to find out more about the mechanisms by which collaborative European Big Science facilities come into being. The chapter builds on a recent book that chronicles and analyzes the ESS story in detail (Hallonsten, 2020a) and several other secondary sources that have documented the politics of European collaborative Big Science from the perspective of their institutional history (Cramer, 2017, 2020a; Hallonsten, 2014, 2016, 2020b; Hallonsten and Heinze, 2012, 2015; Krige, 2003, 2006).

The chapter begins with a brief orientation to the institutional perspective on collaborative Big Science in Europe, with some use of an enlightening comparison between Europe and the United States. Thereafter, Section 3 provides a brief review of theory. In Section 4, the interplay of individual actors and institutional power, resilience, and inertia in the history of the ESS is analyzed, followed by a concluding discussion.

European Big Science: An Institutional View

The 20th century saw an unprecedented expansion of national and international science in the Northern Hemisphere, driven in no small part by the post-World War II bipolar world order and the role that science and technology had played in World War II and was playing in social and technical development. Besides the general growth in expenditure on science in Europe and the United States after the war, a specific feature of the science system was soon to emerge, namely, *Big Science*, a phenomenon with many facets but a term most often used to denote large, governmentally sponsored projects with high visibility and prestige and at the cutting edge of technology (Hallonsten, 2016, pp. 1–21).

In the United States, the mobilization of Big Science was a natural consequence of the wartime mobilization of science in the service of the military, most famously the Manhattan Project which produced the world's first nuclear weapons in a secret undertaking that organized thousands of scientists and laypeople. After the war, the collected assets of the Manhattan Project were transferred to the Atomic Energy Commission (AEC) and its system of National Laboratories, initially eight, which got a default monopoly on the big and growing machines that were needed to undertake cutting-edge peaceful and military research in atomic energy and its sibling disciplines in physics and chemistry (Westwick, 2003). This soon included the construction and operation of particle accelerators for the study of the inner structure of atoms, which moreover became its own branch of Cold War competition, with foremost the United States and the Soviet Union, and also eventually Japan and (Western) Europe, competing to build the largest and most energy-intensive accelerators and to use them to discover subatomic particles and energies with increasingly distant military usefulness but significant chances to win the Nobel Prize in physics and thus render prestige and acclaim to the nation's abilities in Big Science (Hallonsten, 2016, pp. 45–50).

In Europe, several countries pursued national nuclear and particle physics programs after World War II, with both military and peaceful intentions, but in the 1950s, it became increasingly clear that the development of the scientific study of the inner structure of the atom would push sizes and costs of particle accelerator facilities beyond what individual countries could typically afford, and that

Europe's continued global competitiveness in the area therefore hinged upon its ability to build intergovernmental collaboration. The 1950s was also an era of intensifying efforts of bringing (Western) Europe together in attempts to prevent future conflict, and to keep the newly created Federal Republic of Germany in check through multilateral collaborations. The European Coal and Steel Community (ECSC), established in 1951, was the first such attempt and the start of the political integration process in Europe that produced today's EU as an "ever closer union" encompassing 28 countries. In 1954, the European Organization for Nuclear Research (*Conseil européen pour la recherche nucléaire, CERN*) was founded by 12 Western European countries in a parallel effort of scientific integration in Europe, with significant involvement by the United States which thereby sought to secure its influence in Western Europe (Krige, 2006). In the decades to come, several other collaborative European projects were established with CERN as a model: the European Southern Observatory (ESO), the European Space Research Organization (ESRO), and the European Launcher Development Organization (ELDO), all in 1962, the neutron facility Institute Laue-Langevin in 1967, the European Molecular Biology Laboratory (EMBL) in 1973, the European Space Agency (ESA) in 1975, the fusion research facility Joint European Torus (JET) in 1977, and the European Synchrotron Radiation Facility (ESRF) in 1984 (Krige, 2003, p. 899). But while this collection of facilities and projects constituted a progression of ever more intense integration of scientific communities in Europe, it happened entirely outside of the mainstream political integration process that led to the European Community (EC), and eventually the European Union (EU), and thus constituted what has been called "the Other Europe" (Cramer, 2020a). This means that the political work to realize these facilities could never rely on the institutional and regulatory framework of the European Community/Union and that pan-European Big Science therefore developed in a radically different institutional context than its US-American counterpart. In the United States, the Big Science projects conceived, promoted and realized in the 1950s and 1960s within the system of National

Laboratories set a precedent for how such project would be shepherded through the federal political system, and this procedure was, furthermore, formalized in the 1980s by the implementation of a strict review system within the Department of Energy (DOE), which had taken over stewardship over the National Labs from the Atomic Energy Commission upon its dismantling in the 1970s (Hallonsten, 2016, p. 64).

No similar procedure exists for European projects, which means that all of them have had to start with a blank sheet, politically, and mobilize the necessary support of national and regional governmental bureaucracies, and ultimately the political level of governmental decision-making, in direct competition with other diplomatic priorities and agendas and with no central coordinating function to bring projects to the next steps toward realization. This has created a varied landscape, to say the least, with facilities operated as international treaty organizations (e.g. CERN and ESO), limited liability companies operating under the law in the countries they are established (e.g., ILL, ESRF), and most recently, European Research Infrastructure Consortiums (ERICs), a new type of organization form with a legal framework laid down in an EU regulation (Moskovko *et al.*, 2019). Treaty organizations and ERICs have national governments as signatories, but the limited liability companies that operate ILL, ESRF and, more recently, the European X-ray Free Electron Laser (XFEL), may have as shareholders any national organization that the participating countries decide, or even consortia of organizations from several countries that thus participate jointly, through separate bi- or multilateral agreements. Countries participate on vastly different levels and make investments toward the construction and operations costs of facilities based on the outcome of the negotiations that precede agreements and the launch of projects. These negotiations follow no established precedent but are largely shaped by the initiative of individual countries and their leverage, and they mirror the plethora of national preferences and arrangements for participation. Some countries enter into negotiations on the basis of a self-sufficient guarding of their own interests in a specific scientific field, some out of a desire to pool resources with other countries to achieve a competitive advantage in another scientific area, some in order to gain access to a resource they would otherwise not be able to provide to their scientific communities, some with the ambition to be able to get major procurement contracts for high-tech instrumentation and similar to their domestic knowledge-based industry, and some on the basis of a desire to put their own country "on the map" by "winning" the competition for hosting a European Big Science project (Krige, 2003; Hallonsten, 2014; Cramer, 2020b).

The institutional diversity is huge. European countries have very different political systems with very different logics according to which science and innovation policy is shaped and a very different setup of organizations and institutions to shape it: Ministries, research councils, research funding bodies, private foundations, national organizations for Big Science, and powerful academic organizations (universities and national academies of science) are, to varying degrees, involved in priority-setting and the promotion of projects and ventures. Case studies of European collaborative Big Science projects have, moreover, demonstrated that given the lack of precedent and joint institutional structures, the fate of these projects is often decided or strongly influenced by the currents of political relations in Europe: Big Science projects have been used as bargaining chips in larger negotiations, as trial projects for new and bold initiatives, and as symbols of reconciliation in times of political tensions (Krige, 2003; Hallonsten, 2014; Cramer, 2020a, 2020b).

Institutional Change, Institutional Logics, and Bounded Rationality

The analysis of complex historical processes is an undertaking well served by a flexible understanding of theory and a similarly flexible use of it. This means that theory is not taken to mean a set of postulations to be verified or refuted, but instead is given the role of a tool or, preferably, a *toolbox* of complementary tools from which those can be chosen that assist in the undertaking and thus enhance the explanatory value of the analyses.

While this chapter adopts an institutional perspective on the processes under study, it also reports on a vast empirical study of a political campaign in which several actors stand out as decisive for the outcome — "powerful individuals who are in the right place at the right time and manage to turn events in their favor, or in accordance with the agendas they represent" (Hallonsten, 2020a, p. 29). What unites such individuals in spite of their different backgrounds and different rationales is their purposeful and deliberate efforts, which enable and necessitate a conceptualization of *rational actors* as important in the story. Needless to say, absolute or perfect rationality does not exist, and the critique usually issued toward rational choice theory (or public choice theory, as it is usually called in political science) is unnecessary (e.g., Opp, 1999). There is no such thing as a perfectly rational "homo economicus" — it is well-established knowledge in studies of management and decision-making that rationality is "bounded" or "limited" by institutional and structural constraints and by circumstances that are both unpredictable and beyond the view and control of the individual (Simon, 1957; March and Simon, 1958). But individuals can still be conceptualized as acting rationally in the sense that they make decisions and take actions on the basis of the evaluation of the circumstances of specific situations and the prospects for achieving desired outcomes in the long and short term (Esser, 1993; Goldthorpe, 1998). This enables the analysis to discard the endless philosophical debate over rationality and what it means and instead focus instead on the rational actor as a model for understanding and explaining how and why individuals make decisions and act in specific ways, in specific situations. Especially individuals in power, and acting out of an agenda, are simultaneously resourceful, restricted, expecting, evaluating, and maximizing (Lindenberg, 1985, p. 100). In the story of how the ESS ended up in Sweden, it is quite clear that several resourceful individuals with high ambitions pushed developments in certain directions. Ambitions clearly varied between these individuals, and possibly also differed between time periods and situations, and individuals perhaps also hosted dual or multiple ambitions at the same time. Neither in politics nor in science is this unusual:

Personal achievement and the advancement of joint interests, to greater benefits, are commonly mixed with dealmaking and mutual backscratching.

The institutional perspective on the story is complementary to the rational actor perspective: No individual, no matter how powerful, acts in a vacuum but must attach their decisions and actions to structures and procedures that have precedent, credibility and predictability. The concept of "institutions" is ubiquitous in the social sciences but has no commonly agreed definition. Giddens's (1984, p. 24) definition that institutions are "the more enduring features of social life" that provide "solidity across time and space" and Scott's (2001, p. 48) definition that institutions are "social structures that have attained a high degree of resilience" are both imprecise but useful to provide the theoretical countermeasure to the rational actor perspective. Moreover, society is shaped by history's major events and by the slow and inert development that is captured in the French concept, *longue durée*, and the logics of events and processes are always partly shaped by the institutional composition of a society and the sector where the events chiefly take place. In European collaboration in Big Science, and specifically in the case of the ESS, the non-precedent set by the absence of a common legal and procedural framework is a key explanation for the incredibly long time period that passed between the launch of the ESS as an idea and its eventual realization. This is undoubtedly an institutional constraint. Similarly, Swedish politics and the logics by which the parliamentary system and governments obtain and build their mandates clearly influenced the prospects of mobilizing political initiative for a national campaign to win the hosting of the ESS. The strongly institutionalized features of international scientific communities and how they promote and develop projects and initiatives across borders and on the basis of common ideals of scientific and technological edge also played an enormously important role in shepherding the ESS project from a political decision (in 2009) to groundbreaking (in 2015).

Two key concepts are *institutional logics* and *path dependence.* The normative function of institutions — guiding and setting the constraints for behavior — differ between different parts of society and between national and organizational entities. Science has a reward system and a means to solve conflict that differs from politics and the market economy, and as already noted in a previous section, different countries have different institutional setups that greatly influence the nature and level of their engagement in international collaborations. The theory of institutional logics was launched by Friedland and Alford (1991) in an attempt at "bringing society back in" to organization studies and emphasizing the need for attention to other norm systems and guides for behavior and organizing besides that of private enterprise and the market economy and demonstrate how such alternatives may clash with each other and with the profit motive of businesses. The exhaustive cataloguing of the components of specific institutional logics is not the key purpose of the framework; rather, its merit lies in the capacity of identifying and highlighting contrast and analyzing why it exists and creates conflict or tension, or affect decisions and actions, in given situations (Thornton, 2004; Reay and Hinings, 2009; Pache and Santos, 2013). Complementary to the institutional logics perspective is the theory of historical institutionalism, which emphasizes path dependence and the nature of institutional change: Contrary to public images of history as shaped by major disruptive events, historical institutionalists have shown that most change is gradual, incremental, and cumulative and takes place within institutional contexts where stability breeds change, and change enables long-term stability (Streeck and Thelen, 2005; Mahoney and Thelen, 2010; Fioretos *et al.*, 2016).

The remainder of this chapter will be devoted to the essentially empirical questions, explored in much more detail in the book, that the analysis is built on (Hallonsten, 2020a), how and when individuals manage to mobilize institutions in their favor, how and when institutions hinder them or enable them in their pursuits, and how and when the inertia and resilience of institutions work in favor or disfavor of those who have set out to accomplish change.

The European Spallation Source: A Long Story

The use of neutrons for the advanced study of materials was pioneered during World War II and made into a small but reliable feature of the post-war expansion of the scientific use of reactors, both in the United States and in Europe. Prominent scientists in France, the United Kingdom, and the Federal Republic of Germany made progress on the side of instrumentation and their use in parallel with the geopolitical developments that eventually enabled the governments of these three countries to agree on the funding and construction of a dedicated reactor-based neutron source in Grenoble, France, which opened to scientific use in 1973 (Cramer, 2017). The Institute Laue Langevin (ILL), as it was eventually called, became a hotspot for European neutron use and is generally regarded as the key factor for the shift in global leadership in the area from the United States to Europe from the 1970s onward. Meanwhile, smaller national neutron sources provided scientists with important complementary capacity but relied on old reactor technology, and so a looming "neutron gap" became a topic of concern in the 1980s, as national facilities closed and the demand for neutrons exceeded what the ILL could deliver (Hallonsten, 2020a, p. 88ff).

In the 1980s, therefore, plans were made for a new pan-European neutron source that would succeed the ILL and secure European scientific strength in the area in the very long term. In 1990, the European Spallation Source was first proposed, and in 1993, several European research institutes began the work of jointly drafting a design and a proposal. In 1997, the first ESS technical design and science case were presented at an estimated total cost of €1,500 million and an estimated completion in 2007. Political interest was lukewarm, but an endorsement of the project by the Organisation for Economic Co-operation and Development (OECD) in 1999 raised the hopes of the ESS proponents, and a time plan for the political process was laid out, with a 2003 meeting of the European Neutron Scattering Association (ENSA) in Bonn as the point in time when European governments would agree on the financing, siting, time plan, and organization for the ESS facility. None of this happened, however, and in the year after the Bonn meeting, countries instead dropped out of the site competition and relinquished their support of the project — most notably Germany, which was expected to contribute significantly and probably also host the ESS, but which declared that it had other priorities (Hallonsten, 2020a, pp. 89–97).

During the years 2003–2006, therefore, the ESS project was more or less dead on the European stage and only rescued by the work of a devoted few, most notably Peter Tindemans who had chaired the ESS council during the years 2000–2003 and managed to maintain a loose network of ESS stakeholders on the European stage. But the withdrawal of major site contenders — France, Germany, and the UK — also opened a window of opportunity for smaller countries within which especially Sweden managed to act. In 2004, the local ESS Scandinavia campaign in Lund managed to get former Swedish Finance Minister and Brussels bureaucrat Allan Larsson appointed by the government to investigate the prospects of a Swedish bid for the ESS, and in Larsson, the campaign got a formidable ally, champion, and spokesperson who would prove enormously valuable in the years to come. The result of the investigation was meager — though Larsson advocated a strong showing of interest on behalf of the Swedish government, the 2006 election drew closer and put a halt to the initiative. By then, the local campaign in Lund had mobilized support among local politicians and bureaucrats, and built a rather strong power base of individuals in positions of power: Lars Börjesson, who took an active part in launching the ESS Scandinavia initiative back in 2000, had become a high-ranking official in the Swedish Research Council and the Swedish delegate in the European Strategy Forum for Research Infrastructures (ESFRI); Peter Honeth, Director of Administration of Lund University and one of the early non-scientist supporters of ESS Scandinavia, was eventually selected as deputy to the incoming minister for education and research; and both Colin Carlile and Christian Vettier — veteran neutron scientists and former Director-General and Scientific Director of the ILL in Grenoble — had both been recruited to guest professorships at Lund University (Hallonsten, 2020a, pp. 107–152).

But the shift in government was also timely. Majority governments are unusual in Swedish national politics, but when they are formed, they get a significant leeway for ambitious reform agendas. The dynamics of center–right four-party coalitions also goes back decades: Each party tend to have their profile areas, and in 2006, the Liberal Party had settled on education and research, having already envisaged an unprecedented increase in public expenditure on science. In this context, the ESS worked really well as a "moonlander project" (this was the expression used by the incoming Minister of Education and Research, Lars Leijonborg) (Hallonsten, 2020a, p. 157). One of the main champions of the ESS, Peter Honeth, became the highest-ranking civil servant in Swedish research policy and could mobilize power bases in the government to make the campaign a national priority. It took only a few months before the government declared that it was officially seeking to host the ESS in Lund and appointed Allan Larsson as the special negotiator. By this time, the competition for hosting the ESS had become a rather low-intensity race between three contenders — Lund in Sweden, Bilbao in Spain, and Debrecen in Hungary — and the challenge seems to have been to mobilize support in scientific communities, and politically, to get a significant number of European countries to agree to invest in the ESS. The major players — France, Germany, and the UK — were still hesitant, at best, and needed convincing that the ESS was at all worth an investment. A common wisdom at the time was that the host country would have to provide at least 50% of the estimated construction costs to be able to get other countries on board and that there was still a long way to go to build a credible technical and scientific case (Hallonsten, 2020a, p. 153ff).

After an *ad hoc* site review put in place by the European Strategy Forum for Research Infrastructures (ESFRI) in 2008 to give the ESS renewed momentum and evaluate the three site candidates, the campaign intensified and a decision of sorts drew closer. Intensive lobbying on behalf of the Swedish, Spanish and Hungarian campaigns in the winter of 2008–2009 was concluded in a meeting with representatives of interested governments in Brussels on May 28, 2009, where a sufficient number declared their support for Lund. But strictly speaking, "nothing much was really decided, except that it was more or less clear after May 28, 2009, that if the ESS was to be built at all, it would be built in Lund" (Hallonsten, 2020a, p. 183). What ensued was, therefore, the strenuous parallel processes of updating the technical design and building the proper scientific, technical, and project management organization in Lund, and the negotiation between the Scandinavian duo of Denmark and Sweden and the rest of Europe, over the financing of the facility and the structure of its organization. These processes were, naturally, intertwined, but they also took place in different institutional spheres and thus followed separate logics of politics, science, and the management of a large construction project. They involved the recruitment of key personnel to the ESS organization to update the technical design and build a scientific case, most notably a director with experience in project management on the level necessary, and the high-level negotiation between government representatives at the European level (Hallonsten, 2020a, pp. 187–245).

Five years after the site decision, in 2014, the ESS organization in Lund had expanded from a handful of people to over 200 employees (Hallonsten, 2020a, pp. 185–186) and accomplished a comprehensive design update, built a scientific case involving thousands of neutron scientists across Europe, and pushed the plans through the necessary permit processes, and the Swedish government had managed to get 11 other European countries to pledge financial support for the construction of the ESS. In September 2015, one year after the start of construction on site in Lund, nine of these countries plus Denmark and Sweden signed the first legally binding agreement concerning the funding and construction of the ESS, when they entered as members in the European Spallation Source, ERIC. Current estimations (2020) are that the ESS facility will open to scientific use in 2023, with 16 operating instruments. Cost increases have been severe and resulted in cutbacks on technical performance and the need for the Swedish governments to make additional cash contributions (Hallonsten, 2020b, pp. 225–226).

The Actors and Institutions of (Big) Politics and (Big) Science

Lund was, indeed, an unlikely candidate to host the ESS. Sweden has been a reliable (but minor) partner in European collaborative Big Science after World War II, and though Lund is the home of Sweden's largest university and is very close to (or, some would say, even part of) the metropolitan area Greater Copenhagen, Lund had practically no neutron users and Sweden had practically no experience of planning and building large scientific facilities. The framing of the story of how the ESS ended up in Sweden, laid out over hundreds of pages of text in the book that forms the basis for this chapter, is therefore the major question of *how*. Its final chapter notes that "the success of the campaign was therefore both impressive (from a layman's point of view) and puzzling (from the perspective of a historian and sociologist of science)" and cites the several memoirs of central actors in the story that confirm that the ESS "ended up in Lund in spite of all the odds" (Hallonsten, 2020a, pp. 278–279).

Strong individuals emerge everywhere in the story. They are evidently resourceful, evaluating, maximizing, and certainly restricted in their actions by institutional constraints that they are both aware and unaware of, meaning that they make decisions and take action based on ambitions, competences, and what is best described as rational evaluation of the conditions of the situations they are in. These, in turn, depend on the individuals' backgrounds, experiences, personal networks, as well as their formal roles and the rules and conventions that apply in their fields of action. Institutions, with different logics, enabled and restricted them in their pursuit of success.

Scientifically and technically, the ESS was a project firmly grounded in a strong tradition and the mobilization of several cutting-edge research environments across Europe. These employed the top scientists and engineers in neutron scattering instrumentation and use both in the 1990s when the first ESS project was designed and in 2009–2014 when the current ESS design was drafted. The failure of the Bonn meeting is one of the events in the story that most evidently demonstrates how the technical and scientific excellence of the people behind the ESS in the 1990s was matched by a political naivety — the national governments of Europe were not ready to embark on the ESS project, and the predominantly scientific ESS lobby was unable to realize this. For almost 20 years, from the first conception of the ESS to the site decision in 2009, the scientific opportunities that the leaps of performance of the ESS project would bring mattered little — politics had other plans. Put differently, at the time, the ESS project existed in three institutional logics: (Big) science, European science policy, and national politics, with the latter taking precedence and creating a deadlock. This deadlock was only broken by the work of a few powerful individuals who managed to build on their respective power bases and not least make use of the window of opportunity that had opened, being "in just the right place at the right time" and both keeping the hope alive and pushing developments in their desired directions. They came from different backgrounds and entered into arenas where they got influence and made clever use of it.

The original ESS Scandinavia campaign was a grassroots movement of scientists in Sweden, mobilized around one of Sweden's very few internationally competitive neutron users, Lars Börjesson and Danish colleagues. Soon after its start, however, crucial political support was lent to the campaign by not only Lund University Director of Administration Peter Honeth (who would later become a key government official in science policy) but also local and regional politicians around Lund. Their support was vast and decisive in the first few years, but waned after the Bonn meeting and the non-decision it produced: the rather impressive delegation of local and regional politicians that accompanied the ESS Scandinavia enthusiasts in Bonn almost vanished completely thereafter, as other regional priorities took over. The apparently smaller stamina of local and regional politicians, compared to devoted scientists with their eyes on a prize in the distant future, is perhaps also a matter of differing institutional logics. During 2003–2006, ESS Scandinavia was more or less held floating by the work of three people — physicists Karl-Fredrik Berggren and Patrik Carlsson, and administrator Carina Wickberg, with some minor financial support from Lund

University and private foundations in Sweden. That this core group of enthusiasts managed to get the ESS project in front of perhaps the story's greatest campaigner of all, Allan Larsson, was probably its most decisive turn. Larsson's decades of experience in politics within the Swedish Social Democratic Party, the Swedish government, and the European Commission became a major asset and Larsson himself is probably best described as the ultimate mix of a resourceful, restricted, expecting, evaluating, and maximizing individual mastering the institutional logics of politics and mobilizing any necessary institutional arrangements in his favor.

The institutional logics of politics worked in favor of ESS Scandinavia in several ways — Larsson was not able to single-handedly move the incumbent Social Democratic government to act in favor of the ESS but had to wait for the shift of governments in 2006, when the first majority government in decades was installed and the energetic Lars Leijonborg was installed as the Minister for Education (and Research), determined to make a mark and write himself into Swedish science policy history. By this turn of events, electoral politics and the dynamics of party coalitions created formidable conditions for betting on a "moonlander project" favorable, and once Peter Honeth advanced to the ranks of the ministry, it was rather evident what that project was going to be. Suddenly, therefore, through an interplay of the work of purposeful individuals and institutional conditions, the ESS became a top priority of the new government's research policy offensive.

Lund University, an ancient academic organization in Sweden, used the tools at its disposal to bolster the ESS Scandinavia's prospects of getting ahead in the European competition, recruiting internationally reputable neutron scientists Colin Carlile and Christian Vettier to prestigious positions as guest professors. With both coming directly from key leadership positions at the ILL, and with both brandishing impressive resumés and formidable networks, they became formidable *de facto* campaign workers for ESS Scandinavia. With Carlile at the helm, ESS Scandinavia developed into an impressive campaign organization, and with Allan Larsson as chief negotiator and the Swedish government officially backing the efforts, Lund got ahead in the European competition. Institutional conditions mattered: The window of opportunity for smaller countries that had opened in 2003 meant that ESS Scandinavia and its competitors did not have to fight unequal rivals in the shape of France and Germany, and their formidable neutron instrumentation and neutron scattering institutes, but instead keep the idea of an ESS going, build momentum, and eventually present a site candidate with favorable properties.

The ESS story is a story where several actors turn up just in the right place at the right time. The ESS Scandinavia initiative was born out of a desire to have a major research facility located in Scandinavia and gained strength from the wave of cross-border integration and optimism for the future in the region at the time. Lund University and local and regional politicians were able to lend crucial support at the early stages in part because of this dynamic. The scientists that campaigned during the years 2004–2006 seem to have had just the right devotion and energy to continue when others fell off. Lars Börjesson, an early champion, became a high-ranking civil servant in the Swedish Research Council and ESFRI, and certainly carried the ESS Scandinavia cause with him, promoting it whenever he had the chance. Allan Larsson's role has already been mentioned several times — retiring from a long career in public administration and known for his vigor and fervor, he came in at the best possible moment when ESS Scandinavia was running out of steam and the government was dragging its feet. Only a year later, a new government took office and a central figure in the ESS fan club, Peter Honeth, was recruited to its inner chambers. Colin Carlile and Christian Vettier choose to come to Lund when retiring from the ILL.

Several of these people defied the prevalent institutional logics of the Swedish and European research policy and funding systems by mobilizing complementary or competing institutions for their cause. Put differently, the ESS became a reality in Sweden due not to the stability and predictability of known institutional logics in politics and science but by the purposeful work of individuals tapping into other power bases and other institutions and, importantly, by being in the right place at the right time. If the established practices of the Swedish political system would have had their way, and if the prevalent way of promoting initiatives and pushing agendas in science would have been followed by the ESS Scandinavia campaign, it is fair to say that not much would probably have happened at all.

In later stages, after the 2009 site decision, the campaign followed different institutional logics: First, the development of a technical design and scientific case, the creation of organizational structures and an international network of collaborators, and the work to establish a proper project organization. This involved science and engineering as well as corporate governance and project management. Second, negotiations on the European stage, where high-level politics and diplomacy obviously had their way. For the most part, we neither know nor have the ability to systematically map out the logics that rule in politics, other than in a general and schematic sense, i.e., pointing out that power, persuasion, self-interest, and precedent have central roles. In some cases, science beat politics, and in some cases, it was the reverse.

The procedures and regulatory mechanisms, and the norm systems and behavioral patterns that have been given conceptual firmness in institutional theory, including institutional logics, are, of course, only part of the explanation for why things unfolded the way they did in the story of the ESS Scandinavia campaign. Many of the turns of events in the successful campaign also seem to have had an element of chance. Historical institutionalism is useful to analyze such processes that involve seemingly haphazard events and place them in the proper context of long-term, gradual and cumulative change. Lars Börjesson has reflected that the effort to locate the ESS in Lund was the work of "getting a snowball to grow to an avalanche" (Hallonsten, 2020a, p. 288), a notion that fits well with historical–institutionalist analysis. Seemingly insignificant events can indeed alter the course of history.

Another important feature of the ESS story that historical–institutionalist analysis can help identify is the importance of resilience and adaptability. To a great degree, it was the resilience of the ESS Scandinavia campaign that eventually led to victory. No other site contender lasted all the way. Although its key champions and enthusiasts changed, the idea of an ESS in Lund remained. Historical institutionalism teaches us that long-term survival and success depend on long-term stability and adaptability, and the ESS Scandinavia campaign certainly displayed both over its many years of existence and beyond, treating the current ESS organization as its successor and heir. From a local and regional campaign to national and government-sponsored, to international negotiation and the design update phase, and finally to construction, ESS Scandinavia and its successors demonstrated remarkable stability and adaptability. If viewed as a continuum, in spite of the changes in organizational structure, legal status, funding portfolio, top management, and workforce, the campaign was remarkably adaptive. Through several shifts in focus — from the competition with other site candidates to a fight to keep the ESS project alive at all, from science lobby to high-level politics, from campaigning to real project management, from politics to technology and back, and so on, and in a back-and-forth movement between regional, national, and international (European) levels, and between science, politics, and diplomacy — it did not change its overall aim, but eventually succeeded in achieving it. This is also all the more impressive given the lack of precedent and the lack of established procedures for collaborative Big Science in Europe and for European Big Science in Sweden.

References

Cramer, K. (2017). Lightening Europe: Establishing the European Synchrotron Radiation Facility (ESRF). *History and Technology* 33(4), 396–427.

Cramer, K. (2020a). *A Political History of Big Science: The Other Europe* (London: Palgrave Macmillan).

Cramer, K. (2020b). The role of European Big Science in the (geo)political challenges of the twentieth and twenty-first centuries. In K. Cramer and O. Hallonsten (eds.), *Big Science and Research Infrastructures in Europe* (London: Edward Elgar).

Esser, H. (1993). The rationality of everyday behavior. *Rationality and Society* 5(1): 7–31.

Fioretos, O., Falleti, T. G. and Sheingate, A. (eds.) (2016). *The Oxford Handbook of Historical Institutionalism* (New York, NY: Oxford University Press).

Friedland, R. and Alford, R. R. (1991). Bringing society back in symbols, practices, and institutional contradictions. In W. Powell and P. DiMaggio (eds.), *The New Institutionalism in Organizational Analysis* (Chicago: University of Chicago Press).

Giddens, A. (1984). *The Constitution of Society. Outline of the Theory of Structuration* (Cambridge, UK: Polity).

Goldthorpe, J. H. (1998). Rational action theory for sociology. *British Journal of Sociology* 49, 167–192.

Hallonsten, O. (2014). The politics of European Collaboration in Big Science. In M. Mayer, M. Carpes and R. Knoblich (eds.), *The Global Politics of Science and Technology — Vol. 2* (New York, NY: Springer).

Hallonsten, O. (2016). *Big Science Transformed: Science, Politics and Organization in Europe and the United States* (London: Palgrave Macmillan).

Hallonsten, O. (2020a). *The Campaign. How a European Big Science facility ended up on the peripheral farmlands of Southern Sweden* (Stockholm: Arkiv Academic Press).

Hallonsten, O. (2020b). Is there an 'Iron Law' of Big Science? In K. Cramer and O. Hallonsten (eds.), *Big Science and Research Infrastructures in Europe* (London: Edward Elgar).

Hallonsten, O. and Heinze, T. (2012). Institutional persistence through gradual adaptation: analysis of national laboratories in the USA and Germany. *Science and Public Policy* 39, 451–463.

Hallonsten, O. and Heinze, T. (2015). Formation and expansion of a new organizational field in experimental science. *Science and Public Policy* 42(6), 840–854.

Krige, J. (2003). The politics of european scientific collaboration. In J. Krige and D. Pestre (eds.), *Companion to Science in the Twentieth Century* (Oxfordshire, UK: Routledge).

Krige, J. (2006). *American Hegemony and the Postwar Reconstruction of Science in Europe* (Cambridge, MA: MIT Press).

Lindenberg, S. (1985). An assessment of the new political economy: Its potential for the social sciences and for sociology in particular. *Sociological Theory* 3(1), 99–114.

Mahoney, J. and Thelen, K. (eds.) (2010). *Explaining Institutional Change: Ambiguity, Agency, and Power* (Cambridge, MA: Cambridge University Press).

March, J. and Simon, H. (1958) *Organizations* (Hoboken, NJ: Blackwell).

Moskovko, M., Astvaldsson, A. and Hallonsten, O. (2019). Who is ERIC? The politics and jurisprudence of a governance tool for collaborative European research infrastructures. *Journal of Contemporary European Research* 15(3), 249–268.

Opp, K.-D. (1999). Contending conceptions of the theory of rational action. *Journal of Theoretical Politics* 11(2), 171–202.

Pache, A.-C. and Santos, F. (2013). Inside the hybrid organization: Selective coupling as a response to competing institutional logics. *The Academy of Management Journal* 56, 972–1001.

Reay, T. and Hinings, C. R. (2009). Managing the rivalry of competing institutional logics. *Organization Studies* 30, 629–652.

Scott, W. R. (2001). *Institutions and Organizations* (2nd edn.) (Thousand Oaks, CA: Sage).

Simon, H. A. (1957). *Models of Man* (Wiley).

Streeck, W. and Thelen, K. (eds.) (2005). *Beyond Continuity. Institutional Change in Advanced Political Economies* (New York, NY: Oxford University Press).

Thornton, P. (2004). *Markets from Culture. Institutional Logic and Organizational Decisions in Higher Education Publishing* (Redwood City, CA: Stanford University Press).

Westwick, P. J. (2003). *The National Labs: Science in an American System 1947–1974* (Cambridge, MA: Harvard University Press).

Chapter 2.3

Brief History of Synchrotron Radiation Sources in Brazil

Liu Lin

This is a brief history of the development of synchrotron radiation sources in Brazil from a personal perspective. I tried to recollect facts showing how, in a time span of about 40 years, X-ray and VUV physics in Brazil took a big leap from small individual laboratory light generators to the successful completion of one of the most brilliant synchrotron light sources in the world. Over the years, many people contributed to this success story. But one person was very special for his leadership, deep multidisciplinary knowledge, infinite creativity and enthusiasm, and immeasurable kindness. This article is dedicated to the memory of Ricardo Rodrigues, who passed away in January 2020, at the age of 68.

The first discussions to develop synchrotron light science and technology in Brazil date back to the early 1980s, when the president of CNPq, the main federal science support organization in Brazil, Lynaldo Cavalcanti de Albuquerque, asked the Director of the Brazilian Center for Research in Physics (CBPF), Roberto Lobo, for proposals for a new laboratory equipped with a big scientific instrument to leverage experimental science in Brazil. It should be a national laboratory, accessible to researchers from all over the country, that should deeply change the science funding procedure in Brazil: instead of individual proposals for small research equipment, the scientific community would be presented with a large facility that would induce the demand for scientific projects. Lobo sought the advice of Roberto Salmeron (1922–2020), a Brazilian researcher from the Ecole Polytechnique in France, who proposed the construction of a synchrotron radiation facility, which Lobo and de Albuquerque embraced enthusiastically.[1] The discussions that followed resulted in the decision to fund a preliminary design effort, and a small team of four young physicists was sent for a short period to the Stanford Synchrotron Radiation Lightsource (SSRL), early in 1985, to learn about synchrotron source design with Helmut Wiedemann (1938–2020), an enthusiastic supporter of the Brazilian project from its very beginning. Helmut Wiedemann came several times to Brazil in the early 1980s to give lectures about accelerator physics. The Brazilian team was led by Ricardo Rodrigues, who later became the Project Leader for the two light source facilities built in Brazil. I was part of this pioneer team that elaborated a draft conceptual design on a challenging project at the time, a 2–3 GeV storage ring with full energy injection. The resulting design report (later known as Project 1) was the basis for cost estimates and initial planning of the project, as well as for further discussions within the Brazilian government and scientific community.

The Brazilian Synchrotron Light Laboratory (LNLS) was finally created by the newly established Ministry of Science and Technology by mid-1986, with Cylon Gonçalves da Silva, a Physics Professor at the University of Campinas, as its Founding Director. The choice of a synchrotron light source as the first modern national laboratory in Brazil arose from a perception of the need to develop accelerator technologies, enhance research in basic and applied experimental sciences, and establish a new model of scientific organization in the country. The rationale behind the idea of this project was therefore based on a three-fold strategy[2]:

---

[1] Velho and Pessoa Jr (1998).

[2] Gonçalves da Silva *et al.* (1988).

(a) to develop a large challenging engineering project that could attract a critical mass of young scientists and engineers to be trained in key technologies, thus improving the expertise in accelerator physics and engineering in Brazil,

(b) to construct an experimental facility that could provide state-of-the-art resources to a wide spectrum of researchers, including physicists, chemists, biologists, crystallographers, materials scientists, as well as medical and industrial users, and

(c) to introduce the concept of a national laboratory, providing open access facilities to be used by researchers mainly from other institutions, with only a reduced number of in-house scientists who are expected to build up a number of up-to-date beamlines, assist external users and produce high-quality scientific results.

In 1987, with the acquisition of an industrial shed by CNPq, in Campinas, words became action, and the activities for the light source implementation really started. Ricardo Rodrigues was in charge of accelerator engineering development and started recruiting staff, mostly motivated young people looking for their first job and willing to start hands-on activities (Fig. 1). At the time, the Brazilian community of synchrotron light users was very small, and Aldo Craievich passed away in April 2023 (1939–2023), who was actively involved in the early proposal discussions and later was in charge of the LNLS scientific division, took on the task of enlarging the user community. He did it with great enthusiasm by organizing courses, lectures, workshops, training abroad, etc., in a continuous effort that persists at LNLS to this day. The first LNLS Annual User's Workshop took place in December 1990.

The first technical challenge was to have a linear accelerator (Linac) working. An extension of the LNLS shed was built on a rented land next door for the in-house construction of a 50 MeV Linac, about half the design energy, but containing the most technically challenging parts, such as the electron gun, magnets, RF, vacuum, diagnostics. The needed infrastructure was slowly built, including a mechanical workshop, an electronic workshop,

Fig. 1. Some of the LNLS young staff during the second International Machine Advisory Committee meeting in July 1991, in front of the shed that served as the LNLS first headquarters. *Photo*: Roberto Medeiros.

a computer center, a magnetic measurement lab, a high-voltage power supply lab, and so on. The first beam from the Linac was seen on the 19th of December 1989. Observing the first relativistic electron beam produced at LNLS was an important milestone that assured the technical team's self-confidence. The event was marked by an official inauguration on the 19th of February 1990, with the presence of many authorities, including the Nobel prize winner Carlo Rubbia, Director-General of CERN, who had provided important logistics support to our challenging project.

In parallel, beamlines were also being designed and built. In 1992, the first TGM beamline was sent to the Center for Advanced Microstructures and Devices (CAMD), in the University of Louisiana (USA), to be tested. The activities at LNLS were only definitively transferred to the present campus in October 1995, in a 380,000 m^2 land area, donated by the state of São Paulo in 1990, where the construction of the building to house the first LNLS synchrotron light source started in 1993. The "portable" shed extension that before housed the 50 MeV Linac was installed in the new campus and served first as a restaurant, and later, until today, as the campus warehouse.

The difficult economic circumstances in Brazil in the late 1980s and early 1990s led the LNLS board of directors to decide on a more modest facility design, with a storage ring energy lower than that of Project 1, but keeping it scientifically interesting and preserving the original schedule. The new design converged to an electron storage ring with the energy of 1.15 GeV and a circumference of 93 m. The storage ring was named VUV-III and had 12 dipoles and six straight sections, four of which were designed for devices that were starting to be used in storage rings to generate harder X-rays and higher photon flux, the wigglers and undulators. The electron beam energy was later increased to 1.37 GeV by pushing the dipole field to 1.67 T, an important upgrade to better suit the machine to the mostly hard X-ray user community in Brazil. The 1.37 GeV machine was named UVX. To reduce costs, a risky decision was made: to inject electrons into the storage ring at low energy, directly from a 120 MeV Linac, and leaving the possibility of adding a synchrotron booster for a later upgrade. Despite financial difficulties, the in-house activities to develop prototypes for the storage ring components continued. For the magnets, an unusual solution was adopted: a laser cutter machine was used to cut different types of laminations one by one. This was a slow and labour-intensive solution, but cheap, flexible, and capable of producing high-quality magnets.

Step by step, all needed components were produced, and the accelerators were ready to start commissioning in May 1996. Beam accumulation at 120 MeV as well as efficient energy ramping in the storage ring were a big but finally accomplished challenge, and thus the 1.37 GeV UVX light source was opened for users on the 1st of July 1997, with seven beamlines in operation. Figure 2 shows people in the UVX control room during commissioning in 1996. A 500 MeV synchrotron booster was later designed and built, and started operation in 2001, increasing the initial electron beam current from 150 to 250 mA.

Fig. 2. UVX control room during commissioning in May 1996. From left to right, Ricardo Rodrigues, Ruy Farias and Cylon Gonçalves da Silva (front), and Pedro Tavares, Liu Lin, Lucia Jahnel, Aldo Craievich, Fernando Rafael and Carlos Scorzato (standing). *Photo*: Nelson Chinalia.

On the 22nd of November 1997, just a few months after the start of user operations, LNLS embarked on a major institutional change as a presidential decree qualified the Brazilian Association for Synchrotron Light Technology (ABTLuS) as a Social Organization, a non-profit legal entity governed by private law, which could run LNLS under a contract with the Ministry of Science and Technology. This was the result of several years of effort by the Director Cylon Gonçalves da Silva, who realized early on the peculiar needs of an open national laboratory, in particular the establishment of clear finalistic goals that could be followed up by the funding bodies as well as by society in general in a transparent way. The social organisation concept aimed at allowing for the needed flexibility and agility in decision-making while remaining open to scrutiny to guarantee adequate control of the use of public funds.

From 1997 to 2008, LNLS transitioned to a phase of consolidation as a National Laboratory, becoming truly national (and international), integrating several laboratory facilities to the Brazilian scientific community. During this period, José Brum, who became LNLS Director in July 2001, promoted not only a substantial growth in the number of users but also several upgrades in the machine, such as the installation of several insertion devices (a 2 T hybrid wiggler, a superconducting wiggler, and an APPLE-II undulator). In 2003, the machine operated for the first time in single-bunch mode.

UVX was the first synchrotron light source built in the Southern Hemisphere and, after 10 years of construction starting from scratch and 22 years of regular operation for users, UVX was shut down on the 2nd of August 2019. By then, the experimental hall had 17 beamlines serving about 1,500 users per year with more than 97% beam time availability. Despite continuous upgrades over the years, the machine ceased to be competitive anymore. A new light source, capable of delivering much brighter beams, was needed.

The mostly in-house design and construction of UVX, as well as its operation and continued upgrades over the years, served as a grand school for an initially young and inexperienced team of physicists and engineers. The continued hands-on training of human resources had no doubt paved the way for the development of a much more daring endeavour in the 2010s, the construction of Sirius, a state-of-the-art 3 GeV low emittance machine, one of the first fourth-generation storage rings to start operation in the world after MAX-IV in Sweden and with ESRF-EBS in France.

Sirius is a green-field facility based on a 3 GeV, 518 m circumference storage ring with 20 straight sections and 20 five-bend achromat (5BA) cells, making up a light source with bare lattice emittance of 250 pm.rad, a fourth-generation storage ring light source that can provide extremely brilliant VUV and X-ray beams from undulators.

The initial discussions about the need for a new light source in Brazil date back to 2006. In 2008, the LNLS Director, José Brum, submitted a proposal for a new source to replace UVX and keep the LNLS international competitiveness to the Ministry of Science and Technology. As a result, a conceptual design study was funded. In

2009, two prospective workshops to discuss the scientific cases and machine specifications were organized at LNLS. The new LNLS Director, José Roque da Silva, gave the highest priority to the development of the new synchrotron source. Following these workshops, the LNLS team focused on designing a new light source based on a third-generation storage ring. José Roque invited Ricardo Rodrigues back to LNLS to again lead the new project. Ricardo Rodrigues, who had left LNLS a few years before to start a private company, accepted the new challenge in November 2009. His leadership was fundamental to the success of the project. The initial funding for the conceptual design phase was spent to improve the in-house engineering and technical infrastructure in the campus in preparation for the R&D needed for the new machine components.

In 2009, a change in the management structure of LNLS and the associated research labs took place. The whole campus was named Brazilian Center for Research in Energy and Materials (CNPEM) that included LNLS and two additional national laboratories. CNPEM now had a General Director and an individual Director for each national laboratory. Today, in 2022, CNPEM comprises four national laboratories, namely, LNLS for synchrotron radiation, LNBio for biosciences, LNBr for bio-renewables, and LNNano for nanotechnology. Figure 3 shows an aerial view of the CNPEM campus taken in February 2020.

In 2012, the design of a 1.7 nm.rad emittance electron storage ring, a third-generation machine, was mature enough to be presented to an International Machine Advisory Committee (MAC). The MAC's most important recommendation, to push the machine for future brightness standards with a much lower emittance storage ring, in view of the Swedish MAX-IV plans, gave the final incentive for the team to start over a new project design despite having already invested three years in the previous one. In a month's time, after intense design and evaluation work, the official decision to change the project completely to a much more challenging — and therefore exciting — design based on a 5BA lattice was taken on July 2012. This year and the next one represented a turning point for the project. Besides the completely new design, the campus incorporated 150,000 m^2 of adjacent land in 2013, again donated by the state of São Paulo, where Sirius

Fig. 3. Aerial view of CNPEM campus in February 2020. Sirius is on the top left. UVX was housed in the big square building close to the bottom right. *Photo*: Renan Picoreti/CNPEM release.

could be built. The new technical challenges initiated an intense phase of lattice optimization and R&D work for the accelerator components to achieve the required parameters and tolerances. It's interesting to note that only a few years later the term "fourth-generation storage rings" began to be used for these kinds of machines.

From the start, special attention was given to the project of the building and floor stability requirements. Different prototype floor foundations were built, and vibration parameters were measured and modelled, including geological survey data of the land. A 68,000 m^2 building was designed and built from 2014 to 2018, with a special floor for the accelerators and the experimental hall, with an approximately 500 m circumference monolithic concrete tunnel for the accelerators and with building extensions for up to 150 m long beamlines.

As mentioned before, the new design of the light source presented many challenges, requiring every available new technology in the field and opening the possibility for creativity and innovation. In Sirius, the 5BA magnet lattice uses a high field short insert in the center dipole to generate hard X-ray sources while reducing the emittance and only slightly increasing the needed RF power. Special straight sections have their optics set to quite a strained condition to optimize the matching between electron beam and undulator photon beam phase spaces, thus increasing the effective brightness. This condition also allows for the installation of small horizontal as well as vertical gap restrictions in these regions, thus opening the possibility for new types of insertion devices. One problem that arises with the fourth-generation light sources is the high focusing forces required to achieve very low emittances and consequently the need for small magnet apertures. This condition in turn requires small vacuum chambers, implying a challenge to the pumping mechanism. To be able to implement the idea of distributed pumping using NEG coating, a technique not easily available at the time, especially from Brazil, a license was obtained from CERN that allowed LNLS to develop the technique internally. Almost 100% of Sirius vacuum chambers have been NEG coated in-house, including curved dipole chambers with small light extraction channels. To produce the magnets needed with the required quality, LNLS established an R&D partnership with a major international industrial engineering company based in Brazil (WEG S.A.). After several years and innumerable prototypes, high-quality magnets are now guiding and focusing the electron beam in the accelerators. In fact, stimulating the development of local industries is one of the goals of the Sirius project. These are only a few examples of the solutions adopted throughout the project.

In April 2018, the first beam from the 150 MeV Linac was produced. The Linac was acquired from Shanghai Institute of Applied Physics (SINAP), in China, and its installation and commissioning occurred while the construction of the storage ring tunnel was still being finished. In the case of Sirius, the decision was to buy the Linac so that the team could concentrate on the much more challenging task of designing and constructing the storage ring. On the 14th of December 2019, the first stored beam was achieved with on-axis injection from a 3 GeV full energy booster. Two days later, the first X-ray microtomography experimental result was obtained. Figure 4 shows the Sirius commissioning team in the control room to celebrate the first beam. The first beam accumulation with off-axis injection, a process that allows increasing the stored current, was accomplished on the 20th of February 2020.

Fig. 4. First stored beam in Sirius on the 14th December, 2019. *Photo*: Reberson Ius/CNPEM release.

Today, in February 2022, Sirius is operating with six beamlines in science commissioning mode, the other five are under different assembling stages, and three are in the component fabrication phase. The present LNLS Director, Harry Westfahl Jr., now faces the challenge of consolidating the use of this unique facility by the scientific community in Brazil and worldwide, which is expected to address a wide spectrum of new applications needing highly brilliant synchrotron radiation from IR to X-rays generated by Sirius.

Sirius components design started about 25 years after UVX. During this period, many changes took place in the country. Local Brazilian industry can now produce many of the components needed for Sirius, albeit in most cases in a collaborative way, such as magnets, power supplies, girders, ceramics for pulsed supplies and many others. About 40 local companies have been working on technological developments, especially for the Sirius project, and about 85% of the Sirius budget was spent in Brazil. The in-house technical team has not only grown in number but also gained expertise in most areas of accelerator technology, pushing the frontier in many cases with innovative solutions. The Brazilian scientific community is now unanimous in supporting the project, which was not the case for UVX when funding for Big Science projects did not exist in the country.

The successful commissioning and operation of Sirius is the result of the dreams and dedication of many people. For many of us, it has been a lifetime dedication. An entire cycle of scientific and technological development is now complete. A new cycle is underway, with a new generation of experienced people with the mission to continue innovation into the future.

Acknowledgements

I would like to thank many colleagues for pleasant moments remembering good times from the past and for contributions and text revisions, especially Pedro Fernandes Tavares (Accelerator Director of MAX IV Laboratory, Sweden, and former Technical Director of LNLS), Aldo Craievich, Guilherme Franco, Roberto Medeiros, Regis Neuenschwander, Ximenes Resende, Carlos Scorzato, Helio Tolentino, Harry Westfahl Jr., and Cylon Gonçalves da Silva who encouraged us to write this article.

References

Gonçalves da Silva, C. E. T., Rodrigues, A. R. D. and Craievich, A. F. (1988). The National Laboratory for Synchrotron Radiation of Brazil. *Synchrotron Radiation News* 1, 52.

Velho, L. and Pessoa Jr, O. (1998). The decision-making process in the construction of the Synchrotron Light National Laboratory in Brazil. *Social Studies of Science* 28(2), 195–219.

Part III
How Laboratories Evolve: Industrial, National, and Individual

The three chapters in Part III, "How Laboratories Evolve: Industrial, National, and Individual" illustrate the nature and evolution of institutions in different domains of materials research, focusing on materials science laboratories in the United States but pointing beyond them.

"Materials science, more so than some other fields, owed its character to the institutions that supported it," writes Joseph D. Martin in "The Age of Industrial Laboratories." As American industrial laboratories were "some of the key sites in which the materials science developed" in the mid-20th century, understanding the history and character of industrial laboratories up to and during that time "is crucial for situating materials science within the institutional networks that defined its purpose." Companies that invested significant resources in materials science from the late 19th century into the 20th century included General Electric (GE), the Westinghouse Electric Corporation, and American Telephone & Telegraph (AT&T), whose Bell Labs "would remain the epicentre of American solid state physics research for decades." Their programs, enhanced by competition among them, established "a powerful template for American industrial research." Key features of this template were that the research labs were kept separate from production facilities, that credentialed scientists and engineers staffed the labs, and that the labs were largely insulated from the demand for short-term profit. This template was also used by the labs of Eastman Kodak, DuPont, IBM, and the Radio Corporation of America (RCA). But Martin also points out the limitations of focusing on these companies and that template. Many smaller corporations also engaged in materials research and developed crucial instruments for that purpose, while collaborations between industry, academia, and government were also crucial. Nevertheless, Martin's focus highlights three principal patterns of materials science research in industrial laboratories: (1) these labs were important incubators of new materials and training grounds for a new generation of researchers; (2) the research in these labs collapsed the difference between basic and applied research; (3) the institutions involved were highly diverse. These labs, Martin writes, "were as diverse as the materials they produced."

The next essay, Catherine Westfall's "The Age of the National Laboratories: The Advent of the New Big Science," shows how US national laboratories largely took over materials science research in the latter part of the 20th century and into the 21st century, playing a major role in the transformation of Big Science to what Westfall (and I) have dubbed the New Big Science. The US national laboratory system was born right after World War II to support instruments — most notably, reactors and accelerators — whose construction, maintenance, funding, and operation were of too large a scale for individual universities or industries to afford. The subsequent focus on high-energy physics at accelerators big enough to be part of the landscape soon became known as "Big Science." While some small amount of materials science research existed alongside high-energy physics at national laboratories, that research began to grow in the 1960s and 1970s, especially with the development of synchrotron sources. In 1984, the growing interest in materials science facilities was ratified by two Department of Energy panels, the "Planning Study for Advanced National Synchrotron Radiation Facilities" ("The Eisenberger-Knotek Report") and "Major Facilities for Materials Research and Related Disciplines" ("The Seitz-Eastman Report"). The next year, the DOE's director of the Office of Energy Research, Alvin Trivelpiece, forged a plan in which four national laboratories would each get a major new facility. Three of the four projects focused on materials science. After the US Congress terminated the Superconducting Supercollider in 1993, these three materials science projects became the centerpieces of a transformation in which many national laboratories shifted to materials science at their marquee facilities.

Katharina C. Cramer's chapter, "Becoming Materials Science? Turning Points in the History of Materials Science at the ESRF, 1994–2020," illustrates the evolution of materials science inside an institution created for it, and how that evolution changed the institution in turn. Materials science played a major role in research at the European Synchrotron Radiation Source since its founding in 1988. But Cramer traces how the evolution of materials science transformed the "discursive understandings and definitions of materials science," and she pays particular attention to the transformation of materials science as an organizational category of the ESRF's research groups, and to how the transformation affected the career focus of scientists at the facility.

Chapter 3.1

The Age of Industrial Laboratories

Joseph D. Martin

The American industrial laboratory had its heyday through the middle decades of the 20th century. In this era, corporate interests were committed to the value of research and development for the health of industry, and so were willing to invest, sometimes lavishly, in sites where the promise of scientific inquiry could be converted into profitable, patentable discoveries. These laboratories were some of the key sites in which materials science developed. This chapter reviews the history of some of the most important industrial laboratories that nurtured the growth of materials science research in the United States.

Before turning to these individual cases, however, it will be useful to review the origin of the industrial laboratory as a distinct and recognizable site of research, and interrogate the historical circumstances in which it emerged. Materials science, more so than some other fields, owed its character to the institutions that supported it. Industry was one of those institutional settings, and so understanding the industrial laboratory as a site of scientific research in the 20th century is crucial for situating materials science within the institutional networks that defined its purpose.

The Birth and Character of the American Industrial Laboratory

The origins of the American industrial laboratory were electric. The late 19th century saw an explosion of electrical technology and infrastructure, which by the end of the century was consolidated in the hands of three corporations: General Electric (GE), the Westinghouse Electric Corporation, and American Telephone & Telegraph (AT&T). Through the early 20th century, each would invest in extensive research and development (R&D) operations, driven in part by competition among them, which would establish a powerful template for American industrial research.

The idea that industrial concerns should invest systematically in research in order to improve their market position was not novel. It was in some measure an import of a longstanding German tradition of industrial research, particularly in chemistry.[1] Many of the labs that emerged to support the American electrical industry nevertheless exhibited distinctive and novel features. As we will see, these features were not universal, but their prevalence did contribute to the success of industrial laboratories through the middle decades of the 20th century, which is often identified as the golden era of American industrial research.[2]

First, industrial labs were often kept strictly separate from production operations. The function of these laboratories was not to refine existing industrial processes, but to generate new insights, inventions, and materials, and to uncover new principles, which could potentially expand those operations. Second, credentialed scientists and engineers were hired to staff the laboratories. The growth of American industrial research coincided with the growth of the scientific and technical workforce in the early 20th century, which created a glut of trained scientists and engineers, attracted by the salaries, practical rewards, and considerable intellectual freedom industrial employment offered. Third, the labs were typically

[1]Jackson (2011).
[2]Smith Jr (1990).

insulated from the pressures of short-term profit. They presumed a stable corporate setting with a long-term vision, in which the activities of the labs were expected to inform business operations not just five or ten years out, but decades down the road.[3]

The context these features created favored the development of new materials. The technological infrastructures of 1900 were largely wrought of metal, with a smattering of natural fibers. By the end of the 20th century, the material world was transformed by new semiconductors, plastics and polymers, and glasses and ceramics, as well as novel metals and alloys. Many of those new materials — as well as new uses for existing materials — sprung from well-staffed, research-oriented laboratories with the freedom to think beyond the constraints of short-term profit. We meet some of those laboratories, and the materials developed within them, in the following sections.

General Electric

The General Electric Research Laboratory was established in 1900. General Electric had formed eight years earlier, a merger of the Edison General Electric Company (itself an amalgamation of Thomas Edison's earlier endeavors) and the Thomson-Houston Electric Company.[4] The new research laboratory consolidated the lively research tradition of both its antecedent corporations, under the guidance of Willis R. Whitney, a chemist who GE managed to lure away from the Massachusetts Institute of Technology (MIT).

Other academics would follow. The transition was not always smooth; despite the relative latitude GE offered compared to a job in a production laboratory, the differences between academic and industrial culture were often jarring for those used to a self-directed university life.[5] GE in the early 20th century was interested in improving incandescent lighting, and new members of the laboratory were hired to direct their energies in that direction.

---

[3] Reich (1985).

[4] Carlson (1991).

[5] Wise (1983).

The efforts paid off in the field of incandescent lighting. William D. Coolidge, who, like Whitney, had begun his research career at MIT, joined GE in 1905. He developed a procedure for producing tungsten filaments, which, when placed in gas-filled bulbs, solved longstanding problems with incandescent lights, such as gradual blackening of the bulbs, as well as resulting in a product that was both efficient and cheap to produce.[6]

The wealth of trained researchers GE hired, however, meant that its laboratory's focus would not long remain squarely on the problem of incandescent lighting. Another new PhD, Irving Langmuir, would be crucial for expanding the laboratory's scope, particularly into the area of vacuum-tube electronics. Vacuum tubes, and most notably the triode tube patented by Lee de Forest in the United States in 1906, were integral to the emerging field of radio, given their ability to amplify weak electrical signals. By helping GE push forward in vacuum tube research, Langmuir helped GE remain competitive with AT&T, which had secured the patent on de Forest's invention.[7]

By the mid-20th century, GE had secured a leading position as a provider of consumer electronics, a position that owed much to its pioneering industrial laboratory. Not all of the laboratory's efforts bore fruit. Its program aimed to control plant breeding via radiation-induced mutation, for instance, came to naught.[8] But such endeavors are nevertheless indicative of the ethos that animated it: the corporation could be expected to endure, and its long-term health was expected to depend on the timely exploitation of the very latest discoveries in science.

Westinghouse

The Westinghouse Electric Corporation was GE's fiercest competitor in its early years. The two companies sparred to gain control over

[6] Reich, *Making of American Industrial Research* (Ref. 3, Chapter 4).
[7] Noble (1977).
[8] Curry (2013).

the burgeoning electrical infrastructure, with Westinghouse winning its share of bouts, including the lucrative contract for a hydroelectric dam at Niagara Falls, which it completed in 1895. Westinghouse was to General Electric as George Westinghouse was to Thomas Edison. Whereas Edison was a showman, continually engaged in the assembly of his own myth, Westinghouse was a hardheaded engineer and entrepreneur, less interested in flash and novelty and more invested in the prickly details of technical problems.[9]

This in part explains the subtle differences in research culture at the two corporations. Westinghouse came later to the research laboratory game, opening its lab outside the steel city of Pittsburgh, Pennsylvania, in 1916. Established just in time for the entry of the United States into World War I, the Westinghouse Research Laboratory cut its teeth on wartime problems, such as studies relevant to the erosion of gun barrels and other munitions, and analysis of gas-mask materials. Many of its early contributions came in the field of metallurgy, metals being the stuff of the infrastructure Westinghouse sought to build.[10]

After the war, it followed GE in pursuing the radio craze. However, whereas in GE's case, the move to radio represented an expansion of its research mission, in the case of Westinghouse, it represented a narrowing. The laboratory had been founded with a broad vision, rooted in science, in parallel with the mission of GE's research operations. But in the early 1920s, the emphasis shifted to engineering, and to exploiting technologies that were in the process of maturing, such as radio.[11] These efforts led to notable progress in magnetron research, which would become crucial for radar development during World War II.[12]

The research laboratory at Westinghouse remained insulated from the production arm of the business, but it continued to

---

[9] Usselman (1992).

[10] Coltman (2004).

[11] Lassman (2018).

[12] Stephan (2001).

experience tension between a vision driven by a commitment to basic research and one guided by more practical engineering concerns. The pendulum would swing back toward basic research in the 1930s under the directorship of the engineer Lewis Chubb, as nuclear research caught the eye of the American scientific community. The fascination with the nuclear led Westinghouse to recruit the physicist Edward Uhler Condon, to Pittsburgh, poaching him from an associate professorship at Princeton University.[13]

Condon's tenure at Westinghouse is emblematic of broader trends in its research laboratory. As a physicist with an academic pedigree but a fascination with practical problems, he arrived in East Pittsburgh with a vision for expanding fundamental research at the laboratory in a way that erased what he saw as an artificial distinction between the pure and the practical. He succeeded to an extent, luring the solid-state physicist Frederick Seitz to Westinghouse and convincing the company to support the construction of a new 3 MeV Van de Graaff generator for nuclear research, which it used to exploit an emerging market for radioisotopes — presaging the company's postwar move into nuclear power. But his dreams of attracting a large, highly talented scientific workforce to the laboratory never met his ambitions.[14]

The peripatetic Condon would leave Westinghouse for a government position during World War II, in part a reflection of his frustration with not having been able to bring his vision of a fundamental research laboratory linked to practical concerns to full flowering. Westinghouse thus offers both a comparison and a contrast with GE. Its successes in the materials arena, such as commercializing radioisotopes and alloy development for generators and reactors, were indicative of the advantages offered by a long-term mindset and a trained workforce. But it cleaved more closely to practical concerns, and efforts to push into more speculative areas were often met with skepticism by corporate management.

---

[13] Lassman, *Edward Condon* (Ref. 11, Chapter 3).

[14] Ibid., Chapter 4.

AT&T

If the American industrial research laboratory can be said to have reached an apotheosis, then it did so in AT&T Bell Laboratories, or Bell Labs. Whereas Westinghouse followed GE's lead but stuck more closely to practical concerns, Bell went the other way, creating a laboratory renowned for the academic-style freedom of inquiry it afforded its researchers. Bell established its laboratory in 1925, and that ethos would bear fruit in 1937, when Clinton Davisson was awarded the Nobel Prize in Physics for work conducted at Bell Labs showing the wave nature of electrons.

Unlike GE and Westinghouse, which competed directly with each other for control of American electrical infrastructure, AT&T had secured a monopoly over the telephone network. The relevance of Bell Labs to its business mission was therefore not so much to fend off competition as it was to secure its position — that is, to produce developments that would ensure the reliability of the telephone network, and to anticipate major scientific or technological changes that might destabilize AT&T's monopolistic position. Bell therefore had a distinct incentive to invest in basic research. Its position in the marketplace was secure, its investment in basic research served in some ways to keep government regulators at bay, and the potential threats and opportunities scientific research represented were mostly long-term.

These features help account for why it was Bell Labs that hosted the invention of the transistor in 1947. Bell had been an early adopter of quantum theory, developing an expertise in the field at a time when most American universities still lacked a quantum theorist.[15] By the 1940s, it boasted one of the most impressive collections of solid-state physicists in the world. Stefan Machlup, who held a postdoc at Bell after World War II, wrote to William Shockley, the co-director of the Bell solid-state division and co-inventor of the transistor, "I think everybody on my hall was doing exactly what he wanted to do," and marveled at the fact that "if you've got an

[15] Hoddeson (1980).

obscure technical problem, chances are that somewhere in the two-mile network of corridors sits *the* expert on this particular specialty."[16]

The transistor became possible because of some very abstract developments in the understanding of the electrical properties of solids, in particular semiconductors. By preparing "doped" samples of silicon, germanium, or another semiconducting material, in which a pure crystal has some impurities introduced to it in the form of another element, it is possible to create materials with extra electrons or "holes" (places where electrons would otherwise be, which behave like particles inside solids) that can be more easily moved into the substance's conduction band. Exploiting these properties makes it possible to create a device that mimics the behavior of a triode tube: in particular, amplifying and rectifying electrical currents, and acting as a switch.[17]

Such a device had obvious advantages for a company responsible for a national network built with vacuum tubes, which were large, slow to warm up, and subject to breakage. But nor was it an exercise in simple gadgeteering — it rested on cutting-edge theoretical understanding, as provided by the likes of John Bardeen, who would go on to become the only individual to win two Nobel Prizes in physics, following his role in developing the BCS theory of superconductivity.[18]

Bell Labs would remain the epicenter of American solid-state physics research, both basic and applied, for decades, before AT&T's monopoly was broken by the US federal government in 1984. It made many of the key strides in semiconductor physics, as well as in studies of magnetic and amorphous materials, with an intellectual output (at least as measured by papers in top journals) that rivaled or surpassed the top university physics departments.[19] It was in some sense sui generis, the protection AT&T's monopoly afforded

[16] Stefan Machlup, letter to William Shockley, May 30, 1955, William Shockley Papers, University of Illinois Archives, box 2, folder Correspondence 1954 and 1955.
[17] Hoddeson (1977).
[18] Hoddeson and Daitch (2002).
[19] Martin (2018).

meaning it could take risks that would deter other corporations, but in another sense, it embodied the commitment to long-term risk-taking that drove materials development in industrial contexts through the middle of the 20th century.

Eastman Kodak

The laboratories we have considered thus far were all established to support the operations of industries managing large-scale infrastructures. The Kodak Research Laboratories, in contrast, founded in 1912, focused on smaller scales. As a photography company, its interests were mostly chemical, and patents were its bread and butter. In the early 20th century, this meant patents on the products and processes involved in both commercial and consumer technologies, but later in the 20th century, the laboratories at Eastman Kodak found other opportunities for the chemical expertise they contained. Their influence was particularly strong in the field of semiconductor manufacture.

The laboratory grew rapidly under the guidance of its first director, the British photochemist C. E. Kenneth Mees. From a staff of just 20 shortly after its opening, the laboratory's workforce grew to 1,175 by 1955, occupying a laboratory of nearly 300,000 square feet.[20] It naturally focused on photo emulsions, with the goal of improving commercial photographic processes. Because it pursued a patent-driven strategy, Kodak valued secrecy, drawing what researchers referred to as a "silver curtain" between different departments of the research and production operations.[21]

Mees, however, was convinced that successful development could not be programmed and encouraged a culture of free enquiry at the laboratory. He was convinced that the function of an industrial laboratory was to encourage researchers, who might, if left to their own devices, choose to focus on narrowly practical aims, to engage with fundamental questions, publish in the open scientific literature, and

---

[20] Le Guern (2017).

[21] Le Guern, *Contributions* (Ref. 20, Chapter 15).

generally keep abreast of the latest developments in their fields of specialty.[22] Kodak therefore represents yet another mode of industrial research in mid-20th century America. It put considerably more emphasis on secrecy, even internally, than many other major laboratories, but it nevertheless maintained a strong emphasis on fundamental research and contributions to science at large, which it regarded as a wellspring of its most lucrative product — patents.

Kodak's strategy of pursuing chemical patents brought them new relevance during the semiconductor boom when finding new resist materials for the lithographic processes used to manufacture integrated circuits became a pressing priority.[23] As pressure mounted to maintain the progress of Moore's Law — the phenomenon of the regular doubling of transistor density on microchips — that pressure manifested in a demand for better machines and processes for semiconductor manufacture.[24] Kodak filed a patent for one of the first photoresists, KPR, in 1951.[25] It would go on to become a crucial developer of new resist materials, its industrial laboratory thus dedicating itself both to generating new materials and to reconfiguring the processes used to manufacture other materials.

DuPont

Patents were among the most coveted spoils of World War I. Germany, through the 19th century, had developed a world-leading chemical industry. Following the cessation of hostilities, the Woodrow Wilson administration established the Chemical Foundation, Inc., which seized German chemical patents, with the mission to sell or assign them to American corporations. The corporation that benefited the most from these seizures is one that would become synonymous with the American industry over the coming decades: DuPont.[26]

---

[22] Shapin (2008).

[23] Martin and Mody (2020).

[24] Mody (2017).

[25] Minsk *et al.* (February 23, 1954)

[26] Steen (2001).

In the wake of this windfall, DuPont established its fundamental research program. It launched in 1927 and it was inspired by General Electric, and by the German chemical industry from whose patents it now benefitted. Charles M. A. Stein, the program's director, was convinced that the program would bolster the firm's public relations, increase morale and productivity among its technical workforce, and lead to potentially commercializable discoveries. Among the scientists, Stein was able to attract with this scheme was Wallace H. Carothers, a theoretical organic chemist who joined DuPont with the promise that he would not be compelled to undertake practical work.[27]

In an emblematic example of the way the fundamental and the practical tended to bleed into one another in industrial contexts, Carothers's curiosity-driven work on synthetic organic compounds soon lead to one of the most iconic material innovations of the 20th century: nylon. Carothers was interested in the general processes leading to polymerization, but in an era when the microworld constituted a new frontier of investigation, opened up by new instruments and techniques, the quest to understand polymers quickly led to the generation of new polymers for which practical uses could be readily found.[28] Carothers's team synthesized neoprene in 1930, and by the mid-1930s was in the process of perfecting what would become nylon, which would go on to have an astoundingly successful commercial life.

The creation of nylon is in many ways indicative of some of the most productive habits of industrial research that emerged in interwar America. It grew from fundamental research, though crucially fundamental research undertaken in a context where the possibility of eventual applications, however distant, was never far from the mind.[29] It also involved the coordination of different types of expertise — theoretical chemistry, to be sure, but also expertise in high-pressure apparatus, the chemistry of ammonia, and the

[27] Smith and Hounshell (1985).
[28] Cerveaux (2013).
[29] Cerveaux, "Taming the Microworld" (Ref. 28).

knowledge of the production process. The invention and production of nylon pioneered a process that would become the foundation of the discipline of chemical engineering.[30]

IBM

The opening of the microworld that permitted DuPont's polymer research program to thrive also laid the foundations for a new type of industrial laboratory in the postwar period, one dedicated not to emulsions, polymers, and chemical processes, but to solids, surfaces, and electrical processes. That is, industrial laboratories with an interest in materials sprung up around the computer industry. Indicative of this trend is IBM, which invested heavily in commercializing semiconductors and parlayed that expertise into the dominance of the computer industry through the 1960s and 1970s.

Although IBM was representative of a mid-century shift toward an emphasis on computing in American industrial laboratories, it was unusual in other respects. Not for nothing did IBM come to be known as "Big Blue." In the decades after World War II, it blossomed into a massive, multi-national corporation. Its research infrastructure followed suit. Its first laboratory, the Watson Scientific Computing Laboratory, was established at Columbia University in 1945. Just over a decade later, IBM launched its Zurich Research Laboratory in Switzerland. It would move to its Yorktown Heights, NY, headquarters in 1961, and before the end of the century would establish branches established in Haifa, Israel (1972), Tokyo, Japan (1982), Almadén, Spain (1985), Beijing, China (1995), and New Delhi, India (1998).[31] This expansion made IBM the world's most expansive industrial research operation, and although it was unusual for its size, it was representative of the increasingly international character of industrial research during this period.

---

[30] Ndiaye (2007).

[31] More recently, IBM has also opened research operations in Brazil, Ireland, Australia, Kenya, and South Africa.

Both the size and distributed nature of IBM research were conducive to fundamental research with an applied sensibility of the type that had become the hallmark of American industry by the post-World War II era. IBM's labs were well-resourced, and researchers at IBM's satellite facilities often felt as though no one was watching them too closely. In the 1980s, IBM's Zurich lab would be recognized with two Nobel Prizes in Physics. In 1986, Gerd Binnig and Heinrich Rohrer were recognized for their work on scanning–tunneling electron microscopy. The following year, J. Georg Bednorz and K. Alexander Müller would win for their discovery of high-temperature superconductivity. When asked what accounted for such a small, out-of-the-way lab achieving such success, one IBM executive reportedly replied, "poor management."[32]

The scanning–tunneling microscope would become particularly successful for IBM. This was not, all kidding aside, on account of poor management, but because IBM had the operations in place to quickly fine-tune and produce the instrument, in an off-the-shelf, black-boxed form that could be marketed to other laboratories.[33] This was the case with many of IBM's developments. Like Kodak, it was involved in the production of resist materials. It developed the first electron-beam resist material in 1967 and was positioned to deploy it in the developing field of electron-beam lithography in the 1970s and 1980s.[34]

Although unusually large and distinctively distributed, IBM nevertheless indicates how the ethos of American industrial research forged in the early 20th century carried over into the late-20th century. It focused on emerging scientific fields that were ripe for exploitation, empowered its researchers to explore those areas without much top-down guidance of their research programs, and remained ready to pounce on commercializable discoveries — though by the 1950s and 1960s, those growth areas were dominated by semiconductors and the tools used to manipulate and manufacture them.

---

[32] Mody (2011).

[33] Mody, *Instrumental Community* (Ref. 32).

[34] Martin and Mody, "Lithography" (Ref. 23).

RCA

In 1941, the Radio Corporation of America (RCA) began to centralize its research operations. Like AT&T, RCA was a company that made the transition from vacuum tubes to transistors. But like IBM, its concerted focus on in-house research was largely an outcropping of the mid-century. RCA is also a useful contrast point to the other laboratories considered because its management was more circumspect about the operational value of basic research.

Especially in the 1950s, RCA departed from the model pioneered by GE and Bell and maintained connections between its research and production operations.[35] Once the transistor was invented, its commercializable potential was widely evident, but the problem of manufacturing semiconductors at scale was notoriously difficult. Silicon or germanium crystals had to be grown to the right purity with just the right distribution of dopants. Contacts between components had to be precise. Even small faults in the materials production or assembly process could result in costly production failures.[36] The links RCA maintained between its research and production operations made them more adept at overcoming such challenges.

RCA would restructure its operations in the late 1950s, adopting a format more in line with its competitors, with production and research held at arm's length.[37] Its president, David Sarnoff, held fundamental research in high regard and sought to strike a balance between the two in RCA's Princeton laboratory, the David Sarnoff Research Center, named in his honor.[38] RCA's career as a semiconductor company, though, would be indicative of its operations in the early 1950s. It excelled as a producer but lagged when it came to the sort of laboratory research that regularly generated new developments in semiconductor materials through the 1960s and 1970s.

Curiously, the balance between research and production would be reversed in the case of RCA's development of the liquid crystal

---

[35] Choi (2007).

[36] Leslie (2001).

[37] Choi, "Boundaries" (ref. 35).

[38] Gross (2011).

display (LCD) in the 1960s and 1970s. RCA had been a wildly successful color television pioneer, and it aimed to follow up that success. The principles behind the LCD involve the sort of productive overlap that industrial laboratories were built for. It involved developing new materials that could exist in the liquid crystal phase and taking advantage of their responses to electric and magnetic fields to manipulate their optical properties. The ingenious combination of chemical synthesis with knowledge of the mechanical, electromagnetic, and optical properties of molecules made flat-screen displays possible.[39]

But RCA, content with the smattering of patents to flow from this research, failed to effectively commercialize it. An ill-fated move into the computer manufacturing sector made RCA management more risk-averse, and its support for the ambitious LCD program waned. After selling off the relevant patents, RCA withdrew from LCD development, leaving it to be commercialized by other firms, mostly in Japan.[40] In 1985, RCA would be acquired by General Electric, and the Sarnoff Labs were phased into a contract research facility.

Conclusion

"This is the most remarkable company I have ever run across," the University of Chicago chemist Thorfin Hogness wrote in 1948. "Originally it started as a manufacturer of sandpaper. Then it branched out into products which require coatings. As a result of this, they invented scotch tape, which put the company on the map." Hogness went on to list a number of other recent accomplishments of the Minnesota Mining and Manufacturing Company, better known as 3M, and concluded that "the relationship between management and employees in this company must be very good."[41]

---

[39] Gross (2018).

[40] Gross (2012).

[41] Thorfin Hogness, confidential report of an interview with W. L, McKnight, R. P. Carlton, C. W. Walton, and Nelson Taylor, Minnesota Mining and Manufacturing Company, May 7, 1948, University of Chicago. Division of Physical Sciences Records, University of Chicago Special Collections, Chicago, IL, Box 8, Folder 15.

Chicago was targeting 3M as a patron for its new, industry-funded Institutes for Basic Research — among them the Institute for the Study of Metals. Although Chicago did not manage to enlist 3M funding, they remained impressed with its operations and appreciative of what they saw as the company's high regard for basic research.[42]

This anecdote is useful because it highlights some of the limitations of the approach this chapter has taken thus far. It has highlighted aspects of the histories of a few prominent industrial laboratories, but many smaller corporations also supported materials research; some of the most crucial instruments for materials science, for instance, were developed and commercialized in small companies, often those that spun off of academic research projects. Materials research occurred within companies that did not invest heavily in "fundamental" research, but still regarded their work into using existing materials to generate new products as "basic research." Collaborations between industry and academia — as well as industry and government — were crucial for the development of American industrial research, particularly around materials, which were often developed in academic settings and deployed in military settings.

Focusing on a few remarkable institutions nevertheless allows both patterns, and the notable exceptions to them, to emerge. First, the American industrial research laboratory was a potent incubator of new materials, as well as the discipline of materials research. Many polymers, semiconducting materials, plastics, emulsions, and other useful assemblages of matter were first synthesized in American industrial labs, and these labs in turn gave a home to a growing cadre of researchers who began to self-identify as materials scientists.[43]

Second, the distinction between basic (or pure, or fundamental) and applied research, advanced for political purposes in various eras,[44] quickly collapsed into incoherence within the confines of an industrial laboratory. Different sites differed in the degree of

[42] Martin (2020).
[43] Bensaude-Vincent (2001).
[44] Daniels and Krige (2018).

managerial oversight, and the latitude researchers felt to push into areas tangential to the business mission of the corporation, but they were similar insofar as they regarded "basic" or "pure" or "fundamental" as part of their mission, and nurturing that research as a potential wellspring of new technologies, which often meant new materials and new uses for them.

Finally, whatever similarities make it sensible to treat the 20th-century American industrial laboratory as a distinct historical phenomenon — such as the features discussed in the introduction — these institutions themselves exhibited remarkable diversity. GE provided an influential exemplar, but other successful industrial research operations used different relationships between research and production, gave researchers different degrees of latitude, and employed scientists with different ranges of expertise. The mid-20th century might indeed have been a golden age of industrial research, but that does not mean that "industrial research" meant the same thing to every lab conducting it. These labs were as diverse as the materials that they produced, the variety of which is a large part of why we label this age a golden age.

References

Bensaude-Vincent, B. (2001). The construction of a discipline: Materials science in the United States," *Historical Studies in the Physical and Biological Sciences* 32(2), 223–248.

Carlson, W. B. (1991). *Innovation as a Social Process: Elihu Thomson and the Rise of General Electric. 1870–1900* (Cambridge: Cambridge University Press).

Cerveaux, A. (2013). Taming the microworld: DuPont and the interwar rise of fundamental industrial research. *Technology and Culture* 54(2), 262–288.

Choi, H. (2007). The boundaries of industrial research: Making transistors at RCA, 1948–1960," *Technology and Culture* 48(4), 758–782.

Curry, H. A. (2013). Industrial evolution: Mechanical and biological innovation at the general electric research laboratory. *Technology and Culture* 54(4), 746–781.

Daniels, M. and Krige, J. (2018). Beyond the reach of regulation? 'Basic' and 'applied' research in the early cold war America," *Technology and Culture* 59(2), 226–250.

Furfari, F. A. Tony and Coldman, J. W. (November/December 2004). The Westinghouse research laboratory — In memoriam. *IEEE Industry Applications Magazine* 10(6), 9–11.

Gross, B. (2011). The quest for "Magnalux": Redefining technological success and failure at RCA, 1951–1956. *Bulletin of Science, Technology & Society* 31(6), 435–447.

Gross, B. (2012). How RCA Lost the LCD. *IEEE Spectrum* 49(11), 46–52.

Gross, B. (2018). *The TVs of Tomorrow: How RCA's Flat-Screen Dreams Led to the First LCDs* (Chicago: University of Chicago Press).

Hoddeson, L. (1980). The entry of the quantum theory of solids into the bell telephone laboratories, 1925–40: A case-study of the industrial application of fundamental science. *Minerva* 18(3), 422–447.

Hoddeson, L. H. (1977). The roots of solid-state research at Bell Labs. *Physics Today* 30(3), 23–30.

Jackson, C. M. (2011). Chemistry as the defining science: Discipline and training in nineteenth-century chemical laboratories. *Endeavour* 35 (2–3), 55–62.

Lassman, T. C. (2018). *Edward Condon's Cooperative Vision: Science, Industry, and Innovation in Modern America* (Pittsburgh: University of Pittsburgh Press), pp. 30–31.

Le Guern, N. (2017). *Contribution of the European Kodak Research Laboratories to Innovation Strategy at Eastman Kodak* (PhD dissertation, De Montfort University), p. 120.

Leslie, S. W. (2001). Blue collar science: Bringing the transistor to life in the Lehigh Valley. *Historical Studies in the Physical and Biological Sciences Historical* 32(1), 71–113.

Martin, J. D. (2018). *Solid State Insurrection: How the Science of Substance Made American Physics Matter* (Pittsburgh: University of Pittsburgh Press), pp. 142–143.

Martin, J. D. (2020). The simple and courageous course: Industrial patronage of basic research at the University of Chicago, 1945–1951. *Isis* 111(4), 697–716.

Martin, J. D. and Mody, C. C. M. (2020). Lithography. in: Joseph D. Martin and Cyrus C. M. Mody (eds.), *Between Making and Knowing: Tools in the History of Materials Research* (Singapore: World Scientific), pp. 327–340.

Minsk, L. M., Van Deusen, W. P. and Robertson, E. M. (1954). Photosensitization of polymeric cinnamic acid esters, US Patent 2,670,287A, filed January 1, 1951, issued February 23, 1954.

Mody, C. C. M. (2011). *Instrumental Community: Probe Microscopy and the Path to Nanotechnology* (Cambridge, MA: MIT Press), p. 49.

Mody, C. C. M. (2017). *The Long Arm of Moore's Law: Microelectronics and American Soceity* (Cambridge, MA: MIT Press).

Ndiaye, P. A. (2007). *Nylon and Bombs: DuPont and the March of Modern America* (Baltimore: Johns Hopkins University Press).

Noble, D. F. (1977). *American by Design: Science, Technology, and the Rise of Corporate Capitalism* (Oxford: Oxford University Press).

Reich, L. S. (1985). *The Making of American Industrial Research: Science and Business at GE and Bell, 1876–1926* (Chapter 1). (Cambridge: Cambridge University Press).

Shapin, S. (2008). *The Scientific Life: A Moral History of a Late Modern Vocation* (Chicago: University of Chicago Press), pp. 138–39, 151.

Smith Jr, J. K. (1990). The scientific tradition in American industrial research. *Technology and Culture* 30(1), 121–31.

Smith, J. K. and Hounshell, D. A. (1985). Wallace H. Carothers and fundamental research at Du Pont. *Science,* (New Series) 229(4712), 436–442.

Steen, K. (2001). Patents, patriotism, and 'skilled in the art': USA v. The Chemical Foundation, Inc., 1923–1926.

Stephan, K. (2001). Experts at play: Magnetron research at Westinghouse, 1930–1934. *Technology and Culture* 42(4), 737–749.

Usselman, S. W. (1992). From novelty to utility: George Westinghouse and the Business of Innovation during the age of Edison. *The Business History Review* 66(2), 251–304.

Wise, G. (1983). Ionists in industry: Physical chemistry at General Electric, 1900–1915. *Isis* 74(1), 7–21.

Chapter 3.2

The Age of the National Laboratories: The Advent of the New Big Science

Catherine Westfall

At the beginning of the 21st century, the largest projects in the US national laboratory system were devoted to materials science, signaling when this field, with its focus on practicality and industrial participation, was the top national research priority. By the same token, the fate and fortunes of the laboratory system were closely tied to materials science research, which encompassed widescale investigation into the properties of materials as well as the search for new commercial and medical material applications. However, the national laboratory system did not begin with a close connection to the investigation of materials. In the beginning, it was the World War II projects of radar and the atomic bomb that joined scientists, engineers, and military and industry officials. The success of these projects and wartime contacts led to abundant funding opportunities for a vast postwar research enterprise. Prewar and wartime institutions at Los Alamos, Chicago, and Berkeley were transformed into federally sponsored institutions, and new national laboratories were built.

Like the wartime projects, many of these institutions were multidisciplinary, with teams and instruments devoted to a wide range of specialties. However, starting in the 1940s, the largest national laboratory projects were devoted to particle physics, with large particle physics accelerators becoming emblematic of what came to be called

Big Science, that is, science that depended on instruments too big to be sponsored by a single university. Continued funding and growth — of individual laboratories as well as the laboratory system as a whole — were justified by the development of ever-larger and more expensive accelerators needed to advance the search for ever-smaller particles. Premiere accelerators were built at two celebrated multidisciplinary laboratories. In the early 1950s, Lawrence Berkeley Laboratory and Brookhaven National Laboratory built the Bevatron and the Cosmotron, respectively, which kickstarted the advent of particle physics; later in the decade, Brookhaven built the Advanced Gradient Synchrotron. To continue particle discoveries through the 1960s and beyond, two laboratories were built that specialized in particle physics, the Stanford Linear Accelerator Center in Palo Alto, California, and Fermilab near Chicago, Illinois. But a new age was ushered in near the end of the millennium. By this time, particle physics fell into decline just as the fortunes of the newly formed field of materials science were on the rise, with large materials science projects becoming the hope for justifying the continued funding and existence of national laboratories. This chapter describes the events that led to this new age, the New Big Science,[1] the time when large materials science projects reshaped both the field of materials science as well as the US national laboratories themselves.

The Roots of the New Big Science: Through the 1970s

Research Begins

Even as Big Science was flourishing in the post-World War II era, an alternate type of research began to develop. This alternative rose in the 1950s when military and industrial leaders desired new and improved materials for communication apparatus, weapons and energy systems, consumer technologies, and water and aircraft. Given the fundamental scientific uncertainties as well as the multitude of practical considerations they were convinced that creating

---

[1] Crease and Westfall (2016).

and testing the necessary materials could not be accomplished with either traditional engineering methods or scientific investigations alone. With the aim of promoting an approach that combined the skills, expertise, and knowledge of those in a variety of disciplines they created industrial–university–government partnerships with interdisciplinary teams that worked on projects that included the development of new glass and other technologies and materials with commercial potential such as synthetic rubber.[2] To further develop progress in the interdisciplinary development of new materials the military funded a series of interdisciplinary laboratories at universities that began operation in the early 1960s. Initially, these efforts were dominated by engineers, but as the decade progressed teams also included chemists, metallurgists, and physicists.[3] As time went on, more federally sponsored interdisciplinary laboratories were established and materials science developed as a specialty at universities as well as in industry.[4]

Materials science also took root at the national laboratories. At Argonne National Laboratory, the effort grew from wartime work. The postwar laboratory was the successor to the Metallurgical Laboratory near Chicago that spearheaded work on the first reactor that created fissionable material for the atomic bomb. After the war as part of the effort to develop more efficient and safe civilian nuclear reactors, Argonne's Metallurgy Division devised fuels for new civilian nuclear power reactor designs. This work led to a series of reactors that started with improvements to Chicago Pile 1 (CP-1), the reactor used by Enrico Fermi and his team to demonstrate the first sustainable nuclear chain reaction in 1942. Interdisciplinary teams of engineers, chemists, metallurgists, and physicists subsequently developed a series of CP test and research reactors including experimental breeder reactors built at an Argonne test site in Idaho. These efforts were aided by the study of corrosion and radiation effects on reactor materials. In the meantime, Argonne's

---

[2] Lassman (2018).
[3] Martin (2018).
[4] Johnson (1987).

Metallurgy Division, which gave rise to the Solid State Sciences Division in 1959, focused on neutron and X-ray damage to materials over a wide range of temperature ranges. By 1969, Argonne hosted a Materials Science Division. Building on the budding field of materials science, this Division used an array of table-top instruments as well as CP-5 reactor to study various materials, investigating magnetic properties and crystal structures.[5]

By the mid-1970s, materials science was also taking root at other national laboratories. Starting in the early 1950s Oakridge National Laboratory had a Physics of Solids Institute that became a separate division; Oakridge materials science efforts were further expanded in 1961 when the Metals and Ceramics Division was established. At Lawrence Berkeley Laboratory work in high-temperature chemistry and thermodynamics led in the early 1960s to the Inorganic Materials Research Division. Work there was organized into sections: chemistry, metallurgy and ceramics, and reactor materials. Materials science research also began at Brookhaven in the 1950s with groups in the Physics and Metallurgy department, who investigated such topics as radiation damage.[6]

Building Tools and Infrastructure

The emerging field of materials science stimulated the development of new, large instruments. This development had its own dynamic. Unlike particle physics, which required new, larger accelerators to advance, materials science could progress with piecemeal evidence from a large array of instruments of various scales. Thus, the development of large-scale materials science devices lacked the urgency and constituency of particle physics accelerators, which continued to justify the existence of the largest national laboratories through the 1970s. Nonetheless, the new materials science tools, which were developed at smaller laboratories and alongside large projects at the national laboratories, provided the new research opportunities that would usher in the age of the New Big Science.

---

[5] Frost (2003).

[6] Westwick (2003).

Materials science instruments are developed along two separate lines: neutron scattering devices and synchrotron light sources. In the 1960s, the multidisciplinary arsenal of scientific equipment included research reactors that exploited neutron scattering to investigate the relationship between the properties of materials and their structure at Brookhaven, Oak Ridge, and Argonne: the High Flux Beam Reactor or HFBR, the High Flux Isotope Reactor or HFIR, and CP-5, respectively. As the decade progressed, Argonne researchers requested funding for a larger, more powerful device dubbed the Argonne Advanced Research Reactor. When this project failed to attract funding, Argonne began plans for another ambitious project, an accelerator that produced neutrons through spallation that could create short bursts of very high-flux neutrons over a very broad range energy spectrum. Although this project, the Intense Pulsed Neutron Source, was reduced in scale before it was built and faced severe budgetary pressure as it was being completed in the late 1970s, its new technology paved the way for future neutron scattering devices.[7]

The other line of scattering research was synchrotron light sources, which exploit the radiation caused when fast-moving particles follow a curved path. Synchrotron radiation was first observed as an unexpected and unwanted complication by researchers using particle accelerators in the 1930s and 1940s. By the 1950s, however, some were finding ways to create accelerators that would exploit this very energy source as a spectroscopic tool. Starting in the 1960s researchers found they could use synchrotron light sources for materials science, for example, to investigate absorption spectra and for diffraction studies of proteins and crystals. As a result, numerous synchrotron radiation facilities were built both in the US and abroad.[8]

The first device used just for a synchrotron light source was Tantalus, which began operation at the University of Wisconsin in 1968. Like its successors, Tantalus drew researchers from the

---

[7] Holl, *Argonne National Laboratory*, p. 235, Crease (1999) and Westfall (2010).
[8] Westfall (2012).

growing cadres of those studying materials at universities as well as industrial companies such as IBM and AT&T. In the early 1970s, the Stanford Radiation Light Source began operation as a side project at the Stanford Linear Accelerator Center, the national laboratory built for particle physics in the previous decades. Like other so-called first-generation light sources, the Stanford Synchrotron Radiation Light source and Tantalus used components designed for other purposes as did the Cornell High Energy Synchrotron Source, which was constructed in the late 1970s. When upgraded with newly developed beam insertion devices, this type of recycled facility provided a powerful probe of molecular and atomic resonances as well as a way to gain information on the dynamics and structure of atomic, chemical, and biological materials.[9]

As materials science took root with a growing research community and new, large instruments, the federal funding infrastructure was also in flux. After World War II, the major funding agency for research, and the sponsor of the national laboratories, was the Atomic Energy Commission, established in 1946. In response to the energy crisis in the 1970s, President Nixon reorganized the management of the nation's energy and research efforts, separating nuclear regulation from research efforts, creating first the Energy and Research Development Administration and then the Department of Energy (DOE). Shortly after the formation of the new department, a division was formed that focused on materials science research, Basic Energy Sciences. For the first time, materials science had a federal office specifically focused on supporting its development, and the advocacy made a difference. In 1980, the DOE budget for Basic Energy Sciences (BES) was almost $200 million. Although this was less than the $300 million or so allotted for particle physics, this was more than DOE support for nuclear physics. In addition, materials science also continued to be funded at interdisciplinary university laboratories and in industry.[10]

---

[9] Ibid.

[10] Holl (1997).

Big Science Gains its Foothold: From 1980 to the Mid-1980s

By the early 1980s, Argonne's Intense Pulsed Neutron Source began operation, maintaining its funding with the help of DOE Basic Energy Sciences officials. This federal support also helped maintain funding for a similar pulsed source at the Los Alamos National Laboratory, which came to be called the Los Alamos Neutron Science Center. In the meantime, the Brookhaven and Oak Ridge reactors continued materials science neutron scattering research.[11]

By 1980, construction had also begun at two second-generation light sources, that is, devices designed and built expressly to produce and exploit synchrotron radiation: The National Synchrotron Light Source at Brookhaven and another University of Wisconsin light source, Aladdin.[12] Although these devices were experiencing technological difficulties, Lawrence Berkeley Laboratory director David Shirley began plans for a third-generation light source. Shirley, a member of Berkeley's highly respected team at the Stanford Synchrotron Radiation Lightsource and former head of Berkeley's materials science division, was desperate to begin a new large project. The Berkeley laboratory — and indeed the entire DOE national laboratory system — seemed in jeopardy due to Ronald Reagan's promise to cut the size of government. Determined to sidestep such jeopardy, and aware of a new breakthrough in beam insertion technology, Shirley launched the first materials science project aimed at justifying the continued existence of a US national laboratory. By exploiting the new technology, the Advanced Light Source, as it came to be called, promised to provide a tunable beam and thus unprecedented access to ultraviolet and soft X-ray extremes of the electromagnetic spectrum (and thus the device was sometimes referred to as a UV source).[13]

Plans for Berkeley's light source began with optimism after Shirley did some investigation of the new insertion device with his colleague, Exxon researcher Peter Eisenberg. Shirley and his team

[11] Westfall (2010, 2012).

[12] Crease (2008).

[13] Westfall (2008).

felt certain that he would easily obtain funding and proceed quickly and with great success. After all, the project enjoyed numerous advantages. In addition to a promising new accelerator idea, the Berkeley laboratory had a group of experienced users, a sterling research reputation, shared goals, and backing from the highest levels of the US government. However, to Shirley's surprise, Congress voted down the project's funding proposal. Part of the problem was that Shirley had failed to convince would-be users of the comparative research value of the device. In fact, stakeholders critical of the project gathered 75 negative letters addressed to the very legislators in charge of authorization as the DOE funding request was in the balance. Critics noted that the second-generation light sources had still not proven the viability of light source technology, arguing that it was best to stick with the tried and true devices they already had. Underlying their arguments was the fact that materials science, by its nature, needs a variety of devices to advance and concern that an expensive project would bleed more resources than it was worth for the materials science community as a whole. A contributing difficulty was that the project faced such criticism without DOE defenders because Shirley failed to prudently navigate the evolving bureaucratic funding and policy apparatus. By the 1980s, it was a mistake to go straight to the top without cultivating support from the newly developed lower layers of the bureaucracy.[14]

In the end, Shirley got lucky with solving both problems and materials science got a boost that helped propel the field into a new age. The working level managers he faced when he redoubled efforts for Berkeley's light source were Louis Ianniello, the head of the Materials Section of BES, as well as his boss, Alvin Trivelpiece, who headed the DOE Energy Research Office. Both became supporters after being persuaded that the new device would provide money and research opportunities, a winning prospect both for materials science, DOE, and the reputations of managers themselves.[15]

---

[14] Westfall (2012).

[15] Ibid.

Once on board, Ianniello got busy, doing what policymakers do when a desired project is blocked: convene a committee of experts to represent and articulate the interests of stakeholders as part of a reassessment of the project. Ironically, this effort to secure approval for the arguably too big and expensive Advanced Light Source, a project that would cost about $100 million, unexpectedly ushered in a competing light source that would cost about $1 billion, the Advanced Photon Source. In the process, DOE also developed a lineup of materials science projects at numerous laboratories that BES officials would champion as the decade wore on.[16]

The outlines of the new array of materials science devices began to take shape in the fall of 1983. To ostensibly assess the status of light source development, Ianellio convened a committee led by Eisenberg and another respected scientist with experience with large-scale materials science devices, Michael Knotek. They organized a meeting in Albuquerque, New Mexico in October 1983 to bring together all those interested in light sources. Even though the meeting was meant to highlight the Berkeley project, the discussion instead led to considerable excitement for another third-generation light source. This higher-energy light source would cover the hard X-ray spectral region, thus opening the possibility of studying material structure directly through imaging rather than the soft X-ray and extreme UV region covered by the proposed Berkeley machine. Following a pattern that fits with materials scientists' preference for a variety of tools and tried and true technology, the committee gave top priority to completing upgrades at the first-generation light source at Stanford and as well as completion of the second-generation light sources at Brookhaven and the University of Wisconsin. They then judged that if a single facility would be built it should be a 6-GeV hard-ray source. Berkeley's light source came in as the third priority. The committee's recommendations — which were after all to build two expensive new machines given sufficient financial resources — represented an underlying, fundamental shift in the materials science community. For the first time materials scientists

[16] Ibid.

had an energized, sizable interest group committed to the idea that large materials science devices could provide sufficient research resources to be worth significant cost and technological risk. In addition, they had DOE allies poised to successfully finesse the necessary funding arrangements. The necessary initial resources for large-scale materials science were at hand.[17]

Even though the Eisenberger–Knotek Committee opened new possibilities for materials science, it posed challenges for Ianniello and Trivelpiece. Third priority meant poor prospects for the Berkeley light source, which was important to the future of that laboratory and thus DOE. In addition, although the 6-GeV light source had excited considerable enthusiasm, it did not have a specified location. Trivelpiece then did what Washington bureaucrats tend to do when they get an unsettling answer from a review committee: they requested another review. Accordingly, in November 1983, the Office of Science and Technology Policy asked the National Research Council of the NAS to convene a committee to review not only the emerging ideas for light sources but also all other plans for large-scale materials science facilities. It would be chaired by Eisenberger–Knotek Committee member Dean Eastman from IBM and the eminent physicist and statesman Frederick Seitz.[18] The Seitz–Eastman Committee deliberations launched intense negotiations as researchers, laboratory managers, and DOE officially devised plans for new devices and ironed out arrangements for locations and funding priorities. In the process, the outlines of the New Big Science emerged at the US national laboratories.

The Age of New Big Science Takes Shape: After the Mid-1980s

Berkeley researchers reiterated their scientific case for their light source and found ways to reduce cost estimates in the hopes of getting the committee's approval for the project. In the meantime,

[17] Westfall (2008).

[18] Ibid.

proposals for the 6-GeV light source were drafted by experienced researchers at Brookhaven, Cornell, and Stanford. In addition, researchers at Argonne Laboratory threw their hat into the ring. Although Argonne lacked light source experience, it did have a sizeable materials science group and some members had used light sources at other facilities. Perhaps most important, Argonne director Alan Schriesheim was as desperate as Shirley had been to find a new project to justify continued laboratory funding in the harsh political and funding environment of the time. Funding for the civilian nuclear reactor projects that Argonne had depended on was in decline, and the $9 million Intense Pulsed Neutron Source, though productive was a thin reed to depend on to secure the laboratory's prospects. The Argonne pulsed source faced funding challenges, with competition from a similar source at Los Alamos as well as the Brookhaven reactor and a bold new proposal to revamp the Oakridge reactor into a state-of-the-art intense neutron scattering device.[19]

The Seitz–Eastman Committee issued its report in July 1984. It gave the first priority to the 6-GeV light source, asking for a site selection process without specifying which competing laboratory should build the device. The committee gave second priority to the new Oakridge reactor, third priority to Berkeley's light source, and fourth priority to Argonne's Intense Pulsed Neutron Source.[20]

In the wake of the Seitz–Eastman Committee recommendations in the summer of 1984, competitive forces were raging as laboratory directors fiercely lobbied for materials science and other projects needed to secure financial prospects for their respective institutions. Schriesheim was particularly aggressive in his efforts to get the 6-GeV light source for Argonne. In the meantime, DOE officials had an overriding concern: the fate of the $1 billion Superconducting Supercollider for particle physics. Trivelpiece was worried that competitive tensions worsened prospects for the Supercollider and opened the possibility that none of the projects could be funded.

---

[19] Westfall (2010, 2012).

[20] Energy Research Advisory Board (1984).

This would weaken not only individual laboratories but also DOE as a whole.[21]

At this juncture, competition threatened to be lethal and cooperation promised salvation. Agreement on this point secured support for what came to be called the Trivelpiece Plan in early 1985. In this deal, laboratory directors agreed to support the particle physics project and Argonne got the 6-GeV light source, Oak Ridge got the new reactor, and Berkeley got its UV light source. In addition, Brookhaven got a heavy-ion facility for nuclear physics later named the Relativistic Heavy Collider.[22]

The Seitz–Eastman recommendations and the Trivelpiece Plan set the course. But the new age of materials science and the national laboratories could not begin without funding — and a twist of fate. The necessary funding came, slowly but surely, as the 1980s wore on. Berkeley's light source, now slated to cost $100 million, got construction funding in 1986 and its construction began in 1987. Argonne researches had a harder time finding a place in the budget for their much more expensive light source, but in 1989, they did manage to obtain the first installment of their nearly $1 billion budget.[23]

Through the early 1990s, materials science projects grew in the shadow of DOE's premiere project, the multi-billion dollar Superconducting Supercollider, which particle physicists had begun to build in Waxahachie, Texas.[24] Then this project was canceled in 1993. Despite support from within the national laboratory system, the giant particle physics project fell victim to an escalating budget and eroding Congressional support for esoteric science in the Post-Cold War. Thus, when Argonne's light source began operation in 1995 it became DOE's largest, most expensive basic research project. The title was taken over in 2006 by a powerful intense neutron source, the $1.4 billion Spallation Neutron

---

[21] Westfall (2012).

[22] Ibid.

[23] Westfall (2008, 2012).

[24] Riordan *et al.* (2015).

Source at Oak Ridge, a project that replaced the state-of-the-art reactor project canceled due to rising public concerns about the dangers of nuclear power. By that time, the mission of the Stanford Linear Accelerator Center, or the SLAC National Accelerator Laboratory as it was later called, shifted from particle physics to materials science. Its premiere accelerator was a fourth-generation light source, or free electron laser, that allowed an exponential increase in overall radiation intensity than its predecessors.[25] In 2009, Brookhaven began construction of an upgraded version of its light source, the National Synchrotron Light Source II, which began operation in 2014. This collection of large-scale devices provided powerful new tools that provided unprecedented opportunities for studying and creating materials for every industry in the modern world. The age of the New Big Science with its focus on the practicality and industrial participation of materials science had become the raison d'être and intellectual and financial engine fueling the US national laboratories.

References

Crease, R. (1999). *Making Physics: A Biography of Brookhaven National Laboratory, 1946–1972* (Chicago: University of Chicago Press), pp. 316–347.

Crease, R. (2008). The national synchrotron light source Part I: Bright idea. *Physics in Perspective* 10, 438–467.

Crease, R. and Westfall, C. (2016). The New Big Science, *Physics Today*, 69, 30–36.

Energy Research Advisory Board (July 24, 1984). Review of the National Research Council report 'Major facilities for Materials Research and Related Disciplines,' DOE-/S-003.

Frost, B. (June 1996). A history of materials R & D at Argonne. *Advanced Materials & Processes.*

Heinze, T. and Hallonsten, O. (2017). The reinvention of the SLAC national accelerator laboratory, 1992–2012. *History and Technology: An International Journal*, 33, 1-33.

[25] Heinze and Hallonsten (2017).

Holl, J. (1997). Argonne national laboratory, 1946–1996 (Urbana: University of Illinois Press), pp. 278–279.

Holl, J. *Argonne National Laboratory*, p. 235.

Johnson, H. (1987). Ensuring contributions to materials science from small, intermediate-, large-scale science. in: P. A. Psras and H. D. Langrod (eds.), *Advancing Materials Research* (Washington, DC: National Academy Press), pp. 321–334.

Lassman, T. C. (2018). *Edward Condon's Cooperative Vision: Science Industry, and Innovation in Modern America* (Pittsburgh, University of Pittsburgh Press), pp. 159, 173,

Martin, J. D. (2018). *Solid State Insurrection: How the Science of Substance Made American Physics Matter* (Pittsburgh: University of Pittsburgh Press), pp. 121–124.

Riordan, Michael, Hoddeson, Lillian and Kolb, Adrienne (2015). *Tunnel Visions: The Rise and Fall of the Superconducting Super Collider* (Chicago: The University of Chicago Press).

Westfall, C. (2008). Retooling for the future: Launching the advanced light source at Lawrence's laboratory, 1980–1986. *Historical Studies in the Natural Sciences*, 38, 569–609.

Westfall, C. (2010). Surviving to tell the tale: Argonne's intense pulsed neutron source from an ecosystem perspective. *Historical Studies in the Natural Sciences*, 40, 350–398.

Westfall, C. (2012). Institutional persistence and the material transformation of the US national labs: The curious story of the advent of the advanced photon source. *Science and Public Policy*, SCS054, pp. 1–11.

Westwick, P. (2003). *The National Labs: Science in an American System, 1947–1974* (Cambridge: Harvard University Press), pp. 260–261.

Chapter 3.3

Becoming Materials Science? Turning Points in the History of Materials Science at the ESRF, 1994–2020

Katharina C. Cramer

Introduction

Materials science has received much attention within the European science and policymaking landscape since the late 20th century.[1] Several European collaborative Big Science facilities, such as the Institut Laue-Langevin (ILL) or the European Spallation Source (ESS) have made materials science a priority.[2] The history of the European Synchrotron Radiation Source (ESRF), founded in 1988 as a collaborative effort among several European countries, is another point in case: From its pre-founding phase in the early 1980s to its current upgrade program, materials science has always played a major role. Historical records of ESRF's pre-foundation phase already illustrate the close connection between emerging instrumental techniques in materials science and ESRF's scientific mission. Indicative of the importance of materials science

[1] See, e.g., Bertrand and Bensaude-Vincent (2011) and Hallonsten and Gribbe (2017).

[2] See, e.g., Hallonsten and Gribbe (2017, p. 460), Cramer (2020), Hallonsten (2016, p. 211).

throughout the history of the ESRF is perhaps also a bold quote from ESRF's annual report in 1999 stating that "[m]aterials research in its different aspects is the ultimate goal of 90% of all the investigations with synchrotron radiation at the ESRF."[3]

While ultimately important to ESRF, it has no definite definition of materials science. This chapter argues that, although materials science has been and remains a cornerstone of ESRF's scientific program, discursive understandings and definitions of materials science and its manifestations and appearances within the organizational structure of the ESRF and the career foci of leading ESRF scientists varied over time. This chapter hence chronicles and analyzes important turning points of materials science in the history of the ESRF along two dimensions: First, it explores the gradual discursive manifestation of materials science as an *organizational category* among ESRF's research groups (the so-called beamline groups) and reviews committees. Second, it identifies materials science as a key theme in the *career focus* of leading scientists at the ESRF. Through its analysis, the chapter provides insights into institutional changes in Big Science facilities. Moreover, it adds to historical and organizational perspectives on photon sources, such as synchrotron radiation sources or free-electron lasers,[4] and connects to the emerging scholarly understanding of the contemporary history of materials science and related disciplinary approaches and scientific fields.[5]

To that end, this chapter looks at pivotal moments of institutional change at the ESRF. Specifically, it defines three periods of discursive shift as well as organizational and intellectual change throughout the ESRF's history: 1980–2000, 2001–2008, and 2009–2020. This approach draws on existing analysis of institutional change and renewal at Big Science facilities, such as the German Electron-Synchrotron (DESY) or the US-American Stanford Linear Accelerator Centre (SLAC).[6] It also considers complementary

[3] ESRF Highlights (1999, p. 76).

[4] e.g. Hallonsten (2016) and Cramer (2020).

[5] e.g. Martin (2018), Bensaude-Vincent (2011a) and Cahn (2001).

[6] e.g. Hallonsten and Heinze (2012, 2015b).

approaches to the role of discourses and textual bodies in institutionalization processes.[7] Against that background, institutional change is understood as the aggregate of processes that leave the framework (here: the ESRF) stable and intact, but added and/or changed some of its elements.[8] Applied to the ESRF, change of elements occurs at the level of the organization of ESRF, the intellectual and professional backgrounds of its leading scientists, and specific bodies of texts, such as annual reports or pre-founding documents, through which new and emerging understandings and definitions of materials science became institutionalized.

Institutional change has frequently been described as slow and costly, particularly for research organizations and universities.[9] However, recent scholarship revealed that contemporary Big Science facilities, such as the ESRF, are particularly open for institutional change on short time scales.[10] To the extent that the ESRF is a service facility that provides experimental resources to external user groups, it is also dependent on incorporating the innovative demands of its users to remain on the cutting edge of scientific research. This ability to adapt to changing scientific purposes and emerging research techniques is what Hallonsten and Heinze described as the "intrinsic potential to serve many purposes."[11] They further observe that "[t]he labs have been refined and adjusted to multidisciplinary, and partly open-ended, use by purposeful work to integrate all the technical components and organizational elements in comprehensive efforts to lower the barriers to new utilizations."[12]

The investigation and analysis of this contribution are based on different sources. First, historical sources and archival material from the French National Archives and the internal archive of the ESRF provide important insights into ESRF's pre-foundation phase. Second, ESRF's annual reports (*ESRF Highlights*) disclose important

---

[7] e.g. Phillips *et al.* (2004).

[8] e.g. Hallonsten and Gribbe (2017, p. 461, 466) and Hallonsten (2016, pp. 143–147).

[9] e.g. Heinze and Münch (2012, pp. 17–18).

[10] Hallonsten (2016).

[11] Hallonsten and Heinze (2015a, p. 847).

[12] Ibid.

experiments, organizational changes, and ESRF's key financial positions and investments throughout the years. Third, the curricula vitae of leading scientists at the ESRF, such as the directors general or the research directors, are accessed regarding the direct and indirect connections to materials science throughout their careers.

What is the History of Materials Science?

The history of materials science is a history of how, throughout the second half of the 20th century, the study of dissimilar materials such as paper, wood, metals, or ceramics was gradually grouped closer together. While the materials and their applications changed over the years and with it the groups, the aim stays the same: to find new ways and combinations to utilize materials. Bernadotte Bensaude-Vincent identified the "generic concept of materials"[13] as a central aspect within the history of materials science arguing that "[t]he commensurability between such dissimilar materials is based on a functional equivalence for specific uses."[14] She highlights how the study of various materials could find common ground through thinking together towards applications, performances, and functionalities across different disciplines and fields. Through their collaboration, researchers of different backgrounds could form a sense of community and identity among themselves. Similarly, Joseph D. Martin argued that materials science lacks conceptual stringency, but emerged in the early Cold War as a "discipline by convention, assembled in large part to capitalize on the stated preferences of powerful funders."[15] Klaus Hentschel characterized materials science rather as the "overarching science of workable materials."[16] According to Hentschel, the historical processes cumulated around the 2010s in the formation of a self-standing scientific discipline that self-identifies as such through coherent and specific institutions,

[13] Bensaude-Vincent (2011b), p. 108

[14] Ibid., p. 115.

[15] Martin (2018, p. 203f).

[16] Hentschel (2011).

journals, study programs, and professional associations. Despite these differences among experts, there is an agreement amongst them that the important point is that materials science made its way through the history, organization, and politics of contemporary science. As a "disciplinary narrative,"[17] it collapsed the study of different materials and their research techniques under a single umbrella term, providing a unique identificatory category for scientists, policymakers, and funding agencies.

The historical emergence of materials science in the mid-20th century was triggered by changing economic needs that shifted the conceptual characterization of materials. Till then, materials have primarily been seen in contrast to matter.[18] Matter has historically been conceived as a general and abstract term, which leads to an ideal notion of matter. In direct contrast, materials have often been defined by their singularities and imperfections. This conceptual characterization of materials limited them to their functionalities. Dissimilar materials hence remained to be treated as individual entities, such as metals or ceramics, with distinct institutional settings, organizational contexts, and scientific communities.[19] That conception changed with the economic needs and strategic considerations in the early Cold War. Suddenly, researchers had a new incentive to challenge the functionalities and applications of existing materials for innovative uses in communication and consumer technologies and weapon systems.[20]

The application orientation of materials science research and its close connections to industrial production and commercial research and development is key to understand the history of materials science throughout the 20th century in Europe. Around the 1970s and 1980s, materials science also gained prominence in several European countries such as France, Sweden and Germany. Hallonsten and Gribbe stress the important role of new national funding systems in

---

[17] Bensaude-Vincent 2011a, p.46.

[18] Bensaude (2011, p. 112).

[19] e.g. Martin (2018, p. 130).

[20] Martin (2018, p. 122), Mody and Choi (2013, p. 126).

Sweden in the 1980s and a subsequent change in Swedish science policy strategy in the 1990s for materials science to become an important pillar of the Swedish science system.[21] However, these changes alone were not a guarantee for success. Another important factor was the willingness of the scientific community to embrace the new discipline. While France implemented similar policies as Sweden, the outcome was different: Bertrand and Bensaude-Vincent argue that "in stark contrast to the USA and other industrial countries, there was no 'materials boom' in France."[22] The strong political will of French policymakers to make materials science a top priority of the French science system remained largely unheard of among the scientific communities. Although France did not experience a similar tidal wave of new materials science departments and funding initiatives, materials science continued to exist in disciplinary laboratories albeit not as visible as in the United States.[23]

The example of Germany highlights that materials science was especially welcomed into the scientific work of a country when there was a strong commercial incentive. As in Sweden and France, materials science gained importance within German national policy strategies in the 1980s. In 1985, the German Federal Ministry for Research and Technology implemented its first nationwide research program on materials. Specific programs in the federal countries followed around the 1990s. With regard to the increasing importance of materials science for, among others, the German automobile and metal industries, the German Science Council (*Wissenschaftsrat*) in 1993 and 1996 issued fundamental recommendations on the future evolution of materials science in Germany and its competitiveness in global comparison.[24]

After taking off in some European countries, materials science emerged as one of the focal topics in a European context within the third (1991–1994) and fourth (1994–1998) common framework

---

[21] Gribbe and Hallonsten (2017, pp. 491–493).

[22] Bertrand and Bensaude-Vincent (2011, p. 212).

[23] Ibid.

[24] Wissenschaftsrat (1993, 1996).

programs (FPs) of the European Commission.[25] In 2000, the European Commission launched its European Research Area (ERA) initiative which was part of a general strategic turn of European science policy towards greater coherence and coordination efforts among individual initiatives and the national activities of the European member states. This period also saw the increasing interest of science policymakers in technological innovation and added value for European economies.[26] It seems as if science policymakers have also begun to fully acknowledge — at least rhetorically — the importance of materials science for the future of Europe's economy and innovation potential strengthening ties between national efforts and common European activities coordinated by the European Commission. In the FPs that followed in the 2000s and 2010s, materials science increasingly positioned together with other strategic and highly politicized research fields such as nanotechnologies or circular economy.[27]

Materials Science: Making and Shaping the ESRF

Materials science has always been a cornerstone of the European Synchrotron Radiation Facility (ESRF), not least because few European projects have offered more experimental opportunities than the ESRF as a collaborative synchrotron radiation source. Located in Grenoble, France, the ESRF offers extremely brilliant X-rays as multidisciplinary experimental resources to external users that come to the facility on a short-term basis. The synchrotron radiation source is based on a circular-shaped accelerator that accelerates electrons close to the speed of light that are then redirected into beamlines. Since the ESRF opened to external users in 1993, some of these beamlines were dedicated to materials science. This is important to consider because the number of beamlines and the purposes of their experimental stations are crucial questions during the construction work that need to be negotiated and agreed upon in advance.

[25] See, e.g., Tamtik (2016), Wissenschaftsrat (1993).
[26] See, e.g., Ulnicane (2015).
[27] See, e.g., Banchoff (2002, p. 15), European Commission (2015).

The ESRF is embedded within the broad and competitive European landscape in materials science, which hosts a considerable number of national research centers. Examples include the German Helmholtz Centres or Max Planck Institutes, the French National Centre for Scientific Research (CNRS), but also collaborative large-scale research facilities, such as the ESRF itself, ILL or ESS, and research-intensive companies. Furthermore, the ESRF is also co-located with the ILL, a collaborative neutron research reactor project, founded in the 1960s by a French–German effort that was later joined by the United Kingdom. The ILL offers experimental research opportunities with neutrons for external user groups that can be considered as complementary to research with synchrotron radiation.[28] All these different institutions shape the European materials science landscape through complementary experimental resources, scientific cross-fertilization through collaborative projects and circulation of individual skills and knowledge by highly mobile researchers with changing professional affiliations. As a user facility within that research landscape, the ESRF benefits from a constant flow of scientists from and to the ESRF. New cohorts and generations of scientists with their specific expertise, knowledge, and personal networks continuously join and/or leave the facility. This creates changing environments for exchange and synergies across disciplines or scientific fields.

Moreover, the ESRF has close ties to materials science due to the timing and purpose of its establishment. That point becomes clear when looking back at the facility's history. In the late 1980s, the implementation of so-called dedicated third-generation synchrotron radiation sources was underway. Synchrotron radiation, first observed in the late 1940s in the United States, has long been an unwanted and parasitic[29] research activity at particle physics accelerators (*first generation*) that dominated large-scale Cold War research in the United States and Europe. This setting only changed in the 1970s and 1980s with the development of storage rings and

---

[28] Jacrot (2006).

[29] Hallonsten (2015).

the construction of dedicated synchrotron radiation sources (*second generation*) that were independent of particle physics research. Another turning point was the invention and implementation of insertion devices at dedicated synchrotron radiation sources (*third generation*).[30]

The establishment of dedicated third-generation synchrotron radiation sources, such as the ESRF, the Advanced Photon Source (APS) in the United States or the Super Photon Ring-8 GeV (SPRING-8) in Japan signified a broader shift in the scientific activities of major Big Science facilities. Indeed, recent scholarship described this development as the emergence of "New Big Science"[31] or "Big Science Transformed"[32] at the end of the 20th century whereby the once dominant particle physics research (old Big Science) declined at the end of the Cold War. This is where materials science enters the stage. As old Big Science declined, research with synchrotron radiation and related scientific fields that serve experimental research in materials science and life sciences took off considerably.[33] This period has been a crucial point in time at which the use of X-ray techniques at synchrotron radiation sources was raised as one of the central research technologies that considerably advanced the investigation of materials.

Having listed key reasons for why materials science has been and remains a cornerstone of ESRF's scientific program, the chapter now turns to the organizational structures of the ESRF to begin its analysis of changing discursive understandings and definitions of materials science. From an organizational point of view, the ESRF is structured as follows: At its top, the facility is headed by a general director. Then there are a number of important positions regarding ESRF's scientific activities that directly link to the general director. That includes the two research directors (one for the physical sciences and one for the life sciences), the machine director, the

[30] Cramer (2020, pp. 60–62).
[31] Crease and Westfall (2016).
[32] Hallonsten (2016).
[33] Crease and Westfall (2016) and Hallonsten (2016).

1996 1997 1998 1999 2000	2001	2002 2003 2004 2005 2006 2007 2008	2009 2010 2011 2012 2013 2014	2015 2016 2017 2018 2019 2020 2021
PETROFF - Director General (1993-)	**STIRLING - Director General**		**SETTE - Director General**	
KUNZ	**SETTE**		**REICHERT**	
XAFS (1996-1998), Inelastic Scattering, Troika, Surface Science, Nuclear Resonance, Magnetic Scattering, X-Ray Microscopy, X-Ray Microfluorescence, Optics, since 1999: High Energy	Surfaces and Interfaces, Materials Science and Engineering, X-Ray Absorption and Magnetic Scattering, Dynamics and Electronic Structure	Surfaces and Interfaces, Materials Science, X-Ray Absorption and Magnetic Scattering, Resolution and Resonance Scattering, affiliated: Theory Group	Dynamics and Extreme Conditions, Electronic Structure and Magnetism, Structure of Materials	Electronic Structure, Magnetism and Dynamics, Matter at Extremes, Structure of Materials

Fig. 1. Overview of ESRF's Directors General, Research Directors, and Beamline Groups related to Physical Sciences/Materials Science

Source: Organizational charts as displayed within the ESRF Highlights 1996–2020; no organizational charts were provided before 1996.

review committees, and an external scientific advisory committee. While organization is one side of the coin, the other is personnel. It is key to remember that personnel in any organization matters and particularly those that occupy leading positions. The ESRF has seen four different directors general (see Fig. 1): Ruprecht Haensel (1986–1992), Yves Petroff (1993–2000), William "Bill" Stirling (2001–2008), and Francesco Sette (since 2009). Since 1992, the ESRF has seen three research directors for the physical sciences: Christof Kunz (1993–2000), Francesco Sette (2001–2008), and Harald Reichert (since 2009) (see Fig. 1).

Making Materials Science Through Its Instrumental Techniques, ca. 1980–2000

It is worth pointing out that materials science is only marginally used as a discursive concept in historical sources during ESRF's pre-foundation phase. That is largely due to the political context of that phase. The historical sources available, such as bilateral correspondences or short briefings on multilateral meetings, testify to the political compromises among potential members of the ESRF in the pre-foundation phase. Negotiations on political commitment and funding for the ESRF rather hinged on broader political trade-offs and package deals involving not only the establishment of the ESRF but also other Big Science facilities such as the British SNS or the German HERA.[34] Within these political contexts of ESRF's pre-foundation phase, the scientific descriptions of the ESRF project rarely went beyond its characteristic as an X-ray source. Therefore, materials science as a concept received little mention.

However, that changed once the focus shifted to ESRF's scientific program. The Feasibility Study (1979)[35] and the Foundation Phase Report (*Red Book*, 1987)[36] highlighted the use of X-rays as central experimental resources to advance the investigation of materials in the near future.[37] Each of these two reports contained a section on materials science and chemistry that put particular emphasis on individual techniques: Next to the importance of diffraction and scattering techniques, the reports put emphasis on extended X-ray absorption fine spectroscopy (EXAFS), which is a specific subcategory of X-ray Absorption Fine spectroscopy (XAFS) and the more general approach of X-ray absorption spectroscopy (XAS) — in the following summarized as E/XAFS.[38] These reports mattered greatly for the institutionalization process of materials science at ESRF. Indeed, as will be shown below, there is much to suggest that during ESRF's pre-founding phase and early history, materials science institutionalized within the organizational structure of the ESRF through its instrumental techniques, most importantly XAFS, that featured prominently among ESRF's beamline groups in the 1990s.

From all the research areas discussed in these two reports, E/XAFS ended up being of particular importance for the role of materials science in the pre-foundation phase and early history of the

[34] Cramer (2020).

[35] European Science Foundation (1979).

[36] ESRF (1987).

[37] See, e.g., European Science Foundation (1979, p. 15) and ESRF (1987, pp. 3–4).

[38] See European Science Foundation (1979, pp. 27–34) and ESRF (1987, pp. 34–35). For details on EXAFS, see Mottana and Marcelli (2015).

ESRF. XAFS emerged from X-ray absorption spectroscopy (XAS), a technique that has long been in the shadow of X-ray diffraction (XRD). Both techniques, XAS and XRD, apply X-rays for scientific studies of materials. XRD became an important and powerful technique for solid-state physics and the study of ordered matter throughout the 20th century. However, this technique reached its limits when scientists turned to study disordered systems, liquids, and amorphous structures. XAS was indeed able to address disordered and amorphous structures. But the performance of this technique was limited by the unavailability of highly brilliant and powerful X-ray sources, such as synchrotron radiation sources. Only with the emergence of storage rings in the 1970s and 1980s, XAS could make full use of brilliant X-rays to deliver cutting-edge scientific results. In a sense, E/XAFS, as a subset of XAS, became the gold standard for the study of materials around the 1980s and was widely used in many different synchrotron radiation laboratories around the world.[39]

XAFS also played a major role in ESRF's early history, which is in large part due to one person, the physicist Ruprecht Haensel. Haensel had been the Director General of the neighboring ILL since 1985 before he became the first Director General at the ESRF from 1986 to 1992. He has not only been characterized as a pioneer of research with synchrotron radiation.[40] But Haensel has also been promoted as an expert in XAFS.[41] There is much to suggest that his expertise played an important role when names of potential candidates for ESRF's first Director General appeared within the correspondences among governmental representatives and policymakers of ESRF's future members. Haensel's expertise in XAFS was explicitly stressed.[42] The importance of XAFS for the scientific portfolio

---

[39] See Mottana and Marcelli (2015).

[40] See, e.g., Freund (2007) and Heinze *et al.* (2015).

[41] See, e.g., Haensel (1980) and Thulke *et al.* (1983).

[42] See, e.g., *Choses Importantes Que ESRF Demarre Vite et Bien* (25 October 1984) from the French Ministry for Industry and Research, Archives Nationales de France No. 1992 0550/3; *Compte Rendu de la Réunion du 15 Février 1986 du Conseil d'Administration Provisoire de ESRF* (25 February 1986) from the French Ministry for Research and Technology, Archives Nationales de France No. 1998 0422/3.

of the future ESRF was then also resonated by the organization and composition of the first beamline groups at the ESRF, headed by Christof Kunz. One of these groups was dedicated to the application of XAFS. The other groups under Kunz also focused on specific instrumental techniques (inelastic scattering, nuclear resonance, magnetic scattering, X-ray microscopy, and X-ray microfluorescence) rather than scientific disciplines or fields (optics, surface science).

Materials Science Enters ESRF's Organizational Stage: ca. 2001–2008

During the 1980s and 1990s, materials science had shaped ESRF's organizational structure merely through its instrumental techniques. This changed in the early 2000s when new review committees and beamline groups were established, explicitly dedicated to materials science. These developments can be regarded as pivotal examples of how materials science, as a specific terminology, is institutionalized within the organization of the ESRF.

This change did not happen overnight. The initial review committees that were established in the mid-1990s represented rather traditional scientific fields and disciplinary demarcations, such as chemistry, hard condensed matter, and soft condensed matter. Yet, materials science gradually gained traction. The first important shift occurred in 2000 with the establishment of a review committee called *Materials Engineering and Environmental Matters.* In 2006, this committee was anew split into two distinct committees: *Environmental and Cultural Heritage Matters* and *Applied Materials and Environment* (Fig. 2). Under the auspices of the new Research Director for the physics sciences, Francesco Sette, a new research group dedicated to *Materials Science and Engineering* was formed in 2002. The group was later renamed to *Materials Science.* Under Sette's leadership, the ESRF also established the Facility for Materials Engineering (FaME38) on the joint ILL–ESRF site in 2003 in collaboration with several universities and research centers from the United Kingdom. The explicit purpose of this new facility was to support the use of

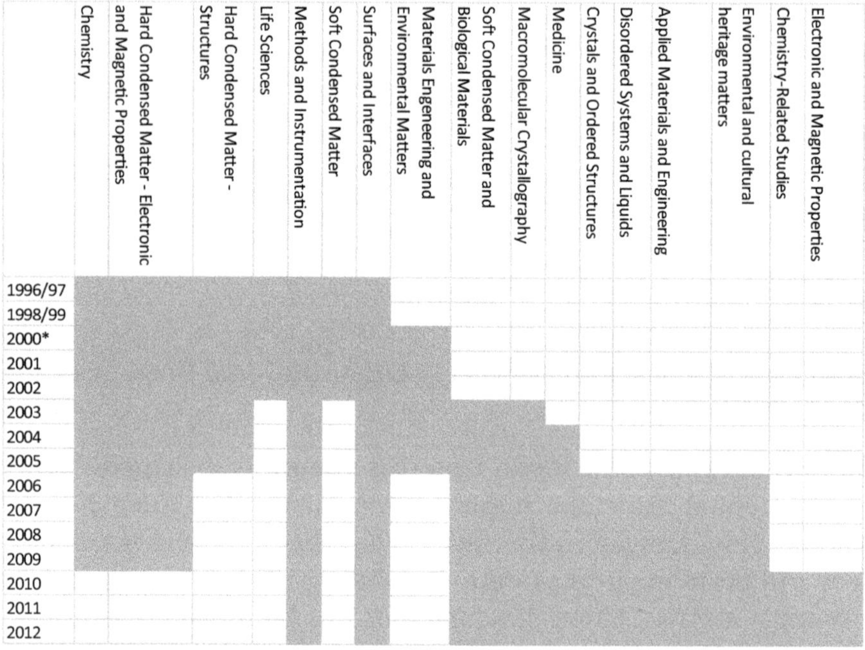

Fig. 2. Overview of Review Committees (Sources: ESRF Highlights, 1996–2012. No precise information on review committees was provided before 1996. After 2012, Review Committees were changed to Beam Time Allocation Panels that relate to the different beamlines and not scientific themes).

X-rays (ESRF) and neutrons (ILL) for materials science providing a supportive environment for all stages of research.[43]

While materials science thus became more and more institutionalized at the ESRF, it seemingly received competition from nanotechnology in the early 2000s. Many countries worldwide started to establish national nanotechnology initiatives, such as the United States, Canada, Russia, Japan, China as well as several European countries, including France, Italy, Germany, and the Netherlands.[44] The increasing political role of nanotechnology and the importance of scientific inquiry on a nanoscale also posed

[43]Webster *et al.* (2003).

[44]See, e.g., Kehrt (2016).

important challenges for ESRF's scientific activities, thus demanding great attention and resources that could have otherwise been spent on materials science. For one, the facility responded and adapted to these developments and the increasing demand from scientists and policymakers to study nanostructures by developing new technologies and refining existing instruments.[45] Moreover, nanotechnology became an important cornerstone of so-called ESRF's Purple Book, published in 2007, which laid out the facility's long-term scientific program.[46] That meant that in the years and preparatory work leading up to the publication, the ESRF, together with its European user community and SAC (Scientific Advisory Committee) had identified several "highlight fields" as "key areas of future scientific development"[47]: nanoscience and nanotechnology, pump-and-probe experiments and time-resolved science, science at extreme conditions, structural and functional biology and soft matter, and X-ray imaging. Materials science — as a discursive category — did not make the cut.

However, it would be an oversimplification to simply depict nanotechnology and materials science as direct competitors. On the contrary, both disciplines benefitted from the other. Indeed, it can be concluded that, although nanotechnology, strictly speaking, is not a direct derivate of materials science, materials science played a key role in the institutionalization and emergence of nanotechnology as a key scientific concern among ESRF's long-term scientific strategy. These connections and their importance to the facility's scientific activities had been repeatedly highlighted in ESRF's annual reports in the mid-2000s.[48] The picture that emerges here bears great resemblance to what Bensaude-Vincent and Hessenbruch have argued in a 2004 article in *Nature Materials,* namely, that the future development of materials science in the early 2000s might be read retrospectively as "the prehistory of a complex system of

---

[45] See, e.g., ESRF Highlights (2003, pp. 4, 30)

[46] ESRF (2007).

[47] Ibid., i.

[48] See, e.g., ESRF Highlights (2006, p. 16; 2007, p. 24).

converging technologies"[49] including nanoscience, biotechnology, or information technology.

Beyond Materials Science: ca. 2009–2020

For the history of materials science, few personal decisions have been as impactful as the nomination of Francesco Sette as ESRF's Director General in 2009. At face value, Sette's nomination can be characterized as the culmination point of a remarkable career at the ESRF. Sette began his work at the ESRF in 1991 as the Head of the inelastic scattering group. As mentioned above, he became the Research Director for the physical sciences in 2001.[50]

For materials science, it marked the end of its short period of high organizational visibility. Already a year after Sette's promotion, the ESFR's new Research Director Harald Reichert oversaw the formation of new beamline groups in 2010. Especially interesting are the names of these groups. While beamline groups in the early 2000s made specific references to materials science terminology, these new groups were remarkably broad and abstract, such as *Dynamics and Extreme Conditions* or *Structures of Materials* (see Fig. 1). In contrast to earlier groups, these names neither point to disciplinary demarcations nor specific techniques or methods. This observation links to broader contemporary tendencies in the general organization of science and scientific inquiry. Specifically, it shows how the increasing growth and complexity of science around the turn of the century have blurred traditional disciplinary boundaries resulting in an ever-growing number of specialized and differentiated journals and publication formats, institutions, instruments, and techniques.[51] That change is reflected in ESRF annual reports around the 2010s. The reports stressed the importance of materials science for other disciplines by adding prefixes such as *meta* materials or *nano* materials.

---

[49] Bensaude-Vincent and Hessenbruch (2004, p. 347).

[50] ESRF (2020).

[51] Marcovich and Shinn (2011).

This chapter has repeatedly stressed the importance of personnel. What is interesting about Harald Reichert is that his professional background stands in considerable contrast to the observed differentiation of traditional disciplinary boundaries that happened under his watch and became translated into ESRF's organizational structure. Indeed, before coming to the ESRF, Harald Reichert worked for 11 years at one of the leading, and perhaps most traditional, German materials science research institutes, namely the Max–Planck-Institute for Metal Research in Stuttgart. The institute was renamed in 2011 as Max–Planck-Institute for Intelligent Systems, to reflect a shift to a truly cross-disciplinary approach.[52] The ESRF described Reichert as "one of the leading experts for the exploitation of high energy X-rays for materials science including the development of dedicated instrumentation."[53] During his time as Research Director, the ESRF completed the construction of two new beamlines in 2016 dedicated to materials science, namely, the *Interfaces and Materials Processing Beamline* and the *Materials Chemistry and Engineering Beamline.*

Ultimately, the changes that took place during the 2020s caused materials science to disappear from the organizational structure of the ESRF as a specific terminology. Instead, it was replaced by extremely broad and abstract beamline groups. Thus, it is not without irony that in 2020, the ESRF announced a job opening seeking a Director of Research for Condensed Matter and Physical and Material Sciences,[54] which had been called until this time the Research Director for Physical Sciences.

Conclusion

This chapter provided new insights into the different institutionalization processes of materials science throughout ESRF's history.

---

[52] Max-Planck-Gesellschaft (2011); see also the biography of Harald Reichert at the ESRF Website.

[53] See the biography of Harald Reichert at the ESRF website.

[54] Physics World Jobs (2020).

This chapter regarded materials science as an umbrella concept that provided a basis upon which scientists can build collaboration and increase political visibility as a coherent, large, and powerful scientific community for science policymakers and potential funding agencies.

The history of materials science at the ESRF and its institutionalization processes within the organizational structures of the ESRF, the individual career paths of leading scientists, and the textual bodies, such as ESRF's annual reports, is complex. It is marked by ambivalence and asynchronicity. Even though materials science has always been and was always imagined to be a decisive part of the ESRF and its work, its visibility in official documents has not always reflected that. That was especially obvious during the pre-foundation area. Back then, documents of the ESRF rarely contained references to materials science *as such*. Instead, materials science was first shaped through its instrumental techniques, such as XAFS, and later through its manifestation and recurrent use of a specific terminology. That ambivalence only changed periodically, during a short period of visibility between 2002 and 2008, when a beamline group dedicated to materials science existed. This group disappeared in 2009, and with them, the use of specific terminology referring to materials science. Instead, it was replaced by beamline groups with very abstract titles, such as *Dynamics and Extreme Conditions*. These developments relate to the preparatory work and the implementation phase of a major upgrade program (2009–2018). In sum, the upgrade program put emphasis on the emerging role of nanoscience and nanotechnologies for the future of ESRF's scientific activities. Materials science, as a specific terminology, certainly became superseded at this time by other terms and concepts such as *meta* materials, *smart* materials, or *nano* materials. But its process of institutionalization was and remains to be sustained through the professional background and expertise of Harald Reichert, the then Research Director for the physical sciences and ESRF's announcement in June 2020 to recruit a new Director of Research for Condensed Matter and Physical and Material Sciences.

This chapter provides a brief, preliminary analysis that needs, however, to be amended by further systematic investigations to obtain a nuanced understanding of the role of materials science throughout the history of the ESRF. Further research on this topic could include systematic, bibliometric analysis of publication activities and the affiliation of leading authors with regard to materials science. Another possibility includes the distant reading of specific textual corpuses such as annual reports, newsletters, and beam time applications with the help of emerging text mining methods.

Acknowledgments

This chapter was finalized with support from the German Academic Exchange Service (DAAD) PRIME Fellowship.

References

Archives Nationales de France

19920550/3 Ministère ou Secrétariat d'État Chargé de la Recherche.

19980422/3 Ministère ou Secrétariat d'État Chargé de la Recherche.

Internal Archive of the ESRF

ESRP (October 1984). *European Synchrotron Radiation Project: European Synchrotron Radiation Facility. Report* of the ESRP (presented by B. Buras and S. Tazzari).

ESRF Foundation Phase Report (February 1987). (Grenoble: European Synchrotron Radiation Facility).

Literature

Banchoff, T. (2002). Institutions, Inertia and European Union Research Policy. *JCMS: Journal of Common Market Studies* 40(1), 1–21.

Bensaude-Vincent, B. (2011a). Building multidisciplinary research fields: The cases of materials science, nanotechnology and synthetic biology. in: M. Merz. and P. Sormani (eds.), *The Local Configuration of New Research Fields* (Springer: Cham), pp. 45–60.

Bensaude-Vincent, B. (2011b). The concept of materials in historical perspective. *N.T.M.* 19, 107–123.

Bensaude-Vincent, B. and Hessenbruch, A. (2004). Materials science: A field to explode? *Nature Materials* 3, 345–347.

Bertrand, E. and Bensaude-Vincent, B. (2011). Materials research in France: A short-lived national initiative (1982–1994). *Minerva* 49, 191–214.

Cahn, R. W. (2001). *The Coming of Materials Science* (Pergamon: London).

Cramer, K. (2020). *A Political History of Big Science: The Other Europe* (Cham: Palgrave Macmillan).

Crease, R. and Westfall, C. (2016). The New Big Science. *Physics Today* 69(5), 30–36.

ESRF (1997–2021). *ESRF Highlights 1996–2020* (ESRF: Grenoble).

European Commission (2015). *Horizon 2020: Work Programme 2014–2015* (European Commission: Brussels).

Freund, A. (2007) Personal view: Hot photons and cool optics. *Synchrotron Radiation News* 20(1), 31–34.

Gribbe, J. and Hallonsten, O. (2017). The emergence and growth of materials science in Swedish universities. *Historical Studies in the Natural Sciences* 47(4), 459–493.

Haensel, R. (1980). Comparison of laboratory and synchrotron radiation EXAFS facilities. *AIP Conference Proceedings* 64(1), 73–83.

Hallonsten, O. (2015). The Parasites. *Historical Studies in the Natural Sciences* 45(2), 217–272.

Hallonsten, O. (2016). *Big Science Transformed: Science, Politics and Organization in Europe and the United States* (Cham: Palgrave Macmillan).

Hallonsten, O. and Heinze, T. (2012). Institutional persistence through gradual organizational adaptation: Analysis of national laboratories in the USA and Germany. *Science and Public Policy* 39(4), 450–463.

Hallonsten, O. and Heinze, T. (2015a). Formation and expansion of a new organizational field in experimental science. *Science and Public Policy* 42(6), 841–854.

Hallonsten, O. and Heinze, T. (2015b). From particle physics to photon science: Multidimensional and multi-level renewal at DESY and SLAC. *Science and Public Policy* 40(5), 591–603.

Heinze T. and Münch R. (2012). Intellektuelle Erneuerung der Forschung durch institutionellen Wandel. in: T. Heinze and G. Krücken (eds.), *Institutionelle Erneuerungsfähigkeit der Forschung.* (Wiesbaden: VS Verlag für Sozialwissenschaften), pp. 15–38.

I: 1962–1977. *Historical Studies in the Natural Sciences* 45(3): 447–492.

Jacrot, B. (2006). *Des Neutrons pour la Science: Histoire de l'Institut Laue-Langevin, une Coopération Internationale Particulièrement Réussie* (Les Ulis: EDP Sciences).

Kehrt, C. (2016). *Mit Molekülen spielen: Wissenschaftskulturen der Nanotechnologie zwischen Politik und Medien* (Bielefeld: Transcript Verlag).

Marcovich, A. and Shinn, T. (2011). Where is disciplinarity going? Meeting on the borderland. *Social Science Information* 50(3/4), 582–606.

Mottana, A. and Marcelli, A. (2015). The historical development of X-ray absorption fine spectroscopy and of its applications to materials science. in: R. Pisano (ed.), *A Bridge Between Conceptual Frameworks* (Springer: Dordrecht), pp. 275–301.

Tamtik, M. (2016). Institutional change through policy learning: The case of the European Commission and research policy. *Review of Policy Research* 33(1), 5–21.

Thulke, W., Haensel, R. and Rabe, P. (1983). Versatile curved crystal spectrometer for laboratory extended X-ray absorption fine structure measurements. *Review of Scientific Instruments* 54(3), 277–283.

Ulnicane, I. (2015). Broadening aims and building support in science, technology and innovation policy: The case of the European research area. *Journal of Contemporary European Research* 11(1), 31–49.

Webster, P., Hughes, D., Malard, B, Sitepu, B., Ratel, N. and Dale, A. (2003). FaME38: A new approach to materials science at the ILL-ESRF. *Journal of Neutron Research* 11(4), 263–266.

Wissenschaftsrat (1993). *Empfehlungen zur Förderung materialwissenschaftlicher Forschung und Lehre an den Universitäten* (Köln: Wissenschaftsrat).

Wissenschaftsrat (1996). *Stellungnahme zur Außeruniversitären Materialwissenschaft* (Köln: Wissenschaftsrat).

Miscellaneous

Biography of Harald Reichert. https://www.esrf.eu/home/about/organisation/directors/bio-harald-reichert.html (last accessed 1 March 2023).

ESRF (2020). ESRF Council confirms Francesco Sette as director general of the ESRF. *Press Release,* 23 June. https://www.esrf.fr/home/news/general/content-news/general/esrf-council-confirms-francesco-sette-as-director-general-of-the-esrf.html (last accessed 1 March 2023).

Max-Planck-Gesellschaft (2011). Neuer Forschungsschwerpunkt “Intelligente Systeme.” Press Release, 15 February. https://www.mpg.de/1157498/intelligente_systeme (last accessed 1 March 2023).

Physics World Jobs (2020). Announcement of the ESRF on the recruitment of a new director of research for condensed matter and physical and material sciences. https://brightrecruits.com/physicsworld-jobs/job/director-of-research-for-condensed-matter-and-physical-and-material-sciences/ (last accessed 10 April 2021).

Part IV

University Centers, Professional Societies, Publications, and Conferences

The four articles in Part IV deal with different aspects of professional life and communication.

As Brit Shields describes in "Government Funding of University Materials Science and Engineering Research Centers," after 1960, a variety of federal agencies began to sponsor materials science research. One of the most successful in the US is the Defense Advanced Research Projects Agency (DARPA), which *The Economist* recently called the agency "that shaped the modern world." In the 1960s, DARPA's predecessor agency ARPA established materials science research centers at several universities, including Cornell, Northwestern, and the University of Pennsylvania, each of which itself has become an institution in the materials science research network. Shields describes how the US federal government's funding and organization of materials science and engineering research centers evolved over the second half of the twentieth century in response to changing national needs, economic contexts, and ideologies of how best to organize such centers. The article presents the founding of the Interdisciplinary Laboratories (IDL) program, as funded by ARPA, in the midst of the Cold War with a focus on meeting the demands of the space race and the development of nuclear weapons. Shields also explores later iterations of the materials science and engineering research centers, such as the National Science

Foundation's Materials Research Laboratories (MRL) in the 1970s with their focus on civilian and industrial applications, as well as the NSF's Materials Research and Engineering Centers (MRSEC) in the 1990s, which fostered cooperation between academia and private industry.

Linn Hobbs's lengthy article, "Materials Science Professional Organizations," begins with accounts of early social organizations needed to create and promulgate specialized materials and materials-based technologies, such as guilds, mining schools, and academic programs and departments that covered glass, metals, polymers, ceramics, and cementitious and electronic materials. He then describes how industry-based materials societies sprang up to promote research in these materials. His article culminates with the turning point in professional societies that came about in the 1960s with the impact of the consolidation of materials science research on professional societies. Particularly important was the formation of the Materials Research Society (MRS) in the wake of which came a reconstruction of traditional societies as well as the formation of new societies and types of societies.

Daniel Ucko, the author of "Materials Science Journals and Journals that Publish Materials Science," is the author who actually formulated what this volume has adopted as the "Rule of Three." As he writes, "To paraphrase Shakespeare, some are born as materials science journals, some achieve it, and some have materials science thrust upon them." Ucko, for instance, discusses journals that grew from metallurgy to materials science, journals that spun off others, and some that started as materials journals from scratch. He also discusses the vast range of materials science journals from specialized journals dedicated to a subfield to journals with a broad materials science appeal, to journals that publish materials science articles that do not have the word "materials" in their name. Finally, Ucko points to likely changes in materials science journals due to the interest in design and discovery studies alongside experimental characterization and synthesis studies as well as to emerging liminal fields, such as nano- and biomaterials.

Meetings are yet another kind of communicative institution closely tied to both professional societies and journals. Guillermo Solòrzano's article, "History of the Brazilian MRS Meetings," provides a detailed picture of this in the founding of the Brazilian Materials Research Society (MRS). A meeting in 2002 at a hotel and convention center in Rio de Janeiro inaugurated the Brazilian MRS; the meeting included participants from US, European, and international materials research societies, and emphasized international materials science collaborations. A second meeting was held the following year, and the proceedings of the first two Brazilian MRS meetings were published in two materials research journals. The number of sessions and participants then steadily increased.

Chapter 4.1

Government Funding of University Materials Science and Engineering Research Centers

Brit Shields

Introduction

Over the course of the second half of the 20th century, the United States federal government sustained significant investment in materials science research and training among universities. The impetus, ideology, and structure for the various iterations of the funding programs reflected the evolving demands of national needs, availability of economic resources, and conceptual frameworks for how to increase research in materials, connect academic research to military and industrial needs, and train new scientists and engineers.

In response to the growing technological demands of the Cold War era, spanning from nuclear weapons to the space race, the 1960s saw heavy government investment in materials-related disciplines.[1] In urgent response to Sputnik, the US military invested in the construction of new laboratory buildings, increases in faculty, and enlarged graduate training programs to foster interdisciplinary research and training within materials-related disciplines.[2]

[1] Mody and Choi (2013), Bensaude-Vincent (2001) and Mody and Martin (2020).

[2] Choi and Shields (2015), Bensaude-Vincent (2016) and Shields (2020).

Largely sponsored by the Department of Defense's newly created Advanced Research Project's Agency, this network of Interdisciplinary Laboratories (IDL) on university campuses was complemented by several similar labs sponsored by the National Aeronautics and Space Administration (NASA) and the Atomic Energy Commission (AEC). Shifting political, economic, and cultural tides by 1970 called for a reexamination of heavy military funding for scientific research and training and the IDL program was moved under the purview of the National Science Foundation. For the following several decades, the NSF responded to fluctuating economic constraints, international competition, and growing demands of private industry. The IDL program was reconceptualized by NSF as the Materials Research Laboratories (MRL) program, oriented toward more civilian and industrial needs. The MRLs were meant to pivot away from the arms and space race and toward medicine, the environment, and industry. The NSF's Division of Materials Science continually adapted its strategy over time in response to new demands, introducing later iterations of organizing university materials research centers to become more nimble, economical, and applied. Thrust groups were introduced to narrowly focus several investigators on a specific interdisciplinary topic with direct applications to industry. Seed funding was meant to spur research in newly established lines of research, with the objective of private industry eventually subsidizing the work. The Reagan administration's 1980s agenda of pivoting from governmental support to private industry support of scientific research and development led the NSF to support research through a network of Engineering Research Centers, meant to closely tie researchers to private industry. By the mid-1990s, the NSF had converted all MRLs into Materials Research Science and Engineering Centers (MRSEC). The MRSEC program not only continued the block funding structure and interdisciplinary research but also propelled researchers to address a specific goal through cooperation with private industry and optimize knowledge transfer. The MRSECs were designed to be applications-focused and adaptable to the research and training demands of private industry. At the turn of the 21st century, the NSF then introduced its Nano Science and

Engineering Centers, which focused on research and education for materials on the nanoscale, which ran in tandem with the MRSEC program.

Interdisciplinary Laboratories

In the years following the Second World War, government officials and academic leaders recognized the importance of materials for national defense and economic growth. Various governmental and military organizations, such as the National Research Council, AEC, and the Office of Naval Research (ONR), established advisory boards or branches that made policy recommendations to support materials research.[3] Further spurred by the watershed moment of the Soviet launch of Sputnik in 1957, the federal government operated with a greater sense of urgency to establish a robust national program to support materials research and education. Eisenhower's Presidential Science Advisory Committee raised alarm that materials for aerospace had become a national security issue and expressed concern that the current university departmental system and lack of equipment were not prepared to address this demand.[4]

Shortly after its establishment in 1958, ARPA took the helm to address the materials bottleneck inhibiting faster innovation in aerospace and nuclear weapons. ARPA sought to address both the immediate concern of lack of materials as well as the longer-term goal of training a future generation of materials scientists and engineers. They sought to foster interdisciplinarity, theorizing that the academic strengths in separate materials-related disciplines such as physics, metallurgical engineering and chemistry could be better harnessed toward applications if those researchers had exposure to the other disciplines. Rather than support individual researchers or even departmental agendas, ARPA wanted to spur interdisciplinary research and collaboration by physically aligning the materials-related researchers together. The proposed solution came in the

[3] Martin (2015) and Fine and Marcus (1994).

[4] Eisler (2013).

form of the Interdisciplinary Laboratory (IDL) program which sponsored the construction of new materials laboratories in university campuses that featured a set of well-equipped and staffed central facilities as well as individual private offices from members of the various materials-related departments. The researchers were to be arranged in the new buildings by research interests and equipment needs rather than departmental affiliation.

ARPA launched the first three IDL contracts in 1960 with Cornell University, the University of Pennsylvania, and Northwestern University. The contracts included a 10-year commitment to pay for building use fees and were largely managed by the local faculty at the home institutions.[5] Shortly after, the program expanded to the University of Chicago, Harvard University, the Massachusetts Institute of Technology, Purdue University, Stanford University, Brown University, the University of Maryland, the University of North Carolina, and the University of Illinois Urbana-Champagne.[6] Beyond the ARPA IDL program, complementary materials science laboratories were also sponsored by the AEC and NASA in the early 1960s.

In the mid-1970s, the National Academy of Sciences organized the Committee on the Survey of Materials Science and Engineering (COSMAT) which produced a study titled "Materials and Man's Needs." The COSMAT report evaluated the ARPA, AEC, and NASA interdisciplinary laboratories and enumerated their institutional and organizational strengths and weaknesses.[7] They were praised for their effective blending of materials science and engineering as an interdisciplinary field, demonstrating block funding as a means of institutional support, fostering excellent research groups, and training new students.[8] However, the report also indicated that there were not a great number of joint publications emerging out of the IDLs.

[5] Stickley (1996).

[6] National Research Council (2007).

[7] Committee on the Survey of Materials Science and Engineering (COSMAT) (1975).

[8] Schwartz (1987).

Rather, the interdisciplinary nature of the IDLs was less in joint publications and more in creating a larger community of researchers working under the umbrella of a block grant with access to central facilities, shared resources, and seminar series.[9] By 1972, ARPA had invested $158 million into the IDL program and saw an increase of 75% in PhDs in materials-related disciplines.[10] At its peak in 1969, the program operated on an $18.7 million budget with 600 faculty, 2,385 graduate students, and 360 PhDs awarded.[11] Ultimately, the legacy of the IDL program, as described by Mody and Choi, was "the creation of a community in the field of materials science with a common institutional experience" as well as a strong tradition of establishing interdisciplinary, university centers with federal block funding.[12]

Materials Research Laboratories

After over a decade of generous ARPA funding and dramatic increases in the numbers of materials-related faculty and graduate students, the political, economic, and societal context in which the materials science IDLs were established shifted. With rising tensions over the Vietnam War, public scrutiny of military-funded scientific research and development, and the passage of the Mansfield Amendment, the IDL program's sponsorship by ARPA was reevaluated. As federal funding of defense and space agendas plateaued, public sentiment shifted toward tackling more civilian ends, such as environmental concerns and the needs of private industry. Ultimately, the decision was made to transfer the IDL program to the National Science Foundation.

Following several years of study, the NSF accepted the task to continue supporting materials laboratories and centers but introduced its own framework. The NSF's approach to managing the

[9] Ibid.

[10] Stickley (1996).

[11] Ibid.

[12] Mody and Choi (2013).

program was largely shaped by the economic context of the 1970s as well as its own ideology to better address the needs of civilian and industrial sectors. As Mody and Choi describe, the NSF sought to support research focusing on more civilian applications, such as environmental remediation, transportation systems, and biomedicine.[13] Mody and Choi further elaborate that by the 1970s, the NSF wanted to spur more direct interdisciplinary research between individual researchers and have them address this agenda of specific problems addressing societal and civilian needs.[14] Bensaude-Vincent emphasizes that during the mid-1970s, the NSF was reacting to broader economic concerns including international competitiveness within private industry, such as the microelectronic market.[15] This confluence of factors pushed the NSF to seek more connection between sponsored science and private industry.[16]

Based on the NSF's aims to foster more direct interdisciplinary collaborations between individual researchers, address broader civilian needs, and promote closer ties between scientific research and private industry, the NSF made significant changes to the IDL program. Schwartz describes the transfer of the IDLs from ARPA to the NSF as a complex, two-year process.[17] This transition involved a newly created NSF Division of Materials Research which transitioned the IDLs into Material Research Laboratories. The MRLs received less investment than the IDLs. While they still received block grants, rather than individual investigator awards, they no longer included funds for the construction of new buildings. Additionally, the MRLs called for more specialized, focused research agendas. Rather than just sponsoring a very large umbrella of funding for an entire IDL, the new MRLs pushed for more direct collaboration between individual researchers within the framework of larger block funding support.[18] According to Schwartz, employing block funding and

[13] Mody and Choi (2013).

[14] Ibid.

[15] Ibid.

[16] Bensaude-Vincent (2001).

[17] Schwartz (1987).

[18] National Research Council (2007).

shared central facilities and equipment which were locally managed by faculty deviated from the NSF's traditional single PI and discipline-oriented funding structures.[19]

As Bensaude-Vincent argues, the MRL program did not fundamentally change the policy of core funding. The most significant changes rather were in the NSF's enforcing more direct interdisciplinary research on the individual researcher level.[20] To promote more direct collaboration and address specific research needs, the NSF's MRL program introduced "thrust areas," which "focused on some broad problem requiring a multi-disciplinary team of researchers."[21] The thrust groups however did not replace the MRLs, which continued to maintain the central research facilities meant to foster collaboration.[22] Schwartz recognized that the workings of the various thrust groups differed between universities, but mainly started at the impetus of a faculty member or the MRL director with an idea for a specific research program and then was followed by the formation of a small team of interdisciplinary faculty to "attack complex problems."[23] The NSF's Division of Materials Research also introduced "seed groups" which could be sponsored with three-year contracts.[24] The seed funding was meant to quickly respond to faculty seeking support for new projects.[25]

At the time of the move from ARPA to the NSF, the IDL program had 600 faculty at 12 universities.[26] The IDL budget for FY 1971 was $17.2 million, but under the MRL program, the budget for FY 1974 was $12.1 million.[27] All 12 IDLs moved from the ARPA to the NSF in 1972 and were later renamed and established as Materials Research Laboratories (MRL). Several more MRLs were added in

---

[19] Schwartz (1987).

[20] Bensaude-Vincent (2001).

[21] National Research Council (2007).

[22] National Research Council (1984).

[23] Schwartz (1987).

[24] National Research Council (2007).

[25] National Research Council (1984).

[26] Schwartz (1987).

[27] Ibid.

the following years at other universities, including the University of Massachusetts, Carnegie Mellon University, Pennsylvania State University, and Case Western Reserve University.[28] By 1974, the NSF's MRL annual budget increased to $14.6 million.[29] By the mid-1980s, the NSF had added a total of five new MRLs, phased out seven others, and operated on a budget of $27 million in FY 1985. The budget was largely allocated to thrust group research (60%), followed by investments in facilities (30%) and seed projects (10%). At each MRL, the average number of faculty was about 30, with six post-doctoral researchers, 25 graduate students, and six technicians.[30]

As per Mody and Martin, the "center model" which emerged in the Cold War era came to be the primary organizational model for interdisciplinary research and was formative for later iterations of interdisciplinary research programs. By the mid-1970s, Mody and Martin add, the track record of the MRLs to weather political and economic changes "cemented the center model as *the* go to science policy instrument."[31] The MRLs served as the model template for other NSF programs, including the Science and Technology Centers and the Engineering Research Centers established in the 1980s, although with modified operating structures and differing program missions.[32]

A large, multi-year study commissioned by the NSF and conducted by MITRE between 1976 and 1978 evaluated the MRLs. A review of the report indicated that the 16 MRLs each had core grants from the NSF which ranged from $200,000 to $1,900,000 for the 1977 FY and had received cumulative funds ranging from $740,000 to $35,800,000 each.[33] The study compared the 16 MRLs with research projects conducted by faculty on individual grants or those at university research centers sponsored by the Department of Energy or NASA. The study indicated that the MRLs and non-MRL

[28] National Research Council (2007).
[29] Ling and Hand (1980).
[30] National Academy of Sciences (1987).
[31] Mody and Martin (2020).
[32] National Research Council (2007).
[33] Ling and Hand (1980).

materials research centers had similar timescales of support and specialized equipment.[34] Ling and Hand reported that the study did indicate however that the MRLs were unique as "sole contributors to certain research areas and in their emphasis on experimental work as well as in their productivity in "terms of major achievements."[35] They continue that the report also indicated that the MRLs were more successful in "attracting researchers with reputations measured as high in terms of scientific honors and esteem among peers."[36] Additionally, it was found that the core grants issued to MRLs were administratively less costly.[37]

Engineering Research Centers and Materials Research Groups

The NSF made another shift in its funding structures for university materials research centers in the wake of the political and economic changes of the 1980s. As the Raegan administration called for closer ties between university researchers and those in private industry, they maintained the "view to weaning universities off federal support."[38] There was a push toward channeling investment in scientific research and development into technological progress and economic development.[39] To alleviate concerns held by the industry about intellectual property ambiguity with federally funded research and to further foster academic–industrial cooperation, the Bayh–Dole University and Small Business Patent Act of 1980 was passed, creating a more consistent intellectual property policy for federally funded research and development.[40]

As the NSF responded to new market needs and the economic context of the 1980s, it also recognized growing international

[34] Ibid.
[35] Ibid.
[36] Ibid.
[37] Ibid.
[38] Mody and Martin (2020).
[39] National Research Council (2007).
[40] Bensaude-Vincent (2010).

competition. Bensaude-Vincent describes Sony's success as contributing the "same humiliating effect as Sputnik" and pushed the US to compete with Japan.[41] At the same time, university facilities and instrumentation in materials were aging and needed heavy government support for modernization.[42] A contemporary study conducted by the NRC remarked on the impact of inflation on instrumentation costs, estimating a need of nearly $1 billion and that only 16% of academic equipment in the physical sciences were considered state of the art, expressing this "crisis in university instrumentation is a matter of national urgency."[43]

By the mid-1980s, the NSF responded to these demands by introducing the Engineering Research Centers (ERC). George A. Keyworth II, the Director of Office of Science and Technology Policy under the Raegan Administration, looked to the ERCs as an important solution.[44] The ERCs, which were generally modeled on the MRLs, were meant to produce academic research applicable to specific industries, such as microelectronics, telecommunications, and car manufacturing.[45,46] The ERCs were composed of interdisciplinary teams who partnered with private industry and focused on a specific problem with industrial applications.[47] By 1989, 13 ERCs were established, 10 of which were related to materials science.[48] The ERCs were established with a 10-year award, renewed at the five-year mark, and then were meant to be supported by private industry.[49]

Also, in the 1980s, the NSF established another program meant to operate on an even smaller and more specialized scale. The Materials Research Group (MRG) program was founded in 1985 to address "rapidly developing opportunities and developments

[41] Bensaude-Vincent (2001).
[42] National Research Council (1984).
[43] Ibid.
[44] National Research Council (2007).
[45] Mody and Martin (2020).
[46] National Research Council (1989).
[47] National Research Council (2007).
[48] National Research Council (1989).
[49] National Research Council (2007).

within thrust areas."[50] While universities with MRLs also could host MRGs, the MRGs were largely intended to reach universities without an MRL.[51] The MRGs were targeted to have about 5–10 researchers and operate with a grant ranging from $500,00 to $750,000/year.[52]

Materials Research Science and Engineering Centers

The NSF continued to reorganize its support of materials research in the mid-1990s in response to new budget demands and to renew the infrastructure for materials research. Seeking to maintain shared experimental facilities, sponsored research, and training of future researchers, while also fostering collaboration between universities and private industry, the NSF introduced the Materials Research Science and Engineering Centers (MRSEC). In the program's solicitation, the NSF described the MRSEC program as fostering "active cooperation with industry to stimulate and facilitate knowledge transfer among the participants and strengthen the links between university-based research and its applications."[53] The MRSECs were required to be interdisciplinary with a specific applications focus, close ties to industry, and provide training. Like its predecessors, the MRSECs were block-funded but differed in that they were also designed to be even more agile, with quick responses to new industrial demands. They were seen as being able to take on higher-risk projects than an individual investigator grant could handle. MRSECs were also expected to engage in educational and outreach activities as well as to foster diversity within the materials science research community.[54]

By the end of 1996, the NSF had converted all of the MRLs into MRSECs. A total of 24 MRSECs were in operation with a $44.28 million annual budget in FY 1996. An NSF report indicated that while there was a 124% increase in the number of centers, there was only a 54% increase in the budget.[55] The report further indicated that

[50] Ibid.

[51] Ibid.

[52] NRC, Materials Science and Engineering for the 1990s, 1989, p. 170.

[53] National Research Council (2007).

[54] Ibid.

[55] Ibid.

the MRSECs were intended to be smaller, with trimmed staff in the shared facilities and decreases in equipment investment. The MRSEC budget slowly increased and then plateaued in 2003–2006 at $53.4 million/year in FY 2006.[56]

At the turn of the century, the NSF also introduced the Nanoscale Science and Engineering Centers, a new block-funded program.[57] With the introduction of the NSEC program, the Materials Research Groups were eliminated.[58]

In FY 2003, the NSF's Division of Materials Research was operating on an annual budget of about $250 million/year, with over half of that funding going toward individual investigators, about a quarter going to centers, such as the MRSECs and Nano Science and Engineering Centers, and the rest split between large facilities and focused research groups.[59]

References

Bensaude-Vincent, B. (2001). The construction of a discipline: Materials science in the United States. *Historical Studies in the Physical and Biological Sciences* 31(2), 223–248.

Bensaude-Vincent, B. (2010). Materials as machines. In A. Nordmann and M. Carrier (eds.), *Science in the Context of Application* (Dordrecht: Springer), pp. 101–114.

Bensaude-Vincent, B. (2016) Building multidisciplinary research fields: The cases of materials science, nanotechnology and synthetic biology. In M. Merz and P. Sormani (eds.), *The Local Configuration of New Research Fields. Sociology of the Sciences Yearbook*, Vol. 29 (Cham: Springer).

Choi, H. and Shields, B. (2015). A place for materials science: Laboratory buildings and interdisciplinary research at the University of Pennsylvania. *Minerva* 53(1), 21–42.

Committee on the Survey of Materials Science and Engineering (COSMAT) (1975). *Materials and Man's Needs* (4 Volumes) (Washington, DC: National Academy of Sciences).

---

[56] Ibid.

[57] Ibid.

[58] Ibid.

[59] National Research Council (2006).

Eisler, M. (2013). "The ennobling unity of science and technology": Materials sciences and engineering, the department of energy, and the nanotechnology enigma. *Minerva* 51(2), 225–251.

Fine, M. and Marcus, H. (1994). Materials science and engineering, an educational discipline. *Annual Review of Materials Science* 24, 1–17.

Ling, J. and Hand, M. (1980). Federal funding in materials science. *Science,* 209(4462):1203–1207.

Martin, J. (2015). What's in a name change? *Physics in Perspective* 17, 3–32.

Mody, C. and Choi, H. (2013). From materials science to nanotechnology: Interdisciplinary center programs at Cornell University, 1960–2000. *Historical Studies in the Natural Sciences* 43(2), 121–161.

Mody, C. and Martin, J. (June 2020). Materials Science. In *Encyclopedia of the History of Science,* accessed 18 September 2023. https://doi.org/10.34758/6afy-w006.

National Academy of Sciences (1987). *Advancing Materials Research* (Washington, DC: National Academies Press).

National Research Council (1984). *Major Facilities for Materials Research and Related Disciplines: Presentations to the Major Materials Facilities Committee* (Washington, DC: The National Academies Press).

National Research Council (1989). *Materials Science and Engineering for the 1990s: Maintaining Competitiveness in the Age of Materials* (Washington, DC: The National Academies Press).

National Research Council (2006). *Midsize Facilities: The Infrastructure for Materials Research* (Washington, DC: The National Academies Press).

National Research Council (2007). *The National Science Foundation's Materials Research Science and Engineering Centers Program: Looking Back, Moving Forward* (Washington, DC: The National Academies Press), pp. 52–53.

Schwartz, L. (1987). Materials research laboratories: Reviewing the first 25 years. In *National Academy of Sciences. Advancing Materials Research* (Washington, DC: National Academies Press).

Shields, B. (2020). Buildings. In J. Martin and C. Mody (eds.), *Between Making and Knowing: Tools in the History of Materials Research,* World Scientific, Singapore (pp. 121–128).

Stickley, C. M. (1996). Uses of DARPA materials sciences technology in DoD systems. Final Report, August 1994–July 1996. https://apps.dtic.mil/sti/pdfs/ADA311508.pdf. Accessed 18 Sep 2023.

Chapter 4.2

History and Sociology of Materials Science & Engineering Professional Societies and Associations*

Linn W. Hobbs

Introduction

Humanity, from its earliest beginnings, has been characterized by its intimate relationship with materials, derived from the natural world, and transformed into objects of social and cultural utility. That relationship has always involved human social interactions — from prehistoric family group specialists to medieval trade guilds to late 19th-century departmentalized academic foundations to

*This chapter is dedicated to the memory of Arne Hessenbruch. Dr. Arne Hessenbruch (1958–2023), the co-creator with Dr. Bernadette Bensaude-Vincent of these four volumes on the history of materials, died unexpectedly on 12 January 2023, just as this contribution was being written. It has been this author's privilege to have spent countless hours, for more than seven years, talking with Arne about materials archaeology and anthropology, development of academic materials fields, materials professional societies and associations, and the personalities that have dominated the 20th- and early-21st-century history of the great human materials endeavor. Arne held a lecturer appointment in MIT's department of Materials Science & Engineering, so he was an academic colleague as well as an intellectual sparring partner and a dear friend, one whose contribution to understanding the materials science field of endeavor will be sorely missed, here, and throughout the world. It is an honor to dedicate this contribution to his memory.

today's professional societies and associations of scientists and engineers, the subject of this contribution. How these human organizations have evolved today must be considered in light of predecessor academic foundations, still defining a wide range of materials fields of endeavor, that in turn trace their origins to traditional craft specialties, whose boundaries still persist in the sociology of these associations, though a move to more catholic integration of materials fields has been occurring over the last 60 years.

Materials in Human Experience

Humans have evolved as highly social species, whose common goal of survival has depended upon individuals associating with each other to share the burdens of surviving in highly competitive and ever-changing environments. The interests of individuals have been well served by collectively promoting the success of the social group, resulting in the formation of human societies and — from a larger perspective — the evolution of human cultures that reflect broad commonalities in solutions to the problems of providing for the common material needs associated with food, shelter, clothing, transportation, protection, and procreation. Domestication of animals (starting *ca.* 13,000 years ago) and invention of agriculture (beginning *ca.* 11,000 years ago) developed more stable food resources and promoted larger fixed settlements — and eventually cities ("civilization") — which increasingly replaced nomadic groups with their traditionally gender-biased task specializations, and permitted broader employment segmentation within urban societies and the rise of specialized trades. Larger markets provided the impetus for geographically wider trading between settlements, facilitated by improved transportation routes and technologies, record keeping, and (eventually) writing, and provided wider cultural and linguistic exposures through which raw materials sources and materials processing technologies became more widely shared.

The history of materials is inextricably entwined with our human evolution. What defines humankind as a species is *not* the ability to employ or construct *tools* (many species do this, well down the evolutionary tree) but the *genius* to *transform materials* available in the natural world into novel socially useful constructs that in turn transform human societies and shape human cultural patterns. In studying the relationship between humans and their material constructs, too often archaeological or anthropological focus has been on the objects themselves (commonly adjudged from art-historical perspectives) or on what they are made of (stone, clay, metals, fibers, pigments, etc., as assessed by modern analytical science) rather than on the human-invented *materials-transformation processes* responsible for the production of these material objects, or on the curiosity-driven empiricism and imaginative insight at work in what is fundamentally an *intellectual* process of re-creation. In fact, it could be argued that this intellectual challenge is what drove the biological and physiological evolution of the human brain that distinguishes the most recent of *Homo* species, *Homo sapiens*, from earlier hominins and certainly enabled the genius that has transformed our present-day material environment out of all recognition from its prehuman manifestation. So, too, did the sociology of the human associations that accompanied and abetted this creative endeavor.

Materials-specialty craft trades were established quite early in the history of human civilizations, to a significant level of sophistication. Utilization of natural polymers for clothing — at first presumably whole animal hides — is inferred[1] to have been necessary for human survival in the known multiple migrations of archaic hominins out of the African continent during several interglacial periods into parts of Europe and Central Asia with increasingly colder climates during subsequent glaciations, beginning approximately 1.2 million years ago (1.2 Ma before present) — well after the loss of hominin body hair, estimated as

[1] Gilligan (2010).

early as 3 million years ago, and well before documented human control of fire (*ca.* 800,000 years ago). Evidence for the use of stone tools to scrape hides dates from at least 780,000 years ago. Anatomically modern humans (AMH, *Homo sapiens*) evolved in Africa much later, some 300,000 years ago, and DNA studies of body lice evolution peculiar to our species, from head lice to clothing lice,[2] suggest that *H. sapiens* began to clothe themselves sometime between 83,000 and 170,000 years ago, coincident with their migration out of Africa and the onset of the last glacial period (beginning approximately 120,000 years ago). Discovery of eyed-needles, dating from 40,000 years ago, suggests that *H. sapiens* constructed more complex *pieced* clothing cut from larger hides and sewn together, presumably with animal sinew. The use of natural polymers as a raw material for textiles came later, with both collagenous animal fibers (wool, hair, later silk) and cellulosic plant fibers (grasses, sedges, milkweed, and other stalk basts, later cotton and flax) being matted (felted, like pounded-bark barkcloth) or twisted (spun) to form threads that were woven into cloth, or knotted, plied or braided into nets and ropes. Impressions of woven textiles (baskets, cloth) date from as early as 27,000 years ago, and of ropes from 19,000 years ago. It has been argued persuasively[3] that the initial impetus for agriculture may have been more about providing food for domesticated animals and supplying plant fibers (e.g., flax in Europe, cotton on the Andean littoral) for cloth.

Firing of aluminosilicate clays to produce hard ceramics (pots, later bricks) dates from at least 14,000 years ago and was common by the tenth millennium BCE. Further experimentation with rudimentary pyrotechnologies led to the discovery and refinement of lime- and gypsum-based cementitious mortars used for the construction of domestic housing before the ninth millennium BCE. Extraction and casting of elemental gold

[2] Toups *et al.* (2022).
[3] Gilligan (2007).

began at least by 5000 BCE. Silver (as well as lead) was produced from low-temperature roasting and reduction of silver-containing lead sulfide ores sometime in the fourth-millenium BCE. Smelting of copper and arsenical bronzes from Cu and Cu–As oxide and sulfide ores began at least by 3300 BCE, followed soon after by alloyed tin bronzes produced by dissolving copper in the molten smelted tin. Smelting of Cu–Fe sulfide ores to produce iron — starting in the second millennium BCE initially in Caucasia, China, India, the Balkans, and West Africa, required higher-temperature pyrotechnologies and was likely driven initially by the geographical scarcity of tin ores for tin-bronze production. Chemically sophisticated ceramics began with the production of high-fired translucent porcelains from kaolin clay, quartz, and feldspar admixtures, during the Shang dynasty in China beginning in the 16th-century BCE, which pioneered very high-temperature (>1500°C) furnace pyrotechnologies later important for casting iron (beginning in 8th-century BCE in China), and much later for Portland cement clinkering. A translucent, even transparent (and colorable) material, now known popularly as "glass," was discovered to emerge from the reaction, at modest temperatures (~700–800°C), between silica (SiO_2, common sand) and alkali (typically sodium and potassium) oxides from the ash of burned plant matter — replacing hogged-out rock crystal for perfume bottles by the late Bronze Age, and metal drinking vessels in the 1st-century BCE after glassworkers in the Levant discovered it could be blown into arbitrary shapes.

Observation (which is the first part of investigation) of materials and their transformations were likely in place in prehistoric eras, but not recorded. By the early classical period certainly, even by the time of early Mesopotamian, Persian, Egyptian, or Minoan cultures of the Old World, observations, investigation, and even deliberate experimental study of materials and their transformations were recorded — by later Greek and Roman times in complete texts, widely available in translation. In this entry, *science* will refer to directed investigation, characterization, or study of structure, attributes or properties of materials, though such a study may involve manipulation of those structures, attributes, or properties of materials through transformation; *engineering* will refer to making something with them. Materials *technology* can involve both science and engineering but generally involves investigation of the *processes* of manipulation with an end product in mind. The current generic phrase used to invoke the discipline is usually *Materials Science & Engineering* (MSE), by use of which the distinction between science and engineering is often unavoidably blurred.

Specialization and Continuity

It is certain that promulgation of these specialized materials-based technologies required from the beginning three facets of social association: (1) *education* (to acquaint the novice with raw materials sources and accumulation, provide an understanding and appreciation of the desired product and its properties, and teach specialist vernacular appropriate to the specialization); (2) *practice* (to emulate existing — and with time to improve — materials processing and fabrication technologies); and (3) *regulation* (to ensure and uphold standards of materials performance and trade credibility and ensure its continuity of production). In early (and smaller) societal groups, a specialty trade was most likely to be propagated through family structures, passed down from parent to child (until the 20th-century CE, and except for fiber technologies, more often from father to son), or as a tribal specialty. The vast Roman concrete industry was, for example, almost exclusively the domain of freedmen (freed slaves) and their progeny,[4] laying the groundwork for the (free) stone masons of the early middle ages in Europe and a proliferation of materials-related craft guilds of the later middle ages and Renaissance and into the 19th century in Europe and its colonial

[4]Lechtman and Hobbs (1987).

outposts. The same familial-to-professional group evolution is likely to have held true for the complex materials technologies evolving in China and for Arabic technologies promulgated throughout the rapid expansion of the Muslim world in the Middle East, Asia, and Africa. It is important to recognize that initial support (financial and structural) for early specialty craft groups came from royal or elite patronage, and not from wider public commerce, and both the regulatory oversight ("quality control") and innovational impetus emanated from the elite demand. Craft- and merchant-based guild structures, involving skilled craft workers starting in the 12th century, evolved in medieval England (100 guilds, mainly in London and Oxford); France (more than 350 craft guilds, known as *corps de métiers,* most centered in Paris, Lyon or Toulouse by the 15th century); Flanders; Italy (Milan, Florence); in Spain (*gremis* in Catalan Barcelona, *gremios* in Valencia and Toledo); and in major cities in individual German-speaking kingdoms before German unification (100 *Gilden* in Hamburg, 80 in Köln, 70 in Lübeck by the 15th century) — most responding to royal court or military pressures. A few north-German guilds unified as the *Hanse,* foreshadowing the unification of more than 200 northern European merchant cities into the more commercially driven *Hanseatic League* trading organization.

Guilds

The guild system as it evolved served all three aforementioned functions of social associations. Novice practitioners were *instructed* in an apprenticeship structure, a guild freeman taking on one or more *apprentices* (typically for 7 years, from age 14 to age 21) who were provided room and board (but, to take them on, an entry fee usually had to be paid in advance to their freeman master). After serving for 7 years and, often, being required to produce a "masterpiece" adjudged up to the standard of the guild, apprentices were granted *freedom* of their specialty guild or city and were free to become *journeymen* (*fahrenden Gesellen,* in German — a label familiar to Mahler devotees; *compagnons*, in French) who subsequently sought paid employment, with the same or a different master, often with the added charge of instructing their employer's apprentices, before in time taking on their own apprentices and establishing their own businesses. The system fostered and ensured a gradual (though not precipitous) continual *improvement* of technology and refinement of products. The guilds were by law granted considerable *regulatory* authority, to oversee and regulate their members' practices, remuneration, and product; to fine members for irregularities and confiscate inferior products; and to prosecute rogue specialist makers operating outside the guild. A guild could strictly limit the number of apprentices (often to as few as one or two at a time) to maintain high levels of employment and remuneration for their profession through an engineered scarcity of practitioners (not unlike the practice of modern physicians, through their professional associations, examination structures, and limitation of class size in medical schools). The guild governance structures were almost identical to those of modern trade or professional associations or societies, with a governing body comprising an elected (head) Master who had more or less automatically elevated through a succession of graduated underling positions with specified sub-responsibilities. Indeed, guilds (along with their less formalized famial, tribal, or court-sustained predecessors) can be thought of as professional mini-societies. In medieval Christian Spain, officials known as *veedores*, inspectors chosen from among guild heads, were put in charge of maintaining overall city craft standards and served as the interface between craftsmen and city councils — an Arabic regulatory concept borrowed from the Muslim bazaars of southern Iberia, under the titles *sahib al-suq* or *muhtasib*.[5]

In England (particularly in London, where they are collectively known as *livery companies*), the system of professional guilds became especially strong and well prescribed in law by Royal charter, persisting well into the 19th century. Ten of the historic

[5] Córdoba de la Llave (2016).

London livery companies are materials-anchored, associated with the products of natural polymer fibers, metallurgy and metallurgical engineering, cementitious mortars, and inorganic glasses.[6]

---

[6]Notable *materials*-anchored London livery companies addressing three principal materials classes: (1) **Natural polymer fibers:** *Weavers* (textiles, established by 1130, chartered 1155); *Hatters* (hatmaking, est. 1347); *Drapers* (thread & cloth production and merchandizing, founded 1361); *Haberdashers* (clothing, haberdashery, regulation of silk and velvet merchants, est. *ante* 1371, inc. 1394, chartered 1448); *Mercers* (cloth and fabric merchants and vendors, chartered 1394), *Clothmakers* (textiles from wool and flax (linen), later silk and cotton, chartered 1528 from union of *Fullers*, est. 1480 and *Shearers*, est. 1508); (2) **Metals:** *Blacksmiths* (iron smithing, est. by 1299, chartered 1571 (included many clockmakers until mid-18th century, since English mechanical clocks before ~1630, were comprised predominantly of iron components)); *Goldsmiths* (gold & silver production, fabrication and merchandizing, founded 1327); *Pewterers* (pewter alloying (~90 wt% Sn–5–10 wt% Sb–2 wt% Cu–tr. Bi, Ag) and pewterware fabrication, established by 1348, chartered by 1451); *Cutlers* (knife and sword making, iron heat treating, est. 13th c., chartered 1461); *Armourers & Braziers* (originally ferrous armor production, est. 1322, chartered 1453, joined by brass workers in 1708); *Ironmongers* (iron workers, originally *Ferroners*, incorporated 1463); *Plumbers* (watertight cisterns, lead smelting, lead sheet and pipe, est. by 1365, chartered 1611); and (3) **Ceramics and glass:** *Masons* (separate entity from Freemasons) (stone masonry, mortars and construction, est. 1367, chartered 1472); *Plaisterers* (lime- and gypsum-based plasters and plastering, chartered 1501); *Glaziers & Painters of Glass* (glass makers, stained-glass workers, est. 14th c., chartered 1638); but surprisingly no potters. Pottery was largely a cottage industry until the 18th century in England, centered mostly in Worcestershire, where there were lias clays, then later in Staffordshire where porcelain manufacture from kaolin clays had become a large industry by the 18th century. Membership in London guilds was surprisingly not exclusive to the trades they represented; craft workers could earn their freedom within their fathers' guilds (freedom by patrimony) even if they were being trained in a different trade, and once freed by a company, a craft worker was free to practice any trade within the city. As examples, before the formation of the Clockmakers Company in 1631, makers of iron tower clocks might have been members of the Joiners or Blacksmiths companies, while many makers of early brass clocks were members of the Milliners Company; many illustrious late-17th-century clockmakers continued in the Blacksmiths Company or were freed in other companies by patrimony. Membership could also be obtained on recommendation and interview and by paying an entry fee (freedom by redemption). **Dutch** potters and tile-makers (*porcelynbackers*) in Holland were members of the Delft guild of St. Luke (*Lucasgild*), founded in 1611, which also included tile painters and glass painters, and formed a separate association within that guild in 1689, controlling the production of Delftware. The trades represented by **French** guilds (*corps de métiers*) were

Most of these London-based guilds persisted in their original form only into mid- to late-19th century, as materials-related production (iron, steam engines, railroads and locomotives, pottery and porcelain, ship building, textiles) became increasingly industrialized and invention-driven, moved predominately to the midlands or the north of England or Scotland, and became dominated by market forces far beyond control by a guild membership. Most of these livery companies still exist today, but instead with charitable or educational agendas; and some have affiliated with university departments or, notably, modern professional societies or institutions (such as, for example, the Institute of Materials, the Society of Glass Technology, etc.), building trades organizations, or even trade unions.

Materials Education

In order to explore the professional and societal aspects of associations of materials practitioners, the subject of this entry, it is imperative to examine the *educational* programs for materials practitioners that arose in the various materials fields in the wake of the final demise of the guild system, which occurred in the 19th century. These academic educational programs themselves came laden with significant elements of the two further requirements for the promulgation of specialized materials-based (or, for that matter, any) technologies: *practice* and *regulation.* The academic institutions that evolved also promoted an early associational identity that preceded and presaged the formation of, and subscription to, professional associations later in the careers of practicing materials professionals. The academic experience, while it did not provide quite the

---

more narrowly specified and exclusive: e.g., separate metal-working guilds were formed for ferriers, knife-makers, locksmiths, chain-forgers, nail-makers, with the armorers subdivided into helmet-makers, harness-makers, harness polishers, and so forth. A single object representing a variety of skills in its production, may have needed to pass through multiple guilds, subject to each guild's regulations (and fees) — an early triumph of the French penchant for bureaucracy.

guild-like intensity of *practice*, provided a much larger peer *social* environment than the apprenticeship experience and thus nevertheless fostered a community identity much earlier, one that later was expressed in a desire for associational belonging. The human products of these educational programs comprised (and still do) the potential membership of subsequent professional societies.

Mining schools

Both metallurgical and ceramics industries have always relied heavily on mining and ore extraction for their raw materials. Technologically oriented schools of mines were established throughout the European colonial period, the first established in Potosí, Bolivia (founded in 1557 by Carlos I of Spain, who became the Holy Roman Emperor Charles V), on the site of the rich Cerro Rico silver mine that supplied Spain most of its silver through an efficient mercury-extraction process utilizing proximate cinnabar deposits. Many others appeared in the 18th century, in Hungary (during the Habsburg reign), Saxony, the Harz Mountains, Prussia, Prague, Norway, Paris, London, and Turin.[7]

---

[7] Besides the very early (1557) foundation at the Imperial Villa de Potosí, overseen by the Spanish viceroy of Peru, many other European schools of mines were established in the 18th and 19th centuries, including: in Hungary (1735, under the Austrian Habsburgs, as the Faculty of Mining and Geotechnical Engineering at University of Miskolc); at Chemnitz (now in Slovakia) (1762, where Georgius Agricola (born Georg Bauer (1449–1555) in Saxony) wrote the first tract on metal ore extraction, metallurgy and metallurgical processing, *De Re Metallica*, which was translated into Ming-dynasty Chinese between 1636 and 1640 by a German Jesuit missionary, and into English from its medieval Latin by Herbert Hoover, who had trained as a metallurgist); at Königsberg, Norway (1757); in Prague (1762); at Freiberg (*Kurfürstlich Sächsische Bergakademie zu Freiberg*, 1765); in Paris (*École des Mines*, established in 1783 by Louis XVI, now *Mines Paris Tech*); in Prussia-Brandenburg (*Märkische Bergschule Bochum*, 1816); in London (1851, as the Government School of Mines and Science Applied to the Art, founded with the support of Queen Victoria's husband, Prince Albert, becoming the Royal School of Mines 1862, incorporated in 1917 into the Imperial College of Science & Technology, and now part of its Department of Materials there); and by similar establishments in the Harz

In Australia, the South Australian School of Mines and Industries was established in Adelaide in 1889. In the United States, the Henry Krumb School of Mines was established at Columbia University in 1864, followed by Geology and Mining at MIT in 1865 (later Mining and Metallurgy), and (after the Morrill Act of 1862 authorizing the formation of state colleges) at many of those state institutions.[8]

By the middle of the 19th century, the growing importance of industrial production of metals, ceramics, and glass for building construction, rail transportation, shipbuilding, and other heavy industries (not least the automobile in the 20th century) shifted the materials-educational and technical innovation roles of craft-based apprenticeship systems to higher educational institutions and professional institutes and societies, and ceded the regulatory responsibilities from self-policing guild structures to governmental oversight, professional licensing boards, and what became national standards-setting organizations[9] — with much of the standardization and

mountains (at Rammelsberg in 1810, established by Count Hans von Bülow, and at TU Clausthal in Clausthal-Zellerfeld (1864, *Bergakademie*, established by George V of Hanover); in Aachen (1865, within the *Polytechnikum* founded by Friedrich Wilhelm of Prussia (later King Friedrich Wilhelm III, to whom Beethoven's 9th Symphony was dedicated), now part of RWTH Aachen), and in Italy at the *Politecnico di Torino* (1907).

[8] Following the Morrill Act of 1862, which established state universities in the US, mining schools were additionally founded at: Missouri School of Mines and Metallurgy (1870), Colorado School of Mines (1874), South Dakota School of Mines (1885), Michigan Mining School (1886, later Michigan College of Mining, now incorporated into Michigan Technological University); New Mexico School of Mines (1889), Montana State School of Mines (1900); Mackay School of Mines (1908) at the University of Nevada; Wisconsin Institute of Mining and Technology (1908).

[9] Prominent among standardization and regulatory bodies established are: in the US, the National Bureau of Standards (now National Institute for Standards & Testing, NIST), American Society for Testing and Materials (ASTM), and Underwriters Laboratory (UL); in Britain, the British Standards Institute (BS); in Germany, the *Deutsche Institut für Normüng* (DIN); in France the (private) *Association française de Normalisation* (AFNOR), in the EC, the *Comité Européen de Normalisation & Standardisation* (CEN), and the International Organization for Standards (ISO). See Yates and Murphy (2019).

regulatory impetus coming from the industries themselves and ultimately, of course, the consumer public, whose (very often inexpert) opinions are now readily amplified through social media internet platforms.

Metallurgical Academic Programs and Departments

Following on from the foundation of schools of mines, many *colleges* or *universities* established departments of mining, geology, and mining, or mining & metallurgy, with most eventually dropping the mining portion and relegating geology to a separate department. In the US, Columbia and MIT were the first with Mining & Geology, in 1864 and 1865, respectively, subsequently altered to Mining & Metallurgy, then just Metallurgy, with other private and state colleges and universities soon following suit.[10] In the UK, metallurgy was a founding discipline in the Government School of Mines and Science Applied to the Arts, established in London in 1851 with support from Prince Albert (renamed the Royal College of Mines (RCM)) in 1862 and becoming a component institution associated with the department of Metallurgy when the RCM was folded into the Imperial College of Science & Technology in 1917. Other UK educational institutions followed suit, particularly in the industrial midlands (Birmingham), in the northern counties of Yorkshire

[10] Following Columbia's lead, US departments teaching metallurgy were subsequently established at: University of California (at Berkeley), department of Mining & Metallurgy (1868); Missouri School of Mines & Metallurgy (1870); Ohio State University, department of Geology, Mining & Metallurgy (1873); MIT, Metallurgy added to department of Geology & Mining (1884); University of Michigan, department of Metallurgy (1904); Pennsylvania State University, department of Metallurgy (1907), Texas State School of Mining & Metallurgy (1913, El Paso, TX, incorporated in 1919 into the University of Texas at El Paso as the college of Mines and Metallurgy); Carnegie Institute of Technology (Pittsburg, PA), bureau of Metallurgy (1924); Rensselaer Polytechnic Institute, course in metallurgical engineering (1933); and increasingly at other institutions, particularly following World War II.

(Leeds and Sheffield) and Lancashire (Manchester), and in Scotland (Glasgow).[11] Sheffield Technical School, founded in 1884 with a

[11] Mining Engineering was a founding discipline (1874) for the West Yorkshire Technical School in west Yorkshire, which on becoming the University of Leeds in 1904 added Metallurgy to Mining. In the south Yorkshire city of Sheffield, famous for its cutlery and silver-plating, Sheffield Technical School was founded in 1884 with a professor of Metallurgy & Mechanical Engineering, the School morphing into University College of Sheffield in 1887 and ultimately Sheffield University. Glasgow College of Science & Arts (later Glasgow & West of Scotland Technical College, then Royal Technical College Glasgow, now University of Strathclyde) established a course in Metallurgy also in 1884. Mason Science College in the important west-midlands manufacturing city of Birmingham established a metallurgical option in chemistry in 1881, becoming a department of Metallurgy in 1887, the college later integrating into what became the University of Birmingham in 1902, with a department of Metallurgy, much later broadening into a department of Metallurgy & Materials in 1981. Metallurgy came somewhat later to England's academically more conservative ancient universities, not unexpectedly defined by a more esoteric scientific content. Metallurgy in Cambridge University began in the 1880's with experiments on metal alloys conducted in the laboratory of Sidney Sussex College, Cambridge by chemists C. T. Heycock and F. H. Neville, who published their first metallurgical paper (1889) on lowering the freezing point of tin by additions of other metals; they were subsequently invited to give the 1903 Bakerian Lecture at the Royal Society. Heycock and Neville's paper, on the constitution of the copper-tin series of alloys, recognized the existence of intermetallic compounds and served as the foundation for the modern science of constitutional phase-equilibria in materials. The Worshipful Company of Goldsmiths endowed a readership in Metallurgy for Heycock in the University Chemical Laboratory in 1908, who immediately introduced a lecture course on the "metallurgy of gold, silver, lead, platinum, copper, tin and other metals." In 1920 the Goldsmiths Company again contributed generously towards the building of the Goldsmith's Metallurgical Laboratory "for the study of metallurgy." The Department of Metallurgy (later becoming the Department of Metallurgy and Materials) established in 1938 a Part II undergraduate curriculum in Metallurgy, developed a close relationship with a second London guild, the Armourers & Brasiers Company, and contributed early important seminal studies of precipitation hardening (Anthony Kelly FRS, Robin B. Nicholson FRS, using electron microscopy), composite materials (Anthony Kelly), and superconductivity (the late Archibald M. Campbell and Jan E. Evetts). In Oxford, William Hume-Rothery FRS (1899–1968), George Kelley Reader in Inorganic Chemistry with an interest in the chemical thermodynamics of solid-solution alloying of metals, established a metallurgical chemistry option in his

professor of Metallurgy & Mechanical Engineering, eventually became Sheffield University, whose Faculty of Applied Science split in 1916 into two parts, a Faculty of Metallurgy and a Faculty of Engineering, highlighting the importance accorded metallurgy in Sheffield at that time.

Academic inclusion of other materials in the UK also occurred at the beginning of the 20th century. In Manchester, a Faculty of Technology was established in 1905 within the Victoria University of Manchester (founded 1824), whose initial engineering curricula included mining, but also textiles — which had always been an important industry in Manchester — were notably added in 1919. The Faculty of Technology also oversaw the Manchester Municipal School of Technology (later Manchester College of Technology, ultimately evolving into the University of Manchester Institute of Science & Technology [UMIST]), to which the Faculty of Technology added a department of Textiles and Paper in 1924. A need for better glass science and improved glass production technologies was noted at the outbreak of World War I in 1914. William E. S. Turner,[12] then a chemistry lecturer at University College Sheffield, founded a department of Glass Manufacture there in 1915 (later subsumed into a department of Ceramics, Glass, and Polymers) and

---

Inorganic Chemistry department at Oxford in the 1950s and in 1958 was appointed to the first Isaac Wolfson Chair in Metallurgy, heading up a new Oxford department of that name. The fledgling Department of Metallurgy forestalled mounting purist doubts about the academic legitimacy of its subject matter by, upon the retirement of Professor Hume-Rothery in 1966, appointing an eminent Cambridge physicist to the vacated Oxford metallurgy chair, P. B. Hirsch (1925), later Sir Peter Hirsch FRS), by then a world pioneer in electron microscopy of metals and Reader in Physics in Cambridge's Cavendish Laboratory (at that time headed by Professor W. Lawrence Bragg FRS who, together with his father William H. Bragg FRS, had pioneered X-ray diffraction methods and were awarded the 1915 Nobel Prize in Physics). Hirsch quickly established a world center for electron microcopy in Oxford's Metallurgy department and immediately broadened the department's scope by adding nonmetallic materials (semiconductors, ceramics, polymers, biomaterials), precipitating a name change to the Department of Materials, a world leader, with most other UK metallurgy departments eventually following suit with their own similar nomenclatures.

[12] Turner (1945).

simultaneously campaigned for the creation of a Society of Glass Technology (established 1916), also in Sheffield and still loosely associated with Sheffield University, and later with the UK Institute for Materials, Mining, and Manufacturing (IOM3) also. Sheffield's department of Ceramics, Glass, and Polymers merged with its Department of Metallurgy in 1978 to create the present department of Materials Science & Engineering.

Ceramics academic programs and departments

Though technologically sophisticated ceramics were being manufactured two-and-a-half *millennia* ago in China (presumably with some kind of learning system to pass down the requisite accumulated empirical know-how), their appearance in Europe spans only two-and-a-half *centuries.* Since the late 17th century, manufacture of ceramics was a historically important industry in England, mostly concentrated in six North Staffordshire pottery towns centered around the town of Stoke-on-Trent, made famous in the 18th- and early 19th centuries by such celebrated firms as Wedgwood (1759), Mintons (1793), Spode (1770), Royal Doulton (1815), and others, and in Worcester (Royal Worcester) near the Welsh border. Stafford University in Stoke-on-Trent was founded in 1915 as a polytechnic institution with an unsurprisingly strong program in ceramic science and manufacturing. The British Ceramic Society was founded in Stoke, from where it has published its *Transactions* and *Journal* since 1974. Other institutions offering historically strong ceramics-oriented university programs were those at Sheffield, Birmingham, Manchester, Imperial College London, Cambridge, and Oxford, now all integrated into Materials departments. Sheffield once had a department of Ceramics, Glass, and Polymers — a not altogether illogical combination — before it was subsumed into their Materials department.

A substantially larger intrusion of ceramic science and engineering into university curricula took place in the US. A department of Clay Working & Ceramics (later Ceramics Engineering) was established at Ohio State University (Columbus, OH) in 1894, chaired by

Prof. Edward Orton, Jr. who had perfected and commercially marketed the earlier German invention of pyrometric cones for measuring high furnace temperatures. That academic foundation, and its proximity to ceramics and glass industries in Ohio and neighboring Pennsylvania, encouraged the formation of the American Ceramic Society (ACerS) in 1899, headquartered in Columbus and championed by Orton and area ceramics companies. The Ceramics Engineering department and Ohio State's counterpart department of Metallurgy (dating from 1873) merged in 1988 to form the department of Materials Science & Engineering. In New York State, a School of Clay-Working and Ceramics was established in 1904 (by an act signed into law by (then) Governor Theodore Roosevelt) in response to an unlikely petition from the small private institution of Alfred University in Alfred, NY (founded as a Baptist teaching college in 1836), on whose campus the public State school was, and remains, housed. Charles Fergus Binns, a British ceramist, served as the School's first director, following his long career with the Royal Worcester Porcelain Works. The School later became known as the New York State College of Ceramics at Alfred University, officially part of the State University of New York (SUNY) system, and continues to turn out notable ceramists and important ceramics research. At the urgings of the American Ceramic Society and local industries, ceramic technology programs were established at Rutgers College (NJ, now Rutgers University) in 1901, University of Illinois (Urbana-Champaign) in 1905, and Iowa State University (Ames) in 1905. Many other notable ceramic engineering or technology programs or departments[13] were soon after established elsewhere, most of

[13] Significant ceramics or ceramic engineering programs in North America were additionally established at University of Washington (Seattle, 1918), University of Saskatchewan (1921), Georgia Institute of Technology (Atlanta, 1924), North Carolina State University (Raleigh, 1924), Pennsylvania State University (University Park, 1924), University of West Virginia (Morgantown, 1924; in Chemical Engineering), University of Toronto (ONT, 1925; within Metallurgy), Missouri School of Mines (Rolla, 1926; now Missouri Science & Technology University), MIT (Cambridge, MA, 1926; a graduate-only ceramics program in Mining & Metallurgy, later subsumed into Metallurgy & Materials Science, now Materials

which have subsequently joined with metallurgy/metallurgical engineering departments and been subsumed into departments of Materials Science & Engineering.

The US undergraduate ceramic engineering programs are particularly relevant in recounting the history of materials societies, because they had additionally spawned two once-robust professional ceramic engineering organizations, affiliated with the American Ceramic Society: *Keramos* (founded 1915),[14] an undergraduate *fraternal* organization, with chapters still existing at 12 US universities that retain some remnant of ceramic engineering in their undergraduate programs; and the *National Institute of Ceramic Engineers* (NICE), a largely r*egulatory* body, at one time responsible for writing and administering the Principles and Practice of Engineering

---

Science & Engineering), University of Texas (Austin, 1945), Clemson University (Clemson, SC, 1948), University of California–Berkeley (1948), University of Florida (Gainsville, 1964; Ceramics division in Materials Science & Engineering). Strong ceramics programs subsequently also formed at Colorado School of Mines (Golden, CO), Case Western Reserve University (Cleveland, OH), Catholic University (Washington, DC), University of Washington–Seattle, Cornell University (Ithaca, NY), Northwestern University (Evanston, IL), University of California–Santa Barbara, and many others, nearly all initially or later merged into departments of Materials Science & Engineering. There remain only *three* ABET-accredited US undergraduate (bachelor's degree) academic programs in *ceramic engineering*: NY State College of Ceramics at Alfred University, University of Missouri (Rolla) and Clemson University (Clemson, SC), while the ceramic engineering program at Pennsylvania State University (University Park) is still a separately accredited option within Materials Science & Engineering.

[14]Kriegel (1982). Keramos chapters were formed at University of Illinois (1914), Ohio State University (Columbus) (1925), NY State College of Ceramics (Alfred) (1932), North Carolina State University (Raleigh) (1932), Virginia Polytechnic Institute (Blacksburg) (1940), Iowa State University (Ames) (1941), Rutgers (New Brunswick, NJ) (1947), University of Missouri (Rolla) (1947), Pennsylvania State University (University Park) (1948), Georgia Institute of Technology (Atlanta) (1950), University of Washington (Seattle) (1952), University of Texas (Austin) (1955), Clemson University (Clemson, SC) (1958), University of Florida (Gainsville) (1976). Only four of these institutions retains an accredited bachelor's degree program in ceramics engineering: Alfred, Missouri (Rolla), Clemson, and Penn State.

professional engineering exam in Ceramic Engineering, and also working with the Accreditation Board for Engineering and Technology (ABET) in accrediting undergraduate degree programs in ceramic engineering in US colleges and universities. Because only four exclusively ceramic engineering programs still survive (at Alfred University, Clemson University, Missouri Science & Technology University, and Pennsylvania State University) — the rest having been subsumed into curricula of materials science & engineering departments — NICE instead collaborates with The Minerals, Metals, and Materials Society (TMS) in advising ABET on accreditation of undergraduate degree programs in Materials Science & Engineering.

With the 1942 publication of *Refractories* by MIT ceramist Frederick H. Norton (1896–1986) and his textbook *Elements of Ceramics* a decade later, the ceramics field began to be put on a more scientific footing. Norton subsequently became chair of a division of Ceramics in the department of Metallurgy at MIT. In 1960, the field of ceramic *science*, as distinct from manufacturing or ceramic engineering practice, was given a huge boost by the publication of MIT professor W. David Kingery's (1926–2000) seminal book *Introduction to Ceramics* (John Wiley & Sons, 1960), which established ceramic science as a broad field more generally encompassing the science and technology of inorganic nonmetallic solids — harnessing the solid-state chemistry and physics, thermodynamical chemistry, and defect science that had been earlier co-opted and applied several decades earlier by the metallurgical community to achieve a more fundamental understanding of the metals and alloys. Joined in the 1960s by glass specialists Alfred R. Cooper (1924–1996) and Donald R. Uhlmann (1938–) and with by then with a strong component in ceramics and glass, MIT's department changed its name to the Department of Metallurgy and Materials, and in 1975 to the Department of Materials Science and Engineering.

Glass academic programs and departments

Modern academic study and engineering of the glassy state, as we know it, did not begin until the 19th century, largely in post-unification Germany — especially in the Thuringian town of Jena — though of course the requisite technologies for the production of art glass had reached an apogetic high in and around Venice in the 13th–15th centuries. The vastly profitable Venetian glass industry was largely centered on the island of Murano in the Venetian lagoon, where glassworkers there adhered strictly to the proscriptions set by the super-secret Muranese glass guild, especially that against the revelation of the "art and mysteries" of glass, on pain of death.

The French were the first to break the Venetian monopoly on the manufacture of plate-glass mirrors in Europe. Louis XIV's stupendous Hall of Mirrors (*Galerie des Glaces,* built 1678–1684) was outfitted by a glass factory with the royal warrant, the *Manufacture royale de glaces de miroirs,* founded by Jean-Baptiste Colbert, with which the Venetian monopoly on the manufacture of plate glass in Europe was broken. Historically, flat glass for European windows was manufactured by flattening a spinning blown-glass balloon, then cutting out square panes, but the method yielded panes neither sufficiently large, nor perfect enough, to provide the king his mirrors. The alternative process involved casting a molten viscous glass charge onto large iron plates, then finely polishing each side of the cast slab, before coating one side with colloidal silver. The first director, Nicolas de Noyer, the appointed tax receiver (collector) for the town of Orléans, was granted a 20-year monopoly. The first blemish-free plate glass was produced in 1666 from what became known as the *Compagnie de Noyer.* In 1695, Noyer's business, and a rival company, both in financial distress, were forced to merge but, retained the monopoly for a further 100 years, until the French Revolution, after which they were obliged to become a commercial company without monopoly protection; the entity that emerged became known as the *Compagnie de Saint-Gobain.* More than a century later, in 1937, the Saint-Gobain company began manufacturing a spun-glass (fiberglass) product, adding a new division, Isover®-Saint Gobain, which began selling in North America (under the trade name CertainTeed). The Company also devised a dipping process (a surface alkali ion exchange) process for manufacturing shatter-proof automobile windshields and added a line of gypsum-plaster drywall products (Gyproc®). In 1990, Saint-Gobain purchased a long-established US ceramics company, Norton Co., in Worcester, MA (founded in 1885 by potter Frank Norton, a forebear of MIT's late professor of refractories F. H. Norton); the company has become a leading manufacturer and vendor of abrasive ceramic materials (silica, emery (alumina + magnetite), silicon carbide, boron carbide, boron nitride, zirconia) for grinding and polishing.

In early 17th-century England, Sir Edward Zouch started a glass works in Vauxhall, a section of London then removed from the City of London (now the Albert Embankment in the Vauxhall section of Lambeth). The enterprise was taken over by the then Lord High Admiral, Sir Robert Mansell, in 1615, who was able to procure a royal monopoly for glass manufacture in England, though plate glass was not produced until the 18th century. The monopoly was void during the Cromwellian period, but shortly after the 1660 restoration of the monarchy by Charles II, a self-styled amateur "chymist," George Villiers, 2nd Duke of Buckingham, took over the Vauxhall site (helped by a former employee glassworker, John Bellingham), and recruited Venetian glassmakers to commence production of blown flat glass. They were joined by George Ravenscroft, who had been commissioned by the Worshipful Company of Glass Sellers to research better materials for glassmaking to improve quality of glass then available. Ravenscroft substituted litharge (PbO) for some of the alkali source potash (K_2O) and invented *lead-crystal* glass (also known as "flint glass"), which proved much more durable than Venetian glass, but abandoned his patent in 1679 (which had proved not then profitable) and joined Bellingham at the Vauxhall glassworks until his death in 1683. Buckingham ended up locked in the Tower of London in 1678 for treason; and, deprived of his patents, the glassworks was inherited by John Dawson (a former apprentice) who entered into partnership with John Bowles, whose respective families operated glassworks until 1780. By the 18th century, the earlier French process of casting flat glass thinly on an iron surface, then polishing both sides, had been adapted for Ravenscroft's leaded glass product and provided the first challenge to the supremacy of Louis XIV's mirrors at Versailles — their plate-glass product known in England as Vauxhall mirrors.

While most glassmaking in Britain migrated to the northeast (Yorkshire), near to where coal mines were located, glassmaking across the Pennines in Lancashire had also begun in the small town of St. Helens, in the Merseyside east of Liverpool, by 1700. What grew to become an international glass-industry giant, Pilkington Brothers, started with a group of six industrious investors in 1826 in St. Helens as the St. Helen's Crown Glass Company; successive departures of early investors left only two, the Pilkington brothers William and Richard, who renamed the company accordingly. In the mid-19th century, Henry Bessemer (of steelmaking fame) unsuccessfully experimented with an alternative for flat-glass manufacture, drawing glass sheet through rollers, with a glass charge still sufficiently plastic while within its glass-transition region (~700–800°C for conventional soda-lime-silicate glass). Meanwhile, subsequent experiments in the US and the UK with floating molten glass on liquid tin ("float glass") proved initially unworkable. A process for *continuous* casting of molten float-glass, initially at 1000°C on top of a six-cm-thick bath of molten tin — under a nitrogen-hydrogen atmosphere to keep the tin bath from oxidizing and finally cooling to a rigid state at 600°C at the end of the bath — was finally successfully perfected by a much later Pilkington heir, Sir Alastair Pilkington, and Kenneth Bickerstaff at Pilkington Bros. glassworks between 1953 and 1957, and dubbed the "Pilkington process." The process was quickly adopted by every manufacturer of flat glass in the world — in the US, principally Pittsburgh Plate Glass (PPG) and Libbey-Owens-Ford (LOF) sheet glass, headquartered in Toledo, OH (formed by successive mergers of Libbey Glass, Owens-Illinois Glass, and Edward-Ford Plate Glass companies), becoming the principal US maker of automobile glass. In 1986, LOF sold its plants and name to Pilkington plc, still headquartered in Lancashire, who were in turn acquired by the Japanese glass-making conglomerate Nippon Sheet Glass Company (NSG) in 2006, making it, together with Asahi Glass in Japan, Saint-Gobain in France and the US-based Guardian Industries (since 2017 owned by Koch Industries), one of the four largest glass producers in the world.

The strongly family-centered industrialization of glassmaking in Britain and the nearly two centuries of corporate monopoly in France conspired to preclude the development of academic programs in glass technology in those two countries until the early 20th century. The first British academic foothold took place across the Pennines, again, in Sheffield when William E. S. Turner (See footnote 12), then a chemistry lecturer at University College Sheffield, founded a department of Glass Manufacture there in 1915. The lecture syllabi as they evolved gave equal coverage to the physics (optical properties), chemistry (glass melt compositions, durability), and physical chemistry (the glass transition, toughening, glass ceramics), each group of topics covered in 10-lecture units. Turner simultaneously campaigned for the creation of a Society of Glass Technology (established 1916), also in Sheffield and still loosely associated with Sheffield University, and later with the UK Institute for Materials, Mining, and Manufacturing (IOM3) also. As noted earlier, Sheffield University's department later became a Ceramics, Glass, and Polymers department and was further broadened into a department of Materials Science & Engineering in 1978. In the US, the NY State College of Ceramics at Alfred University began offering degrees in Glass Technology in 1936, and in the 1990's introduced a PhD program in Glass Science, the first recipient of which was Kathleen A. Richardson, later 2015 president of the American Ceramic Society, followed by John C. Mauro, inventor of "Gorilla® glass" for cell-phone screens at Corning Glass, for which he received the US Department of Energy (DOE)'s Merit Award and election to the US National Academy of Engineering.

In Germany, glass technology took a very different, and decidedly academic turn, largely in the hands of physicists and physical chemists in the late 19th century. The interest was in the optical properties of glasses, specifically their transparency and refractive indices and the consequent ability to form focusing lenses from curved glass surfaces, long an important application in telescopy and microscopy. Otto Schott (1851–1935), son of a Ruhr valley glassmaker, studied chemistry in Aachen, Würzburg, Leipzig, and the Friedrich-Schiller-Universität in Jena (the second largest city in

Thuringia), from which he received his doctorate in 1875, specializing in glass fabrication science. Ernst Abbe, then professor of physics in Jena, had formulated the theoretical resolution limit for lens-based microscopes (about half the wavelength of the light) and was at the time pursuing improvement of microscope glass-lens optics, in collaboration with instrument maker Carl Zeiss. Schott, celebrated for his systematic and meticulous characterization of physical, and especially optical, properties of silicate glasses as a function of composition, began a lifelong collaboration with both Abbe and Zeiss that produced, *inter alia*, borosilicate glasses (like Pyrex®) and the glass compositions used in the manufacture of achromatic microscope lenses, as well as the standard *Jenaer Glas*[15] used for domestic glassware, and the medical glass used in the now ubiquitous COVID-19 vaccine vials produced in the billions. Two enduring academic legacies remaining in Jena are the Otto Schott Institute and Abbe School of Photonics, both at *Friedrich-Schiller-Universität.*

There are, of course, many applications of even only silicate glasses beside windows, containers, and optics (and many different glasses beside silicate glasses — such as borate glasses, chalcogenide glasses, and even metallic glasses). Three late 20th- and early 21st- century applications of glass science and technology — electronic device packaging (including touchscreens), glass-fiber telecommunications and holographic wave guides, and photonic devices (including lasers and optical interconnects in nanoelectronics) — continue to owe much to Schott's meticulous work and to subsequent pioneering industrial research at Bell Labs and Corning, Inc. These commercial drivers have generated an explosion in glass-related innovation (370,000 glass-related patents in the 50 years between 1963 and 2013, a *quarter* of them in the last 4 years of that period alone), with contributions from the expected players (Corning, Inc., Saint-Gobain Glass, Schott AG) figuring prominently but dwarfed by those of Asian competitors, led by Asahi Glass

[15] Hovestadt (1900).

Co. Ltd and Nippon Electric Glass Co.[16] The industrial innovation has in turn spurred explosive growth of the academic glass sector, one of whose quantifiable output metrics — journal publications — now exceeds 10,000 *per year* worldwide (excluding metallic glasses, polymer glasses, and amorphous semiconductors), led by publications in *Journal of Non-Crystalline Solids, Journal of Applied Physics* (and *Letters*), *Physical Review* (and *Letters*), with *Journal of the American Ceramic Society* a distant fourth (See footnote 16). The academic glass league table is led by Tohoku University (Sendai, Japan), but aggregated Chinese institutional outputs now exceed those of the US, Japan, and Germany combined. While neither the nature of the glass transition,[17] nor detailed atom configurations[18,19] for even the simplest non-crystalline silica, vitreous silica, are known or completely understood, these remaining "mysteries" of the glass "art" will no doubt eventually succumb to an intellectual onslaught of such magnitude. The more prominent current or historically important academic programs in glass science and technology in Europe, the US, and Japan are listed separately.[20] Foundation of academic

[16] Mauro and Zanotto (2014).

[17] Ojovan and Tournier (2021).

[18] Marians and Hobbs (1988); Hobbs *et al.* (1998); Hobbs (2008).

[19] Gupta and Cooper (1990); Gupta and Heuer (2008); Gupta (2021).

[20] In addition to the Otto Schott Institute and Abbe School of Photonics, both at *Friedrich-Schiller-Universität* in Jena, dedicated German academic programs in glass science developed also at TU Dresden, *Leibniz-Institut für Festkörper–und-Werkstofforschung* in Dresden, TU Clausthal (Institute for Non-Metallic Materials), *Friedrich-Alexander-Universität* in Erlangen-Nürnberg (Institute for Glass & Ceramics), RWTH-Aachen University, and others. Notable British programs are found in University of Sheffield, Imperial College of Science, Technology & Medicine (London), University of Cambridge, and in France at the *Université Pierre-et-Marie-Curie* (Paris 6), *Université de Montpellier* and others. Italian glass programs exist in the *Universitá degli Studi di Padova,* Czech Republic programs in the *Univerzita Karlova* (Prague). Notable US glass programs include those at MIT, Catholic University (Washington, DC), Case Western Reserve University (Cleveland, OH), Lehigh University (Bethlehem, PA), Pennsylvania State University (University Park), NY State College of Ceramics (Alfred), University of Illinois (Champaign-Urbana), University of Michigan (Ann Arbor), Cornell University (Ithaca, NY), University of Arizona (Tucson, AZ), Ohio State University (Columbus), University of Florida (Gainsville),

glass science programs was notably encouraged by the publication in 1974 of *Glass Science* (John Wiley & Sons, 1974) by Robert H. Doremus (1928–2008), a ceramist and professor of glass science at Rensselaer Polytechnic Institute (Troy, NY), whose popular textbook did for glass science what David Kingery's *Introduction to Ceramics* had done for ceramic science 15 years before.

Many futuristic applications for glasses are envisaged in a promotional video produced by Corning, Inc. entitled *Day Made of Glass*,[21] produced at the beginning of the 21st century, the most pervasive of which is already a reality, the cleverly compositionally engineered (using machine learning) "Gorilla® glasses" of the multifunctional cell-phone viewing screen. Perhaps the most impactful of these future projections, and nearly as close to reality, is the projected use of glass nanoscale interconnects and optical devices in nanoscale electronic circuitry, for example substituting for copper interconnects in microchip technology, providing significant advantage in data processing speed because the speed of light is many orders of magnitude faster than electron transport in copper or semiconductors.

Another significant early 21st-century achievement of the glass scientific community was the development of a suitable glass wasteform for the storage of high-level nuclear wastes — for both defense legacy wastes and accumulating commercial wastes from electrical power-producing nuclear plants — that has been pioneered in France, at the US DOE Savannah River, SC facility, and elsewhere. Alternative proposals for the matrix material in which to accommodate long-lived radioactive isotopes involve using crystalline

---

and others, most now in departments of Materials Science & Engineering. A surprisingly innovative undergraduate research program in glass science is still run out of the Physics department of Coe College, a small liberal arts college founded as a religious training institution in Cedar Rapids, IA in 1851, two notable graduates of whom are S. Donald Stookey (1915–2014, PhD MIT 1940), the inventor of glass-ceramics (and Corningware®) and William Shirer (1904–1993), author of the *Rise and Fall of the Third Reich* (unrelated, but perhaps useful for understanding the ups and downs of Ceramic Engineering programs in the US).

[21] Corning, Inc. (2015).

oxide ceramics with specific complex crystal structures that provide the appropriate substitutional accommodation, particularly, of long-lived actinide isotopes, such as ^{239}Pu (half-life 24,000 years), with decay times ranging from kiloyears to truly geological time scales, and their accommodation in suitable ceramic wasteforms is the study domain of both ceramists and mineralogists-geologists, who are acquainted with the phases and phase stabilities of suitable crystalline systems under conditions of significant radiation damage and element transmutation.[22] Two difficulties attend the employment of *crystalline* wasteforms, though: first, the well-prescribed atomic site environments in crystalline materials limit the range of chemical elements that can be accommodated within stable crystalline phases — not least for when isotopes undergo a *chemical* atom change (transmutation) in nuclear decay reactions; and second, radiation-induced amorphization of most of these oxide compounds to a non-crystalline state introduces unacceptable dimensional changes in the wasteform, leading to cracking and compromised integrity in environments that may not always remain dry. The arguments for glass wasteforms are, first, that they are already non-crystalline (though radiation-induced dimensional changes do occur, and radiation- or thermally induced crystallization or decomposition can also occur, with attendant dimensional changes); and second, that the differing environments at every atom site in the non-crystalline structure increase the probability of accommodation of a very wide range of chemical elements of isotopic concern. Both highly sophisticated science and state-of-the-art glass melting engineering have been required for the glass wasteform approach to nuclear waste stabilization. Both French efforts and the US Savannah River facility have been pouring glass waste "logs" for two decades now, though the

[22] The seemingly eternal saga of the quest for optimal high-level nuclear wasteforms can be followed in the long-running series of proceedings books for the *Scientific Basis for Nuclear Waste Management* symposium, held almost yearly at the Materials Research Society meetings, and elsewhere, beginning in 1979 and now (2023) in its 44th iteration. The very first volume was *Scientific Basis for Nuclear Waste Management*, ed. Clyde J. M. Northrup, now reprinted in *Advances in Nuclear Science and Technology* (Springer, New York, 2013).

long-term isolation of these highly radioactive glass logs in a remote repository has remained an unresolved, partly scientific but largely political issue.

One critical pedagogical issue for glass science is maintaining a pipeline of interested students. In departments of ceramics, or their expansion into departments of materials science and engineering, taught glass science remains often something of an *afterthought* to instruction in ceramic science, and conceptually more difficult than for crystalline ceramics because the huge simplification afforded by crystallography cannot be applied to glasses, given that their structures lack an identifiable unit cell. On the face of it, that reality implies that describing glass structure is a formidable 10^{23} problem (the number of atoms in a cubic cm) instead of a 10^1, or at most a 10^2 problem (the number of atoms in a unit cell, which in a crystal can be infinitely repeated by crystallographic translation) for even the most complex crystal structures encountered in ceramic solids. It is, nevertheless, possible to apply the concepts of connectivity topology (See footnotes 18 & 19) to show that the local atom arrangements in a glass cannot (and do not) differ all that much, locally, from those in an isocompositional crystal, as might be expected given the closeness of the free energies of the two forms, though the differences are statistically distributed.

Efficacious heuristic approaches to education in glass science have been the subject of much recent discussion (e.g., in a session on glass education at the recent *National Day of Glass* symposium[23] held in April 2022 at the National Academy of Sciences in Washington, DC as part of the North American inauguration of the UN's 2022 *Year of Glass* celebration). Many of the educational issues discussed there and elsewhere are summarized in a short article published shortly after, authored by six prominent glass scientist educators from the US, the UK, and Finland,[24] though the article's

[23] *National Day of Glass* symposium, 5–7 April 2022, (Washington, DC: National Academy of Sciences).

[24] Ballatto *et al.* (2022).

appeal to artificial intelligence and machine learning for glass compositional design and properties prediction hardly represents a pedagogic approach. The truly encyclopedic online *Encylopedia of Glass Science, Technology, History & Culture* (Wiley, 2021)[25] is a rich trove of information for reference and graduate instruction in glass science, as is the nearly as encyclopedic *Springer Handbook of Glass* (2019)[26] together with the curricula of the several "summer schools" on glass science that are now offered; but none is quite pitched at a level appropriate for proselytizing impressionable undergraduates.

Cementitious materials academic programs

The historical development of ancient cementitious materials,[27] and of the modern cement industry,[28] are well-discussed elsewhere. Until 1790, the two most authoritative treatises on cementitious materials were by Roman engineer and architect Marcus(?) Vitruvius Pollio(?) (known as Vitruvius, *ca.* 90–20 BCE) who wrote in detail[29] about Roman hydraulic cement in his 10-volume compendium *De Architectura* (re-discovered in 1414 CE by Florentine humanist scholar Poggio Broccolini and, republished 1486 *ff*, after which it remained the principal source for mortar technologies until the 19th century); and the French engineer-chemist Louis-Joseph Vicat (1786–1861), whose inspired research into the origin of hydraulicity in lime-clay mortars[30] enabled him to successfully build a 180 m-long bridge spanning the Dordogne River near Souillac. A key development in cementitious materials was the invention of Portland cement in 1824 by Yorkshire bricklayer Joseph Aspdin, made by high-temperature (> 1200°C) co-calcination of clay and limestone,

[25] Richet *et al.* (2021).
[26] Musgraves *et al.* (2019).
[27] Hobbs and Siddall (2011); Artioli, *et al.* (2019).
[28] Francis (1978).
[29] Vitruvius (1486).
[30] Vicat (1818). See also: Boyet (2020).

replicating in high-temperature solid-state reactions, but in anhydrous form, the principal (hydrated)components developing in the room-temperature aqueous setting of Roman cement — the latter constituted by room-temperature mixing of volcanic tuffs with quicklime that was calcined from limestone at much lower temperatures < 800°C. At first used only as a stucco, or as a water-resistant cement in damp applications (Carlton Terrace, the present home of the Royal Society in London, is one of the first stucco employments, producing a surface finish that resembled the expensive limestone quarried from off-shore Portland island on the Devon coast, with which most of the early 19th-century fashionable London buildings were faced), Portland cement rose in prominence in the early 20th century to become the go-to building construction material, combining with steel skeletons to carve iconic canyons into New York City's hitherto untrammeled airspace (the Empire State Building was the first "skyscraper" so constructed, completed in 1931).

Compared to Roman mortars, ordinary Portland cement (OPC) is substantially stronger (by factors of 10–50) and quicker setting (days instead of months), and its subsequent chemical and mechanical engineering to form modern concretes came to dominantly occupy much of the curricula of Civil Engineering departments after 1930. As the next 90 years were to demonstrate, however, OPC's ubiquity came with co-pathologies and at a considerable environmental cost. With age, concretes tend to crack, not least from stresses imposed by internal corrosion of iron reinforcing bars (rebar), and subsequently degrade catastrophically. Calcination of limestone ($CaCO_3$) to reactive quicklime (CaO) in the co-calcination process emits large quantities of CO_2, while the high furnace temperatures required to generate the requisite solid-state anhydrous reactions require large amounts of fuel, initially supplied by burning coked coal, later methane gas, and more recently by electricity, much of it still generated in coal-fired or hydrocarbon-burning power plants. Grinding up the glassy co-calcined product ("clinker") to form OPC powder requires the expenditure of generous dollops of additional mechanical energy. It is estimated that more than one-third of the present electricity consumption of China goes into the production of Portland cement to modernize its infrastructure, most of its electric power still predominantly generated by burning coal.

Recent environmental awareness of Portland cement's downsides has spurred investigation of alternatives, including additions to OPC of pozzolanic materials (aluminosilicate materials, like Roman tuff — which originated eponymously near the Roman Neopolitan village of Puteoli, now Pozzuoli — or industrial flyash with similar chemistries and surface areas) that spontaneously react in aqueous media with quicklime; and even forays into replicating some forms of Roman concrete, many of which still survive two millennia later. Utilizing pozzolans as raw-material feeds requires less thermal and mechanical energy, since prior volcanic processes have done a lot of the thermochemistry and morphological engineering beforehand.

One landmark decade-long archaeological investigation of Roman *hydraulic* concretes used in underwater construction of harbors, docks, and breakwaters — dubbed ROMACONS[31] — has brought together a multidisciplinary team, including an analytical materials scientist, Marie D. Jackson, whose analysis has revealed many hitherto unappreciated characteristics of Roman concrete chemistries and microstructures, including a useful ability to heal developing cracks. These materials-based investigations have rekindled a research interest within materials science & engineering departments, as well as an enhanced appreciation for (and sometimes appending of) archaeological science. A renown longtime champion of cement studies, mineralogist Della Martin Roy (1926–2021), the first female materials scientist to be inducted into the US National Academy of Engineering, headed a decades-long cement and concrete research program at Pennsylvania State University and was co-founder of the *Journal of Cement and Concrete.* Roy carried out the first study, commissioned by Los Alamos National Laboratory, to explore the origins of the longevity of Roman concrete, whose formulation was then considered as a potential long-term medium for

[31] Brandon *et al.* (2014).

storage of medium-level nuclear waste.[32] Some of the durability of Roman mortar derives from the residual clasts of unreacted quicklime (CaO), first observed by Roy and her students, that more recent investigators[33] have established arose from a dehydrating hot-mixing of the ingredients (the hydration of quicklime with water is highly exothermic). The unreacted lime provides a healing mechanism when water enters into developing cracks in the mortar and reacts to reform the mortar product and fill (and stop) the crack.

Many universities (including Penn State, University of Illinois, MIT, and others) have established interdisciplinary concrete consortia or initiatives to develop better optimized — or other cementitious alternatives to — Portland cement. One example of progress emerging from such efforts is the recent discovery that *forced* carbonation during the earliest stages of Portland cement setting can sequester (like lime mortars) significant amounts of CO_2 without affecting the mechanical properties conferred in the later stages of setting.[34] A few universities have instituted academic efforts to explore the archaeology and underlying ethnology of early materials development, not least of early cementitious materials — first among them MIT, who established an eight-institution Boston-area consortium *Center for Materials Research in Archaeology and Ethnology* (CMRAE) in 1977, founded by the late Cyril Stanley Smith (1903–1992), plutonium metallurgist for the Manhattan project and godfather of materials archaeology studies.[35] Until her recent retirement, CMRAE was directed by Heather N. Lechtman (1935–), a physicist and noted materials archaeologist, specializing in Andean metallurgy and textiles, and a MacArthur Award recipient. She and an MIT colleague (the present author, a ceramist) co-founded in 1982 the MIT undergraduate lecture-plus-laboratory course *Materials in Human Experience,* still going strong after 40 years, the first study unit of which (still being taught) was on replicating ancient and

[32] Langton and Roy (1982, 1984).
[33] Seymour *et al.* (2023).
[34] Stefaniuk *et al.* (2023).
[35] Smith (1960, 1988).

modern cementitious materials, including Roman concrete (See footnotes 4 & 27).

One intriguing alternative to Portland cement was invented decades ago by an innovative French solid-state chemist, Joseph Davidovits (1935–) who pioneered low-temperature aqueous reconstruction of aluminosilicate structures (like platelet clays) in highly alkaline conditions to form feldspathoid-like three-dimensional network structures (similar to zeolites) that form the basis for an ecofriendly cementing technology. These he has dubbed *geopolymers*,[36] which (together with other alkali-activated cementitious processes) have formed a new generation of low-temperature geopolymeric cements that represent a family of alternatives to Portland cements for forming concretes. An entire Australian airport (runway and buildings) has even been constructed with these new cementitious materials, the *Brisbane West Wellcamp Airport in Toowoomba, Queensland,* in 2014. An active geopolymer world community has developed, much of it in materials departments, such as the two-decade research effort of ceramist professor Waltraud Kriven at the University of Illinois's Department of Materials Science & Engineering, accompanied by attendant activity within various professional society divisions (such as the ACerS cement division), even spawning an ECI conference series on geopolymer technologies.[37] An exciting development from the Illinois effort has been to demonstrate that ceramics, more conventionally manufactured by high-temperature sintering of powders, can be more economically and advantageously produced at room temperature *via* aqueous geopolymeric

[36] Davidovits (2020).

[37] ECI Conference series on geopolymers: *Alkali-activated materials and geopolymers,* 28 May–2 June 2023, Calabria, Italy, conference chairs: Waltrude M. Kriven, Cristina Leonelli, J. L. Provis and Aldo R. Boccaccini; *International Conference on Alkali-Activated Materials and Geopolymers: Versatile Materials Offering High Performance and Low Emissions,* Tomar, Portugal, 27 May–1 June 2018, conference chairs: J. L. Povis, Cristina Leonelli, Waltraud M. Kriven, Aldo R. Boccaccini, Arie van Riessen; *Geopolymers,* Hernstein, Austria, 24–29 May 2015, conference chairs: Cristina Leonelli, Waltraud M. Kriven, Aldo R. Boccaccini, Arie van Riessen.

reactions.[38] For some years now, Prof. Kriven has entered her students' geopolymer mugs in the American Ceramic Society's annual mug-drop contest, and left the competition in the dust, so to speak (literally).

Polymer academic programs and departments

Macromolecular materials comprise atoms bonded into molecular units with at least *one* dimension much larger than inter-atomic dimensions. *Inorganic* ordered atomic assemblies (inorganic crystals) and non-crystalline inorganic atom assemblies lacking translational order (glasses, and other disordered forms of inorganic solids) formally qualify as macromolecular materials but are not usually considered as such. *Organic* macromolecular materials include long-chain polymers in crystalline or (more usually) more random assemblages, self-assembling proteins (like *collagen* in hair, hooves, bone or ivory, s*ericin* in silk fibers, or *polysaccharides* in plant cellulose), three-dimensional disordered macromolecular networks (like rubber or asphalt), and organic molecular crystals (like naphthalene in mothballs), and have been historically the province of chemists, organic chemists, biologists, biochemists, chemical engineers, and textile engineers, but only more recently of materials scientists and materials engineers.

People have been using polymers for a long time, probably earlier than any other natural materials resource. As indicated at the outset of this entry, use of natural polymeric fibers for clothing, cord, and rope — both animal-derived (hides, wool, silk) and plant-derived (flax (*Linum usitatissimum*), cotton (*Gossypium hirsutum, G. barbadense*), jute (*Chochorous olitorius, C. capsularis*), hemp (*Cannabis sativa*) fibers) — was critical to human survival and the direction of human evolution in prehistory. Academic study of textiles and ropes was mostly carried out within engineering schools, beginning in the 19th century and continuing well into the 20th century, in which fiber engineering was chiefly carried out in

[38] Bell *et al.* (2009).

mechanical engineering departments. The Clothmakers Company, one the twelve foremost Livery Companies (guilds) in London (founded 1508), helped found the Yorkshire College of Science in 1874 in the West Riding town of Leeds, by then an international center for textile industries and cloth manufacture in Victorian England; the College was folded into the University of Leeds in 1904. (The brief inclusion of the Yorkshire College and its program of textiles into the mining and metallurgy curriculum of Victoria University, predecessor of the University of Manchester, has already been mentioned.) A notable US textile educator was MIT's Stanley Backer (1920–2003), a textile engineer elected to the US Academy of Engineering, who organized a Laboratory for Fibrous Materials in the Quartermasters Company of the US Army in 1944 and for four decades headed the Laboratory for Fibers and Polymers he founded in 1962 in MIT's department of Mechanical Engineering.[39]

Synthetic emulation of natural macromolecular materials, like rubber, ivory or silk, had long been a commercial goal of early organic chemists. Rubber derives from the sticky viscous sap (called *latex*) of the Central and South American rubber tree (*Hevea brasiliensis*), rich in long-chain polymers such as isoprene (the C_5H_8 monomer stoichiometry established empirically by Michael Faraday in 1829). Early Mesoamerican cultures (Olmec, later Mayan and Aztec) had discovered more than 2000 years ago that latex could (in our modern understanding) be cross-linked to form a highly elastic solid, which they fashioned into large rubber balls for an iconic ritual ball game, by mixing it with the juice of the morning glory vine (*Ipomoea indica* and others) that contains sulfur compounds.[40] Two *millennia* later, American chemists Nathaniel Hayward and Friedrich Ludersdorf, working in a latex factory in Connecticut, observed in 1832 that rubber treated with sulfur lost its stickiness and patented their observation, sharing the information in 1837 with self-taught chemist Charles Goodyear

[39] Backer (2002).

[40] Hosler *et al.* (1999).

(1800–1860), then working in his father's ivory manufactory in Naugatuck, CT), selling him their patent. Goodyear patented a process of heating a mixture of latex, sulfur, and lead oxide that subsequently hardened into a hard elastomeric solid, in 1845 — 3 weeks *after* a similar British patent had been filed by self-taught British engineer Thomas Haycock, who had been working with rubber solutions for water-proofing. Goodyear later claimed, in his 1853 autobiography *Gum-Elastica* (written in the 3rd person to enhance its authenticity), to have discovered the sulfur-addition process himself in 1837. The apt name *vulcanization*, by which the process is now known, was coined, however, by a friend of the earlier patentee, Haycock. The first *synthetic* rubber was synthesized by British chemist William Tilden (1842–1926, later Sir William Tilden, FRS) in 1908 by obtaining isoprene from cracking of turpentine (itself still a natural substance, but more locally and strategically available) — a product of the distillation of tree resins comprising mostly the monoterpenes α- and β-pinene, later found by others to vulcanize more quickly with sodium additions.

Ivory is a collagenous solid (as are bone, horn, and fingernails), much used for the classical fabrication of billiard balls. Synthetic ivory has its roots in nitrocellulose chemistry, a compound, which when dissolved in ether (ethoxyethane) additionally diluted with alcohol, yields a thin polymeric film (collodion), first used as a flexible wound dressing in 1847 and later in photographic plates before the invention of gelatin emulsions in the 1870s. Birmingham (UK) chemist Alexander Parkes discovered in 1855 that a solid residue remaining after evaporation of the solvents in collodion photographic films could be used as a flexible waterproofing treatment for fabrics (patented as Parkesine in 1862). American entrepreneur John Wesley Hyatt subsequently discovered that adding camphor (a distillation product from the roasting of camphor trees, rich in α-pinene) to Parkesine yielded a hard, horn-like substance (dubbed *celluloid* by Parkes's brother Isaiah) that they jointly recruited to attempt manufacture of artificial ivory-like billiard balls. Manipulation of the camphor treatment yielded more plastic film-like versions of celluloid that were successfully used as the support for gelatin emulsion in photographic film and later cinema film.

The next two commercially successful macromolecular synthetics were bakelite (a hard, thermosetting phenol formaldehyde, resembling shellac (an insect-derived hard coating) or hard rubber; and nylon (a thermoplastic polyamide solid that can be spun to form silk-emulating fibers for stockings and other textiles). Bakelite was formulated in 1907 by Belgian-born chemist Leo Hendrik Baekeland and publically announced in 1909 at an American Chemical Society section meeting in New York). Nylon was the invention of DuPont chemists, derived from petroleum and synthesized almost two decades later in 1938; its dramatic public announcement was timed for the opening of the 1939 World's Fair. The subsequent eight decades have brought humankind a cornucopia of synthesized "plastics" for every occasion (many proving unrecyclable) that now increasingly more resembles an organic chemist's version of Pandora's proverbial box.

Because these organic assemblies are solid materials with defined (and manipulable) physical, chemical, and mechanical material properties, they have been more recently co-opted by materials scientists and entered the curricula of materials science and engineering departments, joining the traditional materials classes of metals, ceramics, and glasses. They have even invaded the territory of electronics with flexible (and even biodegradable) conductive polymers (for which the Nobel Prize in Chemistry in 2000 was awarded to polymer chemists Alan J. Heeger, Alan G. MacDiarmid, and Hideki Shirakawa) and organic light-emitting diodes ("plastic" LEDs, the first introduced by Cambridge University physicist Robert Friend). While a majority of polymer scientists is still educated in university chemistry or chemical engineering departments, most major research universities worldwide now have either dedicated schools or departments of Polymer Science or Polymer Engineering—like the School of Polymer Science & Engineering at the University of Akron (OH, home of Goodyear Tire), the department of Macromolecular Science & Engineering at Case Western Reserve University (Cleveland, OH), and the department of Polymer Science

& Engineering at University of Massachusetts (Amherst), or significant polymer efforts within Materials Science & Engineering departments or interdisciplinary centers in many universities. While chemistry-based professional societies, like the American Chemical Society, have divisions of polymer or macromolecular chemical science or manufacture, the major materials societies (as well as most materials science & engineering departments) now integrate polymer science and polymer scientists into their materials-catholic remit.[41]

Biomaterials academic programs

The introduction of materials integrated into the human *body* has a long history in antiquity, whether (cosmetic) prosthetic eye balls[42] (the first known, made from clay and gold, dating from Persia in

[41] In the US, dedicated departments (or institution-wide interdisciplinary academic programs in polymers or macromolecular science & engineering) exist at Case Western Reserve University (Cleveland, OH, Department of Macromolecular Science & Engineering; University of Massachusetts (Amherst) Polymer Science & Engineering), University of Minnesota (Minneapolis, department of Chemical Engineering & Materials Science), University of Akron (OH (home of Goodyear), College of Engineering and Polymer Science), MIT (Cambridge, MA, interdisciplinary Program in Polymers and Soft Matter (PPSM)), University of Southern Mississippi (Hattiesburg, School of Polymer Science & Engineering), University of Connecticut (Storrs, Institute of Materials Polymer Program), and University of Chicago (Pritzer School of Molecular Engineering); in Europe: Netherlands at Groenigen University (*Laboratorium voor Polymeerchemie*), Germany (Berlin, joint program in Polymer Science, *Technische-Universität* Berlin, *Frei Universität* Berlin, *Humboldt-Universität* Berlin, *Universität-Potsdam*), Slovenia at *FTPO Slovenj Gradek* (Faculty of Polymer Technology); in Asia: China at Xi'an Jiaotong University (Institute of Polymer Science), Zhejiang University (Hangzhou, department of Polymer Science & Engineering) and others, Singapore (Nanyang Technological University ((NTU)), Taiwan (National Taiwan University (NTU), Institute of Polymer Science & Engineering (PSE). Academic polymer materials programs exist within most departments or institutes of Materials Science and/or Engineering, Chemical Engineering, and/or Biological Engineering, in the US and in many countries worldwide.

[42] Dyer (1980); see also: Ocular prostheses. https://en.wikipedia.org/wiki/Ocular_prosthesis (last accessed 1 March 2023).

2900 BCE, in-socket glass eyes introduced in 16th-century Venice, poly(methyl methacrylate) (PMMA) used after World War II); cranioplasty[43] (gold patches for missing bone in skull fractures used *ca.* 2000 BCE in the Peruvian Andes, xenographic bone graft patches from the denatured animal bone in 1668, later human allografts or autografts starting 1821, PMMA from the 1940s); or prosthodontics (animal or human tooth prostheses for missing teeth by at least 700 BCE by Etruscans, 17th c. ivory or bone prostheses, 1770s French porcelain dentures). The introduction of materials into human *biology* began in several areas of medical practice: principally orthopedics, dentistry, and cardiology, but also for drug delivery to specifically targeted sites or across the blood–brain barrier.

The *hard tissues* of bone and teeth (both enamel and dentin) are largely comprised of a naturally assembling bi-continuous organic-inorganic composite comprising a protein, primarily type-I collagen (a triple α-helix long-chain poly(peptide) polymer expressed in bone by osteoblast cells, and in dentin and enamel by odontoblast and ameloblast cells), and a crystalline hydrated calcium phosphate ceramic, hydroxyapatite $Ca_{10}(PO_4)_6(OH)_2$ (formed — at least for bone — from a hydrated amorphous calcium-phosphate gel injected by the mitochondria of osteoblast cells into tiny vesicles that transport their contents through the cell wall into the inter-cellular matrix).[44] Expressed collagen molecules stack end-to-end and self-assemble into fiber bundles, the hydroxyapatite then crystallizing — from the exuded contents of the vesicles — with a platelet morphology at the sites of regularly repeating gaps in the collagen molecule stacking in the fiber bundles, forming a biphasic composite hard tissue.[45]

In bone pathologies (e.g., bone defects from infection, bone fractures) or at prosthesis/bone junctions (e.g., hip-joint or knee joint prosthetic implants), an orthopedist desires to induce growth of new bone (osteogenesis) at bone defect sites or at bone/prosthesis interfaces to promote bonding or rebonding. Prosthetic joint

[43] Shah *et al.* (2014).

[44] Boomrungsiman *et al.* (2012).

[45] Rosen *et al.* (2001).

components are typically metals (e.g., Co–Cr, zirconium alloys, though sometimes zirconia or alumina ceramics), which must be biocompatible and non-allergenic, and coating these with bioactive Ca-phosphate compounds (like, but not limited to, hydroxyapatite $Ca_{10}(PO_4)_6(OH)_2$) has been found to accelerate initial bonding of bone to the implant to anchor it.[46] Similar bioactive coating considerations arise with metallic implants into the jaw bone to serve as anchors for artificial tooth replacements.

Physiologically degradable biologically active "bioglasses," incorporating calcium and phosphorous ions, were first explored by ceramist Larry L. Hench (1938–2015, University of Florida and later Imperial College London) and frequently used for filling bone defects to stimulate replacement bone growth for the same reason.[47]

Of course, a variety of metal alloys (impressed mercury amalgams, cast gold alloys) has long been employed for filling dental enamel decay cavities, now augmented by an array of polymeric and notably ceramic[48] substitutes that can be exquisitely tailored to replace the missing tooth material.

Soft tissue pathologies (e.g., tendon rupture, muscle-tendon junction rupture) can be repaired using a wide array of strong fibrous non-biological polymers that substitute for the soft biological tissue. Soft-tissue/hard-tissue junctions (e.g., bone-tendon junctions) pose additional difficulties because such joins are complex graded junctions involving typically at least three different cell and tissue types (e.g., bone-forming osteoblasts, chondrocytes that make cartilage, tenocytes that make tendon) and can take a year to reform; tendon is typically apposed to bone for that time period with resorbable polymer sutures (typically poly(lactide-glycolide) or ε-caprolactone polymers)[49] that are anchored to bone with similarly resorbable polymer anchors, sometimes doped

[46] Porter *et al.* (2001, 2004).

[47] Hench and West (1996).

[48] Hobbs (2007); Doremus (2008).

[49] Pappalardo *et al.* (2019).

with calcium compounds to accelerate bone regrowth after resorption. Synthetic soft materials are employed as replacements for diseased or decayed cardiovascular tissue, e.g., blood vessels or heart valves. Metal or polymeric stents, cleverly mechanically designed or using shape-memory alloys to facilitate insertion, are employed to prop open soft-tissue arterial vessels, such as coronary arteries.

The holy grail of repair is, of course, regeneration of tissues by recruiting the body's own indigenous tissue growth mechanisms. Apart from blood cells, most human body cells are anchorage-dependent and rely on naturally engineered extra-cellular matrices to promote and guide appropriate tissue growth. Appropriate natural or artificial scaffolds, especially bioactive ones, are thus important to construct in order to promote and guide tissue regrowth and morphology after injury or disease.[50] Denatured bone (See footnote 45) has long been used as a scaffold for bone regrowth, as have "artificial skin" scaffolds for promoting the regeneration of new epithelial tissue in burn victims. Tubular guidance channels with appropriate biochemistries can be used to guide and accelerate severed nerve regrowth after trauma or neurodegenerative disease. In addition to resorbable polymers for soft-tissue scaffolding and guide channels, advantageous parallel use of bioactive ceramics, like hydroxyapatite or bioglasses, in soft-tissue repair has also been demonstrated.[51] A recent addition to the biomedical arsenal for soft-tissue repair are hydrogels, three-dimensional soft matrices capable of incorporating and retaining a large fraction of water molecules, common natural examples of which are chitosan, collagen, gelatin, and cellulose. Many schemes have profitably relied on biomimetic approaches.

Drug and other therapeutic agent release delivery vehicles involve a wide range of materials types and morphologies.[52] As an early example, glasses were formulated to be biodegradable, and

---

[50] Chan and Leong (2008).

[51] Mazzoni *et al.* (2021).

[52] Vargason *et al.* (2021).

miniature glass spheres were used as delivery vehicles. A far more sophisticated approach is to recruit nanoparticles as nanoscale therapeutic agent transport vehicles with specifically targeted delivery protocols used to deliver agents to specifically targeted sites (like cancerous tissue, for example) by attaching biomarkers to the delivery vehicle; these have hugely proliferated in the last two decades, and now involve virtually every materials class and a good deal of sophisticated biology.[53] Some nanoparticle agents can be formulated such that they can be externally directed, for example by magnetic fields or ultrasound, and are able to breach the imposing blood–brain barrier for the treatment of brain diseases.

With both biology and medicine embracing an even broader range of naturally occurring, biologically active inorganic and macromolecular organic compounds, materials engineering of these and their synthetic versions — endowed with engineered properties that coordinate with or replace biological tissues — become appropriate targets for informed materials scientists. The field of biomaterials (or biomedical materials, or bioengineering) began in the 1950s. A difficulty well into the 1990s (and to a lesser extent even 30 years later) was that a majority of materials practitioners still lacked adequate knowledge of biology, though by the same token, a majority of biologists equivalently lacked materials know-how, so dual educational tracks have since arisen in a huge proliferation at the outset of the 21st century of dedicated departments of Biomaterials or Biological Engineering or Biomedical Engineering,[54] as well as in materials

[53] Mitchell *et al.* (2021).

[54] A necessarily partial catalogue of active dedicated academic programs in Bioengineering, Biomaterials Engineering, Biomedical engineering, Biomolecular Engineering, etc. *Programs in the* **US***:* University of Alabama (Birmingham), Biomedical Engineering; Brown University (Providence, RI), Center for Biomedical Engineering; Emory University, Biomedical Engineering; University of California (Berkeley), Bioengineering; University of California (Davis), Biomedical Engineering; University of California (Riverside), Bioengineering; University of California–San Diego (La Jolla), Bioengineering; University of California–Santa Barbara, Biological Engineering; Carnegie-Mellon University (Pittsburgh, PA), Biomedical

Engineering; Case Western Reserve University (Cleveland, OH), Biomedical Engineering; City College of New York, Biomedical Engineering; University of Central Florida (Orlando), Biomedical Engineering; University of Cincinnati (OH), Biomedical Engineering; Clemson University (Clemson, SC), Bioengineering; University of Colorado (Boulder), Chemical & Biological Engineering; University of Colorado (Medical) (Denver), Bioengineering; Columbia University (New York, NY), Biomedical Engineering; University of Connecticut Health Center (Faringdon), Biomedical Engineering; Cornell University (Ithaca, NY), Meinig School of Biomedical Engineering; Duke University (Durham, NC), Biomedical Engineering, *and* Center for Biologically Inspired Materials and Materials Systems; University of Florida (Gainsville), Biomedical Engineering; Georgia Institute of Technology (Atlanta), Bioengineering (School of Chemical and Biomolecular Engineering); Harvard University (Cambridge, MA), Bioengineering (Paulson School of Engineering & Applied Science); University of Illinois (Champaign-Urbana) Bioengineering; Johns Hopkins University (Baltimore, MD, Biomedical Engineering; Indiana University (Indianapolis)–Purdue University (Indianapolis) (joint), Biomedical Engineering; Iowa State University (Ames), Biomaterials and Biomedical Engineering; Johns Hopkins University (Baltimore, MD), Biomedical Engineering; University of Kansas (Lawrence), Bioengineering; University of Kentucky (Lexington), Biomedical Engineering; University of Maryland (College Park), NIH/NIBIB Center for Engineering of Complex Tissues; University of Memphis (TN), Biomedical Engineering; University of Michigan (Ann Arbor), Biologic and Materials Sciences (School of Dentistry) *and* Biomedical Engineering; University of Minnesota (St. Paul), Biomedical Engineering; MIT (Cambridge, MA), Department of Biological Engineering; University of Nebraska (Lincoln), Biomedical Engineering; North Carolina Agricultural & Technical State University (Greensboro), Chemical, Biological and Bioengineering; North Carolina State University (Raleigh)–University of North Carolina (Chapel Hill) (joint) Biomedical Engineering; Northwestern University (Evanston, IL), Biomedical Engineering; Ohio State University, Chemical and Biomolecular Engineering; Oregon Health & Science University (Portland), Division of Biomedical Engineering (School of Medicine); Pennsylvania State University (University Park), Biological Engineering; University of Pittsburgh (PA), Bioengineering; Rice University (Houston, TX), Bioengineering; University of Rochester (NY), Biomedical Engineering; Rutgers University (Piscataway, NJ), Biomedical Engineering; Syracuse University (Syracuse, NY), Biomedical & Chemical Engineering; University of South Dakota (Sioux Falls), Biomedical Engineering; University of Texas (Austin), Biomedical Engineering; University of Texas (El Paso), Metallurgical, Materials and Bioengineering; University of Texas (San Antonio), Biomedical Engineering; Texas A&M University (College Station), Biomedical Engineering; Tufts university (Medford, MA), Biomedical Engineering; University of Utah (Salt Lake), Biomedical Engineering; Vanderbilt University (Nashville, TN), Chemical & Biomolecular Engineering; University departments in universities without independent biomaterials departments. At MIT — although it had long had world-leading programs in cellular and molecular biology and toxicology — a proposal to begin a department of Bioengineering was met by initial institutional skepticism in 2005 about whether this field was sufficiently established to supply its graduates adequate employment opportunities. That situation has greatly changed. One dedicated professional association, the Biomaterials Society, discussed later, has existed for the last three decades, but all the major materials societies, as well as a growing number of materials departments, now consider biomaterials science and engineering, and associated practitioners and academic programs, within their purview. A significant help has been the parallel more recent proliferation of available textbooks[55] (some better than others) and curricula geared equally to students with materials background or students with biology backgrounds. Or, increasingly with both: for example, MIT made at least *one* heavy-duty molecular, cellular, or environmental biology course an Institute *requirement* for *all* its undergraduates in the mid-1990s. (MIT undergraduates are not required to complete a *materials* course, though approximately 60% of undergraduates satisfy an Institute chemistry requirement with a freshman course in solid-state chemistry — essentially materials science — taught by materials faculty members since the 1960s.)

of Virginia (Charlottesville), Biomedical Engineering; Virginia Commonwealth University (Richmond), Biomedical Engineering; University of Washington (Seattle), Bioengineering & Chemical Engineering; Washington State University (Pullman), W. M Keck Biomedical Research Laboratory; University of Wisconsin (Madison), Biomedical Engineering; Worcester Polytechnic Institute (MA), Biomedical Engineering; Yale University (New Haven, CT), Biomedical Engineering. *A few Biomaterials/Bioengineering academic programs* ***outside the*** **US:** Chinese University of Hong Kong, Institute for Tissue Engineering and Regenerative Medicine; Imperial College London, Materials Science; McMaster University (Hamilton, ON, Canada), Chemical Engineering; University of Minho (Guimães, Portugal), Research Institute on Biomaterials, Biodegradability and Biomimetics; University of Porto (Portugal), Instituto de Engenharia Biomédico; RMIT University (Bandoora, VIC, Australia), School of Engineering; University of Toronto (ON, Canada), Institute of Biomaterials and Biomedical Engineering.

[55] Ratner *et al.* (1997–2013); Wagner *et al.* (2020).

Electronic materials academic programs

Electrical and magnetic phenomena were noted early in human history and, by the 16th century, formed the basis of three practically important inventions: (1) The *compass* was invented by the Chinese during the Han dynasty 2,000 years ago, using a needle of (ferromagnetic) iron that has been temporarily magnetized by uniaxially rubbing against a chunk of (ferrimagnetic) iron ore (magnetite, Fe_3O_4, known as *lodestone*, exhibiting a much larger remanence than that of iron), and which, suitably suspended, will then orient its induced remanent magnetization axis parallel to that of the earth's magnetic field). (2) The *lightning rod* was invented in the late 18th century (an elevated pointed metal rod connected by a copper wire to the earth to conduct lightning safely to ground), following on from Benjamin Franklin's (1706–1790) 1752 confirmation that lightning was electricity (electricity was similarly implicated in the stimulation of frog leg muscles by physician and physicist Luigi Galvani (1737–1798) in 1780). (3) The aqueous chemical *battery* (two electrodes of dissimilar metals immersed in a salt solution or separated by brine) was invented by physicist Alessandro Volta 1745–1827 in 1799 that provided an electrical source. Observable electrical and magnetic phenomena were sufficiently fascinating that they became popular parlor demonstrations in the early 19th century. Several useful principles emerged: (1) Metals could conduct electrical charge, many other substances (insulators: amber, glass, ceramics, dry wood) did not, while some solids like carbon conducted but only poorly, resisting the charge transport. (2) Electrical charge could be generated by rubbing a non-conducting solid (*triboelectric* effect) and stored in a *capacitor*, essentially comprising two closely spaced metal plates separated by an air space or other non-conductor (originally, a cylindrical metal-foil-wrapped glass jar, called a Leyden jar after the Dutch town of the scientist-inventor, Pieter van Musschenbroek (1692–1761), who invented it in 1746). (3) English chemist Michael Faraday (1791–1867) established that aqueous solutions of soluble salts could also conduct charge. (4) In further experiments, Joseph Henry (1797–1878), an

American, discovered that charge transported through a metal wire generated a radial magnetic field, which when coiled in a helix (physicist André-Marie Ampère (1775–1836), in 1820) produced an axial magnetic field that could be enhanced by inserting magnetizable iron into the helix core (physicist William Sturgeon (1783–1850), in 1825). The relationships between charge, current, magnetic field, electrical field, and resistance (or its reciprocal, conductance) were compiled by Scottish physicist James Clerk Maxwell FRS (1831–1879), a Trinity College Cambridge don, who compressed them into four differential equations, known as Maxwell's Relations, in 1865. Materials were assigned two innate properties, their dielectric constant and their magnetic susceptibility, that determined the speed of temporal propagation of electromagnetic disturbances in that material medium — and even foretold the speed of light (an electromagnetic wave) propagating in the complete absence of any material medium (a vacuum), the latter vacuous medium then (puzzlingly!) assigned non-zero fundamental values for each of these two materials constants.

Resistors, capacitors, inductors, the basic circuit elements of the electrical and magnetic devices that were critical to much late 19th-century industrial growth — motors (Moritz Jacobi, 1834), generators (Samuel Varley, 1856; Werner von Siemens, 1866), telegraphy (Charles Wheatstone, 1837; Samuel F. B. Morse, 1837; Alfred L. Vail, 1840), telephony (Alexander Graham Bell, 1876), electric lights (Thomas Alva Edison, 1878), radio (Heinrich Hertz, 1887; Aleksandr Popov, 1895; Guglielmo Marconi, 1895) — soon transitioned from objects of phenomenological interest to physicists to the contents of electrical engineers' toolboxes. The first academic institutions to introduce an educational curriculum in electrical engineering were MIT, which offered a degree program through the Physics department starting in 1882, followed by the *Technische Universität Darmstadt* (Germany, founded 1877) and Cornell University (Ithaca, NY) in 1883. The first university chair in Electrical Engineering was established at University College London (1885) and the first free-standing department of Electrical Engineering at the University of Missouri–Columbia (1886, now the Missouri University of Science &

Technology). Other institutions throughout the world quickly followed.

Experiments with gas discharge tubes by two university physicists (Julius Plücker and Heinrich Geißler in Bonn, 1856), and the discovery of mysterious "cathode rays" (by Plucker and Johann Wilhelm Hittorf, 1859), led to three important discoveries: (1) Discovery of X-rays (1895), by William Konrad Röntgen (1845–1923) at *Universität München*, made possible the first atomic-scale investigation of atomic arrangements in crystalline materials using X-ray diffraction by father-son duo William H. (1881–1942) and W. Lawrence Bragg (1890–1971) at the Cavendish Laboratory in Cambridge University. (2) Identification of the negatively charged electron was first made by Irish physicist George Johnstone Stoney, who measured the charge (per broken bond) associated with atom-bond breakage during electrolysis (in 1874) and named this fundamental charge unit the *electron* (1891); J. J. Thompson (1856–1914), then director of the Cavendish, measured the charge-to-mass ratio of what he identified as a "corpuscle" (later identified with Stoney's electron) responsible for cathode rays (in 1897); and Ferdinand Braun (in 1897) deduced its collision with the glass tube walls to be the source of Röntgen's X-rays and discovered that a narrow beam of cathode rays could be mechanically collimated and scanned across a phosphor screen by the magnetic field of an electrified external coil, forming the first cathode-ray tube (and parenthetically forming the basis for the invention of the electron microscope by Ernst Ruska in Berlin in 1931, now the most powerful imaging tool for atomic-scale investigation of materials). (3) The *thermionic diode* was invented by John Ambrose Fleming, a sometime consultant to Edison's electric light company, in which electrons thermionically emitted from a heated source in a fully evacuated discharge tube conducted electrical charge (to a metal plate anode maintained at a positive electrical potential) but in only *one* direction, thus providing rectification of an alternating AC current, even at radio frequencies (RF). It was thus deduced that the (negatively charged) electron was the charge carrier in electrical circuits, therefore reversing the direction of charge flow previously presumed by electrical engineers. Fleming (later Sir John Fleming, FRS) was the first Professor of Electrical Engineering at University College London and patented the thermionic diode (later known as a vacuum-tube rectifier, or valve rectifier) in 1904 — later modified in 1906 by American inventor Lee de Forest by addition of an intermediate grid element to form a triode tube (or valve), which was able to amplify audio or radio signals (dubbed the "Audion" tube by de Forest) and established the field of electronics and electronic engineering.

Jagdish Chandra Bose, a professor of Physics in Calcutta, had previously patented the point-contact galena (PbS) rectifying diode (the standby rectifier for generations of young crystal-radio set enthusiasts) in 1901. PbS is a solid-state semiconductor, the first precociously introduced into electrical engineering. The point-contact configuration of another revolutionary solid-state semiconductor device, the three-element point-contact transistor, initially using germanium (Ge) and relying on surface charge effects, was invented by three Bell Laboratory physicists (John Bardeen, Walter Brattain, and William Shockley) in 1947. Beginning in 1954, high-purity silicon (Si) was substituted as the semiconductor of choice, locally doped by other elements, e.g., boron (B) or antimony (Sb), to provide regions with, respectively, an accessible deficit (*p*-type) or excess (*n*-type) of electrons that could be induced to flow across the interface, from one area to another (across the *p-n* junction formed), by application of an appropriately polarized small electrical potential. Three-element solid-state transistors could then be constructed in *p-n-p* or *n-p-n* configurations that, like a three-element vacuum tube, provided the possibility of signal amplification. Another possible way to effect induction of current flow in semiconductors, known as the *field effect*, was first demonstrated and patented in 1926 by Polish-American physicist Julius Edgar Lilienfeld, initially working at *Universität Leipzig*, then in the US at Amrad Corporation in Malden, MA (later bought by Magnavox). Brattain *et al.* had found the field-effect method too unstable and opted instead for their point-contact devices, but an unstoppable migration to field-effect transistors (FETs) began in the 1960s, and finally to the metal-oxide-semiconductor field-effect transistors (MOSFETs). The MOSFET was made possible by thermally forming a passivating SiO_2 oxide layer on a planar silicon surface, first reported by Mohamed Atalla, an Egyptian-American Bell Labs engineer, at a 1958 meeting of the Electrochemical Society. Processing methods developed by Atalla and others enabled huge numbers of MOSFETs to be patterned on a single flat water of silicon, together with other circuit elements, forming integrated circuits (ICs), thus ushering in the age of microchip technology. Most ICs now use paired MOSFETs, known as complementary MOS devices, or CMOS, which since 2007 have used HfO_2 (with a higher dielectric constant than SiO_2) for insulating the gate electrode. Parallel developments led to the use of compound semiconductors (compounds formed from elements in Groups III and V, like GaAs or InP, or from elements in Groups II and VI, like CdTe or ZnSe — or, indeed, even PbS), useful in high temperature or high-frequency applications, and also for solar cells and light-emitting diodes (LEDs), often grown on SiC substrates.

Other classes of "device" electronic materials are ferromagnetic and ferrimagnetic materials (e.g., magnetite, Fe_3O_4; CoCrPt alloys), used for magnetic storage media (magnetic recording tape, computer hard drives); piezoelectric materials (e.g., quartz (SiO_2)) used for frequency control and in accelerometer guidance systems; a subset of piezoelectric-ferroelectric materials, *electrostrictive* piezoelectric compounds (e.g., perovskite-structure lead zirconate titanates, $Pb(Zr,Ti)O_3$) (PZT)), used as actuators (audio speakers, micromanipulators) and sensors (sonar, ultrasound imaging); and solid-state ceramic electrolytes and electrodes in fuel cells and for Li-ion and other battery systems (e.g., Y-doped ZrO_2 (YSZ) or $Li_7La_3Zr_2O_{12}$ (LLZO)). Ironically, PbS, Si, Ge, GaAs, CdTe, ZnSe, SiC, Fe_3O_4, SiO_2, HfO_2, Y-ZrO_2, $Pb(Zr,Ti)O_3$, $Li_7La_3Zr_2O_{12}$, and related inorganic solid compounds are, in fact, *ceramic* materials, while CoCrPt and related alloys are decidedly *metals*. A major reason for relating the foregoing history of electronic materials and devices is to point out that their *initial* investigation and development did *not* take place in traditional departments of ceramics, or by industrial ceramists, nor in traditional departments of metallurgy, or by professional metallurgists, but instead in academic and (mostly) industrial laboratories staffed with physicists, chemists, electrical engineers, and radio/electronics engineers. This seeming oversight by the ceramics and metals fields, two nuclei for the development of materials science, both as a field and as academic curricula, has meant that other academic departments (principally physics and electric engineering, in some cases chemistry) and their associated professional societies (physics; chemical, electrochemical, electrical and electronic engineering societies) have, respectively, trained and represented scientists and engineers practicing in these electronic materials fields, certainly until well into the 1960s when materials science & engineering departments were first being formed. The reluctance of ceramists to take on the important materials class of semiconductors was particularly glaring and has been to their detriment — though electrical ceramics were expertly and enthusiastically embraced (not least by ceramists at Penn State, MIT, NUS Singapore, and at many other institutions world-wide) by the 1970s.

Electronics and radio engineering academic programs had developed largely within traditional electrical engineering departments, many of which later became departments of Electrical Engineering and Computer Science post-1990, when personal computers and the Internet first became conspicuous, then ubiquitous — both enabled by what were really key electronic materials developments. MIT provides one early example: The famous Radiation Laboratory ("RadLab"), formed at MIT during the Second World War, was instrumental in developing radar (along with industrial partners, like Raytheon). After the War ended, the RadLab was renamed the Research Laboratory for Electronics (RLE) in 1946 and, while nominally interdisciplinary, was closely associated from its start with the department of Electrical Engineering and was directed by MIT electrical engineer, Julius Stratton. The next six subsequent directors were either physicists or electrical engineers. Electrical engineering (EE) at MIT was first taught by the Physics department in 1882, enshrined in 1884 as MIT's sixth academic degree course (Course VI), and was given department status in 1902. (Stratton, a 1923 graduate of MIT's EE department, earned the *Dr. rer. nat.* degree in *physics* from the *Eidgenössiche Technische Hochshule* (ETH) in Zürich in 1928 and returned as an EE department faculty member, later becoming MIT's 9th president in 1959.) In 1975, his department changed its name to Electrical Engineering & Computer Science, reflecting its growing involvement with the computer field. The 7th RLE director (Yoel Fink, chosen in 2000) was a *polymer materials* scientist in the department of Materials Science & Engineering. MIT's former department of Metallurgy, which had added a ceramics curriculum in the 1950s, had made its name similarly more inclusive in 1967, becoming the department of Metallurgy & Materials Science, and shortly after — with the inclusion of polymers *and* electronic materials — the even more ecumenical department of Materials Science & Engineering in 1975, a decade before Prof. Fink's arrival as a graduate student in that department.

Graduate Materials Education

With the proliferation of materials-centered university undergraduate programs in the 1930s and the number of former soldiers taking advantage (at least in the US) of government tuition grants (the "GI bill") to attend university, materials-related professionals (metallurgists, ceramists, polymer scientists, and engineers) entering the work force in these fields did so with professional undergraduate degrees, rather than acquiring post-secondary school technical experience on the job as had often been the case earlier. Meanwhile, the nature of what was being taught changed. Metallurgists advantageously applied the principles of solid-state chemistry and solid-state physics, developed in the first three decades of the 20th century, to develop a fundamental solid-state science of metals, with a keen eye for applications and with an engineering twist. The nature of the potential membership for professional societies in materials fields therefore began to change — from trade-based empirically trained to academically and scientifically trained — with reduced stress on engineering practice and more on the origins of fundamental materials properties and their advantageous modification. As a consequence, technical societies had to become more scientifically sophisticated and less oriented toward traditional industrial trade practices.

This change was nowhere more apparent than in the ceramics field, with a sudden shift from traditional ceramics production (whitewares [toilet bowls, crockery], glazes [bath tubs, enameling], refractories for metal smelting and casting, Portland cement, porcelain electrical insulators, abrasives, and window glass) to highly technical advanced ceramic and glass materials with cleverly thermo-chemically engineered properties — in applications such as high-temperature turbine-blade coatings, highly fracture-resistant structural ceramics and glasses, ceramic battery electrodes and solid electrolytes, solar cells, piezoelectric activators, high-temperature superconductors, key components of microelectronic devices, and radiation-resistant components of nuclear reactor and space vehicles. MIT, which had added metallurgy in 1884 and dropped mining engineering by the 1940s, added ceramics to their Department of Metallurgy with two sequential (and consequential) professorial hires: F. H. Norton in 1939, who applied thermochemistry to the characterization and engineering of ceramic refractories and created an interdisciplinary program in ceramic science within a Metallurgy department; and in 1951 his doctoral student, W. David Kingery, who completed the application of solid-state science principles — that had so revolutionized modern metallurgy by the late 1940s — to the study of inorganic *non*metallic solids in his groundbreaking textbook *Introduction to Ceramics* (1960). Departments of Ceramic Engineering and Metallurgy in many other US universities soon followed suit, many eventually uniting both disciplines into materials-catholic departments of Materials Science or Materials Science & Engineering. The result was a growing cadre of very different looking ceramists within the American Ceramic Society by 1970; and the Society, like the profession, was obliged to change its culture in order to survive the predations of other material-centric societies on both its membership and its scientifically gentrifying subject matter.

With the rising popularity of further post-graduate study, beginning in the late 1930s, and after the post-World War II emergence of the "research university" in science and technology fields, an ever-increasing larger proportion of post-graduate students elected to pursue PhDs for 4 or 5 years (or more) of fundamental or applied research. The academic system quickly began to increasingly resemble nothing so much as a reincarnated *guild* system: doctoral students were effectively apprenticed to a professorial advisor in a specialized department (in exchange for payment of their university tuition from their advisor's grant funds), were provided a meager stipend — barely adequate to offset the cost of room and board — and contributed most of the research legwork for their professorial master who was busy competing for research funding (most frequently public funding from government agencies[56]). After 4 or 5 years the student apprentices produced a "masterpiece" in the form of a thesis, adjudged by a committee of their academic superiors comprising professionals from within and without their

[56] In the US, the materials research support sources are mainly the National Science Foundation (NSF), Department of Energy (DOE [formerly Energy Research & Development Administration (ERDA)], Department of Defense [Air Force Office of Scientific Research (AFOSR), Office of Naval Research (ONR)], National Institutes of Health (NIH) (for biology-related research), and a small number of industry-consortia funded institutes, such as the Electric Power Research Institute (EPRI); in the UK, formerly the University Grants Committee (UGC), now the Universities Funding Council (UFC), the UK-wide Engineering & Physical Sciences Research Council (EPSRC), and the Biotechnology and Biological Sciences Research Council (BBRFC) operating under the umbrella of UK Research & Innovation (UKRI), the UK Atomic Energy Authority (UKAEA), (additionally in Scotland) the Scottish Funding Council, and the various societies or academies (such as the Royal Society, Royal Academy of Engineering, Academy of Medical Research) and various endowed legacy trusts, such as the Wellcome Trust (for medical research); in Germany, mainly the *Deutsche Forschungsgemeinschaft* (DFG) and a number of well-endowed private or industry-legacy foundations, such as the *VolkswagenStiftung* and the *Max-Planck-Gesellschaft zur Förderung der Wissenschaften* (with its extensive network of specialized Max Planck Institutes (MPI)); in France, the *Agence Nationale de la Recherche* (ANR), and the *Centre National de la Recherche Scientifique* (CRNS) network, to which many French academics enjoy joint appointments; in the EC, various directorates-general and Joint Research Centers.

institution (but likely in the same specialty). After admission to the cohort of qualified professionals, these newly minted PhDs generally sought post-doctoral employment, often in their former advisor's research group or with those of professors at other universities. Post-docs functioned much as journeymen in the guild system, paid but not handsomely, and often put in charge of the day-to-day supervision of graduate student apprentices and their projects in professors' research groups. The regulatory function exercised by guilds was largely achieved instead through peer review of grant proposals or research publications in journals. As in guilds, the end product (in this case, research publications — though occasionally patents) invariably bore the name of the professorial advisor, identified as the (sometimes nominal) "principal" investigator (or inventor). Productive graduate students would likely be afforded opportunities to participate in local section meetings of a national society, and attend (and possibly present at) a specialist conference or an annual or semi-annual conference organized by a major research society. Ministering to these graduate students (and competition for their allegiance) became an increasingly important role among competing materials-based societies by the late 1970s, each establishing student chapters in many universities; these were later often combined into a single "Materials" student organization.

Materials-Intensive Corporate Laboratories and National Laboratories

Once they become "freed," employment options for these materials PhDs, who now formed a major constituency for all materials societies, now consequently widened in their subject embrace, and the sociological profile of these employees differs considerably from what it was 50 years ago. In the 1950s and 1960s, the US boasted an enviable phalanx of major self-funded in-house fundamental research laboratories maintained by large industrial companies (General Electric, US Steel, Ford Motor Company, Bell Telephone, IBM, RCA, DuPont, Westinghouse, Raytheon, Texas Instruments,

3M, Honeywell and, in Europe, Philips in Eindhoven, Netherlands) who invested significant corporate money in long-term materials research and in sustaining materials research departments (with focus on metallurgy, metallurgical engineering, ceramics, polymers, or electronic materials as appropriate to the company interests, and sometimes all) that were the envy of US universities. These large corporate research centers not only provided ready employment for post-graduate-trained materials scientists and engineers, but for many of these employees served as smaller self-contained in-house scientific and engineering professional societies, with often more interaction between professionals operating in different specialties than normally occurs within universities, because the research necessarily spanned many disciplines all related to company products vital to the success of the company and their employees' continued employment. These much-envied corporate laboratories are now largely gone, victims of an unfortunate pre-occupation with quarterly profits; their roles have, to some extent, been taken up by the similarly large US Department of Energy (DOE)-funded National Labs and industry-funded research collaboratives, like the Electric Power Research Institute (EPRI), that additionally provide funding for related university research projects and — like the former large corporate laboratories — are inherently interdisciplinary and focused on discrete national priorities, so by themselves again serving somewhat as *de facto* scientific professional mini-societies for their employees.

Two more modest (but highly successful) US corporate materials research efforts that *have* survived productively into the 21st century are Dow [formerly Dow Chemical] (Midland, MI), now styling itself as "The Materials Science Company," though continuing its dominant former focus on polymer-based coating and adhesion products and lately investing significantly in biomaterials; and Corning [formerly, Corning Glass Works] (Corning, NY), similarly styling itself "a global leader in materials science technology," though maintaining its long-standing expertise in glass-science and ceramic-science technologies, optics and optical communications.

Both companies maintain strong links to university materials research efforts. Pilkington once enjoyed a similar status in the UK.

The numerous well-funded Max-Planck-Society Institutes in Germany and the CNRS and *Centre d'Etudes Nucleaires* (CEN) networks of national laboratories in France, though similarly large and well-funded, function in a less interdisciplinary fashion and in less broad societal contexts than do the US National Laboratories (or did the former corporate fundamental research centers), because the focus of each laboratory in their systems is largely confined to a single sub-discipline. The UK Atomic Energy Authority's Atomic Energy Research Establishment (AERE) at Harwell functioned, from the 1950s until the early 1980s, as a national fundamental science research laboratory, in many ways more like a research university, with a broad materials remit, while Britain's National Physical Laboratory (NPL) continues to fulfill its intended standards-setting regulatory role, much as does the US National Institute for Standards and Technology (NIST). Japan is well on its way to integrating many major sectors of materials science and engineering into a near-comprehensive national laboratory in a centralized site. It began by establishing a National Research Institute for Metals (NRIM) in 1956 in Meguro, Tokyo, with an office in Tsukuba in 1979 (to help populate the Tsukuba Science City, to which site NRIM had fully removed by 1995); and establishing the National Institute for Research in Inorganic Materials (NIRIM) in Toshima, Tokyo in 1966, moving to Tsukuba in 1972. NRIM and NIRIM were then merged in 2001 to become the National Institute for Materials Science, whose Tsukuba campus now houses centers for biomaterials, superconducting materials, computational materials science, steel research, ecomaterials, high magnetic fields, and materials nanoarchitectonics, with instrumentation sections for materials information technology, materials analysis, and ultrahigh-resolution electron microscopy. It has additionally integrated a materials *education* component through the establishment of a doctoral program in Materials Science & Engineering within the Graduate School of Pure & Applied Sciences at the co-located University of Tsukuba.

Professional Societies, Associations, and Institutes

Associations and societies exist because human beings are social, like their identity affirmed, are curious to learn what advantage can be wrought from the status, effort, and successes of others, and seek to augment and calibrate their own contributions. This somewhat dystopian assessment ignores, of course, the group benefit that such associations can confer to establish schools of thought or further human understanding, not least the scientific corpus. Members may be persuaded to engage in working together, not always altruistically, toward shared goals that benefit a larger population. Communication is invariably more efficient in person than in written dialogue, not least when a shared specialist language is agreed upon and employed, a second benefit of associating with like minds. To an extent, in this sense, present-day scientific or engineering departments, in universities with post-graduate research programs attracting many students and mentors, qualify as mini-associations, or microsocieties (of, admittedly, somewhat limited scope) providing many of these associational functions in microcosm — professional education and social training, and of course regulatory oversight — and serve as practice runs at membership in larger scientific communities. Venerable early scientific associations, like the British Royal Society or the French *Académie Royale des Sciences,* once congenial (occasionally competitive) venues for discussion of scientific findings, novel theories, or promising applications, now instead function largely honorifically and through restricted membership by special election, as indeed do also the US national academies of Science, Engineering, Medicine and Arts & Sciences. Professional societies, as distinct from these honorific academies or institutes, are unrestricted self-elected assemblies of more quotidian professional practitioners that provide wider dissemination of new ideas from a broader array of sources, not least through a frequent sponsored conference or targeted meeting venues (and often in-house journals or proceedings publications).

Materials represent, foremost, a universal *human* experience and therefore are pervasively associated with multiple aspects of human endeavor. Professional societies that have emerged to serve materials-related disciplines accordingly differ from those serving more closely delineated scientific disciplines, such as theoretical or high-energy physics or pure mathematics, in at least three significant ways: First, they tend to be more academically interdisciplinary, and therefore more scientifically catholic, attracting researchers from diverse backgrounds in (or employing the tools of) physics, chemistry, biochemistry, geology, microscopy and microanalysis, engineering, electronics, mathematical modeling, computational science, materials economics, archaeology, anthropology, and still other areas, in addition to those circumscribed by the five conventionally identified materials sub-disciplines of metals, ceramics, polymers, electronic materials, and biomaterials. Second, they invariably and enthusiastically embrace broader ranges of scientific, industrial, commercial, or public import. Third, their focus tends to alight on materials, material processes, or materials production with unique or significant potential for wider application, and by implication with a particularly strong industrial component. Materials societies are generally perceived as adopting mediative roles, both encouraging *and* enabling collaborative and interdisciplinary endeavor, made easier by the broad range of practitioner backgrounds and expertise available in the membership cohort embraced — and perhaps more interesting.

Traditional Industry-Based Materials Societies

Until the 1960s, there was seemingly little need for an overarching professional research society, embracing all materials classes, because the various materials-class constituencies were perceived as being substantially unrelated. Metals were crystalline, adopting only a few (~4) basic crystal structures involving simple chemistries (mostly binary or at most ternary phase equilibria); exhibited geometrically simple polycrystalline boundaries; were mostly ductile with industrially accessible melting points which facilitated processing by casting, rolling or forging; had simple defect structures that provided the required plasticity and diffusion transport required at easily attainable temperatures; could be further reduced to required net shape by machining or surface finish by abrasion (using hard ceramic powders); were electrically conductive. Ceramics, though (with the exception of glasses) also crystalline, adopted complex crystal structures that required a more nuanced appreciation of crystallography; were chemically complex, with multiple atomic species, phases, and stoichiometries requiring higher-dimensional phase equilibria; exhibited complex defect types and chemically complex polycrystalline boundary structures; were brittle and refractory (with often inaccessible melting points), properties that precluded casting or forging to shape, and (for the most part) had to be processed into net shape by sintering (high-temperature solid-state defect-facilitated fusion) of amalgamated ground powders. Organic polymers were an entirely different beast, requiring sophisticated knowledge of structural organic chemistry; were chemically complex with often poorly understood or characterized microstructures and nanostructures; were generally processed (and processible only) from powders and usually under pressure; though sometimes remarkably elastic, were frequently brittle or deformable plastically *via* only poorly characterized deformation mechanisms; were (for the most part) electrical insulators. After 1947 (when the point-contact transistor was invented by three solid-state physicists at Bell Labs), semiconducting solids (such as Ge, Si, and later III–V and II–VI compounds) — though technically ceramic materials — became the province of electrical engineers and physicists and were largely ignored by metallurgists, ceramists and polymer scientists.

Despite clear inherent commonalities between these materials types — cast iron (Fe-4%C) and many metal alloys are also brittle; ceramics (and some brittle polymers) can be made to plastically deform at similar homologous temperatures ($T_H = T/T_{\text{melt}}$) and by similar or analogous mechanisms; powder metallurgy has a lot in common with powder processing of ceramics and even polymers; sintering of powders (ceramic, metal or polymer), and grain growth in both polycrystalline metals and polycrystalline ceramics, proceed by largely identical or analogous defect-enabled mechanisms; ternary and higher-order phase equilibria are necessarily important in many commercially important metal alloy and multicomponent ceramic systems; silicate glass, ceramic, metal and polymer fibers all exhibit some similar rheological properties; some metal systems can — like ceramic glasses — be rendered non-crystalline ("amorphous metals"); under particular externally imposed conditions (e.g., partial pressures of oxygen, temperature) some ceramics can be rendered electrically conducting; sulfide ceramics and some ceramic oxides are semiconducting at room temperature; etc. — these four encampments remained stubbornly immiscible. Ceramic science is, in fact, more closely related to geochemistry, because of the commonality of the chemistries involved (largely oxides) and the high-temperature phases encountered, while polymer scientists continue their allegiance to structural organic chemistry.

Inorganic nonmetallic glasses are non-crystalline network assemblies of atoms with low connectivity to each other, no fixed melting point, and a solid-to-liquid (or *vice versa*) transition that occurs, not suddenly as it does for crystal melting or crystallization in metals or ceramics, but gradually over a wide temperature range. They have traditionally remained in the ceramics camp on account of their familiar chemistries (most commonly based on silica (SiO_2) or alkali silicates, or chalcogenide analogs) and their frequent appearance in high-temperature phase assemblages during ceramic firings and in-service refractories. They are also a common product of geological vulcanism. Though with atom connectivity topologies remarkably similar to those of their isocompositional crystalline ceramic counterparts (See footnote 18), inorganic glasses are fundamentally irregular macromolecular networks that can be said to share a kinship with polymer and other macromolecular networks. Asphalt, for example, is an *organic* macromolecular glass, with a glass transition temperature approximately 50°C, and it is asphalt (not alkali silicate window glass) that is by volume the *most* pervasive glass in our modern environment! Because the structural details of such irregular solids cannot be accessed with the simple diffraction tools (such as X-ray and electron diffraction) that are the mainstay for investigating structures and phase assemblages of crystalline metal and ceramic systems, they remain forbidding *terra incognita* for most metallurgists and at best inscrutable traveling companions for most ceramists (and too much like physics for the average inorganic chemist).

What has been responsible for bringing these long-entrenched encampments closer together are three changes in the materials landscape that have provided increased common ground: First, the resolution of characterization tools has been reduced to atomic dimensions, capable of imaging common nanosized defects: for example, transmission electron microscopy (TEM) can image dislocations (mobile line defects in the regular atom ordering in crystals, not unlike rumples in a carpet that, when (easily) sequentially nudged along, can eventually shift an otherwise immobile carpet across a floor) that are responsible for the plasticity of metals and also present in ceramics, but there mobile only at much higher temperatures (but at similar *homologous* temperatures); atomic-resolution TEM can also directly reveal details of actual atom arrangements and chemistries at interfaces like grain or phase boundaries, using nanometer-scale microanalysis tools. Second, many new materials — such as *metallic* glasses, or the charge-conducting solid-state *ceramic* electrolytes now found pervasively in "lithium"-battery electric cars, or ceramic high-temperature zero-resistivity *super*conductors (the potential dream grid!), or super-plastic nano-grained metals that stretch like poly(ethylene) or taffy (the latter a polysaccharide glass) — come with properties once considered exclusive to only one class of materials but showing up surprisingly in others. Third, novel altered educational strategies brought forth curricula (first at Penn State and Northwestern Universities and soon after in many traditional metallurgy and ceramics faculties) that began to emphasize the commonalities in materials behaviors rather than their apparent differences — sort of like treating different materials classes as second-cousin species instead of unrelated phyla. A 20th-century invention — single-mode optical glass fibers for telecommunication (proposed theoretically by an American Optical Co. scientist in 1961, realized practically by scientists at Corning Glass and Bell Labs) — brought glass science and technology abruptly to public attention. Two notable 21st-century successes — tough "Gorilla® glass" for cell-phone screens, developed at Corning (by an Alfred graduate, John C. Mauro, now a distinguished professor at Penn State), and incorporation of photonic (speed-of-light) components for nanoelectronics (vastly outpacing the slower electron-conduction-limited information-transfer rates afforded by conventional copper microchip traces) — have elevated glass to at least first-cousin status to ceramics, metallurgy *and* electronic materials!

Five key professional organizations in the US that had *historically,* before 1970, dominated the representation of the traditional materials *industries,* particularly through their journals and conferences were as follows: the *American Ceramic Society* (ACerS), the *American Society for Metals* (ASM) [and its UK counterpart, the *Institute of Metals* (IOM3)], the *American Institute of Mining, Metallurgical, and Petroleum Engineers* (AIME) and its adjutant *The Metals Society* (TMS), the *Society of Plastics Engineers* (SPE), and the *Institute of Electrical and Electronic Engineers* (IEEE). On a more fundamental science level, materials interests were additionally represented by the *American Physical Society* (APS) [and its UK counterpart the *Institute of Physics* (IOP)] in condensed matter physics sub-divisions, the *American Chemical Society* (ACS) (with divisions for polymer science and for polymer engineering), the *Japanese Society for Polymer Science* (SPS), the *Electrochemical Society* (for corrosion of metals and solid-state ionic ceramic battery electrodes and electrolytes), the *Optical Society of America* (now *Optica*) for glass science and glass fiber and photonics technologies, and the *Electron Microscopy Society of America* [since 1992 the *Microscopy Society of America* (MSA)], together with its adjutant *Microanalysis Society* (MAS) for materials ultramicroscopy and analysis.

Ceramics Societies

American Ceramic Society (ACerS)

Study of, and research into, ceramic materials in the US grew out of 19th-century needs of commercial industries that produced materials made from fired clay: bricks, tiles, porcelain (china), pottery, sanitary ware, and refractory materials that had to withstand high temperatures, as well as glass products: plate glass, table glassware, enamel, and glaze frits. The first curricula, organizations, and literature to focus on ceramic materials began at the end of that century, in Ohio and Pennsylvania, at the time (and still) two centers for ceramics and glass manufacturing. In 1898, the formation of a society was discussed at a meeting of the National Brick Manufacturers' Association in Pittsburgh, Pennsylvania, and realized a year later at Ohio State University in Columbus, Ohio, where the first American department of Ceramic Engineering had just been founded by Edward Orton. The new society was officially incorporated, in1905 in Columbus, as the *American Ceramic Society* (ACerS). It began with 21 charter members in 1898 and experienced steady (537 members in 1916) then explosive growth (over 11,000 members by 2023). In the dozen years between 1912 and 1923, Corporate membership had grown from 5 to 212, with most US ceramics, pottery, brick, glass and refractories manufacturers represented, including many corporations still prominent a century later — among them AC Spark Plug, Babcock & Wilcox, Bausch & Lomb Optical, Carborundum, Hall China, B. F. Drakenfeld, Engelhard Refractories, Harshaw Chemical, Libbey Glass, Norton Abrasives, Pittsburgh Plate Glass (PPG), Square D, and Western Electric.

The new society's members were diverse — chemists, engineers, geologists, and commercial manufacturers, operating within the ceramics and glass fields and industries. What drove the initial formation of this professional society was the felt commonality among several related industries, particularly those with products involving fired clays, glazes and enamels, and glass, and the clear need to pursue research into the engineering of a wider range of ceramic and glass materials with advantageous ceramic properties. For its members, it brought colleagues with similar interests and problems; for manufacturers, it provided access to information about best practices and new market prospects; for the Society, it defined (and staked out, if not circumscribed) a field of technical endeavor and laid the groundwork for potential expansion. Publication was seen as a crucial unifier, and the first issue of the society's *Transactions* appeared in 1899. *Ceramic Bulletin* and the *Journal of the American Ceramic Society* both followed in 1922. Education was a second important ongoing function, and an Association of Ceramic Educators (later the Ceramic Educators Council (CEC)) was formed in 1935.

The Society's seven initial interests were demarcated by what became its foundational divisions: Art, Enamels, Glass, Heavy Clay Products, Refractories, Terra Cotta, and White Wares (later, Whitewares). In 1929 an exposition of American ceramic products took place in Chicago, Illinois, (repeated at least well into the 1970s at annual Society meetings), reflecting the Society's original clay ceramics origins and its continuing dominance in ceramic technology and industry as the Society's preoccupation. After several decades of societal focus on clay products and glazes, the Society's attention shifted as ever more commonalities were discovered between clay-based ceramics and other commercial ceramic materials, such as refractories, glasses, and cements. Subsequent additions (and deletions) revealed the shifting reach and gradual maturation of the ceramics field. In 1930, a *Materials & Equipment* division was added, with *Structural* Clay Products replacing Heavy Clay Products. In the 1940s, a *Design* division replaced Art, and Terra Cotta was discontinued. The 1950s brought much-needed additions of a *Basic Science* division (1951) and *Electronic Ceramics* (1958). *Ceramic-Metal Systems* replaced Enamels (1961), encompassing the important area of ceramic-metal seals.

Following the publication in 1960 of W. D. Kingery's transformative textbook *Introduction to Ceramics* that established ceramic *science*, as opposed to ceramic engineering, on a firm footing, the Society's founding interests thereafter began to expand gradually to eventually include ceramic materials used in electronics, energy storage, telecommunications, biological applications, nuclear energy and wasteforms, ceramic art, and ceramics in human prehistory. A *Nuclear* division was added in 1965, and 1971 brought the (long overdue) addition of a *Cements* division at the urging of Geoffrey Fronsdorff (a British physical chemist, later chief of NIST's Building Materials Division), who had organized a successful Symposium on Cements at the Society's 1970 annual meeting. The CEC offered its first shortcourse, on kinetics in ceramic processes, in 1975, and the first volume of *Advances in Ceramics* was published in 1980.

The 1980s and 1990s brought some relabeling as the Society's compass was forced to broaden: the *Engineering Ceramics* division replaced Ceramic-Metal Systems (1985), Refractories became *Refractory Ceramics* (1989), the Glass division became the *Glass & Optical* division, Nuclear became the *Nuclear & Environmental* division; and a moniker change, from ACS to ACerS, was introduced to avoid confusion with the American Chemical Society, on whose territory it began to intrude (and *vice versa*). Three new publishing endeavors: *Ceramic Source* (a yearly source book of manufacturing companies, equipment and instrumentation, and materials suppliers), started in 1985; a new (Wiley) journal *Advanced Ceramic Materials,* launched in 1985 (though 4 years later it was absorbed into the longstanding *Journal of the American Ceramic Society*); and ACerS entered into a landmark agreement with the (then) National Bureau of Standards (NBS, now NIST, under the US Department of Commerce) to undertake the monumental effort of collecting and cataloguing equilibrium phase diagrams for nonmetallic inorganic systems and to incorporate these into a regularly updated series of volumes of *Phase Diagrams for Ceramists,* comprehensive graphical compendia of the thermodynamic properties governing phase equilbria in multicomponent ceramics.

As the Society's centenary approached and a new century loomed, concerns about waning membership numbers and finances grew. The Society spun off its first subsidiary society, the *American Ceramic Industries Association* (1995), to better retain its traditional industrial base, and launched the more engineering-oriented *International Journal of Applied Ceramics Technology.* In an attempt to tap into the enormous ceramic artists' community, ACerS acquired the artist community's magazines, *Ceramics Monthly* and *Pottery Making Illustrated,* and established a Potters Council (2001), which later morphed into an international Ceramic Artists Network. In 2014, ACerS spun off an international *Ceramics and Glass Industry Foundation* to provide a home for ceramics and glass industry professionals and provide philanthropic support for various educational initiatives. For most of its first century, the governance structure and culture of ACerS remained dominantly male (as was largely true for most of the traditional, historically engineering-originated materials societies), even as more women entered the ceramics fields, mainly through more gender-neutral materials science departments (at one point, MIT's materials department undergraduate student enrollment was over 70% female). ACerS elected its first women president in 1996 (a prominent National Lab glass scientist from the Nuclear division). There have now been eight more in the intervening 17 years, for a total of nine women presidents, and five of the sixteen 2022–2023 board members are female, an impressive course correction. By comparison, in the same period, ASM International has had three women presidents, TMS two, and AIME one.

The UK counterpart *British Ceramic Society* never quite attained equivalency with ACerS, or their organ, *Transactions of the British Ceramic Society,* with the ACerS *Journal.* In 2016, a noted British ceramist, William E. Lee, at Imperial College became president of ACerS, their first international president, and in 2017 ACerS began instituting international chapters. Forty percent of ACerS's 11,000 members now hail from 70 countries outside the US.

The Metals Societies

Societies devoted to metals research followed a somewhat more convoluted history, In Britain, eight medieval guilds dealt in some way with metallurgy, all well established by the 14th century, with active control for the next four centuries. Ferrous metals (iron, later steel) were the province of the Blacksmiths, Cutlers, Armourers & Braziers, and Ironmongers; non-ferrous metals (copper, tin, lead, gold and silver, and their alloys) that of the Goldsmiths, Pewterers, Founders, and Plumbers (See footnote 6). Similar medieval metallurgical guilds functioned in France, Germany, Italy and Spain. Sienese metallurgist Vanoccio Biringuccio (1480–1539), author of the influential text *De la Pirotecnia* (published posthumously in 1540) and late in his life head of the Papal foundries, was a member of the secretive guild *Fraternitá di Santa Barbara* (patronized by the artillery gunners of the Royal Marine). His influential book did much to dispel the secrecy surrounding the "art and mystery" of the metallurgical trade. Subsequently, Georgius Agricola (1491–1555), a Saxon metallurgist, mineralogist, physician, and humanist scholar, wrote the encyclopedic 12-volume sequel *De Re Metallica librii XII* (published posthumously in 1556), following on from his earlier comprehensive textbook *De Natura Fossilium* (1546) on mineralogy and mining.

Institute of Materials, Minerals and Mining (IOM3) [UK]

During Roman times, tin — for alloying with copper to make bronze, was obtained from Cornish tin mines, but bronze (which was brittle) was increasingly displaced by malleable brass (an alloy of copper and zinc initially produced by the cementation process, involving sublimation of zinc vapor from heated calamine ($ZnCO_3$)). By the early 17th century, large brass foundries (controlled by the *Founders* guild) were located just outside the City of London's northern gates in Moorfields, partly to supply the burgeoning English clock industry with brass (then known by its Latin name *latten*) sheet and castings, but direct alloying of copper and zinc from zinc metal (speltering) was not widely practiced until mid-to-late 18th century, brass production eventually shifting to Bristol, using coal for smelting.

By the mid-18th century, the ferrous metallurgical industry had also largely shifted from control by the London guilds to locations to the north, in the midlands and Pennine districts, where iron ore deposits and remaining forests for charcoal, and later mineable coal, were more proximate. In the midlands, Birmingham became an early center for iron production and manufacturing, using coal, and in the 19th century the Shropshire town of Coalbrookdale, where iron ore was first advantageously smelted by using *coked* coal (a purer carbon form of coal from which hydrocarbons have been burnt out), became the center of the cast-iron (known as pig iron, with 2–5% carbon content) industry, supplying the railroads with cast boilers and cast bridge components. In the 1850s, American metallurgist William Kelly and English industrialist Henry Bessemer (later Sir Henry Bessemer, FRS) experimented with reducing the carbon content of pig iron to make malleable iron (low-carbon steel) by direct oxidation of carbon with pressurized blown air, both likely inspired (and perhaps mutually informed) by early accounts of Chinese and Japanese methods for reducing the brittleness of cast iron by burning out the carbon and silicon impurities (forming volatile CO and CO_2, with SiO_2 joining the slag). Bessemer (whose father was a Hugenot metallurgist who had devised steel dies for the Paris mint) patented his Bessemer converter in 1856 — for which he was later knighted in 1879 — that enabled economic production of steel, replacing the inefficient and labor-intensive piecemeal bloomeries that produced only100-lb spongy, unmelted, low-carbon iron "blooms." These were maintained in a low oxygen environment under cover of burning charcoal and molten flux, but retained substantial slag impurity inclusions that had to be hammered out of the spongy bloom. Coked coal and Bessemer's process soon shifted British steel production further to the north, to the Yorkshire towns of Middlesbrough and Sheffield (the latter a town largely founded by Bessemer), closer to coal supplies, and shifted American steel production to locations with similar access to mineable coal.

A British association of iron and steelmakers and related industrialists in the north of England was organized in 1869 as the Iron & Steel Institute (ISI), first meeting biannually in London (and other European cities, too — Paris, Brussels, Vienna, Düsseldorf — and even New York!). Its first president was Lord William Cavendish (who endowed the Cavendish Physics Laboratory at Cambridge University), Henry Bessemer its second, and it received a Royal charter in 1899. Recognizing that no similar association of *non*-ferrous metallurgists then existed, a group of copper and brass founders resolved to form a similar institute for brass and copper in 1908, equivalently established along the lines of the Iron & Steel Institute, but immediately expanded the purview to include not only copper producers and brass founders but also "shipbuilders, engineers and others interested in non-ferrous metallurgy," the latter embrace circumscribed as including "tin, zinc, nickel, aluminium, silver, gold, platinum and their alloys." The new association was christened the Institute of Metals (IOM), with an eminent naval architect as its first president. The society first met in Birmingham, followed by London, Manchester, and Glasgow, and reported its proceedings in its *Journal of the Institute of Metals,* the first volume of which appeared in 1908. The first "local section" of the Institute was established in Birmingham in 1910. A May lecture series was also begun, the second of which, in 1911, was given by eminent metallurgist G. T. Beilby FRS and presciently titled "The hard and soft states of metals" (which a century later might have been more appropriately styled as "The hard and soft states of matter"). Membership rose from 200 in 1908 to 600 in 1912, with a sixth resident in 20 countries. A council was formed in 1911 to regularize the naming of alloys. Not surprisingly, a growing relationship developed with the ISI, and the two institutes eventually combined as an enlargened Institute of Metals in 1974. Further mergers with the *Institute of Ceramics,* the *British Composites Society,* and the *Plastics & Rubber Institute* in 1993 and the *Institute of Mining & Metallurgy* in 2002 resulted in the now conglomerate *Institute of Materials, Mining, and Metallurgy* (IOM3), which has subsequently voraciously gobbled up the additional specialist associations *Institute of Packaging* in 2005, *Institute of Clay Technology* in 2006, and *Institute of Wood Science* in 2009 — these constituent professional specialties now represented by divisions or groups of IOM3 — completing more than three decades of dizzying consolidation of the materials-based societies. Cognizant of its illustrious ferrous roots, IOM3 still awards an annual Bessemer medal.

American Iron and Steel Institute (AISI)

Iron-making was the first heavy industry established in British colonial outposts. John Winthrop (the younger) was the eldest son of Massachusetts Bay Colony's first governor of the same name, later becoming himself an early governor of the Connecticut Colony. Before departing Massachusetts, he founded (with British capital eager for a larger supply of pig iron) America's first successful integrated ironworks, near Lynn, MA, on the Saugus River just north of Boston, beginning operation by 1646 with its own blast furnace, forge, slitting and shearing mills, and quarter-ton water-driven trip hammers, and drawing on local deposits of bog iron ore.[57] By 1776, 80 iron works had been established throughout the Colonies — mostly in coastal regions of New York, New Jersey, Virginia, and Pennsylvania, with access to bog-ore deposits and charcoal fuel from colonial forests — exporting pig iron to Britain, but prohibited by a 1750 act of Parliament from manufacturing steel or iron plate.

Iron smelting in the first half of the 19th century began to relocate, in larger operations, nearer the more extensive ore deposits found inland, as coastal bog-ore sources became increasingly exhausted, but continued to use charcoal as the primary fuel, though anthracite coal began to be substituted as early as 1827 in Pennsylvania. Coke, from the coking of coal (heating in the absence of oxygen to remove volatile hydrocarbons), provided fuel with a higher specific heat content, and largely replaced charcoal (similarly derived from wood) by about 1840, though the higher sulfur content (primarily pyrites) in bituminous and lignite coals resulted in a detrimental impurity in the pig iron produced. The Bessemer process for reducing the carbon content of the pig iron was quickly implemented after its invention in 1856 and soon begot (with massive infusions of capital from prominent financiers) huge steel mills in Pittsburgh (Dale Carnegie, Carnegie Steel), Cleveland (Cyrus Eaton, Republic Steel), Bethlehem, PA (Charles Schwab and Eugene Grace, Bethlehem Steel); Scranton, PA, in the Lackawanna Valley (George & Seldon Scranton, later moving their plant to Buffalo, NY, Lackawanna Steel); and Gary, IN, on Lake Michigan, adjacent to Chicago (J. P. Morgan, US Steel), all sites with convenient access to iron ore (drawing on abundant ores from Minnesota's Mesabi Range abutting Lake Superior, easily shippable through any Great Lakes port, or those from Pennsylvania's Allegheny Mountains) and to coal (abundant throughout Pennsylvania, Ohio, and West Virginia). Pittsburgh, situated at the confluence of three major rivers, in northwest Pennsylvania and not far from the Lake Erie port of Erie, PA; Bethlehem and Scranton in eastern PA, proximate to

[57] Hartley (1957).

rich deposits of both anthracite coal and Allegheny iron ore; Gary, IN on Lake Michigan; Cleveland and Buffalo on Lake Erie, all satisfied both criteria.

With the rapid expansion of the American railroad system in the mid-19th century, the demand for pig iron increased steadily — to 7 million tons by 1855 — the same year that eastern ironmakers felt a need to establish a trade body to represent their interests, resulting in the foundation of the *American Iron Association.* Following the introduction of the Bessemer process for steel manufacture, the iron industry transitioned into the production of steel — for railroads and munitions, and later for automobile parts and architectural construction — and accordingly modified the Association's name to the *American Iron and Steel Association* (AISA) in 1864, toward the end of the American Civil War, the Association at that time headquartered in Philadelphia. With the explosive growth of the steel industry at the beginning of the 20th century, the largely statistical functions of the AISA were considered inadequate, so an American Iron and Steel Institute (AISI) was created in 1908, with US Steel's chairman Elbert H. Gary (for whom Gary, IN is named) as its first president. Based in New York City, the Institute functioned alongside the Association until the latter was merged into the AISI in 1913. By the 1920s, American steel represented 40% of world production, but the Great Depression following wreaked havoc on the steel industry. As a result of the National Recovery Act, passed in 1935, AISI was placed in charge of representing the iron and steel industry and has persisted into the 21st century as the major trade association for that industry. Among its contributions, it introduced, and for years maintained, a numbering system for identifying stainless steel alloys by their composition and treatment, handing off further maintenance of the scheme to the Society of Automotive Engineers (SAE) in 1995.

American Society for Metals (ASM)

While many US steel manufacturers still preferred to keep their steel-making processes secret, many realized the importance of sharing information and techniques about steel treatments. As a result, in 1913 just as the AISI was consolidated, a Steel Treating Club formed in Detroit, MI, which was by then already a center for automobile manufacturing, a major consumer of sheet and forgeable steels. Ford Motor Company in fact established its own River Rouge steel foundry in adjacent Dearborn, MI in 1921, just upstream of the Detroit River that joins Lake Huron (and through it the great Lakes Michigan and Superior) to Lake Erie. The strategic location provided water access to the massive Masabi Range iron ore deposits adjacent to Lake Superior, the coal of Pennsylvania, the automobile markets of Chicago and the West, the consumer markets of New York and the East Coast through the Erie Canal from Buffalo to the Hudson River, and to the Atlantic Ocean and Europe through Lake Ontario and the St. Lawrence River canals and later seaway). During the First World War — which had vastly expanded the demand for steel for aircraft, tanks, ships, and munitions — the Club was renamed the Steel Treating Research Society, still dominated by Detroit-based members.

After World War I, a group of participants from Chicago (with its own burgeoning steel industry, starting in 1908 with the founding of US Steel's gigantic steel plant in adjacent Gary, IN) withdrew to create their own society, the *American Steel Treaters Society.* Initially rivals, the two groups united to form the *American Society for Steel Treating* (ASST), incorporated in 1921. However, neither would allow the other to be the site of its headquarters and after some negotiation agreed to put it in Cleveland (on Lake Erie, with its own vigorous steel industry). Publication of the Society's *Transactions* began that year, with a *Handbook* containing data sheets for steel heat-treating, published 2 years later. The *Metals Progress* magazine (later *Advanced Materials & Processes*) began in 1930.

Lacking an indigenous guild history during the 17th and 18th centuries, North American colonial metals production, other than of pig iron (which did begin in 1646), did not get started until after Independence — not, in fact, until the middle of the 19th century — not least because it had been specifically prohibited during the colonial era by the British Parliament, who required the colonists to instead purchase UK-fabricated finished goods from Britain, generating multiple levels of middleman profit. Some non-ferrous metalsmithing was nevertheless done, for example silversmithing and domestic brass clockmaking in the 18th century, but the silver came from melting down (British) silver coinage in circulation or recycled silver objects, and the brass came from melting down recycled brass kettles or candlesticks (of original British manufacture) and re-casting. (Recycled copper scrap could be converted — though not often and likely illegally — to lower melting point castable β-brasses (40–55wt%Zn, liquidus temperature $T_\ell < 900$ °C) or more malleable α-brasses (~20wt%Zn, liquidus temperature $T_\ell \sim 1000$ °C) by the calamine process, using $ZnCO_3$ carbonate ores that were available locally.) Eighteenth-century colonial clockmakers did fabricate brass domestic clocks, after English models, the brass sometimes coming from casting of recycled brass re-melts, but a majority was obtained as rough castings from brass founders in Britain, even more so after 1776.

The first major post-independence brass industry was started in the Naugatuck valley of Connecticut (Waterbury, CT) by brothers Abel and Levi Porter, only in 1802, for the manufacture of brass buttons (subsequently gilt) that were punched-out from brass sheet obtained by cold-rolling cast-brass ingots, replacing the prior cast-pewter colonial buttons. The enterprise proved unprofitable for the Porters, who sold out to partners (becoming Scovill Manufacturing Co.). Levi Porter next commissioned 4,000 wooden clocks to be mass–produced by clockmaker Eli Terry during the period 1806–1809. Terry's two clockmaker hires, Seth Thomas and Silas Hoadley, bought out Terry after the Porter contract had been fulfilled. Ironically, Seth Thomas soon after left the partnership with Hoadley to take up manufacturing domestic clocks from punched-out rolled sheet brass, at a fraction of the cost of English clocks — an early US indigenous industry that further spurred sheet brass production in Connecticut and by the late 19th century had decimated the revered English brass clock industry that had previously dominated world clockmaking for over three centuries.

Meanwhile, in neighboring Massachusetts, silversmith-*cum*-revolutionary Paul Revere, who had learnt to cold-roll silver sheet, became interested in rolling copper into sheet form for sheathing the hulls of US Navy ships. He purchased an idle iron-rolling mill in Canton, MA in 1800 and began cold-rolling copper sheet in 1801, using cast-iron rollers, later mastering hot-rolling that produced a more malleable sheet product, though still using imported copper. The Phelps Dodge Company (partly based in Liverpool, UK) was one of the US import-export companies in the metals importation trade; founded in 1834, it exported southern cotton to Britain and imported metals, primarily copper, zinc and tin, to the US.

Domestic smelting of copper metal from ores did not, however, begin until 1844, drawing on native copper and mined copper ores in Michigan's upper peninsula (UP) and along adjacent Lake Superior. The Waterbury and Detroit Copper Company, organized and controlled by Scovill Manufacturing Co., was founded in 1850. Copper production soon burgeoned, thanks to a protective tariff barrier erected in 1846 against importation of foreign copper, and during the 1850s decade domestic production of copper rose from 650 tons to 7,200 tons per annum. During the period 1847–1887, 8 million tons of copper were smelted from ore extracted from the Delaware Copper Mine, just south of Copper Harbor, MI in the UP's Keweenaw peninsula. Spelter zinc (for production of brass from copper by direct alloying) was at this point still largely supplied from Germany and Belgium sources, at least until 1867, and it was not until the 1870s that American zinc production was commercialized as extensively as had been copper two decades earlier. Tin was still obtained partly from Cornish mines and the rest from the East Indies rather than from domestic sources. US domestic production of lead was not commercialized until the 1860s. Western US metal ore sources began to be utilized in the third quarter of the 19th century. The Detroit Copper Mining Company established at Morenci, AZ a copper ore mining and smelting operation, incorporated in 1872 and later purchased by Phelps Dodge & Co. in 1897. Phelps Dodge Corporation subsequently became the world's second largest producer of copper; a subsidiary, Climax Molybdenum, became the largest world supplier of molybdenum in the 20th century. Commerical electrochemical extraction of aluminium from bauxite ore began in 1886 in Pittsburgh, PA, production later migrating to Niagara Falls a decade later (as discussed in footnote 64) under the company moniker Aluminum Company of America (Alcoa).

The Depression of 1929–1941, however, put enormous pressure on the greatly expanded metals industries, leading to a growing resources competition between the critical production needs of these much broadened industries and the parallel pursuit of rapidly expanding scientific and technical research into their metallic systems. In the Depression's pre-war wake, the ASST morphed into the broadened American Society for Metals (ASM), the steel treaters retreating into their own much smaller *Metals Treating Institute.* World War II brought an even greater need for metals engineering and research — particularly into newer alloys involving Be, Al, Cr, Mg, Ni, Ti, Zr, Nb, Ta, and W (and U and Pu, following the Manhattan project) — leading to a World Metallurgical Congress in 1951. In the ensuing years, the disciplinary and global scope of the ASM grew to include international participation and, increasingly, associated nonmetallic materials (starting with refractories and slags — both the province also of ceramists — and extending to still others). In 1986 the society changed its name (and its business model) to ASM International.

ASM is organized on an affiliate member society model, with affiliate societies focused on Heat Treating, Thermal Spray, Failure Analysis, (International) Metallography, Shape Memory and Superelastic Technologies, and *Electronic Device Failure* (*sic*!). Its governance structure draws from academically trained scientists or engineers who have risen to prominence in academic governance, business management, or national government service, while their oversight boards appear to be more drawn from industrial or government managerial ranks. Its flagship publication is *Advanced Materials & Processes* magazine, together with periodicals aimed at its individual industrial constituencies. Of the traditional materials-related societies, ASM is the most directly industrially oriented. It has set up a "Materials Solution Network" service to the industry, and the Society indeed styles itself as "the materials information society," despite its principal continuing focus (with the exception of the electronic device failure community) on the metallic state.

American Institute of Mining, Metallurgical, and Petroleum Engineers (AIME) and *The Metals and Materials Society (TMS)*

The *American Institute of Mining Engineers* (AIME) was founded in 1871 in Pennsylvania. It began to broaden its scope after World War I, adding a Petroleum Division (1922), an Iron and Steel Division (1928), and an Institute of Metals Division (1933). The resulting diversity provoked a name change, and in 1957 the AIME became the American Institute of Mining, Metallurgical, and Petroleum Engineers (though, unlike their UK equivalent IOM3, retaining the same AIME acronym). The diversity also spawned several separate member societies affiliated with the Institute, including the *Society for Mining, Metallurgy and Exploration* (SME), the *Minerals, Metals and Materials Society* (TMS), the *Society of Petroleum Engineers* (SPE), and eventually, in 1974, the *Iron and Steel Society* (ISS). Among the somewhat dizzying array of metallurgical societies extant or merging before the late 1970s, TMS emerged as carrying most of the academic weight and, following the emergence of materials science as an academic discipline throughout the 1960s, renamed itself *The Metals and Materials Society* in 1989 (retaining TMS as the moniker) and changed the name of its flagship publication *Journal of Metals* to just *JOM* (without specification), and of its *Metals Transactions* journal to *Metals and Materials Transactions,* to accommodate the broader intended coverages. Like ACerS, TMS is organized in a divisional structure, though somewhat more generically, with divisions in *Extraction & Processing, Functional Materials, Light Metals, Materials Processing & Manufacturing,* and *Structural Materials* that belie its original and continuing metallurgical focus. Its current (2023) membership stands at around 13,000.

Among the many luminaries in the modern metallurgical field, two merit singling out — both for their materials catholicity, though each for somewhat different illustrative purposes here. German-born Hans (later John) Werner Cahn (1928–2016) was, like many eminent scientists in this review of the materials associations, a refugee from Nazi Germany, fleeing as a 5 year old with his family after Hitlers's *Machengreifung* in 1933, initially to Amsterdam, then in 1939 to New York City, losing most of the remainder of his extended family to the Holocaust. On becoming a US citizen in 1945, Cahn was stationed with the US occupation army force in Japan after the War. He subsequently trained in chemistry and physical chemistry at the University of Michigan (BS, 1949) and the University of California–Berkeley (PhD, 1953) and joined the chemical metallurgy effort at General Electric's Schenectady, NY Research Laboratory, then led by David Turnbull (profiled later), extending Turnbull's pioneering work on thermodynamics and kinetics of phase transformations. With British-born colleague John E. Hilliard (later professor in Northwestern's Materials Science & Engineering department), Cahn explored the thermodynamic factors driving phase separation in solids, particularly with a kinetic and morphological transformation known as spinodal decomposition — resulting in a well-known governing principal described by what is known as the Cahn-Hilliard equation. In 1964, Cahn left GE to take up a professorship at MIT in their Department of Metallurgy and Ceramics (later Materials Science & Engineering), conducting seminal work on the influence of mechanical stress on solid-state transformations, on vector-based thermodynamic treatments of interfaces and capillarity, and on solidification and the solid-liquid interface. With graduate student (later MIT professor) Samuel M. Allen, Cahn investigated order-disorder phenomena in iron-based alloys, resulting in the governing Allan-Cahn equation. Cahn left MIT in 1979, for the National Bureau of Standards (later NIST).

In 1982, an icosahedral phase of an Al-Mn alloy was discovered by Israeli electron microscopist Daniel Shechtman during his sabbatical at NBS, adding a new class of solids, *quasi-crystals*, that exhibited sharp crystalline-looking diffraction patterns that were, seemingly impossibly, characterized by fivefold symmetries, forbidden in crystals. Cahn worked out the theory for the phenomenon, and the resulting 1984 paper introducing quasi-crystals was jointly authored by Shechtman, Cahn, and two others. Until his dying day (ironically the publication date of the Shechtman *et al.* paper publication), Nobelist Linus Pauling (in concert with others) vehemently disputed Shechtman's interpretation and deprecated Shechtman, who was roundly disgraced, dubbed a quasi-scientist and worse, until he was ultimately proven correct and awarded the 2011 Nobel Prize in Chemistry (demonstrating that the scientific method not only serves to dismiss erroneous assertions but also to affirm correct ones). Finally, in 2004 Cahn and a (by then NIST) colleague demonstrated that a non-crystalline isotropic metallic phase (a *metallic glass*) could be grown from a molten metal melt — for which solidification mode, as Cahn then established, a nucleation barrier must exist, implying a discontinuous metal/liquid interface between. Cahn was the recipient of astonishingly large lists of awards, including the Hume-Rothery Award of TMS, the Saveur Award of the ASM, the von Hippel award of the Materials Research Society, honorary life membership in ACerS, the Kyoto Prize and, in 1998, the National Medal of Science and election to the National Academy of Engineering (though surprisingly not sharing in the 2011 Chemistry Nobel prize with Shechtman).

A second exemplar who deserves singling out is Julia Randall Weertman (1926–2018), the first woman fellow and first woman director of TMS, a society that (in common with most scientific and engineering societies) has only recently accorded women the sort of professional recognition afforded its male members. Encouraged by materials physicist Frederick Seitz (later president of the National Academy of Science) to pursue a scientific career, Julia Randall enrolled at Carnegie Institute of Technology (now Carnegie Mellon University, uniting bequests of two 19th-century financial magnates) as the first female student in the College of Science & Engineering. There she earned bachelors, masters, and DSc degrees in physics before embarking on a post-doctoral study in Paris at École Normal Supérieure, afterward joining the US Naval Research Laboratory as a solid-state physicist. She had met her husband, Johannes (Hans) Weertman, at Carnegie Tech, and when he joined Northwestern University's fledgling graduate department of Materials Science & Engineering as a professor in 1959, she came along and raised two children over the next 13 years in Evanston, IL. Along the way, she co-authored the textbook *Elementary Dislocation Theory* with her husband (1964, among the first of Northwestern's Macmillan series of textbooks on materials science & engineering, reprinted by Oxford University Press in 1962). In 1972, she was persuaded by Department head Morris Fine to teach in the department, then joined her husband on the faculty and rose to hold a Walter P. Murphy chair. She was elected head of the department in 1987, the first female head of a materials department in the US — doubling the number of departmental MSE majors in her 5-year tenure and recruiting two further women faculty members. Her research explored deformation and creep behavior of polycrystalline metal alloys, then explored cavitation during creep using small-angle neutron diffraction, and finally pioneered the study of nanocrystalline metals where the metal grain size becomes small enough that the grain boundaries dominate the mechanical behavior and properties of the material. She was awarded a Guggenheim fellowship in 1986, elected to the National Academy of Engineering in 1988, and, in addition to TMS, chosen as a fellow of ASM International, the American Geophysical Union, and in 2017 the American Academy of Arts and Sciences. She was the recipient of the 2003 von Hippel award of the Materials Research Society, the Gold Medal of the ASM in 2005, a Distinguished Engineering Educator award from the Society of Women Engineers, and the John Fritz medal of the American Association of Engineering Societies in 2014. In 2017, TMS renamed its Educator Award the TMS Julia and Johannes Weertman Educator Award. Julia Weertman served as an exemplary mentor for women scientists and a tireless recruiter of girls into science and engineering careers, and especially in materials science & engineering.

Association for Iron and Steel Technology (AIST)

In 1907, a group of twenty-eight electrical and mechanical engineers and plant supervisors from the pre-eminent US steel companies met at a trade exhibition in Pittsburgh, showcasing electrical equipment of potential interest to electrifying steel production, and promptly formed the *Association of Iron and Steel Electrical Engineers,* with district sections established in Pittsburgh (1914), Chicago (1915), Cleveland and Philadelphia (both 1916). An iron and steel exhibition was organized in St. Louis in 1919, and a monthly journal *Iron and Steel Engineer* begun in 1924. In 1936 the Association's name was truncated to *Association of Iron and Steel Engineers* (AISE) to better reflect the increasing inclusion of other engineering disciplines in the steel-making process and steel product manufacturing, and over the next 50 years AISE grew to number 14,000 members in 18 district sections, with 16 technical divisions, including (as well as electrical engineering), energy and combustion engineering, lubrication, maintenance technology, safety, coke production and byproducts, iron production, and oxygen steelmaking (in 1954, following its introduction in Austria in 1952), materials handling systems, bar and shape rolling, electric steelmaking, environmental engineering, health & safety — reflecting the evolution of steelmaking technology and practices. Society publications included, besides numerous sub-committee reports, *The Modern Strip Mill* (1941), *Roll Design and Layout,* and the annual *Directory of Iron & Steel Plants.* There were no academic educational programs, but the Association was a charter supporter of the National Merit Scholarship program in the US. The decline in, and eventual demise of, domestic US combined iron and steel production (from 229 million metric tons in 1973 to 107 million tons by 1984 — much of that loss in production ceded to overseas producers) was reflected in a steadily declining AISE membership, diminishing to 6,000 by 2003, less than half that in 1980. In 2004, AISE merged with the ISS Division of AIME to form the *Association for Iron and Steel Technology* (AIST) that remains a sister society within the AIME umbrella.

Polymer and Materials Chemistry Societies

American Chemical Society (ACS)

The American Chemical Society (ACS) was founded in 1876 at the College of Pharmacy in New York City (NY), among the first US large-scale professional scientific societies, with a 1926 membership 50 years after its founding of 18,000 members, and now — nearly 150 years on — among the largest worldwide, with a membership of 151,000, 28,000 of whom reside abroad in 28 different countries.

Its first president, John William Draper, was born in St. Helens, Lancashire (where Pilkington Bros. had begun their glass company 50 years before) and studied chemistry under Edward Turner FRS, a popularizer of Dalton's atomic theory, at University College London, before moving to Virginia in 1832, following the death of his father. He soon after attended the University of Pennsylvania School of Medicine and became a physician. Like many notable scientists, Draper was something of a polymath — a chemist, physician, photographer, historian, and philosopher — notably investigating the origins of the photochemistry upon absorption of light and improving the daguerreotype photographic process to capture the first clear representation of the human face (his sister's), and also of the moon, as well as taking the first photomicrographs by 1850. Applying the scientific method to history, he wrote three historical treatises: *The History of the Intellectual Development of Europe* (1862), *The History of the American Civil War* (in three volumes, 1867–1870), and a *History of Conflict between Religion and Science* (1874), the first and last of these extending his arguments in an earlier lecture he'd delivered during the famous 1860 Oxford evolution debate, his paper entitled "On the intellectual development in Europe, considered with reference to the views of Mr. Darwin, and others, that the progression of organisms is determined by law."[58] He helped found the New York University Medical School in 1839, served as its president for 23 years and was elected to the NAS in 1876. It is worth remembering that there have always been strong links between chemistry, pharmacology, and medicine, and that pharmacists then (and still now in the UK) were called chemists.

The ACS has long divided itself into disciplinary divisions, 17 in 1930, and 33 in 2023 (all now assigned convenient four-letter identifiers!). Those featuring materials content are the divisions of Polymeric Materials: Science & Engineering (PMSE), Rubber

---

[58] Fleming (1950).

(RUBB), Biochemical Technology (BIOT), Catalysis Science & Technology (CATL), Energy & Fuels (ENFL), and Geochemistry (GEOC). Of these, the Polymeric Materials division is probably the most important for the materials community, since the only other dedicated polymers professional society is the US is the Society of Plastics Engineers (SPE), with much less science content, while worldwide the closest competitor is the Society for Polymer Science (SPS) in Japan. The ACS has two large annual meetings, one in the Fall and one in the Spring, each featuring a "theme," but with extensive divisional content in most divisions. Of recent, materials figure surprisingly commonly in the annual meeting themes: for example, "Materials for Health & Medicine" in Fall 2012, "Chemistry & Materials for Energy in Spring 2014, "Advanced Materials, Technologies, Systems and Processes" in Spring 2017, "Nanoscience, Nanotechnology, and Beyond" in Fall 2018, and "Macromolecular Chemistry: The Second Century" in Spring 2021, reminding how much chemistry still resides in the practice of materials science and engineering.

The ACS is a *huge* publisher, universally known for its longstanding *Chemical Abstracts* service (45 million abstracts to date, which include most materials science papers), and maintains a catalog, *CAS Registry*, of 165 million organic and inorganic substances. Its flagship publication, the *Journal of the American Chemical Society*, was founded in 1879 and is published *weekly*, accruing to some 19,000 pages annually. The Journal has produced a volume of its history, *Celebrating 140 years: Discover 140 years of JACS' History*, for those (like the inaugural president and historian Draper) with a historical inclination; ACS had briefly published a series of monographs in the 1920s, likewise of historical interest. ACS in addition publishes the separate *Journal of Organic Chemistry*, *Journal of Physical Chemistry* (in 4 parts) and *Journal of Physical Chemistry Letters*, *Langmuir*, *Chemical Reviews* and 76 (!) other journals, including some specifically aimed at materials communities or with materials content: *Accounts of Materials Research, ACS Applied Bio Materials, ACS Applied Electronic Materials, ACS Applied Energy Materials, ACS Applied Materials & Interfaces, ACS Applied Nanomaterials, ACS Applied Optical*

Materials, ACS Applied Polymer Materials, ACS Biomaterials Science & Engineering, ACS Catalysis, ACS Materials Au (gold open access), *ACS Materials Letters, ACS Nano, Nano Letters, ACS Nanoscience Au* (gold open access), *ACS Photonics, ACS Polymers Au* (gold open access), *ACS Sensors, Biomacromolecules, Chemical & Biomedical Imaging, Chemistry of Materials, Crystal Growth & Design*, and *Macromolecules* — pretty much covering the waterfront of materials chemistry, notably excepting inorganic glasses and ceramics, hard-core metallurgy, and electronic materials physics.

Society of Plastic Engineers (SPE)

The Society of Plastic Engineers was founded in 1942 as a societal home for professional engineers employed in the plastic and polymers industries (though not academia). Society membership stood at 22,500 in 2023, spread across the US and Canada and in 82 other countries, with 39 university chapters, only four of them outside North America. The Society divides itself into 32 technical chapters, representing the various interests of the plastics industries. The Society publishes three journals: *Polymer Engineering & Science, Polymer Composites*, and the *Journal of Vinyl and Additive Technology*, plus a (gold) open-access journal *SPE Polymers*; it also publishes the *Plastic Engineering* magazine, *SPE News*, the *Plastic Insig*ht newsletter, along with a *Resource Guide* to the plastic industries, and maintains a large online library of relevant articles. SPE holds an annual technical conference, ANTEC. The presidents have come mainly from industry (the 2023 president is from Dow Chemical), but SPE had no women presidents until 1993, then three between 1994 and 2007, and none since.

Society of Polymer Science–Japan (SPSJ)

The largest professional society catering to polymer scientists, after the American Chemical Society, is the *Society of Polymer Science in Japan* (SPSJ), founded in 1951. Now in its eighth decade, SPSJ has approximately 12,000 members worldwide, organized into eight regional chapters from Hokkaido to Kyushu, with a supernumerary chapter, *Kōbunshi Doyukai* (Federation of Technological Management Professionals in Polymer Industries), for management professionals in the polymer industries in Japan. The Society maintains 27 research groups, *inter alia* polymer gels, polymer surfaces, plastic films, biomedical polymers, adhesion and coatings, and supermolecules. SPSJ holds three major meetings each year, an Annual Meeting in May (in 2007, featuring a guest lecture by Nobelist Akira Yoshino on the Li-ion battery), a Symposium on Macromolecules in September, and the Polymers Materials Forum in November. The Society additionally sponsors eight lecture meetings each year, organized through individual research groups, focusing on current topics in polymer science and technology.

SPSJ publishes its flagship international peer-reviewed organ *Polymer Journal* (*Kōbunshi Gakkai*) in English, monthly since 1970 (and including the original articles published by Akira Yoshida on the Li-ion battery), with additionally a Japanese language monthly journal *Kōbunshi* (Polymers) published since 1952, and *Kōbunshi Ronbunnshu* (Japanese Journal of Polymer Science and Technology) in Japanese, with English abstracts, discontinued after 2019.

SPSJ's third president was Ichiro Sakurada (1904–1986), who had by his election in 1961 already risen to international prominence as an early pioneer in polymer science and fiber chemistry. Fascinated as a high schooler by red-colored colloidal gold sols, Sakurada was encouraged by his mother to study engineering (rather than following in his father's footsteps as an itinerant foreign correspondent) and matriculated at Kyoto Imperial University in 1923 in the department of Industrial Chemistry. There he became additionally interested in fibers, investigating cellulose acetate fibers, and stayed on for 2 years of postgraduate study with his mentor Gen-itsu Kita, who afterward sent him to Germany in 1928. At *Universität Leipzig*, Sakurada worked with colloid chemist Wolfgang Ostwald, son of Nobel-prize winning physical chemist Wilhelm Ostwald (of Ostwald ripening systems), then moved to the *Kaiser Wilhelm Institut* (KWI) *für Chemie* in Berlin-Dalhem to study with cellulose chemist Kurt Hess, an ardent national socialist, who was equally ardently involved in the debate about the molecular structure of polymers. Hess favored the idea that polymers were physical aggregates of small molecules, rather than of large, three-dimensional macromolecules — the alternative theory, proposed by Hermann Staudinger (inventor of pyrethrins) who imagined instead long stiff rods (for which he was awarded the Nobel Prize in Chemistry in 1953). Back in Kyoto in 1931 for his doctoral studies, Sakurada, who had been unable to crystallize the supposedly small molecules of cellulose, came to the conclusion that such polymers must instead be long flexible chain molecules and pursued this line of thought as a professor in Kyoto, publishing in 1940 the seminal paper (in Japanese) entitled "The shape of threadlike molecules in solution and the relationship between solution viscosity and molecular weight," that remained largely unknown in the West, though it proposed the correct structure of polymers.

Sakurada then turned his attention to polymer fibers, seeking a synthetic version of silk. When Dupont came out with nylon in 1939, Sakurada was able to get a few milligrams and deduced its structure by X-ray diffraction. He then helped start a department of Fiber *Chemistry* in Kyoto and joined others in creating a research *Association for Synthetic Fiber* (*Nippon Gosei Sen'i Kyokai*). He next focused on making synthetic polymer fibers from polyvinyl alcohol (PVA), with which he synthesized a PVA-derived synthetic fiber, dubbed *vinylon*, subsequently used in agriculture, fishing, and other industries. In 1943, the research association was transformed into the *Kōbunshi Kagaku Kyokai* (Association of Polymer Chemistry), which started publishing a journal *Kōbunshi Kagaku (Polymer Chemistry)*, almost 2 years before the appearance of Herman Mark's[59] *Journal of Polymer*

[59] Herman Francis Mark (1885–1992) was a Viennese-born polymer scientist with a Hungarian Jewish father (who had converted to Christianity) and a Viennese mother. He began university studies in Vienna, but followed his thesis advisor to the *Kaiser-Wilhelm-Institüt* (KWI) in Berlin-Dahlem (which had initially specialized in fiber science under the directorship of Fritz Haber and later became the *Fritz Haber Institüt* of the *Max-Planck Gesellschaft*). There Mark joined the research group of Michael Polanyi and experimented with X-ray diffraction, eventually investigating the structures of silk, cotton and wool fibers. Mark was later thesis

Science in the West. In 1951, Sakurada and others transformed the Association into the *Kōbunshi Kagaku Gakkai* (Society of Polymer Science in Japan, SPSJ), and as the Society's third president, served for *four* consecutive terms 1961–1968.

Materials Physics Societies

American Physical Society (APS)

The *American Physical Society* (APS) was founded by a group of 36 physicists meeting at Columbia University (New York City) in 1899 in order to promote the discipline of physics. Today, it

---

supervisor for Eugene P. Wigner (Nobel prize in Physics, 1963). With a physicist colleague, Raimund Wierl, Mark invented and pioneered the technique of electron diffraction (orders of magnitude higher scattering amplitudes compared to X-rays) from molecules in the gas phase, making possible the measurement of atom-to-atom bond-length distances in the molecule, starting with the Cl-Cl bond length in CCl_4 and extending to other simple molecules. Mark taught electron diffraction to Linus Pauling (1901–1994), who had come to Europe on a Guggenheim fellowship in 1926 and visited Mark; much of the data on bond-lengths in Pauling's 1939 book *The Nature of the Chemical Bond* came from gas-phase electron diffraction. (Pauling won his first Nobel Prize in Chemistry in 1954 for his work on X-ray diffraction of proteins and determination of the structure of the α-helix). Shortly after Pauling's visit in1926, Mark left the KWI Fiber Institute to take up a position with IG Farbenindustrie (IGF), where he began research into cellulose, silk, rubber and synthetic polymers. His X-ray diffraction studies, both at KWI and at IGF, provided incontrovertible proof that polymers were made up of long-chain *flexible* molecules, settling the long-running controversy between Staudinger and those who favored three-dimensional macromolecules. Staudinger never acknowledged the seminal contribution Mark's X-ray work had made to understanding polymer structure, nor Sakurada's, even when Staudinger alone received the Nobel Prize in Chemistry in 1953 for the enunciation of the long-chain polymer motif (but without its flexible elucidation, which was mainly Mark's). Owing to his father's Jewish heritage, Mark moved to Vienna before the *Anschluß* but was afterwards detained by the Gestapo, then released, and made a harrowing escape from Berlin to Switzerland, thence to Canada and finally the US, eventually settling in at Brooklyn Polytechnic Institute in New York. Mark founded the *Journal of Polymer Science* in 1951. A memorial remembrance has been offered by Hungarian historian of science Istvan Hargittal: (2022).

numbers approximately 50,000 members. The second president was Prussian born-American Albert A. Michelson (1852–1931), who emigrated with his parents to the US when he was age 2 and attended the US Naval Academy in Annapolis, MD, where he made early measurements on the speed of light, then later took a position with the Case School of Applied Science (Cleveland, OH, now Case Western Reserve University). In 1887, he and Western Reserve chemistry professor Edward Morley performed the famous Michelson-Morley experiment that disproved the existence of any "aether" as a medium through which light or radio waves were supposed to have propagated. Michelson was awarded the 1907 Nobel prize in Physics. The list of succeeding APS presidents reads like a veritable who's who list of American physics super-stars: University of Chicago's Robert A. Millikan (1923 Nobel Prize in Physics for his oil drop experiment that precisely measured the electron charge); Harvard spectral physicist Theodore Lyman IV (Lyman spectral lines); Princeton physics chair Karl Taylor Compton (later president of MIT); Harvard's Percy W. Bridgman (1946 Nobel Prize in Physics for his experiments in high-pressure physics); Hungarian-American Columbia physicist Isidor I. Rabi (1943 Nobel Prize in Physics for nuclear magnetic resonance measurements of magnetic moment and nuclear spin of atoms, and during World War II a "Rad Lab" physicist); German-born Cornell theoretical physicist Hans A. Bethe (1967 Nobel Prize in Physics for stellar nucleosynthesis, and head theorist for the Manhattan project); Hungarian-American physicist Eugene P. Wigner (1963 Nobel Prize for Physics for his work in Göttingen on group-theory mathematical formulation of quantum mechanics — Wigner also worked for a time at the *Kaiser-Wilhem Institüt* at Berlin-Dahlem); Harvard's John Hasbrouck Van Vleck (who shared the 1977 Nobel Prize in Physics for his establishment of the quantum mechanics of magnetic systems and crystal field theory, shared with Philip W. Anderson and Sir Neville Mott); and Göttingen (and later MIT) theorist Viktor Weisskopf (1980 National Medal of Science and a multiple nominee for a Nobel Prize, co-founder of Union of Concerned Scientists); and many later luminaries.

Two later APS presidents who made *immense* contributions to materials science are Frederick Seitz (1911–2008) and Mildred Spiewak Dresselhaus (1930–2017). A brilliant mathematical physicist, Seitz worked under Wigner at Princeton and had faculty appointments in several universities, most notably the University of Illinois (Urbana-Champaign) and earlier at the University of Pennsylvania (Philadelphia) where at age 29 he wrote his classic *Modern Theory of Solids* (McGraw-Hill, 1940) that firmly established the field of solid-state condensed matter physics, particularly the role of atomic-scale defects in the properties of solid materials; in his later career, he was President of the US National Academy of Science and of Rockefeller University, a National Medal of Science recipient and a key figure in promoting the consolidated field of Materials Science & Engineering.

MIT's Dresselhaus has been called (affectionately, with *never* a hint of misogynist condescension) "Millie, the Queen of Carbon Science" for her encyclopedic work on the many forms of carbon-based solids (graphene, graphite, graphite intercalation compounds (important electrodes in some advanced battery designs), "buckeyball" fullerenes, and carbon nanotubes and nanowires) over her long career, one that began at Hunter College (New York City), then to University of Chicago for her PhD under Nobelist Enrico Fermi. Her 57-year sojourn at MIT started at MIT's Lincoln Laboratory, then as a professor of both Electrical Engineering and Physics in those MIT departments, and finally as a much-revered Institute Professor. A steadfast champion of women in science, she was the first woman to be elected to the National Academy of Engineering and was awarded both the National Medal of Science *and* the Presidential Medal of Freedom, the Enrico Fermi award, and countless other honors. She was also a serious musician, and her family supplied an all-family quintet dubbed "Voice of the Whale." Despite her stellar example, APS has at best a modest record of female presidents: five in the past 40 years, and only one in the past 25; but that sparsity likely more reflects the scarcity of women in physics and not any inherent APS bias.

At first, APS only held meetings, four times a year, but in 1913 it took over the journal *Physical Review,* that had been founded at Cornell University (Ithaca, NY) in 1893. That publication, which much later was divided into five different parts (as essentially separate journals), was joined in 1929 by *Reviews of Modern Physics.* The Society now meets twice a year, an annual meeting in March and an annually designated topical meeting in April, both with large attendances (6,000+), each with divisional sub-structures. APS sustains 10 regional sections and 25 topical forums (including Magnetism and its Applications and Shock Compression of Condensed Matter) and divides itself structurally into 18 divisions, those with materials prominent being: Materials Physics, Condensed Matter Physics, Polymer Physics, Laser Science, Biological Physics, Chemical Physics, and Soft Matter Physics, nearly half the complement. It also supports numerous committees, in fact even a Committee on Committees, to keep track of them all.

Besides the five flavors of *Physical Review,* of which the most relevant to materials is *Physical Review* B, devoted to condensed matter physics and materials physics; there is also frequent material content in *Reviews of Modern Physics* and sometimes in *Physical Review* X (for timely impact, online) and *PRX Life* (life sciences, online).

American Institute of Physics (AIP)

The *American Institute of Physics* (AIP) is a ten-member federation, founded in 1931, of the APS with other scientific societies that promote physics and physics education. Those member societies with some materials content are as follows: APS; AVS: *Science & Technology of Materials, Interfaces & Processing* (formerly *American Vacuum Society*); *American Crystallographic Association* (ACA); and *Optica* (formerly OSA, the *Optical Society of America*) — together with another 15 affiliate societies, including (again, those with varying degrees of materials content) the *Microscopy Society of America* (MSA), the *Electrochemical* Society (ECS), the *American Nuclear Society* (ANS), the *International*

Centre for Diffraction Data, the *American Geophysical Union* (AGU), the *Geology Society of America* (GSA), and the *Biophysical Society.* AIP is mainly known for its publications; those with some materials content are as follows: *Physics Today* magazine, *Journal of Applied Physics, Applied Physics Letters, APL Materials, APL Photonics, APL Bioengineering, APL Energy, Journal of Renewable & Sustainable Energy, Matter and Radiation at Extremes, Journal of Chemical Physics, Chemical Physics Reviews, Review of Scientific Instruments, Nanotechnology & Precision Engineering, Biophysics Review,* together with numerous conference proceedings.

Institute of Physics (IOP) [UK]

The history of the *Institute of Physics* (IOP) begins in 1874 with the founding of the *Physical Society of London,* for much the same reasons the APS was founded, though there was a good deal more physics done (and physicists accumulated) in Britain during the 17th–19th centuries than in North America. Among the diverse group of 29 founders were, not only professional academics but also school teachers and amateur scientists, including women — though the group still self-described as *Fellows* of the Physical Society of London. The fledgling Society met in the West End of London, at the Royal College of Science and Imperial College. Intent on improving the status of physicists, the Physical Society consulted other parallel societies (the *Faraday Society,* the *Optical Society,* and the *Röntgen Society*) and agreed jointly to construct a federation of the societies, incorporated in 1920 as the *Institute of Physics* (IOP), which shared offices with the Physical Society. The inclusion of two materials-related societies, the *Royal Microscopical Society* (RMS) and the *Röntgen Society,* both joining as associate societies, explains why much of the X-ray diffraction work of the 1920s–1950s and electron microscopy from the 1950s onwards were societally channeled through the IOP. The Physical Society had, in 1905, begun an annual exhibition of scientific instruments and related apparatus, and that emphasis on experimental physics was reflected in the inaugural publication of the federated Institute of

Physics: the *Journal of Scientific Instruments,* begun in 1922. The IOP added *Reports on Progress on Physics* in 1934. The Physical Society began to reach out geographically from London, establishing branches first in Manchester, then in Australia and India in 1932. In 1947, merger talks began but did not bear fruit until 1960, after the *Faraday Society*, the *Royal Meteorological Society* and the *British Institute of Radiology* had withdrawn from the IOP federation.

The president of the Physical Society at the time of the merger, Sir Neville Francis Mott FRS (1905–1996), was a highly influential condensed matter physicist, who won the 1977 Nobel Prize in Physics, shared with Philip W. Anderson and John Hasbrouck Van Vleck of the APS, for work on the structure of disordered electrical and magnetic systems, especially amorphous semiconductors. Mott spent 38 years at the University of Bristol, as chair of theoretical physics and director of the H. H. Wills Physical Laboratory, and 17 years as Cavendish Professor of Physics in Cambridge, and is known also for introducing the concept of a Mott insulator in solids (e.g., transition metal oxides, glasses, amorphous semiconductors) that should conduct electricity but don't, and so can undergo a metal-insulator transition, and for the Mott-Littleton method of modeling atomistic defects in ionic solids.

The (merged) IOP now has 20,000 members worldwide and supports 53 special interest groups; those with significant involvement with materials science are as follows: Carbon, Computational Physics, Dielectric and Electrostatics, Electron Microscopy & Analysis (EMAG), Magnetism, Materials & Characterization, Nanoscale Physics and Technology, Nuclear Industry, Optical, Polymer Physics, Quantum Electronics & Photonics, Semiconductor Physics, Structural Condensed Matter Physics, Superconductivity, Theory of Condensed Matter, and Thin Films & Surfaces. IOP also administers a large number of topical conferences across the range of the interest groups and publishes *Physics World* magazine, various conference proceedings, and well over 100 journals. Those with substantial materials content are: *Journal of Physics Materials, Journal of Physics Condensed Matter, Journal of Physics Photonics, Journal of Solid State Science & Technology, Electronic Structure, 2D Materials, Materials Futures,*

Materials Research Express, Modeling & Simulation in Materials Science & Engineering, Nano Express, Nanotechnology, Nanoscience & Technology, Bio-inspirational & Biomimetics, Biomedical Materials, Progress in Biomedical Engineering, Semiconductor Science & Technology, Superconductor Science & Technology, Smart Materials & Structures, and *Reports on Progress in Physics.*

Glass and Optical Materials Societies

International Commission on Glass (ICG)

The first international organization disseminating knowledge about the glassy state and promoting shared industrial technology arose from a close collaboration that developed between Sheffield's *Society of Glass Technology* and the *German Society for Glass Technology* in Frankfurt-am-Main, leading to a joint meeting in 1928. The driving forces were the ubiquitous Prof. William E. S. Turner, head of the (by then) Department of Glass Technology at Sheffield, and Prof. Georg R. Gehlhoff, head of the Glass Laboratory at Weißwasser and a board member of the lighting company Osram in Berlin (now Osram Sylvania in the US). A center of European glass production by 1900, the north Saxony town of Weißwasser boasted eleven glass factories and five glass refineries. The first *International Congress on Glass* was held in Venice in 1933, with Turner as its first president. The Congress morphed into a full-fledged international association of national glass societies and glass-related associations, the *International Commission on Glass* (ICG), replete with executive and committee structures, with 32 member countries participating (including, of course, ACerS, whose Glass and Optical Materials division has supplied five of ICG's 27 presidents).

As well as organizing an International Congress on Glass meeting, held every 3 years (the 26th of which was held in 2022), ICG supports frequent smaller joint conferences and glass summer-schools (the 14th ICG Montpelier Summer School on Glass was held in 2023). It compiles a variety of bibliographic, reference, and topical volumes and collections, and was successful in convincing the United Nations to declare 2022 the International Year of Glass.

Glass and Optical Materials Division of ACerS (GOMD) and *Society for Glass Technology (SGT) [UK]*

Because the *Optical Society of America* (*Optica*, described in the following section) has a very different mission, focused on optics and photonics, the *Glass Division* of the *American Ceramic Society* (GOMD) is the only comprehensive glass-science organization in the US and is paralleled in the UK by Turner's *Society for Glass Technology* (SGT), founded in Sheffield in 1905 at about the same time the ACerS Glass Division was established. A former division head of the ACerS Glass Division, Arun Varshneya, is now President of the *Society of Glass Technology*, and there have been many other links since both were founded. Varshneya and two of his Alfred University glass colleagues spun off a company 26 years ago for chemically-strengthening the glass cartridges of the life-saving epinephrine-injecting EpiPen®, a half-billion of which have been distributed throughout the world.

The ACerS Glass Division has been a particularly active ACerS division in its own right, having been formed at the outset of ACerS and claiming a rich history of intimate divisional meetings that firmly cemented the glass community, not only in the US but also internationally. For many years, from the 1930s to the 1970s, the Glass Division meetings were held in resort settings mostly in Pennsylvania (often Bedford Springs, PA) or the New Jersey shore, featuring Gordon Conference-like ambience, but with 300 members typically attending, including many from the industry. MIT professor Bertram E. Warren, who carried out the first attempts at structural characterization of simple glasses using X-ray diffraction, was a frequent attendee of these early GOMD meetings. W. E. S. Turner even came over from Sheffield for a GOMD meeting in Washington, DC in 1938 to plan the upcoming 1939 ICG meeting scheduled for Germany, which of course never took place.

Another faithful attendee was Norbert J. Kreidl (1904–1994), a refugee from Vienna just after the Nazi-imposed *Anschluß*, who fled to the US in 1938 and ended up at Pennsylvania State University, where he continued his glass career. Trained in Germany, he had worked for 10 years in the glass industry in Czechoslovakia, where he patented some opalescent glasses, before emigrating. In 1943, Kreidl joined Bausch & Lomb (founded in 1853 in Rochester, NY), eventually rising to become their director of materials research, and retiring at the (then) mandatory retirement age of 65. With the supply of German optical glass cut off during the war years, Bausch & Lomb pioneered US production of lens-quality optical glass for the huge number of military instruments needed in the war effort; postwar, the company developed the improved optics needed for CinemaScope in movie theaters and is now one of the world's largest supplier of contact and intraocular lenses. For the 25 years following his forced retirement, Kreidl returned to teaching, first at Rutgers University, then University of Missouri–Rolla for two more decades, and later at the College of New Mexico in Santa Fe. An annual award of the Glass Division, for the outstanding young glass scientist is named for him.

In 1977, Kreidl personally asked L. David Pye (who had worked for him at Bausch & Lomb, and later become Dean of the New York State College of Ceramics at Alfred) to organize at Alfred a conference on *Boron in Glass* (borate (B_2O_3)-containing glasses, like Pyrex®), which both metamorphosed into an on-going university conference series on glass science and inspired an additional series of recurring international conferences on borate glasses. Pye subsequently served as both the 1997–2000 president of the International Congress on Glass (ICG) and the 2007 president of the American Ceramic Society (ACerS).

A fourth Glass division luminary was Australian-born (Charles) Austen Angell (1933–2021), Regents Professor at Arizona State University, who was among the world's most cited authorities on the physics and chemistry of glass and glass-forming liquids. He graduated from Melbourne University, where he read chemical metallurgy, then researched molten salts with John Bockris at the University of Pennsylvania, before heading off to Imperial College London for a

PhD in physical chemistry, studying self-diffusion in metal-molten salt solutions, afterward returning to Australia. In 1964 he was back in the US and joined Argonne National Laboratory, where he explored cation coordination in ionic solvents, then took up a professorship at Purdue University (West Lafayette, IN) for the next 23 years. He made seminal contributions to two phenomena: the first, the anomalous behavior of super-cooled water, predicting a critical point singularity at –45°C; the second, establishing the concept of *fragility* in glass-forming liquids, referring not to mechanical strength but to the extent to which formation of an aperiodic (amorphous) solid upon cooling deviates from Arrhenius behavior with temperature. For a *strong* glass-forming liquid (least fragile) — like molten (liquid) silica (SiO_2) — Arrhenius behavior predicts that the logarithm of liquid viscosity η (viscosity is a measure of resistance to flow) should vary linearly with inverse temperature ($1/T$) as the liquid approaches the glass-transition temperature T_g, implying a unique activation energy governing viscous flow and gradual transition to a glassy solid over a comparatively wide range of temperatures upon cooling. A plot of $\ln(\eta)$ *vs.* T_g/T is known as an *Angell plot.* For a *fragile* glass-forming liquid (which includes many organic and biological glass-forming liquids), the viscosity increases far more quickly as the glass transition temperature is approached, implying a much narrower temperature range over which the liquid solidifies. This way of categorizing the glass transitions revolutionized thinking about the formation of amorphous solids, and an entire 1988 special issue of the *Journal of Physical Chemistry* B was devoted to papers from a *festschrift* for Austen Angell, with an Angell plot on the cover. In 1989, Angell was offered a Regents Professorship at Arizona State University, where he continued his seminal work on establishing fundamental paradigms governing anomalous behavior of liquids and became recognized as perhaps the most versatile physical chemist of his generation. He received enormous recognition for his insights, with major awards from four leading materials-related scientific societies: the Materials Research Society (the 2006 David Turnbull lecture), the American Ceramic Society (1990 George W. Morey Award), the Electrochemical Society, and the American Chemical Society, as well as the W. H.

Bragg lecture from Imperial College, the Otto Schott Research Award from Schott GmbH, the Lise Meitner award from Swedish Gothenburg Physics Centre, and the Galileo Galilei award from the International Society for Polymer Electrolytes in Italy.

Frederick Seitz (profiled earlier, at the time chair of the Research Council of the US National Academy of Sciences (NAS)) in 1957 asked Canadian-born ceramist Van Derck Fréchette (1916–2001), longtime professor of ceramic and glass technology (1944–1987) at the New York State College of Ceramics at Alfred University, to organize a timely conference at Alfred on "Non-Crystalline Solids" — just at the time when float glass was developed by Pilkington in the UK and controlled crystallization of glass (leading to the production of glass-ceramics like CorningWare®) at Corning Glass, and just before metallic glasses had been successfully demonstrated (1960) by Belgian-born physicist Pol Duwez at California Institute of Technology. (Metallic glasses were soon after exploited by a small Rochester, NY company developing a revolutionary dry photocopying process, subsequently renamed Xerox). Frèchette's 1958 *Conference on Non-Crystalline Solids*, whose keynote speaker was Hans Bethe (later 1967 Physics Nobelist), proved a watershed in the study of non-crystalline materials science in North America and shortly after prompted noted UK-born American glass scientist (John) Douglas Mackenzie (1926–2020), then working at GE's Schenectady, NY Research Laboratory to launch the three-volume monograph series on *Modern Aspects of the Vitreous State* that morphed (in 1969) into the *Journal of Non-Crystalline Solids* (JCNS), while Mackenzie was a professor at Rensselaer Polytechnic Institute (Troy, NY). Mackenzie was editor of JCNS for the next 20 years and was later inducted into the National Academy of Engineering. These notable American developments led to the convening of an ICG congress for the first time in the US, in Washington, DC in 1962. In a wry characterization by David Pye (the former ICG president, for whom the ACerS-GOMD glass science lifetime-achievement award is named, the 2019 recipient of which was David Mackenzie), glass science had become "an adopted orphan in the emerging solid-state discipline."

The ICG and the ACerS Glass Division were instrumental in getting the United Nations to declare 2022 as "The Year of Glass." To kick-off this recognition of the importance of the glassy state to humanity, the ACerS Glass Division held a *National Day of Glass* celebration at the National Academy of Science in Washington, DC in April 2022, followed by the *National Day of Glass Symposium* (See footnote 23), organized by North American *Year of Glass* steering committee led by David Pye and others, which featured prominent glass scientists, recounting all aspects of the past and future of glass science and industry, from nuclear waste glass to glass photonics, and even included internationally prominent glass artists, such as the much celebrated Dale Chihuly (the 2020 recipient of the David Pye lifetime-achievement award).

ACerS-GOMD publishes the *International Journal of Applied Glass Science* (founding editor L. David Pye) currently under the co-editorship of another former GOMD chair, Mario Affatigato, a chaired professor at Coe College (IA), and Alicia Duran, head of glass research in the Ceramics & Glass Institute of Spain's *Consejo Superior de Investigaciones Cientificas* and 2019–2022 president of ICG. ACerS has also established a *Ceramic and Glass Industry Foundation,* as well as co-sponsoring an annual *Conference on Glass Problems,* now in its 84th year, organized by the *Glass Manufacturing Industry Council* and Alfred University. Scientific articles in glass science most frequently appear in the Elsevier journal *Journal of Non-Crystalline Solids,* at present co-edited by a former Norbert J. Kreidl awardee Liping Huang, professor and Associate Dean of Research at Rensselaer Polytechnic Institute (Troy, NY).

Optical Society of America (OSA) [since 2021 *Optica*]

The *Optical Society of America* (OSA) was founded in 1916 by a group of 30 optical scientists and optical instrument makers in Rochester, NY (home of Eastman Kodak, which pioneered photography and photographic optics). Since that foundation, the Society has increasingly become the worldwide leading association for the fields of optics and photonics. In 2008, in light of its increasing international presence and membership, the Society changed its name to just the *Optical Society* and in 2021 to *Optica.* Now catering to some 24,000 individual members and 340 corporate members in 183 countries, the Society publishes multiple journals: the flagship publication *Journal of the Optical Society of America,* which first appeared in 1918, and since 1984 has been divided into two parts, *A* for research in optics, image science and vision, and *B* for research in optical physics; *Applied Optics and Photonics* since 1962; *Optics Letters* in 1977 (print and now online); since 1975 a magazine *Optics and Photonic News*; and since 2009 *Advances in Optics and Photonics.* Photonics is, broadly defined, the science, manipulation, and application of light, through research into photon sources, photon-conducting media, and photon detection devices — a scientific field that has grown hugely since the introduction of fiber optics as the basis for telecommunications and will continue to grow as the introduction of photon-conducting pathways into nanoscale electronic circuitry and devices promises greatly enhanced processing speeds. The US *National Research Council* has recently weighed in on the centrality of glass photonics to the our shared future.[60]

At the end of last century, OSA began to broaden its publication base in both topical content and geographical access, starting with the open-access publications *Optics Express* in 1997, followed by *Biomedical Optics Express* in 2010 and *Optical Materials Express* in 2011. *Optica* was added, in 2014, then in 2018 the open access *OSA Continuum,* both aimed at rapid dissemination of high-impact results in all areas of optics and photonics. OSA has also partnered with other societies in a supporting role or in joint publication endeavors, such as *Applied Spectroscopy* (with Society for Applied Spectroscopy, since 1951), the *Journal of Lightwave Technology* (with IEEE, since 1998), the *Journal of Display Technology* (with IEEE, 2005–2016), *Journal of Optical Communications and Networking* (with IEEE, since 2009), *Photonics Research* (with Chinese Laser Press, since 2013), and *Journal of Optical Technology* (translated from the Russian *Opticheskii Zhurnal,* since 1999).

---

[60] National Research Council (2013).

OSA meetings were at first joint with the American Physical Society (APS), starting in 1918. Later, OSA began its own series of meetings, now mostly centered on topical subjects and international meetings, rather than annual Society meetings. Notable OSA presidents include physicist Arthur L. Schawlow (1921–1999), who with Charles Townsend, co-invented the laser and shared with him the 1981 Nobel Prize in Physics; AT&T scientist Tingye Li NAE (1931–2012) who pioneered microwave and lightwave telecommunications methodologists in wide use today; and Canadian optical physicist Donna T. Strickland FRS — OSA's fifth woman president — who shared the 2018 Nobel Prize in Physics for chirped pulse-amplification, providing ultrashort terawatt and petawatt light pulses. In recent years, as OSA/Optica has increasingly internationalized, the Society has elected several distinguished presidents from outside the US, for example, Ian Alexander Walmsley FRS, Provost of Imperial College London and formerly pro-Vice Chancellor for Research at Oxford University, who currently directs the Networked Quantum Information Technologies hub in the UK National Quantum Technology Programme.

Electronic Materials Societies

Electrochemical Society (ECS)

Electrochemistry traces its roots back to Humphry Davy (1778–1829, later Sir Humphry Davy Bt FRS and President of the Royal Society), a self-made chemist who was the first to employ solution electrolysis to extract elements Na, K, Ca, Sr, Ba, I and B from their soluble compounds. Working at the Royal Institution in London and using Voltaic piles (batteries), Davy also investigated the corrosion of iron floodgates by copper-clad ship bottoms and explored cathodic protection of the copper cladding using iron or zinc sacrificial anodes. His laboratory assistant and successor at the Royal Institution, Michael Faraday (1791–1867), introduced the terms *anode* and *cathode* electrodes and the intervening *electrolyte*, for the three constituent components of electrochemical cells, and subsequently deduced that (1) the amount of product (or chemical change) produced by current at an electrode-electrolyte boundary is proportional to the quantity of electricity used, and (2) the amounts of chemical changes produced by the same quantity of electricity in different substances are proportional to their equivalent weights. Meanwhile, a recently appointed 46-year-old professor of chemistry at nearby King's College London, John Frederick Daniell, had eliminated the danger of hydrogen production in Volta-style batteries in 1836 with a clever design, now called a Daniell cell (forerunner of modern battery design), that solved the polarization problem. A 31-year-old professor of experimental philosophy at the London Institution (modeled after the Royal Institution), William Robert Grove — a barrister turned scientist (later knighted) — showed in 1842 that, by *reversing* the polarity of a cell for electrolysis of water (which evolved hydrogen and oxygen), hydrogen and oxygen could be recombined electrochemically, producing water and electricity and so inventing the (hydrogen gas) fuel cell; Faraday was the first to witness Grove's invention.

German chemist and Nobelist Walther Hermann Nernst,[61] who famously formulated the Third Law of thermodynamics, subsequently developed the theory of the electromotive force of the Voltaic cell in 1888, applying the concepts of thermodynamics to the

---

[61] (Walther) Hermann Nernst (1864–1941) was educated in Physics and Mathematics successively at the universities of Zurich, Graz (under Ludwig Boltzmann and others), and Wurzburg (under physicist Friedrich Kohlrausch, who also taught Svente Arrhenius), writing his doctoral thesis on the electromotive force generated by magnetism in heated metal plates. In 1888, he joined German physical chemist Wilhelm Ostwald (Nobel prize in Chemistry, 1909), Dutch physical chemist Jacobus Henricus van't Hoff (winner of the first Nobel prize in Chemistry in 1901) and Swedish physicist Svente Arrhenius (Nobel prize in Chemistry, 1903) at *Universität Leipzig* in 1888, and there elucidated his theory of galvanic cells, inspired by Arrhenius' dissociation theory. He formulated what is now known as the Nernst equation, relating the change in thermodynamic free energy ΔG of an electrochemical cell reaction to the electrochemical cell potential E, $\Delta G = - z\ (N_A\ q)\ E$ where z is the number of electronic charges transferred in the cell reaction, q is the fundamental electron charge, *NA* is Avogadro's number, and $(N_A\ q)$ is known as the Faraday constant F. Nernst formulated his famous heat theorem, now the Third Law of thermodynamics, for which he is perhaps better known, in 1906, and won his own Nobel prize in Chemistry in 1920.

chemical reactions taking place in the battery. Johann Wilhelm Hittorf[62] introduced the concept of ion transport number, the fraction of electric current carried by each ionic species. In 1898, Friedrich W. G. Kohlrausch's seminal book *Das Leitvermögen der Elektrolyte insbesondere der Losungen* (Conductivity of Electrolytic Solutions) (Benedictus Gotthelf Teubner, Leipzig, 1898) established that electrolytic solutions exhibited a specific resistance. In the same year, Karlsruhe chemist Fritz Haber[63] consolidated all these concepts in his definitive textbook on electrochemistry: *Grundriß der technischen Elektrochemie auf theoretischer Grundlage* (Theoretical Basis for Electrochemistry) in 1898, formally circumscribing the field of chemically induced electrical phenomena. By that time, the German electrochemistry society *Deutsche elektrochemische Gesellschaft* was well established and had published the first issue (1894–1895) of its organ *Zeitschrift für Elektrochemie.*

By 1900, North America was no stranger to industrial electrochemistry. Electroplating of gold, silver, copper, and particularly nickel (for the railroads) was widely practiced, especially after the invention of the electric generator in 1873; the rechargeable lead-acid battery (invented by French physicist Gaston Planté in 1859) was in widespread use, though not for self-starting automobiles until

[62]Johann Wilhelm Hittorf (1824–1914) was a German physicist and a professor of Physics at *Universität Münster*, who was the first to observe the mysterious rays (coined "cathode rays," now known to be electrons) extending from the negative electrode of a gas discharge tube and causing fluorescence (and emitted X-rays) where they hit the glass walls of the tube.

[63]Fritz Haber (1868–1934) was a German chemist who spent a large part of his career in Karlsruhe. He devised the Haber-Bosch process for catalytic production of ammonia, important for fertilizers, for which he received the Nobel Prize in Chemistry in 1918. He also devised, with physicist Max Born, the Born-Haber cycle, describing the cohesion energetics of ionic solids. After 1911, he left Karlsruhe for the *Kaiser-Wilhelm-Institüt für Physikalisches Chemie und Elektrochemie* in Berlin-Dahlem, where he later became director. An ardent nationalist, he conducted research on production and effects of poison gas, employed during the First World War (and after), which generated considerable criticism. A Jew, he was deprived of his position at KWI in 1933 and left for England, then Spain and Switzerland. He was offered directorship of the Sieff Research (later renamed the Weizmann) Institute in Rehovat, Palestine by zionist Chaim Weizmann, but died *en route.*

1912); and electrolytic production of alumin(i)um, by electrolysis of molten alumina Al_2O_3, dissolved in cryolite (Na_3AlF_6) — first carried out in 1886 independently by Paul L. T. Héroult in France and recent Oberlin College graduate Charles Martin Hall in his home laboratory in Pittsburgh — was by 1900 well established at Niagara Falls,[64] using the massive hydroelectric power available there. An informal delegation approached the board of the *American Chemistry Society* in 1901, hoping to establish an electrochemistry division, but ACS demurred and the disappointed supplicants decided to start their own society. A particularly enthusiastic young Philadelphia chemist, C. J. Reed, who had been meeting with an eminent electrical engineer, Carl Hering, and a German chemist, E. F. Röber who worked for Hering, was greatly impressed by the new *Deutsche elektrochemische Gesellschaft* and its *Zeitshcrift* publication and was able to convince Joseph W. Richards, a well-known professor of electrometallurgy at Lehigh University, to lend support. Joined by Samuel P. Stadtler, William H. Wahl, and Charles A. Doremus at a formative discussion meeting in 1901, the six hopefuls wrote letters to all members of the ACS and the AIEE inviting them to a foundational meeting for an Electrochemical Society (ECS) in Philadelphia, home of the Franklin Institute. Eighty-seven registered for the meeting, 52 attended, and a total of 350 paid membership dues.

J. W. Richards served as the first president (1902–1904) and the original instigator, C. J. Reed, as secretary. Among those who joined

---

[64] Charles Hall was challenged by his Oberlin College chemistry professor, Frank Fanning Jewett, during his freshman year (Hall then just 17 years old) to come up with a way to produce aluminium inexpensively. Hall tried unsuccessfully throughout his Oberlin years, but after graduation persevered for two further years in a shed laboratory behind his parents' home in Oberlin, OH, patenting his successful process in 1886. Ultimately finding financial backing in Pittsburgh, PA, he joined with metallurgist Alfred E. Hunt, an 1876 graduate of MIT's department of Mining and Metallurgy, to co-found the Pittsburgh Reduction Company, which moved to Niagara Falls, NY in the 1890s and caused the price of aluminium (Hall's original patent spelling) to fall from $12 to $2 a pound. In 1903, Hall's company (later renamed Aluminum Company of America, or Alcoa in 1907) supplied the Wright brothers aluminium with which to cast the engine block and crankcase for their historic flight in Kitty Hawk, NC.

in that first tranche were Charles Martin Hall (founder of Alcoa), Herbert Dow (founder of Dow Chemical Company), Edward Weston (founder of Weston Instruments), and Thomas Alva Edison (who remained a member for the next 28 years). The Society then began meeting biannually.[65] Its second meeting was at Niagara Falls, NY (by then the center of aluminum production), with 117 members present and at which was read a congratulatory letter from Dutch theoretical chemist and Nobel Laureate Jacobus H. van't Hoff, then president of the *Deutsche elektrochemische Gesellschaft* and co-founder of *Zeitschrift für physikalische Chemie.* Industrial participation came from the principal electrochemical industries at the time: electrodeposition, electrolysis, and electrothermic processes, representing approximately $100M of product in the US. (The thirteenth meeting, in 1909, was again held in Niagara falls, but on the Canadian side, the Society's first international foray.) A decade out, the membership stood at 2,300. Twelve decades out, the membership now stands at 8,000 (extending internationally), and the 243rd meeting of ECS is scheduled for May 2023, joint with the 18th *International Symposium on Solid Oxide Fuel Cells.*

Notable among subsequent ECS presidents have been Herbert H. Uhlig (1907–1993), a physical chemist, first at General Electric's Research Laboratory and later professor in MIT's department of Metallurgy (later Materials Science & Engineering), head of MIT's Corrosion Laboratory, editor of the *Journal of the Electrochemical Society,* and founder and editor of the world-renown *Corrosion Handbook*; Ernest B. Yeager (1924–2002), Frank Horvoka Professor of Chemistry at Case Western Reserve University, founder and longtime director of what would become the Yeager Center of Electrochemical Science, an authority on oxygen electrodes, and president (1969–1971) of the International Electrochemical Society; David A. Vermilyea (1923–2022), noted corrosion scientist and expert on anodic thin-films (at GE, for his entire 33-year career), co-author and co-editor of *Structure & Properties of Thin Films* (Wiley, New York, 1959) — and a centagenarian; and Harry C. Gatos (1921–2015),

[65] Burns (1977).

professor in MIT's Materials Science & Engineering and Electrical Engineering & Computer Science departments, and one of the co-founders of the Materials Research Society and its first president.

Local sections were begun almost immediately, in 1905, growing to 18 in 1976, but then eliminated in favor of local student chapters, of which there are now (2023) 69 in the US and 58 in 26 other countries. Society headquarters moved to New York City (to be near Columbia University) in 1921, and then to New Jersey (to be near Princeton University) in 1958. Subject-wise, ECS initially organized itself into interest groups and technical committees. Beginning in 1921, it consolidated these into divisions, ending up with 16. The first two divisions were Electrodeposition (1921), Electrothermic (1922, later Electrothermic & Electrometallurgy (1954), now High-Temperature Energy Materials & Processes), with technical committees remaining for, *inter alia*, batteries, organic electrochemistry, electroanalysis, with corrosion as an interest group. The latter committees and groups later found their way into divisions: Electronics (1931, later Electronics & Photonics), Theoretical (later Physical) & Analytical Electrochemistry (1936), Organic & Biological Electrochemistry (1940), Corrosion (1942), Industrial Electrolytic (1942, merged into Electronics & Photonics), Industrial Electrochemistry and Electrochemical Engineering (1943), Dielectrics and Insulation (1945, later Dielectric Science & Technology), and Batteries (1947). With the rapid developments in new materials and materials science beginning in the 1970s, three additional divisions were added in the early 1980s: Luminescence & Display Materials (1982), Nanocarbons (1983), and Energy Technology (1983).

Per the desire of initiator-founder C. J. Reed, the Society began to publish at the outset, with a monthly journal, *Electrochemical Industry* (later *Electrochemical Technology*) beginning in 1902, and *Transactions* volumes publishing papers given at its meetings (which continue today as *ECS Transactions* full-text conference proceedings). Likely much to the disappointment of Reed, who had envied the German society's *Zeitschrift für Elektrochemie*, the ECS *Journal of the Electrochemical Society* was not begun until 1948, into which the

Electrochemical Technology was absorbed. ECS also began a series of monographs, starting with the *Corrosion Handbook*, founded and edited by H. H. Uhlig in 1948, followed by a series of supplemental corrosion volumes, the first of which was *High Temperature Oxidation of Metals*, as well as a dozen other monographs on other subjects and a score of ECS special *Symposium* proceedings. ECS added a second scholarly journal publication, the *Journal of Solid-State Science and Technology* in 2011, published by the British IOP publishing arm. The Society has also accumulated and maintained an assortment of awards on society, division, and local section/chapter levels.

Until the mid-20th century, electrodes in the batteries, associated electrolysis apparatus, and fuel cells were generally solid materials (metals or carbon), but the electrochemistry was done only in *liquid* (or gel) *electrolytes*, in which charge is conducted by transport of charged ions in solution. The *Electrochemical Society* (ECS) was the first professional society to embrace the importance of *solid* electrolytes, in which charge transport occurred by *solid-state* diffusion of ions, putting these squarely in the playing field of ceramists. The early exploration of solid-state electrolytes for batteries was, however, not pursued by ceramists but instead by physicists and chemists working at Ford Motor Company's Scientific Laboratory in the 1960s, focused on so-called β-alumina (a sodium polyaluminate) *ceramic* compound in which rapid transport of Na^+ ions could be achieved by their diffusion along a two-dimensional network of channels in the structure. The target battery system was the (molten) sodium/(molten) sulfur combination, operating at temperatures of 300–350°C, in which the β-alumina was introduced as the solid electrolyte separating the two molten species. Though commercialized, the Na-S battery and its successors were not suitable for vehicle propulsion. Other solid-state electrolytes, however, found application in sensors, the most widespread of which is oxygen-deficient yttria-doped zirconium oxide ($Y_xZr_{1-x}O_{2-½x}$) [YSZ] in which the missing oxygen is accounted as electrically charged oxygen vacancies (unoccupied oxygen sites in the ionic ZrO_2 crystal structure) that can diffuse at temperatures greater than ~400°C by a hopping mechanism (trading places with next-door oxygens). With two different (partial) pressures of oxygen on either side of this zirconia solid electrolyte — say air (0.2 atm oxygen pressure) on one side, hot combustion gas from automobile exhaust on the other — an electrochemical potential (voltage) develops across the zirconia electrolyte and can be used to determine the oxygen content in the combustion gas, which information is used to adjust the combustion parameters in the engine for best efficiency. These sensors have been installed on every automobile produced for the last four decades.

A second, now equally pervasive, electrochemical device is the Li-ion battery, in which solid electrodes are of primary importance as sources and sinks for migrating Li^+ ions that diffuse more quickly than larger Na^+ ions, and facilitate a large number of recharge cycles. This breakthrough battery was the brainchild of British chemist M. Stanley Whittingham (1941–), starting with his Oxford University DPhil thesis, and John B. Goodenough (1922–2023), a solid-state physicist trained at the University of Chicago under physicist Clarence Zener, subsequently working initially at MIT's Lincoln Laboratory and then for a decade as head of the Inorganic Chemistry department at Oxford. Whittingham explored the intercalation of Li^+ ions in a variety of crystalline hosts without altering their crystallographic structure and constructed the first rechargeable Li-ion battery, with a Li-Al alloy anode, a TiS_2 titanium disulfide cathode, and a polymer gel electrolyte. For his part, Goodenough gave thought instead to what had long been an interest of his, complex transition-metal oxides, and came up with a $LiCoO_2$ oxide cathode in the prototype. Akira Yoshino, who had studied at Kyoto University (Japan), worked afterward with the Asahi Kasei company, later completing doctoral studies at Osaka University. He replaced Whittingham's Li-Al metal anode with a soft-carbon anode (less prone to bursting into flame if exposed to moist air), adopted Goodenough's $LiCoO_2$ cathode, and in 1987 patented a rechargeable Li-ion battery using a carbonate-ester electroltye that was commercialized first by Sony in 1991 and the next year jointly by Asahi and Toshiba. Yoshino's anode was later replaced by graphite, into whose layer structure Li^+ ions readily and reversibly intercalate (a discovery of Rachid Yashimi, a Morrocan doctoral student at *l'Institut Politechnique de Grenoble* (France), *Groupe INP*. Whittingham, Goodenough, and Yoshino shared the 2022 Nobel Prize in Physics for their development of the Li-ion battery, a joint triumph for electrochemistry, solid-state chemistry, physics and electronics, and materials science — which clearly demonstrates how interdisciplinary materials development has become. Subsequent refinements have included replacing the $LiCoO_2$ cathode with lithium iron phosphate ($LiFePO_4$), lithium manganese oxide spinel ($LiMn_2O_4$), lithium manganese oxide (Li_2MnO_3)-based lithium-rich layer structures, or lithium-nickel-manganese-cobalt oxides and the graphite anode with SiC polytypes. These materials are also used in supercapacitors and solid-oxide fuel cells (SOFC), as well as in solid-state batteries.

The most recent and novel development for grid-scale electrical energy storage is the redox-flow solid oxide iron-air rechargeable storage battery (RFSOIAB), announced in 2011 by Xuan Zhao and Kevin Huang at the University of South Carolina's SmartState Center for Solid Oxide Fuel Cells. Huang was a postdoc with Goodenough at University of Texas (Austin) in 1995 and afterwards worked for Siemens-Westinghouse before becoming director of the SmartState Center in 2010. In contrast to scarce lithium, the iron-air battery utilizes the first (oxygen) and fourth (iron) most abundant elements on earth to initiate a reversible oxidation of iron (well-known to users of powdered-iron hand warmers) and can use YSZ or other solid-oxide ceramic electrolytes.

Institute of Electrical and Electronics Engineers (IEEE)

Yet another player in materials science research is the *Institute of Electrical and Electronics* Engineers (IEEE). It, too, grew out of prior, more trade-based societies. Toward the end of the 19th century, industries associated with electric power and illumination, and telegraphy and telephony, led to the formation of the *American Institute of Electrical Engineers* (AIEE), which first met in Philadelphia in 1884, in historic Franklin Hall — with some *living* electrical luminaries present as well, among them Norvin Green, the president of

Western Union, and Alexander Graham Bell, inventor of the telephone, respectively, becoming the first and sixth presidents of AIEE. The badge of the new society — adopted during Bell's presidency — depicted, somewhat anachronistically, Franklin's kite, with Ohm's law in gold letters below. Among other founders were Nikola Tesla, inventor of the AC motor (among many other inventions), and Edward Weston, founder of Weston Instruments. The invention and growing importance of the radio (which Tesla had also worked on) led to the formation of a parallel society, the *Institute of Radio Engineers* (IRE), in 1912. In 1962, the AIEE and IRE merged to form the *Institute of Electrical and Electronics Engineers* (IEEE).

While the IEEE is associated with electrical and electronics research, it owes much of its participation in materials science research to the development of transistors, which caused the IEEE to promote much research into silicon, germanium, and other semiconducting materials, electronic ceramic materials that had been abandoned by ACerS. IEEE also evinces strong activity in magnetic, ferroelectric, piezoelectric, and electrostrictive materials for use in computer memory, frequency control, guidance, and sensor applications, as well as superconducting materials (for potential electrical grid applications) and polymer and ceramic packaging materials for electronic devices. Rather than divisions, IEEE has 39 (sub)-"societies" that represent topical interest groups. Those with some materials focus are: Dielectrics and Electrical Insulation Society, the Electronics Packaging Society, the Magnetics Society, the Photonics Society, and the Ultrasonics, Ferroelectrics, and Frequency Control Society.

The Institute publishes a large number of journals and transactions with materials content, including *Journal of Electronic Materials, IEEE Journal of Flexible Electronics, IEEE Open Journal of Nanotechnology, IEEE Journal of Photonics, IEEE Photonics Technology Letters, IEEE Journal of Photovoltaics, IEEE Sensors Journal, IEEE Sensors Letters, IEEE Open Journal of Ultrasonics, Ferroelectrics and Frequency Control, IEEE Transactions on Applied Superconductivity, IEEE Transactions on Biomedical Engineering, IEEE Transactions on Biomedical Circuits & Systems, IEEE*

Reviews in Biomedical Engineering, IEEE Transactions on Component & Packaging Technologies, IEEE Transactions on Advanced Packaging, IEEE Transactions on Device and Materials Reliability, IEEE Transactions on Dielectrics and Electrical Insulation, IEEE Transactions on Magnetics, IEEE Magnetics Letters, IEEE Transactions on Nanotechnology, IEEE Nanotechnology Magazine, IEEE Journal of Photovoltaics, IEEE Transactions on Energy Conversion, and *IEEE Transactions on Nuclear Science.*

With 450,000 members, 340 local sections throughout the world, and hosting over 2,000 technical meetings annually, IEEE claims (undisputed) to be the largest technical society in the world. It appears that it is among the largest publisher of technical communications as well, though not of materials content. It additionally maintains a standards association. The many student chapters fall under the extensive roster of local sections. Among the sponsored or co-sponsored conferences, meetings or workshops with materials content are: Biomedical engineering, healthcare & sustainability, Electronic components & technology, Device research (co-sponsored with MRS), Electronic devices & manufacturing, Active-matrix flat panel displays and devices, Young professionals in electron devices and materials, Electrical insulation and dielectric phenomena, Applications of ferroelectrics, Frequency control, Lasers and electro-optics, Semiconductor lasers, Electromagnetics, Magnetics (co-sponsored with the *Magnetics Society of Japan*), Magnetism and magnetic materials, Advanced materials and processes for RF and THz applications, Artificial materials for novel wave phenomena (metamaterials), Electrical materials and power equipment, Materials for advanced metallization, Nano/micro-engineered and molecular systems, Nanotechnology materials and devices, Microwave and optoelectronics, Electronics packaging technology, Photonics, Photovoltaics, Semiconductors, Semiconductor interfaces, Biosensors, Sensor & nanotechnology, Solid-state sensors, actuators and microsystems (transducers), Silicon photonics, Soft robotics, and Superconductivity and electromagnetic devices.

In the past 60 years, there has been no IEEE president with a materials background, and only two women presidents.

Biomaterials Societies

Society for Biomaterials (SFB)

The history of the intrusion of materials into medicine, from ancient times to the explosion of the biomedical materials field starting in the 1960s, together with that of the explosion in biomaterials educational programs in the early years of the 21st century, have been recounted earlier. The *Society for Biomaterials* (SFB) was founded in 1974, following a series of international biomaterials symposia, begun in 1969, that focused on the natural and synthetic materials involved in the practice of reconstructive surgery. As a consequence, the Society has maintained strong ties to the medical profession; their surtitled motto, adopted later, reads *Giving life to a world of materials.* In SFB's first decade, its yearly membership fee was 3 or 4 times that for other materials societies, which reflected the large contingent of physician members; it has since reduced that to about twice that of other materials societies, but includes a subscription to its *Journal of Biomedical Research & Applied Biomaterials* and offers reduced fees for students and post-docs. Coinciding with (and partly responsible for) the explosive growth of biomaterials/biomedical materials/bioengineering field, now pursued in over 75 academic departments in the US alone, the society has itself grown explosively in parallel, growing from 500 members in 1977 to a membership of several thousands in 2023. With students presently comprising 34% of its membership, it has established 38 student chapters in US universities. The Society maintains 14 special interest groups, spanning the wide interests represented by its members, *viz:* biointerfaces, biomaterials education, biomaterials & medical products, biomaterial-tissue interaction, cardiovascular biomaterials, dental/craniofacial biomaterials, drug delivery, engineering cells and their microenvironments, immune engineering, nanomaterials, ophthalmic biomaterials, orthopedic biomaterials, surface characterization and modification, and tissue engineering.

As a recent (and successful) foundation as a multi-disciplinary society, like the Materials Research Society and the same age, the backgrounds and affiliations of SFB's five accomplished founders are similarly of both historical interest and societal relevance. Its principal founder, the late C. William Hall MD (1922–1992) was a pioneer in the development of the artificial heart at Baylor College of Medicine (Houston, TX), where he directed that fledgling program, and from 1970 was director of the Department of Bioengineering at Southwest Research Institute in San Antonio, TX. The late Samuel L. Hulburt (1936–2016) was a ceramist, educated at the New York State College of Ceramics at Alfred (BS, PhD), joined the ceramics faculty at Clemson University (Clemson, SC) and in 1973, as Dean of Engineering at Tulane University (New Orleans, LA), helped found the department of Biomedical Engineering there. Hulburt later served for 28 years as the president of Rose-Hulman Institute of Technology (Terre Haute, IN, founded 1879 as an undergraduate engineering institution, the first institution in the US to award a degree in chemical engineering); at Rose-Poly, he continued to teach biomedical engineering, completing 30 years as a materials educator. A veritable polymath, PhD scientist Sumner Norton Levine (1923–2007) was a longtime professor (1961–1990) at the State University of New York at Stony Brook, joining just after its founding, and a member of (what is now) the Department of Materials; but before that, he had been a senior scientist at RCA, Director of the VA hospital at East Orange, NJ, a senior scientist at the atomic energy division of Westinghouse, founded two journals, the *Journal of Biomedical Materials and Biomedical Engineering* (in 1967) and the *Journal of Socio-Economic Planning Science*, and served as editor-in-chief of vols. 1–4 of *Advances in Biomedical Engineering and Medical Physics* (1968–1971). As an avocation, Levine self-educated himself in finance, edited a massive two-volume reference book *The Financial Analyst's Handbook* and edited the *Dow-Jones Business and Investment Almanac* for 15 years. Sigmund A. Wesolowski MD (1921–2009, late in his life going by an alternative name Adam Wesolov) was a longtime surgeon at the Department of Surgery, State University of New York Downstate Medical Center (Brooklyn, NY). He wrote his 1948 master's thesis on the construction of an extracorporeal heart pump and subsequently experimented on an early mechanical heart, achieving long-term survival of animal models on by-pass surgery; in his later career he worked on the development of intracardiac surgery for the correction of human congenital cardiac defects (and served as president of SFB in 1976). Richard S. Woodbury JD served as the Society's counsel, an important role in a society representing physicians and industries with significant malpractice and product liability litigation and patent concerns.

SFB held its inaugural meeting at Clemson University (Clemson, SC) and has met every Spring since. It participates (and helps organize) the series of quadrennial World Biomaterials Congresses, the first of which was held in Vienna in 1980; the twelfth will be held in Korea in 2024. For the past two Autumns (2021, 2022) SFB has maintained a presence at the large multi-society *Materials Science & Technology* (MS&T) conference sponsored jointly by the ACerS, AIST, and TMS. SFB publishes its own *Journal of Biomedical Research* (Parts A & B, now in its 100th volume and incorporating *Applied Biomaterials*); a quarterly newsletter, *Biomaterials Forum*; and an online *Biomaterials Bulletin.* It has also lent its imprimatur to the 4th edition of the multi-authored textbook *Biomaterials Science: An Introduction to Materials in Medicine* (See footnote 55). Although 15% of its membership resides outside the US, it has yet to establish affiliated international outposts, though it has once met jointly with the Japanese Society for Biomaterials (in Hawaii).

Of the 49 presidents of SFB, seven have been women, and several have strong materials science backgrounds or exposure, among them, for example: ceramist Larry Hench NAE (1938–2015), professor of Biomaterials in University of Florida and Imperial College London materials departments, the inventor of bioactive bioglass for hard tissue repair; polymer chemist Buddy Ratner NAE (1947–), endowed chair in Technology Commercialization and director of the Center for Biomaterials at University of Washington (Seattle), authority in natural, synthetic and resorbable polymeric biomaterials for tissue repair and regeneration, and founding co-author and lead editor of *Biomaterials: Introduction to Materials in Medicine* (1997–2013); materials scientist Paul Ducheyne (1948–), professor of Biomaterials and Orthopaedic Surgery at University of Pennsylvania, who has long investigated use of bioactive ceramic, glass and xerogel materials for orthopedic and dental hard-tissue engineering, editor-in-chief of *Comprehensive Biomaterials* (2011–) and co-author of *Functional Behaviour of Orthopaedic Biomaterials*; and bioengineer Myron Spector (1945–), retired director of the Orthopaedic Research for Brigham & Women's Hospital, Boston, emeritus professor of Orthopaedic Surgery at Harvard Medical School, and co-editor-in-chief of *Biomedical Materials: Materials for Tissue Engineering & Regenerative Medicine*, who has long addressed muscular-skeletal and nerve regeneration, with a variety of materials-based interventions, and has co-taught biomaterials-tissue interactions at MIT for 30 years with Ioannis Yannas NAE (b.1935–), professor of Polymer Science & Mechanical Engineering at MIT and developer of the first artificial skin scaffold for burn victims.

ACerS Bioceramics Division, ACS, and *TMS Biomaterials Committees*

Three other US-based materials societies cater, to a lesser extent, to the biomaterials community. The *American Ceramic Society* (ACerS) formed a topical division of their society focusing on ceramic biomaterials, specifically designated Bio*ceramics* (added in 2017). The *American Chemical Society* (ACS) chose not to establish biomaterials as one of its technical divisions, but has featured biomaterials as a major emphasis theme for some of its huge national annual meetings; for example, the 2012 ACS annual meeting emphasis theme was entitled "Materials for Health and Medicine." *The Minerals, Metals & Materials Society* (TMS) maintains a biomaterials committee that is connected to both its Functional Materials and its Structural Materials divisions.

The Great Amalgamation: Materials Science & Engineering

The post-war 1950s were remarkable in that major changes were wrought in the way atom arrangements in the solid state were conceived of, and in particular the critical role that imperfections and disorder at the atomic level played in determining physical, chemical, and mechanical properties of solids. Investigative imaging techniques such as high-resolution analytical electron microscopy and surface probes of all kinds have developed that have made it possible to examine materials at a nanometer scale — ultimately even at the atomic scale — and to study structure, composition, and defects in atomistic detail (and in false color!). It has become increasingly recognized that defects are often the key to understanding the special properties exhibited by many materials, and special attention has been focused (literally) on their characterization and roles across the spectrum of materials using all manner of rapidly developing imaging and microanalytical techniques, whose multiplying acronyms alone now require an encyclopedic guide.

Harvard's David Turnbull (1915–2007, profiled later), in a retrospective commentary on the emergence and evolution of materials science as a discipline,[66] credits the inspirational lead of notable solid-state physicists, mostly theorists, such as (to mention a few) Neville F. Mott (1905–1996) at Bristol and then Cambridge (UK), F. C. Frank (1911–1988) at Bristol, Frederick Seitz (1911–1988) (author of the seminal books *Modern Theory of Solids* (1940) and *Physics of Metals* (1943)) at Illinois, William B. Shockley (1910–1989) at Bell Labs, and Clarence M. Zener (1905–1993) at Chicago after WWII (later Director of Science at Westinghouse). All gained international eminence: Mott and Frank were both elected FRS and knighted; Shockley (co-inventor of the p–n junction transistor) was awarded the US National Medal of Merit; Zener had the Zener diode named after him from his foundational work on the Zener effect underlying it, among many other phenomena in solids (such as anelasticity of metals and internal friction); Seitz became president of the US National Academy of Sciences; Mott and Shockley (along with his Bell Labs colleagues John Bardeen and Walter Brattain) both won Nobel prizes in Physics. As physicists, none of them had parochial educational backgrounds from traditional materials departments of metal, ceramics, glasses, or polymers (though Shockley's father was a mining engineer), yet the breadth of applicability of their work and their interest in explaining a wide

[66] Turnbull (1983).

range of properties of material solids brought a decided interdisciplinarity and materials catholicism to their contributions.

It should be remarked here that a practical model for implementation of interdisciplinarity was supplied by another physicist, Arthur Robert von Hippel (1898–2003), educated at *Universität Göttingen,* where he was taught by a veritable dream-team of early-20th physicists: David Hilbert, Peter Debye, Max Born and Gustav Hertz (the last three Nobel Prize winners) and supervised by another Nobelist, James Franck (of the Franck-Condon principle), for his PhD studies at Göttingen's Institute for Applied Electricity, marrying Franck's daughter Dagmar. Afterward, he studied electrical breakdowns in gases and crystalline solids. Because he was an outspoken critic of the Nazi regime and Dagmar was Jewish, von Hippel fled Germany, with his wife, in 1933, spent a year at the Niels Bohr Institute in Copenhagen, then accepted an offer of employment from another eminent physicist, Karl Compton, then President of MIT (who had done *his* PhD under Nobelist O. W. Richardson at Princeton). At MIT, von Hippel established in 1940 a Laboratory for Insulation Research (important for the war effort) that became a model for an interdisciplinary approach to materials research. He recruited graduate students for his Laboratory that came from chemistry, physics, electrical engineering, and metallurgy (then including ceramics) to synthesize and characterize a wide range of dielectric materials. In 1944, von Hippel made the key discovery of ferroelectricity and piezoelectricity in barium titanate ($BaTiO_3$), a ceramic compound later in widespread use for electromechanical transducers and electronic circuitry. His Laboratory became a major center for developing dielectrics for radar applications during World War II, then being developed at MIT's proximate wartime Radiation Laboratory, and later a model for how a disciplinarily blind approach could be recruited advantageously for materials research.

It subsequently fell, however, to others — such as metals chemist William Hume-Rothery (1899–1968) in Oxford (Hume-Rothery electron rules for alloy formation), metallurgist J. Herbert Hollomon (1919–1985) at General Electric (head of metals and ceramics research at GE), metallurgist Cyril Stanley Smith (1903–1992) at Chicago's Institute of Metals (Smith had pioneered the metallurgy of a formerly non-existent metal, plutonium, for the Manhattan project), and even Harvard applied physicist (originally physical chemist) David Turnbull himself (who *inter alia* established the formal criteria for glass transitions and (correctly) predicted the formation of metallic glasses) — *each* pointedly invested in manipulating traditional materials types and manufacturing new ones, to take the lead in bringing a *common* phenomenological basis to traditional materials studies. They could now *explain* exactly why the historical empirical approaches to traditional materials optimization had worked so well. What followed from their efforts was an increasing ability to utilize the fundamental physics and chemistry underlying materials properties to craft totally *new* kinds of materials — heterotopic in some cases (like polymer car bumpers or ceramic carving knives), wholly novel in others (like metallic glasses or high-temperature ceramic superconductors) — that created the stage for promulgating a "materials science of everything" and creation of the *novel* construct of materials science and engineering. This was, indeed, that ages-old human genius, alluded to at the outset of this entry, for *transforming our material environment out of all recognition from its prehuman manifestation* — transformations that have so demonstrably led, once again, to a profound alteration of human cultures over the last 70 years.[67,68]

The 1960s brought about a substantial attendant consolidation both of materials science research and of the associated professional societies. As early as 1955, the polymath John von Neumann (1903–1957), then a uncontestedly influential commissioner of the Atomic Energy Commission (AEC) and emphatic proponent of intercontinental ballistic missile development, had identified materials as a critical field for the strategic development of US cold-war capabilities and recommended establishment of a materials-intensive institute in an academic setting. Frederick Seitz, who shortly after von

[67] Bensaude-Vincent (2001).

[68] Bensaude-Vincent and Hessenbruch (2004).

Neumann's untimely death chaired an AEC advisory panel in 1956 on the materials needs of the AEC, pushed for materials education programs and research laboratories to be funded at universities — including new buildings to house them — but the AEC was reluctant to commit funds. One deciding factor in re-invigoration of that blunted proposal was undoubtedly the USSR's launch of Sputnik in 1957. Realizing that it was so apparently far behind in missile technology, the panicked US Department of Defense (DOD) called on its Advanced Research Projects Agency (ARPA, later styled the Defense Advanced Research Projects Agency, DARPA) to foster the greater study of a wide range of materials, including metal alloys, ceramics, and other compounds serviceable in extreme environments, especially the tremendous pressures and temperatures to which missile skins were subjected at launch and re-entry, together with the advanced solid-state electronics required for remote radio-frequency guidance and telemetry. A committee on materials to advise ARPA was formed, chaired by Julius H. Harwood (1918–2014). A corrosion metallurgist, Harwood at that time headed up the Metals and Ceramics branch of the Office of Naval Research (ONR); he left ONR in 1960 to join Ford Motor Company as manager of their metallurgy department, then later became overall director of Ford's materials sciences laboratories. It was clear to Harwood, and to all, that this effort would require a novel realignment of science and engineering educational programs in US universities, particularly at the graduate level, which had remained historically balkanized into doggedly separate materials classes. While it recognized what the country fundamentally needed, long-term, was a larger cohort of post-graduate professionals trained in materials sciences across disciplines, the DOD was not set up to fund educational endeavors, so what was specified as needed instead was "research" (hence ARPA's involvement), with the logical expectation that it would be mostly graduate students who would be carrying out that research. An expansion of research (and graduate student populations) and the advanced instrumentation required by both clearly translated into increases in laboratory and office space, but the DOD was not in the business of funding grants for bricks-and-mortar projects; so it was arranged that the cost of any new or reconfigured buildings would be borne by the university but that the DOD would pay ground rent on some proportion of the increased space amortized over 10 years, as reimbursement.

Materials Science & Engineering Departments

In 1959, ARPA solicited proposals to establish several interdisciplinary laboratories (IDLs) and associated graduate programs from US universities; 45 universities ended up submitting IDL proposals, and 12 eventually were selected, over a 12-year period. Accounts of the philosophical background for, and rollout of, the ARPA IDL program have been thoughtfully summarized by physicist Robert L. Sproull (1918–2014)[69] (at the time of the IDL proposal solicitation, director of Cornell University's Laboratory for Atomic and Solid State Physics, Cornell's first IDL director, and later director of ARPA). Cornell University, Northwestern University, and the University of Pennsylvania were the first three universities chosen, in 1960. These three initial choices were made largely based on future potential for integration and, to some extent the interdisciplinary histories of the institutions.

Of the three, Northwestern was the only university that already came with its own integrated materials science department and building, begun in 1959 and replacing an earlier 1954 Metallurgy program.[70] Under the prescient chairmanship of Morris E. Fine (1918–2015) — a phase-transformations metallurgist (with interests also in ceramics) who had worked under Cyril Stanley Smith in the Manhattan project — this novel materials science & engineering department had already embraced four materials classes (metals, ceramics, electronic materials, and polymers); "engineering" was added to better reflect the practical implications of such integration and the inclinations of its new faculty. Northwestern's department, almost immediately, launched an introductory

[69] Sproull (1987).

[70] Fine and Marcus (1994)

textbook series (in inexpensive paperback format) with MacMillan, the department surmising (correctly) that the initial tranches of graduate students were unlikely to have had much background in materials science, more likely coming instead from undergraduate programs in physics or chemistry, or in mechanical, electrical or chemical engineering. The first four introductory texts, respectively, addressed phase transformations, diffraction methods, mechanical properties, and dislocation theory, later Martensitic transformations, and recovery and recrystallization. The newly formed interdisciplinary department also had the advantage of being located within a brand new extension of Northwestern's Technological Institute, a sprawling completely interconnected building complex containing the university's science and technology library and already housing solid-state physicists and chemists and all four of the existing engineering departments and their laboratories. The Technological Institute building (today the main building of the McCormick School of Engineering and Applied Science), and its MSE department footprint, were further enlarged after 1966 when the Northwestern campus was extended into Lake Michigan by cannibalizing sand from the Indiana sand dunes.

The University of Pennsylvania in Philadelphia had had a Department of Metallurgy since 1955 and a School of Metallurgical Engineering, whose director (until 1972) was the eminent metallurgist Robert Maddin (1919–2019). Maddin joined physicist Eli Burstein, chemist Robert Hughes, and chemical engineer Norman Hixon in the proposal for what was grandly christened the *Laboratory for Research on the Structure of Matter* (LRSM) — which was to embrace solid-state physics and chemistry, as well as polymers and metallurgy. They forecast doubling aggregate faculty positions in those disciplines from 20 to 44, and the graduate student cohort from 83 to 188 — but had no building to house their proposed IDL. A new building was eventually constructed in 1965[71] and over time acquired a passel of star-studded residents, 28 of them in turn elected to the

[71] Choi and Shields (2015).

US national academies, and five of them winning Nobel prizes (Robert Schrieffer (1931–2019) for superconductivity (1972), Ahmed Zewail for femtosecond spectroscopy (1999), and Alan Heeger, Alan MacDiarmid and Hideki Shirakawa for conductive poly(acetylene) polymers (2000)).[72] In 1973, the metallurgy department became the department of Materials Science and Engineering, chaired from 1972 to 1983 by Maddin, who then migrated to Harvard to work on metals archaeology in its (then) Peabody Museum.

Cornell University's path to an IDL is related in elaborate detail by the late synchrotron radiation physicist Paul Hartman (1913–2005),[73] who has provided an equally detailed account of the history of the ultimate decision by the DOD to fund an IDL program. Cornell was one of 11 institutions originally approached by ARPA in 1959, a group including also California Institute of Technology, University of California–Berkeley, Carnegie Institute of Technology (now Carnegie-Mellon University), Case Institute of Technology (now Case–Western Reserve University), University of Chicago, Harvard University, University of Illinois, Massachusetts Institute of Technology, Pennsylvania State University, and Purdue University — though the number of proposals eventually submitted reached 45. By 1959, in addition to longstanding departments of Physics, Chemistry, Chemical Engineering, and Metallurgy (the latter initially part of Chemistry), Cornell had established several newer, more interdisciplinary groupings: a Laboratory for Atomic and Solid State Physics (1959), an Engineering Physics department (founded just after WWII, in the School of Applied & Engineering Physics), and a department of Engineering Mechanics & Materials. Like, Penn, Cornell proposed doubling materials faculty, to 55 (by 2022 it had risen to 123) with a

---

[72] LRSM. Timeline history of the LRSM. https://www.lrsm.upenn.edu/timeline/ (last accessed 1 March 2023).

[73] Hartman, Paul L. (1992), with a forward by Robert L. Sproull. *Cornell's Materials Science Center: the Early Years.* Original manscript title: "MSC decade & a half: A brief history over the formative first decade and a half of the Materials Science Center at Cornell University." Available from Cornell University Materials Science Center, Ithaca, NY.

proportionate increase in the graduate materials student population. Space for the proposed anticipated expansion of teaching and research facilities was, however, tight — though Bard Hall was under construction for what was to be a separate department of Metallurgy. Cornell proposed to expand its 1904 Rockefeller physics building with a new adjacent physical sciences building, a seven-floor structure (named Clark Hall for its benefactors, completed in 1965), to house the larger portion of a potential Materials Science Center (MSC); to utilize Bard Hall (then in construction, completed 1963); and to make additional modifications to Chemistry facilities in the Baker Lab, connecting it to the proposed new Clark Hall. In 1962, the second year after the 1960 IDL award, the metallurgy program, in Chemistry, and the department of Engineering Mechanics & Materials amalgamated first into a department of Engineering Physics & Materials, headquartered in the newly completed Bard Hall, and then in 1965 into a department of Materials Science & Engineering, chaired by Walter S. Owen (1920–2007), a Liverpool metallurgist, later Dean of Northwestern's School of Engineering, and still later chair of MIT's Materials Science & Engineering department. Like Penn, Cornell has had its share of illustrious faculty affiliated with the MSC (now known as the Cornell Center for Materials Research (CCMR)) who have been elected to the NAS or the NAE, such as Robert W. Balluffi NAE (1924–2022), a former chair of Cornell's Materials Science & Engineering department, who was elected to both NAS and the American Academy of Arts & Sciences, and some like Hans Bethe (1906–2005) sporting Nobel Prizes. Balluffi spent the last 18 years of his career at his *alma mater*, MIT, before returning to Ithaca to finish his second book *Introduction to Elasticity for Crystal Defects* (his first was on *Interfaces in Crystalline Materials*), and to paint; like Oxford's Hume-Rothery, Balluffi was an equally talented water-colorist.

After Cornell, Northwestern, and Penn, nine more IDLs were awarded is the succeeding year: Brown, Chicago, Harvard, MIT, and Stanford (June 1961); Maryland, North Carolina–Chapel Hill, and Purdue (October 1961); and Illinois (1962), the latter IDL administered by the AEC. As one consequence of the student anti-war protests in the late 1960s, a provision in the 1969 Military Authorization

Act placed restrictions on the funding of long-term research by the DOD (and with it, on ARPA); indeed, Cornell IDL's ARPA-funded Clark Hall had been a target of student protests, and the Charles Stark Draper Laboratory at MIT, which developed *inter alia* missile guidance systems, had had to be divested by the Institute under similar pressure in 1973. The result was that ARPA was more transparently renamed the Defense Advanced Research Projects Agency (DARPA), and the IDL program was transferred to the US National Science Foundation (NSF) in 1970. After its own review, NSF altered the criteria by which recipients of the IDL block grants would be assessed, emphasizing that these, now renamed, *Materials Research Laboratories* (MRLs) must be seen as fostering "coherent, multidisciplinary and multi-investigator projects requiring the expertise of two or more materials-related disciplines." The required reorientation was instituted so as to differentiate the substantial block grant funding for MRLs from NSF's more usual mechanism for funding of individual investigator projects, and so to blunt some of the jealousies increasingly harbored by universities who had not received MRL support. Starting in 1973, NSF awarded six more MRL grants: Carnegie-Mellon University and University of Massachusetts–Amherst in 1973, Pennsylvania State University and Case Western Reserve University in 1974, Ohio State University in 1982, and University of California–Santa Barbara in 1992. Eight of the MRL programs were terminated in the period 1977–1988, and the remainder in the period 1994–1996, after which the NSF instituted a *new* program of *Materials Research Science and Engineering Centers* (MRSECs), funded by 6-year grants oriented to the multi-investigator two-or-more-subdisciplines research-thrust pattern that NSF had previously imposed on the IDLs when they became NSF MRLs.

As might have been anticipated, a sense of elitist unfairness permeated how the IDL/MRL programs were viewed by academic institutions outside those programs — and even by non-MRL faculty within institutions with MRLs.[74] A case in point is that of the

---

[74]When the LRSM faculty at the University of Pennsylvania proposed in 2004 augmenting the (by then) 40–year-old LRSM building with a new structure to

Pennsylvania State University, not chosen as one of the 12 DOD IDLs and awarded one of the four last MRL grants, made by NSF, only in 1974. Penn State — at the urging of Rustum Roy (an energetic ceramist and solid-state chemist, profiled later, and a prominent co-founder of the Materials Research Society), and drawing the long history of its College of Earth and Mineral Sciences (starting as Mining in 1896, then Mining and Metallurgy, when it had been known as the State College of Pennsylvania) — had by 1959 developed its own integrated graduate program and curriculum in solid-state technology, and by 1962 its own independent interdisciplinary Materials Research Laboratory (now going by the name of its successor, the *Materials Research Institute*). Indeed, it would have been difficult in 1962 to find a more interdisciplinarily minded and capable materials faculty dominant across the materials spectrum, then largely dominated elsewhere by metallurgists and physicists. By 1967, that spectrum was consolidated at Penn State into one of the first departments of Materials Science and Engineering. It was no accident that the first formational meeting for the Materials Research Society was held on Penn State's campus in 1972, intentionally posing the question of whether the direction of a field should be decided by its rank and file or by a government committee.

Without speculating on the particulars of this example, which doubtless involved personalities, differing perspectives on the

---

house a newly created Bio/Nano Interface Center (NBIC) — following Penn's successful bid to establish one of the Nanoscale Science and Engineering Centers (NSECs) supported by NSF as part of the National Nanotechnology Initiative (NNI) — there was unexpected opposition from other sectors of Penn's faculty who questioned whether another "interdisciplinary" building would actually foster interdisciplinarity in the sharing of the accompanying research funding (as had been previously argued for the LRSM building erected in 1965, meeting with similar skepticism then). Choi and Shields (2015). In the end, the new NBIC center, led by ceramist-scanning-probe-microscopist Dawn A. Bonnell (since 2013 Penn's senior Vice Provost for Research), was successful in obtaining funding from major donor Krishna P. Singh, and the $91M Singh Center building was opened in 2013, providing new space for Penn's MRSEC as well as the NBIC.

scientific disciplines, and possibly latent prejudices concerning the institution involved, it is at least a fair question to ask whether the IDL/MRL experiment resulted in the intended outcome — which was the advantageous consolidation of the materials sciences and the enhanced production of materials-savvy PhD scientists and engineers capable of improving the safety or welfare of the tax-paying population that funded it, or indeed of humanity as a whole. The answer to the first is yes, at least an acceleration of the consolidation; the answer to the second is yes, enhancement of breadth of education and training of the PhD product in at least 18 institutions, but whether this production has increased either safety or welfare of larger populations involves a longer, larger and still unfinished assessment. Certainly, there cannot be any doubt that human cultures have been altered more rapidly than perhaps ever before in human history — an affirmation of the undeniable connection between technological development and societal change through many positive (and some negative) feedback loops.

There have been several assessments of the IDL/MRL program, one instigated by the US National Academies in 1970, known as the *Committee on the Survey of Materials Science and Engineering* report (COSMAT),[75] grandly surtitled "Materials and Man's Needs: Materials Science and Engineering." The Committee was chaired by eminent metallurgist and MIT Institute Professor Morris Cohen (1911–2005), NAE *and* NAS, Kyoto Prize and National Medal of Science recipient; and co-chaired by physical chemist (and polymer scientist) William O. Baker (1915–2005), NAE and likewise National Medal of Science recipient, then vice-president of Bell Telephone Laboratories (later 5th president 1973–1979), who had earlier helped develop synthetic rubber. In addition to a thorough review of the history and accomplishments of the IDL/MRL program up to the time of its transfer to the NSF, the report provided a comprehensive survey of, and powerful *raison d'etre* for, the materials science field at that time. A second assessment was commissioned by the NSF and carried out under contract by the Mitre Corporation, for 18 months, 1976–1978. The fundamental questions being asked in

[75] Baker and Cohen (1975).

the Mitre report were far less grandiloquent, more pragmatic, and even parochial: Were the materials faculty members within the IDL/MRL umbrella any more productive than were the individually funded materials faculty in universities without this block-grant funding, and how did the administrative costs compare? As a measure of productivity, the Mitre group collected those publications of IDL/MRL researchers that were carried out with IDL/MRL funding and had them assessed for quality by an independent evaluation committee, then compared the result to a collection of publications, similarly assessed, coming from an equivalent number of researchers who were supported by individual NSF grants in institutions without IDL/MRLs which had strengthened their MSE efforts by consolidation in similar ways without the aid of block funding. The Mitre group reported[76] that the differences were not striking, except perhaps for the higher quality of achievements at the IDL/MRLs and the observation that the IDL/MRLs on the whole had attracted more celebrated faculty members. The administrative costs for the MRL research were somewhat lower than those for individual grants (economies of scale), and the overall program cost per publication was a little lower for MRL-generated publications.

There were many obvious fundamental flaws in the Mitre review protocols (despite occupying 7.5 person-years at a cost well over $1M), among which were the assumption that the sought-after "product," to be quantified and qualified, was investigators' research publications, whose lasting impact could not be immediately assessed; or the seeming ignorance of the simple zero-sum (and less easily quantified) fact that if faculty members were relieved of some of the onerous and time-consuming tasks of writing many more individual grant proposals, they could perhaps spend more time thinking about their research, or consulting with their graduate students, or teaching and proselytizing undergraduates in order to popularize a discipline with still almost zero public recognition. The wherewithal to purchase large and expensive pieces of equipment in central facilities, like ultrahigh-resolution electron microscopes or surface analytical tools, vastly beyond the means of any individual

[76] Ling and Hand (1980).

research grant, was an almost incalculable (but not quantified, either, by Mitre) godsend to progress in so many aspects of materials study. But those shortcomings aside, the Mitre survey simply asked (or was asked to ask) the wrong question. The right question was what effect did the concerted infusion of blocks of money into materials education and materials research at an early stage of their evolution have on consolidating the historically disparate constituencies in the materials field, on increasing the number of materials-savvy materials-catholic PhD-level graduates generated, and on promoting a significant multiplier effect when a substantial fraction of these graduates began populating the faculty ranks of other institutions. It is a truism at a place like MIT, or for that matter any leading research university, that its faculty members now compete with their own former graduate students who have gone off to enrich the faculties of other research universities. This was von Neumann's vision for materials: a rising tide to lift all boats, and the standard of education and research volume at *all* major materials departments have been remarkably enhanced over the past five decades. Departments without the advantage of IDL/MRLs simply consolidated their programs, invented their own MRL equivalents, and competed. NSF's present MRSEC program levels the playing field considerably, specifically encourages project interdisciplinarity, and caters to many more players now in the field. But the pump had already been primed.

A more personal assessment[77] 25 years on is provided by Lyle H. Schwartz, a former Northwestern student (NU's first undergraduate student in Materials Science & Engineering '59, PhD '64; this author being the second). A subsequent Northwestern faculty member, and national champion weight lifter, Schwartz was for 13 years the director of the NBS/NIST Materials Science & Engineering Laboratory, and like this author was wooed even as an undergraduate into materials science, properly portrayed as an exciting field with infinite potentialities by a faculty with a broad conception of its mission — that missioned enriched by the largesse and vision of some thoughtful predecessors in the DOD and later NSF.

[77] Schwartz (1987).

Since 1960, a ground-up revolution has clearly "materialized" in materials science, documented still further elsewhere in this series. Materials scientists are now able to do so much more than simply co-opting traditional raw materials and processing them in particular ways that improve the properties of existing materials. In parallel, professional materials societies have moved away from being industry-based organizations and instead have increasingly focused on research-based methodologies that resemble those of other scientific societies and are common to what had been previously maintained as disparate materials fields. Researchers have gleefully examined the new toolbox of analytical and imaging methods available to them, and the prodigious range of new materials and products that could be produced with information gleaned from those tools that often cut across several materials classes at once (for example, in composite structures or devices — among them touch-screens, high-temperature superconductors, LEDs, ferroelectric actuators and speakers, ubiquitous sensors, cell-phone Gorilla® glass, lithium batteries, geopolymeric concretes, custom nanoengineered multi-class structures, photonic devices, bionic body parts, and so many others — their number, acronyms, and impact mounting almost daily.

Academically, while the discipline of materials science & engineering straddles both schools of Arts & Sciences (physics, chemistry, biology, geology, archaeology, anthropology) and schools of Engineering (mechanical engineering, electrical engineering, chemical engineering, civil engineering, industrial and systems engineering, nuclear engineering, aeronautical-astronautical engineering, bioengineering) — and even schools of Business (life cycle analysis, strategic resources, materials economics) — materials departments, almost without exception, have ended up located in schools of engineering; this school placement is typically problematic only for accreditation purposes, when acrobatic similitudes have to be invented: for example, the ubiquitous engineering *design* requirement has had to be creatively interpreted as *microstructural design of component phases in multiphase assemblages,* or as *microchip device topology,* or perhaps *polymer processing schema.* Of course, there are academic materials scientists to be found in any of these schools and departments and their equivalent divisions within industrial or commercial research structures. MSE is not so much subject- or material-oriented as it is *approach*-oriented, so societal representation by a unitary association emphasizing concept and methodologies above composition provides multiple associational benefits for this widely distributed network of materials practitioners. Not surprisingly, such broad professional society representation was not long in coming.

Professional Society Implications

The balkanization of university departmental structures initially associated with separate materials classes had its counterpart in the less-than-flexible specialization boundaries of traditional materials societies, nearly all of which had grown up from strong roots in craft-based trades dating back centuries. By the early 1970s — a decade out from ARPA's MSE educational initiative — it was the time (some would have said *over*time) for professional societies to more broadly reflect the integration now appearing in the educational realm. Formation of the *Materials Research Society* was the direct outgrowth of these materials-catholic ontological and educational developments.

The Materials Research Society (MRS)

The *Materials Research Society* (MRS) is the first professional society in the roster of materials-related societies not to have arisen from a traditional materials type, nor of craft-based origins, so it is worth looking at why, and particularly how, it was founded. MRS was formed in 1973 from a diverse set of academic and industry-based professionals across the materials spectrum, addressing the evident need for a more catholic approach to materials with missionary zeal.

More than anyone else among those who founded this novel interdisciplinary instrument, the charismatic force behind it was Rustum Roy (1924–2010) of Pennsylvania State University. Roy was born in Bihar province in eastern India, bordering Nepal, the 7th of 11 children. A youthful encounter with Mohandas Gandhi left an indelible imprint of the Mahatma's ethical humanism. Roy took a

Cambridge School Certificate from St. Paul's School in Darjeeling, then studied physical chemistry to the master's level at Patna University in Bihar, and earned his doctorate in engineering ceramics at Pennsylvania State University in 1948, the same year he married his lifelong companion, Della (Martin) Roy (1926–2021), a mineralogist who studied the glass transition in silicas, invented porous biomaterials for bone repair, and became a world-renowned expert in cementitious materials. Like her husband, she was inducted into the National Academy of Engineering, only the third woman to be so elected and the first woman in materials science.

Rustum Roy was an early (later tireless) advocate of interdisciplinarity and noted that

> "....universities have been forced into new interdisciplinary patterns, not only by the dollar sign, but also by the inexorable logic that the real problems of society do not come in discipline-shaped blocks."

He and wife Della had indeed been the co-founders of the aforementioned Penn State interdisciplinary solid-state technology graduate program (instituted in 1959) and its counterpart Materials Research Laboratory (begun in 1962), with Rustum as its first director. He edited the proceedings of a 1969 national colloquium[78] on the discipline of Materials Science and Engineering in the US and organized the meeting at Penn State in 1973 that led to the formation of the MRS, the same year he was inducted into the National Academy of Engineering. The next year, he and Olaf Müller co-authored the compendium *The Major Ternary Phase Diagrams.* Ever the promoter of the materials science movement, his 39-page 1991 lecture on *New Materials: Fountainhead for New Technologies and New Science* was published by the NAE in 1994. He delighted in

[78] Roy (1970). It is instructive to compare this earlier assessment of the materials science and engineering discipline to one provided 30 years later by Merton C. Flemings NAE, a metallurgist and former chair (1982–1994) of the Materials Science & Engineering deparment at MIT. Flemings (1999)

creating new ceramic materials using his patented solution sol-gel process, and in his 65-year career, he schooled generations of students in the "art and mystery" of ceramic synthesis. Exceptionally broad in his purview, Rustum Roy also wrote on nuclear waste, US science policy, the fusion of religion with technology, the social impact of materials technologies, integrative medicine, and Ayurvedic and holistic treatment of disease, and began a book series on materials education for materials educators. Unfailingly the ethical humanist, he put a human face on both big and little science. He and his wife Della Roy even co-authored a book on sexual ethics entitled *Honest Sex*.

Several preceding events precipitated the formation of the MRS.[79] The first, of course, was the ARPA funding of the Interdisciplinary/Materials Research Laboratories (IDL/MRLs) at US universities, (eventually 12) starting in 1959. Even before that, as early as 1957, Rustum Roy at Penn State had proposed with the Carnegie Institute of Technology (after 1967 Carnegie-Mellon University) to establish a shared interdisciplinary research program, but there was at the time no interdisciplinary agency to fund it (agencies like NSF, DOE — and its predecessor, Energy Research & Development Agency (ERDA)), and even its National Laboratories, remained still divided along traditional disciplinary lines); it was as a consequence that the Roy began Penn State's own independent interdisciplinary Materials Research Laboratory in 1959.

The *Committee on the Survey of Materials Science & Engineering* (COSMAT) (See footnote 75), convened by the National Academies at the behest of the Government, was started in 1970 and reported in 1974; but its provisional recommendations — sufficiently leaked by 1972 — were understood to advocate the formation of a *federation* of existing traditional-discipline materials societies, as well as favoring the foundation of a dedicated interdisciplinary materials society.

At the time, there were 31 traditional-discipline professional societies whose compass included some form of material, eleven of

[79] Roy (1993). Gatos (1993).

which (*viz* ACerS, ACS, AISI/AISE, APS, ASM, EMSA (later MSA), IEEE, OSA (later *Optica*), SPE, and the *National Association of Corrosion Engineers* (NACE) were singled out for federation; all of these societies, except the last mentioned, NACE, are profiled in this entry. (The *Electrochemical Society* (ECS) has a strong Corrosion division, which is more material science-based than is NACE.) Organization of a federation of these eleven was set in motion by Nathan E. Promisel, chair of the National Materials Advisory Board (NMAB), and a steering committee organized in 1971 for what eventually became a reality in 1977 as the *Federation of Materials Societies* (FMS), under the leadership of Robert Laudise, head of Materials Research at Bell Labs. But FMS could do little to *break down* the disciplinary barriers separating these societies.

Meanwhile, Rustum Roy had organized two small sequential interdisciplinary conferences under the title *International Conference on Characterization of Materials*, a topic that intentionally cut across materials discipline boundaries: the first (ICCM-I) at Penn State in 1966, the second (ICCM-II) in Rochester, NY in 1968. After ICCM-I, Roy sent a letter to Frederick Seitz (then president of the National Academy of Science (NAS), and an eminent materials physicist) stressing the need for a professional organization in the field of materials. Roy further championed this idea as the ACS representative to a National Research Council (NRC) meeting. In 1968, Seitz appointed a *Committee on Intersociety Cooperation in Materials* (CICM)), including Roy, Harry C. Gatos (MIT), N. Bruce Hannay (vice-president of Bell Labs), Robert A. Huggins (ARPA) and Israel Warshaw (NSF) and came up with a first shot at a written proposal that appeared to have cleared the decks of the Naval Research Laboratory (NRL) — joined by DOE (ERDA) and ARPA managers in their firm opposition to the formation of a new society.

Roy put the proposal to a vote of attendees at the ICCM-II meeting in 1968, where it was favored by a 3-to-1 majority of that rank-and-file. Roy then called a "national colloquy on the field of materials" in 1969 at Penn State (his word choice betraying a decidedly British imprint on his Indian schooling), and at that gathering the group of W. O. Baker, I. Warshaw, R. A. Huggins and Roy decided to support the NAS in implementing the COSMAT survey, advocate for a federation of materials societies, and begin the founding of a new interdisciplinary materials society. A constitution was written and circulated to the group of those involved most closely in the lead-up: Rustum Roy (Penn State), Harry C. Gatos (MIT), Mark B. Myers (Xerox), Eric Baer (Case Western Reserve University (CWRU)), Robert A. Huggins (Stanford Center for Materials Research), N. Bruce Hannay (Bell Labs), Rodney Hanneman (GE), Israel Warshaw (NSF) and Leonard R. Weisberg (RCA, who was made treasurer), that group later enlarged — with the addition of Kenneth Jackson (Bell Labs), Robert Laudise (Bell Labs), E. Banks (Brooklyn Polytechnic Institute), Eric Kay (IBM), S. Victor Radcliffe (head, department of Metallurgy and Materials Science at CWRU), B. Sheldon Sprague (Celanese), Richard S. Stein (University of Massachusetts–Amherst), James J. Tietjen (RCA) and S. Stemstein (Rensselaer Polytechnic Institute) — to form the formal founding committee. William O. Baker (Bell Labs VP, later president), whom Roy labels as the "godfather of materials science," kept a watchful eye on the whole proceedings.

Given the novelty of a foundation without a background history linked to any traditional materials type nor earlier craft specialty, it is both instructive and appropriate to profile those most responsible for this transformative formation of a new form of an interdisciplinary professional society. Roy cites five as particularly effective in the early stages of the Society's existence (See footnote 79). They are Harry Gatos, first MRS president, like Roy a very early proponent (and practitioner) of interdisciplinary approaches; Mark Myers, an early doctoral student at Penn State's independent MRL, migrating to Xerox Corporation; Eric Baer, a polymer chemist who had created the department of Macromolecular Science & Engineering at Case Western Reserve University; Ken Jackson, a Canadian crystal grower at Bell Telephone Laboratories; and Israel Warshaw, at NSF during the whole formation period.

Harry C. Gatos NAE (1921–2015) studied Chemistry at the University of Athens, the University of Indiana, and MIT, where he received his PhD in inorganic chemistry in 1950. He worked for

3 years with DuPont, then joined the Electronic Materials Group at MIT's Lincoln Laboratories, becoming head of the Solid State division. In 1962 Gatos was appointed Professor with joint appointments in the departments of Metallurgy and Electrical Engineering (later Materials Science & Engineering and Electrical Engineering & Computer Science), where he pursued research in electrodeposition, metallic corrosion, superconductors, high-pressure phase transformations, and crystalline and amorphous semiconductors, founding the journal *Surface Science*, serving as its editor-in-chief from 1962 to 1996; he became head of the Corrosion division of the Electrochemical Society and ECS president in 1967.

Mark Myers (1938) undertook his PhD with Profs. Rustum Roy and K. Vedam in Penn State's Materials Research Laboratory on high-pressure measurements in solids, then joined the research arm of Xerox Corporation in Rochester, NY in 1964, studying arsenic-antimony-selenium amorphous glasses. He eventually rose to the position of Vice-President of Research at Xerox, from which he retired in 2001 and took a position with the Wharton School of Business at the University of Pennsylvania, retiring again in 2005 to become a gentleman farmer in southeast Pennsylvania and president of the *Philadelphia Society for Promoting Agriculture.*

Eric Baer (1932), like Arthur von Hippel, John Cahn, Norbert Kreidl, and Sir Peter Hirsch (profiled later) before him, was a refugee from Nazi Germany. His father was taken to Buchanwald concentration camp after *Kristallnacht*, but released in 1939, and his family immediately emigrated to England, settling in London. Like Peter Hirsch and countless other London children, Baer was separated from his parents and evacuated north of London during the blitz. He rejoined his parents after the War, and the family moved to Baltimore, MD in 1947. He matriculated at Johns Hopkins University (JHU), and 2 years later, in 1953, took an MA in organic chemistry, transferring to the JHU school of engineering to take his doctorate in chemical engineering in 1957. He worked at E. I. DuPont's Polychemical Division before joining Case Western Reserve University (Cleveland, OH) in 1962 as founder of their department of Macromolecular Science & Engineering, where he held the

Herbert Henry Dow chair in Science & Engineering. His principal research has been in polymer layered systems, for which he won the Paul J. Flory Research Prize and the Charles Goodyear Medal of the ACS Rubber Division.

Kenneth A. Jackson NAE (1930–2022) took his first degree in Physics from the University of Toronto and his PhD from Harvard in 1956, joining the faculty there. In 1962, he was appointed head of the Materials Physics research division at Bell Labs, where he investigated the kinetic processes of crystal growth, thin film growth and characterization, and ion-beam processing of semiconductors. He established (with Robert Laudise) the *American Association for Crystal Growth* and was its founding co-president. He is well known for his treatment of constitutional supercooling as a determinant of many of the surface, morphological, and defect characteristics of crystals and their growth, for which he was recognized — like Roy and Gatos before him — by election to the National Academy of Engineering. He authored a definitive book for the crystal-growing community entitled *Kinetic Processes: Crystal Growth, Diffusion and Phase Transformations in Materials.*

Israel (Izzy) Warshaw was program director of the Engineering Materials program in the engineering division of the National Science Foundation (NSF) during the formative period for MRS, in 1971 becoming acting division director of the Division of Engineering when a separate Division of Materials Research (DMR) was created at NSF.

The new society's inaugural meeting was held at Pennsylvania State University in 1973, attracting 300 scientists, of which 215 signed on as members of the new organization.[80] The carefully chosen single topic, the application of phase transformations in materials science, deliberately cut across all materials disciplines. MRS began publishing a newsletter in 1974 that eventually grew into the monthly *MRS Bulletin* in 1982. The second meeting was held at

---

[80]A first-hand account of MRS from its beginnings to its 40-year mark in 2013 is given by Gail Oare, director of memberships, meetings and publications for 27 of those years. Oare (2013). An account of the growth of MRS in the first 20 years is given by S. Thomas Picraux (1993 MRS president). Picraux (1993).

Princeton in 1975, with again a single generic theme, this time defect-property relationships in solids. The third meeting was held in Cambridge, MA and featured six interdisciplinary symposia, establishing a pattern of parallel symposia as the MRS meeting format that continued for the next nearly 50 years. The fourth meeting was held back at Penn State, but with the fifth meeting, in 1978, the Society moved its meeting to Boston, where it has remained as the annual MRS Fall Meeting in late November, immediately following the US Thanksgiving holiday. At that meeting, the introduction of two subsequently long-running symposia — Scientific Basis for Nuclear Waste Management, and Rapid Thermal Processing — spurred a rapid growth in MRS membership. The first few Boston meetings were somewhat crowded and boisterous, crammed into undersized facilities at an older classic Boston hotel (which the Society's Fall Meeting soon outgrew), but nonetheless enthusiastic and permeated with revolutionary zeal. Later, the Boston meeting venue expanded more comfortably into Boston's principal conference center and adjacent hotels, and the Society added an equally well-attended West Coast meeting, initially in San Francisco, later more peripatetic. By the 2000s, the meeting attendance (and MRS membership, since that came with paid attendance) had climbed into the many thousands, outpacing all the traditional materials society competitors, whose meeting and memberships complements were falling. In 2023, MRS membership stood at 16,000.

Adopted from its third meeting onwards, MRS's unique organizational paradigm for its biannual meetings, built around free-standing individual symposia devoted to specific issues and problems, enabled MRS to be more nimble than other societies, more responsive to new materials developments, and to become immediately responsive to new potential membership sectors. The administrative tasks of undertaking a large, increasingly multidisciplinary conference were at the same time spread among a much larger cadre of rank-and-file enthusiasts: the symposia chairs and their organizing committees. From the outset, MRS did not necessarily focus on any particular class (or classes) of materials, and it had no divisional structure for specific interest groups — though it does maintain an internal committee structure, e.g., publications, government outreach, global outreach, etc. Its meeting agendas were driven by its rank-and-file in the form of symposia that are proposed and organized by individual members or groups of members, and that direct input has greatly facilitated the rapid emergence of new materials research directions. One meeting might host a symposium on soft materials, another might entertain symposia centered on broad topics with significant commonality to all materials classes, like surfaces, or general attributes, or specific properties, such as structural topology or radiation resistance, which cut across multiple materials classes. Other sessions have been organized around particular investigative or modeling tools, biometrics, or the politics of strategic materials that transcend specific material identities, or environmental sustainability.

The symposia for each meeting were initially assigned letter designations (reserving X for special lunchtime lectures highlighting the current frontiers of materials research — named in honor of co-founder Rustum Roy and presented by prominent materials practitioners); those 25 designators were quickly exhausted, and — after reaching into triple-letter assignments by 2012 — MRS moved in 2016 to broad *topical clusters* of symposia, each topical cluster typically represented by 5–12 symposia, each symposium comprising perhaps 20–50 presentations. The most recent (2022 meetings) clusters were: *Broad Impact* (often societal), *Characterization, Materials Computing & Data Science, Electro-Optical & Quantum, Nanomaterials, Soft Matter and Biomaterials,* and *Structural & Functional Materials.* An especially attractive feature of earlier MRS meetings was that individual four- to six-page articles (not just abstracts) from a symposium were soon after published in retrievable book form — at least until the sheer number of symposia proceedings from the bi-annual meetings, a major publishing chore, overwhelmed the publishing arm of the society and were offloaded onto electronically retrievable platforms.

The MRS governing board is drawn from many sectors, both academic and industrial, and almost from the start from many countries of the global materials community. Five of the six presidents following the initial presidencies of Harry Gatos (1973) and Rustum Roy (1974), came from the electronic materials community (mostly from Bell Labs, starting with Kenneth Jackson), which had long been without a societal home, before broadening out to the national laboratories (Sandia, Oak Ridge, Livermore) and other industries (General Motors, Exxon), then settling out subsequently to about 40% academics, with a broad cross-cut through the various materials disciplines. There was an early focus on young materials scientists, and on graduate and undergraduate materials students, with well-publicized presentation awards offered to student authors in its meetings, and free admission for students willing to help out with A–V oversight in the many meeting rooms. One of the MRS directors was specifically tasked with introducing graduate and undergraduate special educational programs and establishing student chapters in universities that, similarly, cut across disciplinary and departmental boundaries in their memberships and attempted to integrate (somewhat successfully, though with blowback from competitor societies) existing student chapters tied to single materials-class metallurgy, ceramic, polymer or electronic materials science or engineering identifiers under a unifying MRS umbrella student-chapter organization. MRS's first short course was held in 1982, and Rustum Roy began a series of educational modules on materials science & engineering topics aimed at undergraduates.

The materials-catholic administrative, meeting, and educational structures rendered the MRS significantly different from the more traditional societies, which were organized for the most part around specific classes of materials, associated with a significant traditional industries bias, and so by contrast provided a strategic temporal advantage in nimble capture of new developments within the materials research community — such as high-temperature superconductors, or advanced battery research, or optoelectronics — early in their evolution. The MRS quickly became the home for electronic materials scientists, as a more applications-oriented venue than APS provided and a more solid-state physics approach than afforded by IEEE. Notably, MRS engaged an effective lobbyist in Washington to keep the importance of materials science and technology to the US's continuing world-leadership position prominently before Congress. It also runs a Congressional Fellows program (jointly with TMS), many of whose participants have gone on to distinguished careers in science policy and continue their materials advocacy.

In common with all professional materials societies, MRS has established awards to honor outstanding individuals in its fields of endeavor. Since the remit of the MRS is so broad, effectively *all* materials, the contributions of individuals whom it chooses to so honor could be in any discipline with materials implications. Rather than choosing one of its founders after whom to name its two most prestigious awards, MRS chose instead to honor (and now memorialize) two individuals whose work has exemplified broad integration across the materials spectrum. The first so honored, for the Society's highest award, was Arthur von Hippel, whose broad working integration of a wide range of fields in his Laboratory for Insulation Research in 1940 provided a demonstrable model 30 years before the MRS was founded, and whose life was profiled earlier. The von Hippel award "recognizes those qualities most prized by materials scientists and engineers — brilliance and originality of intellect, combined with vision that transcends the boundaries of conventional scientific disciplines." The first recipient was von Hippel himself in 1978; the second recipient was W. O. Baker (president of Bell Laboratories, and the *eminence grise* who hovered over the whole foundation of MRS), and the third was Harvard professor David Turnbull. The list of von Hippel awardees reads, again, like a who's who of materials science.

The other major award, a plenary lecture delivered at the annual Fall MRS meeting, is intended to "recognize the *career* contribution of a scientist to fundamental understanding of the science of materials through experimental and/or theoretical research" and bears David Turnbull's name, who was himself the first awardee in 1992. Born to a farming family in northwest Illinois, 100 miles west of Chicago, David Turnbull (1915–2007) pursued his undergraduate studies in physical chemistry at Monmouth College (IL) and his PhD at the University of Illinois, before joining the faculty of Case Institute of Technology (later Case Western Reserve University) in 1939. During the War years at Case, he conducted synthetic rubber research and taught Naval cadets metallurgy. He left in 1946 for a job with General Electric's Research Laboratory, investigating solidification of metals, in particular supercooling of their melts, and later formulated a criterion for the ease of glass formation from supercooled melts with high viscosities (as in silica glasses); he then extended his ideas to metal alloy systems with deep eutectics and predicted formation of metallic glasses for sufficiently fast cooling, and three decades later in 1982 demonstrated that bulk metallic glasses could be obtained with cooling rates as low as 100 K/s — a factor of 10 slower than had been previously demonstrated — using the Pd–Ni–P alloy system. He continued to explore nucleation during solidification, confirming the role of impurities in catalyzing the liquid-solid phase transformation. Turnbull and Frederick Seitz published the first edition of *Solid State Physics* in 1955, now an annual series, the same year that John Cahn came to the GE Research Laboratory, with Turnbull as its director of the Chemical Metallurgy division. Turnbull joined Harvard University as Gordon McKay Professor of Applied Physics in 1962 and taught there for 23 years; among his many successful students are King-Ning Tu (IBM) and Profs. Frans Spaepan (Harvard), Michael Aziz (Harvard), and Carl Thompson (MIT). During his career, he wrote scores of research papers each on a large number of materials topics: solidification, nucleation, glass formation and amorphous solids, diffusion, crystal growth, thermionic emission, gas thermodynamics, organic physical chemistry, corrosion He was, in his successor's Aziz's words in his memorial tribute, "the master (in) conceiving of and performing the simple but critical experiment." Upon coming to Harvard, Turnbull felt he could not compete with GE's sophisticated arsenal of equipment, instead instructing his students that "while we can hardly achieve a distinction with our hardware, we may do so with our wits." He was elected to the National Academy of Science in 1968 and, in addition to his two MRS distinctions, was awarded the Japan Prize in 1986 and the Franklin Medal in 1990.

Of all the US-based materials societies, MRS has proven the most pro-active in achieving gender balance — perhaps not surprising, since it did not have to contend with a legacy of historically male-dominated craft-centered fields — though in the 1980s still only 10% of MRS members were female. Forty years later, two-thirds of the officers of the Society are presently women (including the last two presidents), and seven of the eleven Board members. Six of the last ten presidents have been women. This representation much more closely resembles the gender distribution that has come to characterize the undergraduate student enrollments in many US materials science & engineering departments (though, with a substantial fraction of graduate students in every US department drawn from countries outside the US, and coming from cultures that remain male-dominated in science and engineering fields and education, the gender balance in graduate student enrollments remains tipped substantially in favor of men).

MRS soon became global, as well, with symposium organizers and participants drawn from many nations (probably 50% of MRS members are from outside the US). In 1983, the European Materials Research Society (E-MRS) was formed in Strasbourg (now with 4,000 members), and since then many other similar societies have been established, some like E-MRS retaining a collaborative attachment to the MRS. In 1986, MRS began publishing its own international journal, the *Journal of Materials Research.* It continues to extend its publishing roster, most recently adding two rapid-publication journals *MRS Communications* and *MRS Advances*, and a special topic periodical *MRS Energy & Sustainability*. MRS has also launched several public outreach endeavors, including two science museums/ centers exhibitions — the first, *Strange Matter*, running for the past 10 years, and the second, a more recent sustainability program titled *Strange Matter Green Earth.* MRS partnered with the public television NOVA program to produce the four-part series *Making Stuff* and has designed an undergraduate course *Impact of Materials on Society*, aimed at liberal arts curricula to acquaint non-STEM students (and their instructors) with the strong connection of developments in the materials field to societal change.

Now beginning its second half-century — which will give some of its surviving older members pause — the MRS can trace a history of substantial accomplishments, not least of which is freeing materials from their traditional engineering origins and embracing the fundamental and applied science at their core. Six notable topics with which MRS had been involved at its 40-year mark (and continues to be still, a decade later) were catalogued[81] by former MRS president Gordon Pike (Sandia National Laboratory) and science writer Philip Ball in 2013 as responsible for societal change (and, it should be added, significant cultural shifts). These six topic areas were the clever employment of oxides in electronic materials, organic electronics, carbon nanostructures, multiscale materials modeling, materials for tissue engineering, and materials science in nuclear waste management, all of which are still very much in play a decade later and all mentioned elsewhere in the present article. In 2023, one might want to add still others: materials for energy production, conversion and storage (e.g., solar energy materials, fuel cells, Li^+- and other ion-conduction battery systems); use of light conduction (instead of slower electron conduction) in optoelectronic devices; the systematic engineering of extraordinary materials properties in applications involving multiple requirements (such as the glass cell-phone screen); an evolving fundamental understanding of the structure of non-crystalline solids (a challenging scientific frontier still remaining on the materials horizon); the sequestration of carbon in our degrading climate environment; materials characterization at the atomic scale using multiple analysis techniques in massively parallel employment; and nuclear materials challenges, not only for nuclear waste containment (still a problem, but in large part now mostly political), but more importantly for invention and rapid deployment of a new generation of nuclear energy-production systems (still perhaps the *only* viable medium-term solution to clean energy on the 20 TWatt scale required by the human population of this globe). Perhaps MRS could institute a Symposium X_S in future MRS meetings, the subscript referring to human survival, toward which existential goal materials could yet play a deciding role, doubtless to the post-mortem delight of the late Rustum and Della Roy, who both expressed such consistent concern for the societal challenges — and salvational opportunities — presented by materials technologies.

[81] Ball and Pike (2013).

Post-MRS reconstructions of traditional societies

The impact of the academic shift to departments of materials & engineering (MSE), and of the formation and rapid growth of the Materials Research Society, on traditional departments of metallurgy & metallurgical engineering and ceramics & ceramic engineering and their professional societies was not inconsiderable. At first dismissed, and not long after demonized, the generic materials movement — and especially the MRS — represented unexpected and unwelcome competition. While traditional societies may have policed their borders out of pride or a historic sense of ownership, the MRS did none of that: everything was fair game, and it didn't help the other societies that upcoming students were being doubly proselytized, taught that materials science & engineering was both a historically grand *and* a unified field of human endeavor (if still relatively unrecognized by the general public). Academic ceramic engineering departments — at least those that clung to clay-based end-products or traditional ceramics formation processes based on consolidation of powders to near-net shape, found themselves in an increasingly existential race, competing with materials science & engineering departments and the increasing number of universities that exclusively offered them. US undergraduate ceramic engineering programs exhibited a near-perfect inverse-U shaped population demographic, rising to an apogee (at fourteen programs) by 1960, followed by an immediate mirror-reversal decline (to four remaining programs by 2006) once MSE departments were begun.[82] The *Keramos* fraternity reflected a corresponding downturn, with four of the thirteen active chapters going inactive by 1972 for lack of initiates or discontinuation of an academic program in ceramic engineering (See footnote 14). By 2006, ceramic engineering programs graduated only 50 ceramic engineers out of a total of almost 75,000 engineering graduates from all US colleges and universities; by that same date, there was a total of 65 MSE departments, 9

[82] Readey (2008); a transcribed written version is retrievable at: https://www.mines.edu/faculty-senate/lecture/2007-readey/ (last accessed 1 February 2023). Ready was President of ACerS in 1991, but also is a fellow of ASM, a member of MRS, and served on the TMS Board of Directors.

metallurgical engineering departments, and 4 ceramic engineering departments in the US. One long-time ceramics engineer lamented that, even as the number of ceramic engineering degree programs and their graduates was waning, the volume of academic ceramics research (then being conducted in materials departments) was actually increasing rapidly.

One contingency band-aid was to append "materials" to departmental titles, e.g., as was done at MIT (in 1967, *Metallurgy and Ceramics* changed to *Metallurgy and Materials*, changing again in 1974 to *Materials Science & Engineering*), at Oxford University (*Metallurgy* changed to *Materials*), at Cambridge University (*Metallurgy* changed to *Materials Science & Metallurgy*), or at Clemson University (*Ceramic Engineering* changed to *Ceramic & Materials Engineering*). By the 1980s, virtually all previous metallurgy, metallurgical engineering, ceramics, or ceramic engineering programs worldwide had been subsumed into departments of materials or materials science & engineering. The same was not true for programs of macromolecular science or polymer engineering, which continue to exist as stand-alone departments in some cases, or as divisions of chemical engineering departments, though nearly every department of materials has integrated programs in polymer science and engineering, or else interdisciplinary programs in polymer science and technology (MIT does both), as well as well-integrated inclusion of electronic materials.

Responses of traditional professional societies — both to MRS and to unified materials curricula in university programs — varied and developed more slowly than in academia. A near-universal response was to add other materials classes to divisional or meeting structures. For example, ACerS (which had already added *Basic Science, Electronic Ceramics,* and *Ceramic-Metals Systems* to its divisional complement by 1961) added *Nuclear Ceramics* in 1965 and *Cements* in 1971 — all before MRS was founded — and appended "*Optical*" to the *Glass* division, "*Environmental*" to the *Nuclear* division in the 1980s, and "*Archaeology & Conservation*" to the *Art* division in 2013. Now well into its second century and beginning to experience downward membership pressures, ACerS transitioned the *Whitewares and Materials* division to a broader *Manufacturing* division, added a *Bioceramics* division in 2017 and an *Energy Materials & Systems* division in 2019, and additionally shored up its ceramics and glass industrial base by launching a *Ceramic and Glass Industry Foundation* in 2014.

Responses by the metals societies were complicated by the demise of the steel industry, in both the US and the UK, which had served as both impetus and backbone for these societies for a century or more. The downward spiral began in 1967, and by 1970 American steel producers could no long compete effectively with low-wage producers abroad. Most US steel mills closed or went bankrupt. By 2002, the giant US Steel corporation employed only a tenth the number of workers it had employed in 1902. World production peaked in 2007, but prices fell 40% during the recession of 2008. India's Tata Steel bought out UK producers, then merged with western Europe's major steelmaker to form the world's largest multinational steel producer, capturing 10% of world production. By 2013, however, ensemble production in China dwarfed all others, comprising almost 75% of the world's steel. Of course, there were other high-volume metal industries for ASM and TMS to represent in their various ways (copper, aluminum, titanium, zirconium, niobium, specialty steels, metallic glasses) that were growing faster than domestic low-carbon steel.

TMS (originally styled just *The Metals Society*, one of the four constituent societies of the American Institute for Mining, Metallurgical, and Petroleum Engineers, AIME) added *materials* to its name in 1989, though still retaining the *TMS* moniker and the *JOM* initials only (without specification of materials class) for its former *Journal of Metals* organ. The American Society for Metals affected the same makeover, re-styling itself " ASM (without specification), the Materials Information Society." The Institute of Metals in Britain added ceramics, composites, mining, steel, clay, and wood — by progressive society mergers — to its initial nonferrous metal representation, re-styling itself as the Institute of *Materials*, Metals, and Mining (IOM3) by 2002. In an academic counter-offensive, TMS, ACerS, ASM, and AIST decided to collaborate and *jointly* launch their *Materials Advantage* materials-engineering student-chapter program in universities, as a direct challenge to MRS's student chapter initiative, similarly aiming to consolidate the various local student society organizations (at least those in materials science and engineering departments, particularly those with an engineering bias) into a single chapter entity, though the Materials Advantage initiative has not established quite the same reach into physics, chemistry and polymer science student communities — who were then also engaged in various forms of materials education and research — that MRS has enjoyed. Both ASM and TMS set up foundations to provide student scholarship assistance and awards as a recruiting vehicle and initiated public outreach programs. TMS has also partnered with MRS in their joint Congressional Fellows program. To match ACerS's traditional "ceramic mug drop" student competition at its annual meeting, TMS set up an analogous "bladesmithing" competition at its.

In a bid for increased membership, and to compete with MRS's international embrace and its international affiliate MRS societies, all the traditional societies went international, too. In 2016, ACerS elected its first non-American president, eminent UK ceramist William E. Lee at Imperial College London (now emeritus professor), and the next year established affiliate "chapter" groups in several countries outside the US — first in the UK, followed by India, Italy, Canada, and Germany. It launched two new pointedly international journals, the *International Journal of Applied Ceramic Technology* in 2004, the *International Journal of Ceramic Engineering and Science* in 2019. ACerS was also instrumental in organizing the first International Conference of Ceramics in Toronto in 2006, since held every 2 years in sites around the globe, and with other ICG member societies helped successfully lobby the UN to declare 2022 as "The International Year of Glass" (See footnote 23). ASM renamed itself ASM International in 1986. ACerS and TMS both co-sponsored international meetings series, in particular their respective Pacific Rim conferences series — involving Japanese, Korean, Chinese, and Australian materials-related societies — (for ACerS) *Ceramic & Glass Technology* (now up to PACRIM-15) and (for TMS) *Advanced Materials & Processing* (now at PRICM-11).

But by far the most direct counter-offensive mounted to the MRS incursive challenge was the co-operative consolidation of the annual or Fall meetings of four of the major traditional societies into a *single* huge materials meeting and exposition known as MS&T

(for *Materials Science & Technology*), first held in Chicago in November 2003 and every Fall thereafter. The original collaborating societies were ACerS, TMS, ASM (until 2020), and AIST, and since 2022 the Society for Biomaterials (SFB) has provided a small representation as well. MS&T has, however, replaced society annual meetings *only* for the ACerS, which continues to maintain a vigorous slate of its well-established annual divisional conferences in the Spring, the other societies retaining their traditional annual meetings as Spring venues. ASM withdrew from the MS&T consortium with some irritation in 2020 to concentrate on its own annual meeting, with its own preferred timings and venue, now renamed the *International Materials Applications & Technologies* (IMAT) conference. Before the COVID-19 Pandemic, MS&T had averaged attendances of approximately 3,500 annually, with approximately 1800 papers presented each year. This statistic compares with MRS's Fall or Spring meetings with 6,000 attendees each. MRS currently numbers some 16,000 members world-wide and maintains 134 student chapters in 22 countries.

It is hard *not* to argue that MRS occupies the pole position in materials societies. In numbers of members, meeting attendees, and national and global reach, it is now always the elephant in the room. However, not every electrical engineer is an electronic materials scientist, nor is every polymer engineer a soft-matter scientist. (In fact, MRS has experienced notable difficulty, both at its meetings and in its *Journal*, in attracting polymer researchers from their more traditional homes in ACS or SPE.) MRS focuses on the *research* frontier, which is admittedly very close these days to the industrial vanguard, at least the high-tech vanguard. So there should be still plenty of room for other kinds of professional societies, with different emphases and a more traditional or less academic clientele. MRS deliberately, but presciently, inserted cements and nuclear waste media into its meeting lineups just to offset the high-tech laser processing and electronic materials biases that had set in at its early meetings and could have threatened to overwhelm the originally intended breadth of the Society, simply because there weren't many alternative homes for researchers in these fields. The mistake has been, though — at least in this author's opinion, informed over the past 60 years by multiple society governance positions and academic responsibilities in three universities — for any society to try to do everything in order to remain competitive.

The American Ceramic Society has done better — if sometimes slower on the uptake than some members may have liked and with a few glaringly missed opportunities (like semiconductors) — in developing a broad profile of materials utility, but confining its attention to the explosion of technically sophisticated *ceramic* materials while at the same time attempting to retain traditional ceramic engineering cohort (or what remains of it), from ceramic artists to whitewares manufacturers, in an unbroken 120-year history as a ceramics society. The same cannot be said for several of the diversifying metals societies, who claim to inform about *all* materials classes by simply appending "materials" to their increasingly opaque identifiers. The British Institute of Materials, Metallurgy & Mining (IOM3) has at least widened its breadth a bit more honestly by merging with other materials class societies to enrich its composite portfolio. In contrast, the American Ceramic Society has chosen to continue to style itself as "the home for *ceramics* and *glass* professionals" — not *all* materials professionals. Similarly, the American Chemical Society continues to style itself as "the premier home of chemistry professionals, "despite having divisions of *Polymeric Materials, Science & Engineering*, and *Biomaterials*.

A further internationalization phenomenon has been the formation of regional or global *federations* of materials societies, for example, the eponymous (but short-lived) *Federation of Materials* Societies.[83] Recommended by the 1974 COSMAT report (See footnote 75), FMS was officially co-founded in 1977 and presided over by eminent crystal-grower Robert A. Laudise (1930–1988), head of Bell's Labs Materials Research Laboratory. The Federation had usefully included as members the APS divisions of Condensed Matter Physics and Materials Physics, the University Materials Research Council that represented most major university materials departments, and several major corporations, and additionally had representation from

[83] Laudise (1995).

the US Solid-State Sciences Committee (SSC), from the National Materials Advisory Board, from NAS and NAE, and from other advisory groups to the National Research Council (NRC). A principal aim was to help the materials community to arrive at consensus materials policies that could be used to inform and persuade government policy makers. The Federation in the end, though, mostly functioned as an infrequent consultative body, lacking a pro-active operations side and dedicated government lobby, and ceased operations altogether in 2011.

Still very much alive, though, is the *Federation of European Materials Societies* (FEMS), founded in 2006 and now representing 27 materials societies in 22 European countries, with its own series of Euromat conferences (begun 1989) and control taken of its flagship publication *Advanced Materials* (a Wiley-VCH journal spun off from *Angewandte Chemie* in 1988). Some of FEMS's constituent member societies are quite venerable: for example, the *Deutsche Gesellschaft für Materialkunde* (DGM), a German technical-scientific society for materials science and engineering technology, was founded originally in Berlin in 1919 under the name *Deutsche Gesellschaft für Metallkunde* (German Society for Metallurgy). Another active conglomerate federation is the *International Union of Materials Research Societies* (IUMRS), founded in 1991 with the help and encouragement of the MRS (though without formal links) and variously administered from Singapore or China, as a loose union of the materials societies outside the US, now comprising 12 national or regional materials societies in Africa, Brazil (Sociedade Brasileira de Pesquisa em Materiais), China, Europe (E-MRS), Indonesia, Korea, Mexico (Socíedad Méxicana de Materiálias), Russian Federation, Taiwan, and Thailand. The IUMRS co-hosts annual and topical conferences — for example the IUMRS-*International Materials Research* conference (IUMRS-IMRC) in Cancún, México (which hosts approximately 1,800 conferees annually — in collaboration with MRS), the *International Conference on Electronic Materials* (IUMRS-ICEM), *International Conference on Advanced Materials* (IUMRS-ICAM), *International Conference in Asia* (IUMRS-ICA), and

International Conference for Young Researchers on Advanced Materials (IUMRS-ICYRAM).

Future materials landscapes

One might ask if there are now too many materials conferences or, indeed, even too many materials societies; but it is a hard sociological fact of history that it is difficult to keep human beings with like interests apart (*per* the well-established BFFT associative syndrome[84]). The primary question for an organization like MRS is what kind of society it ultimately wants to become. At present, it is a phenomenally successful *meetings* society, drawing upwards of 6,000 participants to each of its 2 yearly meetings (in largely non-overlapping sets), not only from the US but from many other countries as well. But, as founder and former president Rustum Roy remarked at the two-decade mark, big is not necessarily better ("small remains beautiful"), and the MRS's "unwillingness to tame the growth and grandiosity that affects American institutions has, perhaps led MRS to more of the same, instead of continually experimenting with the new." Apart from the camaraderie apparent in each of the delimited individual symposia, where many participants likely know or know of each other and certainly share commonality of interest, how close to the MRS as a society do participants really feel? Where that kind of allegiance shines is in the robust committee structures that now characterize the MRS organization, where committee chairs and their members feel a very strong sense of belonging to MRS and its

[84] *Birds of a feather flock together.* The origins of this pithy English proverb likely lie with the English naturalist William Turner (the earlier, non-ceramist, one by that name), fellow of Pembroke Hall, Cambridge, cleric, physician and sometime Dean of Wells Cathedral, considered the father of English botany and ornithology. Turner observed in his compendium *Avium praecipuarum, quarum apud Plinius et Aristotelem mentia est, brevis et succincta historia* (1544) — with hardly brief nor succinct, and certainly less poetic, inflection: "Byrdes of on Kynde and color floke and flye allways together." Turner is more remembered for his courageous riposte to Henry VIII, whose 1543 proclamation limited reading of the Bible to men of social standing: "What kind of ungodly belly wisdom, to say that rich men and the nobles are wiser than the poor?"

shared missions. It is there that the allegiance and socialization is felt most strongly. Apart from the growing tradition of its well-established awards lectures, MRS has chosen to eschew the traditionalist trappings of other societies in favor of looking unrelentingly toward the future and the newest developments; changing out the MRS Meeting Chairs every year ensures that forward projection. Embracing the future together is not necessarily the same as socialization, which relies on well-established social cues and long-standing commonalities with roots in the past. Many in MRS feel that the Society has not yet settled on the appropriate — or at least most efficacious — balance that will ensure its dominance into the next half century *and* increasingly provide a societal "home" for the prodigious number of materials scientists being churned out by the grand educational experiment that is the materials science and engineering discipline. To paraphrase what at least two more recent MRS presidents have remarked: "kudos, MRS, but what will you be when you grow up?" Will it end like Gatsby?[85]

What always strikes outsiders to the materials field is how many idiosyncratic materials societies there are, of every stripe and flavor. Would the materials community be better off with just one huge materials society, one "federation of everything" like ACS or APS, with lots of distinguishable divisions, or an IOM*n* — where *n* is large and positive — that gobbles up every conceivable materials field as a sub-society? Or a mega-MRS that solves its paucity of small-group sociability? The problem is ubiquity: materials are all around everyone, everywhere, and any individual sees something of intriguing interest somewhere. Physicists will boldly proclaim their compass is not just what's around everyone in everyday life: their embrace is the *whole universe*! No materials scientist would have the temerity to lay that leveling claim, so the materials community is likely stuck with everyday multiplicities, and a huge field that unavoidably cuts across every traditionalist disciplinary hedgerow, no matter how doggedly it might conspire to jump over or eliminate those persistent dividers.

[85]The last line of F. Scott Fitzgerald's novel *The Great Gatsby* laments: *"So we beat on, boats against the current, borne back ceaselessly into the past."*

Imaging and Analysis Societies

One would be remiss not to include in the materials societal roster two categories of professional societies devoted to the conquest of *inner* space, the imaging and microanalysis organizations, whose broad range of techniques has provided much of the understanding of materials phenomena over the last 80 years. Microscopy has from the start of scientific investigation played a large role in elucidating the "mysteries" of materials, starting with the 17th-century investigations of the Dutchman Antonie van Leeuwenhoek (1632–1723) and the English polymath Robert Hooke (1605–1703), who not only discovered "animalcules" and ultimately biological cells (the former term, a translator's invention in the English translation of van Leeuwenhoek's many letters to the Royal Society in London, represents what would be later known as bacteria; the latter term was first coined in Hooke's 1665 book *Micrographia*), but also some internal material structures, such as the columnar cellular microstructure of cork revealed by Hooke, or surface topography, such as that at the sharpened edge of a razor blade. (Cork is a material with the unusual mechanical property of a negative Poisson ratio, a consequence of its cellular topology: when compressed in a direction across the cell walls, it can unexpectedly *contract* along the lengths of the cells.[86]) The microscopes of van Leeuwenhoek and Hooke were capable of magnifications of perhaps only 500 times, but it was the polymath Hooke who developed a theory of light optics and Ernst Abbe in Jena, who two centuries later, established the origins of the resolution limits of optical instruments that made possible the science of microscopy and led to progressive improvements in microscope instrumentation.[87]

An amateur Sheffield scientist, Henry Clifton Sorby (1826–1908), was responsible for the leap to what became known as the science of *metallography* (See footnote 35). As a young student, Sorby

[86]Fortes and Nogueira (1989); Sergi and Trilio (1989). "Cork" is the bark of the cork oak tree; Fortes' and Nogueira's "backyards" are in Portugal, where most of the world's cork oaks are grown, not least for providing wine corks for Port wine.

[87]For a popular account of early light microscopy and its 20th-century evolution to atomic-scale electron microscopy, see: Hobbs and Revel (1989).

became interested in geology, mineralogy, and meteorites, and taken on a trip to Germany by his mother in 1852 met the German mineralogist Ferdinand Zirkel, who went on to invent the discipline of microscopic geology and mineralogy with the technique of thin-sectioning. Returning to Sheffield and his preoccupation with meteorites, Sorby subsequently became interested in applying some of the petrographic microscopy techniques to metals (given his home in Sheffield), turning his attention to the effects of ornamental etching in revealing underlying substructure of steels, thus beginning the now universal practice of metallography. Zirkel and fellow mineralogists had perfected light microscopy through ground thin sections (~25 μm thick) of transparent minerals, but Sorby was limited to surface observations of *opaque* metals (whose microstructural details had previously been adduced — incorrectly — only from the topography of their fracture surfaces), hence the importance of selective surface etchants. In 1864, he penned the following autobiographical ditty (related by C. S. Smith, see footnote 35) and sent it to the librarian of the Royal Society:

> Hard crystals I anatomize
> And iron and steel, too
> And stubborn rocks I shave and grind
> And look them through and through.

Abandoning an unpromising career as a poet for that of a metallographer, Sorby went on to become president of the *Geological Society*, the *Royal Microscopical Society*, the *Mineralogical Society*, and in 1882 the president of an academic institution, Firth College, that would shortly become the University of Sheffield, with an academic department of Metallurgy.

The resolution of any optical instrument is ultimately limited to about half the wavelength of its investigating probe; in the case of visible light, this limit is approximately 200 nm, a limit closely approached by the 19th-century light microscopes of Abbe and Zeiss using Otto Schott's improved Jena glasses. Further imaging forays had to await the invention of the electron microscope in 1931 by physicist Ernst Ruska (1906–1988), his doctoral thesis project at the *Technische Hochschule Berlin* under his electrical-engineer supervisor Max Knoll. The work of French physicist Louis-Victor de Broglie (1892–1987), later 7ème *duc* de Broglie, had earlier (in his 1924 doctoral thesis) proposed that moving quantum particles, like electrons, should exhibit wave-like properties; the electrons in Ruska's instrument were accelerated through electrical potentials of 10 kV to speeds of 20% the speed of light with a de Broglie wavelength (~12 pm) some 40,000 times smaller than that of visible light, theoretically promising sub-atomic resolution. Ruska's 1933 electron microscope (*übermikroskop*) version, a direct electron-optical analog of the transmission light-optical microscope, exceeded the resolution of its predecessor light microscope with images magnified 12,000 times, but (as with Zeiss's light-optical instruments) the *practical* resolution was limited by aberrations in the lenses, which were sufficiently corrected to achieve the instrument's sub-atomic resolution potential only in the early 21st century.

Ruska's instrument was commercialized (in 1937, with the help of Ruska's brother-in-law Bodo von Borries) by the electricity-industry giant Siemens & Halske, and some of these first 1939 instruments were ignominiously shipped to the Cavendish Laboratory at the University of Cambridge as part of war reparations at the termination of World War II. One of these instruments was employed by the group of Peter Bernhard Hirsch (1925, later Sir Peter Hirsch FRS) at the Cavendish to examine the nanostructure of hammered and chemically thinned metal foils (viewed "through and through," as in Sorby's ditty line), through which thinned foils the accelerated electrons were able to pass. The hammering deformation had filled the foil specimens with atomic-scale line defects, called dislocations, the movement of which is responsible for the plastic deformation of all crystalline solids (including glacier ice and the Earth's mantle rocks); and Hirsch's group were the first to see them move and confirm their postulated role in material deformation. Hirsch and his collaborators assembled lecture notes, from a summer school on electron microscopy they'd organized at the Cavendish, into the first and still seminal book on electron microscopy of materials, *Electron*

Microscopy of Thin Crystals,[88] published in 1965. Formation of an Electron Microscopy and Analysis Group (EMAG) of the UK's Institute of Physics followed soon after.

Because the original form of the electron microscope was an analog of the petrographic (transmission) light microscope, specimens were viewed in transmission (electrons passing through the specimen, hence the "thin crystals" of the Hirsch *et al.* title), and that instrument is known as the *transmission electron microscope* (TEM). A second kind of instrument, the *scanning electron microscope* (SEM), was invented by the German engineer Manfred von Ardenne (1907–1997) in 1937, using instead a very small focused probe of electrons scanned across a thick sample, with images formed from scattered electrons collected by an electron detector that modulated the scan of a cathode ray tube scanned synchronously with the electron probe, but over a vastly larger area — hence producing a magnified image. Von Ardennes' instrument was lost in World War II and commercially developed, by Cambridge Instruments, only in the 1960s (and now ubiquitously produced by all electron microscope manufacturers). An advantage of the SEM is that the incident electron beam also generates (in addition to both back-scattered electrons and emitted lower-energy secondary electrons) X-rays and light photons from the specimen, and induces an electron current in the specimen, so that any of these signals can also be used to modulate the image display, providing information about the identity of atoms present and their immediate surroundings in compositional maps. A hybrid instrument, the *scanning transmission electron microscope* (STEM) produces both a transmission image and all of these analytical chemical signals at atomic resolution.

Two associated societies — both North American in origin, but with wide international subscription and participation — stand out in both their size and their reach: the *Microscopy Society of America* and the *Microanalysis Society of America*. Joined at the hip, sharing much instrumentation in common, and now meeting together annually, the two organizations nevertheless maintain separate histories and

---

[88] Hirsch *et al.* (1965).

governance structures. Well over half of the MSA organization is populated by materials scientists, and probably more than 90% of MAS.

Microscopy Society of America (MSA) [formerly Electron Microscopy Society of America (EMSA)]

Electron microscopes had in the 1930s also been assembled in North America, at Washington State University (Pullman), Washington University (St. Louis, MO), and the University of Toronto. The first truly high-resolution instrument, comparable in resolution to that of the Siemens instruments, was constructed in Toronto by Cecil Hall, James Hillier, and Arthur Prebus in 1938. Hillier was subsequently hired by Alexander Zworykin (father of television) at RCA, which commercialized Hillier's instruments. The growing North American community of electron microscopists led to their formation of the *Electron Microscopy Society of America* in 1942; its first meeting was held at Columbia University (New York, NY) in 1944. Both biological and materials applications of the instrument were demonstrated; and, in fact, the biological applications have revolutionized all of biology and macromolecular protein structure determination.

Sir Peter Hirsch's electron microscopy centers, in the Cavendish Laboratory and later in his Oxford Materials department, spawned a remarkable roster of subsequent distinguished electron microscopists, including: in the UK — in addition to co-authors Sir Robin Nicholson FRS (UK Chief Science Advisor 1983–1985), David W. Pashley FRS (Imperial College London), Michael J. Whelan FRS (Oxford Materials) and Archibald Howie FRS (Cavendish Laboratory, Cambridge) — Sir Colin Humphries CBE FRS (Cambridge Materials), the late David J. H. Cockayne FRS (Oxford Materials), Peter Nellis FRS (Oxford Materials); and in the US the late John C. H. Spence FRS (Arizona State University), John Silcox (Cornell University, EMSA President 1975), Linn W. Hobbs OBE (MIT, EMSA President 1987), the late Terence E. Mitchell (Los Alamos National Laboratory, MSA President 1995), Dale E. Newbury (NIST, MAS

President), Charles E. Lyman (Lehigh University, EMSA President 1991), C. Barry Carter (University of Connecticut, MSA President 1997), the late David C. Joy (University of Tennessee/Oak Ridge National Laboratory, MSA President 1999), J. Alwyn Eades (Lehigh University, MSA President 2003), and the late Gareth Thomas (University of California–Berkeley, EMSA President 1975, an early member of Hirsch's Cavendish group who started a large electron microscopy group at Berkeley from the mid-1960s. A rival TEM group to Hirsch's Cavendish and Oxford groups had also formed at the University of Melbourne in Australia, led by the late John M. Cowley FRS (1923–2004) — who had earlier mentored D. J. H. Cockayne FRS, J. C. H. Spence FRS and David J. Smith (Arizona State University, MSA President 2009) in Melbourne. Cowley was eventually wooed to Arizona State University, where the National Center for Electron Microscopy was established, consolidating the US electron microscopy effort (a second was similarly established at Lawrence Berkeley National Laboratory in the San Francisco Bay area, and another central user facility more recently at Oak Ridge National Laboratory).

EMSA had always represented both the biological and materials sides of electron microscopy (with the election of successive presidents from each camp alternating in their terms), but the explosion of all manner of hybrid microscopical methodologies and their importance to both communities led the society in 1992 to change its name and greatly enlarge its reach as the *Microscopy Society of America* (MSA), the largest such organization in the World. No materials laboratory can now function without multiple versions of these instruments, and it can be said without exaggeration that there are precious few materials scientists or engineers who do not regularly utilize the sophisticated microscopy techniques represented. MSA now hosts (together with its sister society, the *Microanalysis Society* (MAS)) a large annual meeting, appropriately titled *Microscopy & Microanalysis* (M&M), and has launched two publications, the periodical *Microscopy Today* (owned by the Society, with 17,000 subscribers) and the journal *Microscopy and Microanalysis*, formerly published for MSA by Cambridge University Press and more recently by

Oxford University Press. It also publishes proceedings of the M&M annual meeting. Its membership numbers approximately 5,000 scientists, engineers, and technologists worldwide. MSA is a charter member of the *International Federation of Societies for Microscopies* (IFSM) (originally founded in 1951 as the Joint Commission on Electron Microscopy), with member societies in 28 countries.

A variety of new microscopy techniques arrived in the late 20th- and early 21st centuries. Two, in particular, have revolutionized microscopy: (1) *tomography* (from the Greek τόμος, to slice) provides three-dimensional images (formed by light, X-ray or electron microscopy), in which the specimen object is illuminated from a wide range of angles and the resultant images combined by high-speed computational tools to form a three-dimensional rotatable image (and resulting in truly massive data sets!), a method familiar to most people from clinical X-ray computed tomography (CT) scans; and (2) *ptycography* (from the Greek πτύχή, fold), in which (in one method) an illumination-defining aperture is moved relative to the specimen object, and successive interference patterns some distance from the specimen fold (convolute) into each other and can be deconvoluted by computer to form a real-space image. Such reconstructed images preserve phase-difference information induced by the specimen object and are unaffected by lens aberrations (but likewise produce massive data sets of TB dimensions). In the electron microscope, with electron wavelengths vastly smaller than atomic dimensions, the (1873) Abbe resolution limit of about half a wavelength is not limiting; instead it is the lens aberrations that limit image resolution — which before electron ptychography stood doggedly just greater than 0.1 nm. Recent application of ptychographic methods has produced the highest-resolution atomic images to date, less than 0.03 nm (30 pm or 0.3Å), corresponding to nearly the positional uncertainty from the atomic vibrations of the atom.[89]

Eliminating that leading "E" has opened up MSA to a host of exciting new microscopical tools that use illumination sources other

[89] Chen *et al.* (2021).

than electrons (as well as recruiting many more practitioners to its membership), from light microscopies that challenge the Abbe limit (confocal, near-field photon-tunneling, near-field scanning optical, and far-field light microscopies[90]) to synchrotron-sourced X-ray photon (instead of electrons or light) imaging microscopy using Fresnel zone plates[91] and its employment in transmission, scanning transmission, holographic, and even ptychographic X-ray microscopy modes — with < 10 nm image resolution possible in three-dimensions in thick specimens, even in water environments, with crystal structures able to be identified by simultaneous X-ray micro-diffraction, and atomic chemical identities revealed by X-ray absorption near-edge spectroscopies (XANES) (See footnote 91). There is a multitude of inorganic, organic, and biological materials applications for synchrotron X-ray imaging, in every branch and twig of the materials tree; one recent example[92] is time-resolved X-ray microscopy imaging of cement components (in Roman pozzolanic concrete, or modern Portland cement), as they react in *real time* on a 5-nm *dimensional scale* in hydrating environments, that is revolutionizing the fundamental understanding of the setting of cementitious materials.

Microanalysis Society (MAS) [formerly Microbeam Analysis Society (MAS)]

For many years, EMSA had also shared its meetings (every 3 years) with another society, the *Microbeam Analysis Society* (MAS, now the *Microanalysis Society* — but retaining the same moniker) that had inherited its credentials from the geologists, who extensively employed another electron-based instrument, the electron probe microanalyzer (EPMA) — invented over the period 1948–1950 by French physicist Raimond Castaing (1921–1998) — to generate characteristic X-ray emissions from a thick polished

[90] Neice (2010); "Super-resolution microscopy" https://en.wikipedia.org/wiki/Super-resolution_microscopy.

[91] Sakdinawat and Attwood (2010).

[92] Brisard *et al.* (2020); Li *et al.* (2019)

sample and thus perform chemical *microanalysis*. The first two gatherings of *aficionados* of the technique were at an ASTM meeting in 1963 in Atlantic City, NJ and at an Electrochemical Society (ECS) meeting in 1964 in Washington, DC. At the latter, a group of 15 enthusiastic practitioners decided to organize what was first known as the *Electron Probe Analysis Society*, the first meetings of which were at College Park, MD in 1966, and the following year in Boston. The meetings soon broadened to include other emerging microanalytical techniques such as the ion microprobe, scanning electron microscopy, soft X-ray emission microanalysis, and energy-dispersive X-ray microanalysis (XEDS) techniques. By the mid-1970s, still other tools had emerged, among them: surface analysis by the scanning Auger microprobe (SAM), secondary ion mass spectrometry (SIMS) and ion scattering spectroscopy (ISS); analytical transmission electron microscopy (analytical-TEM); electron energy loss spectroscopy (EELS); and proton and heavier energetic ion-induced X-ray emission (PIXE). The Society responsively changed its name to the *Microbeam Analysis Society* (MAS), and still later — when EMSA dropped an "E" in order to include all microscopies — shortening its own name further by dropping the "beam" and becoming the *Microanalysis Society* (MAS, still). MAS now meets annually with MSA at the joint *Microscopy & Microanalysis* meeting, and shares with MSA publication oversight for both *Microscopy Today* and *Microscopy & Microanalysis*. MAS numbers approximately 600 members worldwide, including the many corporate manufacturers of microanalytical instruments.

Other Topical Conference Venues and Formats

What might be classed as "sometime" professional societies, replete with the exclusivity, selectivity, and inbred governance worthy of a guild system, are the more intimate conference series known as the *Gordon Research Conferences* (GRC), together with the similar (if cushier) series of conferences run by the *Engineering Conferences International* (ECI) organization.

Gordon Conferences were the brainchild of Neil E. Gordon (1886–1947), a chemist at Johns Hopkins University, who saw an advantage to getting together knowledgeable practitioners, working at one or more of the frontiers of scientific research, for sociable discourse in relaxed surroundings. The first began in the summer of 1931 at Johns Hopkins University, and expanded into the AAAS Gibson Island (MD) conferences on chemical research. Now held in a variety of intimate venues around the world (most in small private schools or colleges, predominantly in New England, but others in less sequestered venues) and with a predominantly chemical and biological content, the Gordon Conference organization puts on more than 20 four-to-five-day conferences *per week* throughout the North American summer season — and occasionally Antipodal summer venues — drawn from almost 800 separate topics, any one topical conference repeating as an ongoing series at typically 2- to 4-year intervals. Chairs-elect are usually proposed by a panel of previous chairs, sometimes held out for election by conferees present, and progress through a (guild-like) vice-chair succession at the next meeting of the series, to eventual chairing the meeting following that. Attendance is usually (though not always) limited by vetting applications and is typically in the range 75–150 attendees.

ECI conferences have an even earlier (if less well known) history, beginning with the establishment in 1912 of the *Engineering Foundation,* set up by steel magnate Andrew Carnegie on behalf of the five principal engineering societies at the time. One of the Foundation's initial missions was to provide the engineering community with international, interdisciplinary leading-edge engineering research conferences. Re-configured as a stand-alone conference organization in 1962, ECI is a not-for-profit global engineering conferences program that provides opportunities for the exploration of problems and issues of concern to engineers and scientists from many disciplines — with a decidedly engineering or applications bias. Since 1962, ECI has organized over 1600 four-to-five-day conferences in a variety of intimate (and particularly attractive) venues throughout the world. A conferences board perennially comes up with potential new conference topics and seeks out capable conference chairs for each. A series usually develops, at 2- to 3-year intervals, with succeeding chairs selected by attendees of the current conference, much as in the GRC tradition. Conference attendances run from 50 to (in a few cases) 300 attendees, with a typical attendance of 100 participants.

Both GRC and ECI conference topics frequently involve materials-related subjects. An illustrative, though not necessarily complete, compilation records the wide range of materials focus for each.[93]

---

[93] Recent Gordon Research Conference (GRC) materials-related topics include: Additive manufacturing of soft materials, Advanced materials for sustainable infrastructure development, Biomineralization, Computational materials science & engineering, Crystal engineering, Crystal growth & assembly, Defects in semiconductors, Electronic processes in organic materials, Energetic materials, Environmental nanotechology, Fuel cells, Granular matter, High-temperature corrosion, Hydrogen-metal systems, Industrial ecology, Ionic liquids, Liquid crystals, Metals in medicine, Multifunctional materials & structures, Nanomaterials for applications in energy technology, Nanoporous materials & their applications, Neutron scattering, Noble metals nanotechnology, Peptide materials, Physical metallurgy, Polymer physics, Polymers, Quantum control of light and materials, Quantum sensing, Renewable energy: solar cells, Rock deformation, Science of adhesion, Scientific methods in cultural heritage research, Soft condensed matter physics, Solid-state chemistry, Solid-state studies in ceramics, Spin dynamics in nanostructures, Structural nanomaterials, Superconductivity, Three-dimensional electron microscopy, Topological and correlated matter, Tribology, Two-dimensional electronics beyond graphene, Unconventional semiconductors and their applications, Upconverting nanoparticles, Vibrational spectroscopy, Volume electron microscopy.

Recent Engineering Conferences International (ECI) materials-related topics include: Alkali-activated materials and geopolymers, Beyond nickel-base superalloys, biofrabrication for hierarchical *in vitro* tissue models, Carbon-based nanomaterials & devices, Ceramic matrix composites, Colloidal macromolecular and biological gels, Composites at Lake Louise, Design & manufacture of functional microcapsules and engineered particles, Electric-field assisted sintering and related phenomena far from equilibrium, Electrophoretic deposition, Environmental damage in structural materials under static/cyclic loads at ambient temperatures, Fibrous nanocomposites, Field-enhanced processing of advanced materials, Functional glasses — properties & applications for energy & information, Innovative materials for additive manufacturing, Green process engineering, Life cycle assessment and other assessment tools for waste management and resource optimization, Nano- and

Such gatherings develop an almost-family (or at least guild-like) character, with multiple generations of investigators attending and becoming mutually acquainted over years of attendance.

Materials Publications

The publications of the materials societies, the materials-related divisions of other scientific societies, or the organs of non-materials societies like *Nature, Science, Physical Review, Philosophical Magazine*, or *Proceedings of the Royal Society* or the *National Academy of Sciences* and countless others that publish materials-relevant articles, have always served an important associational function and are addressed in the well-informed entry by Daniel Ucko in this volume.[94] The major scientific publishers (Macmillan, Springer, Elsevier-North Holland, Taylor & Francis, Kluver, *et seq.*) continue to spawn new materials-related journals, or spin-offs of their long-established parent journals (e.g., *Nature Materials* or *ACS Nano*) in ever-increasing numbers, with surprisingly little of the immense consolidation that has marked the recent evolution of the major materials societies — or perhaps not so surprisingly, since publishers are just hawking their wares to an ever-expanding and diversifying market. Materials articles appear now in a staggering 300 or so different journals. This corpus of journal contributions provides at least two of the important roles of human associations — *education* and *practice.* They also, like conferences, play an important (if virtual) socialization role that creates a sense of community.

---

micro-mechanical testing in materials research & development, Nanotechnology in medicine, Nature-inspired engineering, Non-stoichiometric compounds, Pathways to carbon capture utilization & storage deployment, Polymer reaction engineering, Rare-earth minerals/metals sustainable technologies, Semiconductor technology for ultra-large-scale integrated circuits & thin-film transistors, Shotcrete for underground support, Syntactic & composite foams, Thermal & environmental barrier coatings, and Ultrahigh-temperature ceramics.

[94] Ucko, Daniel. Materials science journals and journals that publish materials science. This Volume.

Perhaps the most *useful* kind of publication is the review article or the summary (invariably invited) articles (or chapters) appearing in journals (or books) that document "advances in" or "progress in" a specialty field, because such contributions periodically provide a welcome clarifying filter for the plethora of publishable outputs from those practicing some sort of materials science or engineering — well over a million highly cited materials-related articles in the last 100 years! These select summary authors represent the *guild elites* of our era, and their sage assessments provide in some way the aforementioned *regulatory* function important in communities of associating individuals (a function once provided by the guilds system but curiously absent in present-day scientific societies), together with an *evaluatory* standard-setting function once also provided by guilds — and, 50 years ago, by the careful scrutiny of august thesis committees, a measure certainly preferable to (if less quantifiable than) the spurious indices presently marshaled out by various convolutions of citation numbers with journal impact factors. Despite their availability on microfiche or on-line data bases, no one reads doctoral theses any more, which still often provide the best current (if transitory) information about a specialty field and its progress. Currency is, alas, a highly evanescent quantity nowadays in materials fields, with at best a half-life less than 1 year. But as a sage physicist pointed out a few decades ago, as the march of *Physical Review* volumes across the bookshelf approaches the speed of light, ultimately no information becomes transmissible.

Social Function and the Greater Community

The greater materials science & engineering community (accounting for overlaps in society memberships, since their memberships are not entirely orthogonal) probably numbers upwards of 100,000 individuals, exclusive of Chinese and Indian materials scientists and engineers for whom demographic information is not readily available. That is about twice as many as in the physics community, half as many as chemists, and a quarter of electrical and electronic engineers — a respectable but far from overwhelming presence, especially considering that there are 1.3 million licensed attorneys, 500,000 elected political office holders and 400,000 financial analysts in the US alone. Given the impact materials have had, and increasingly continue to have, on the world economy and on many now near-universal cultural norms, materials science & engineering remain significantly underrepresented in the general population — and largely unknown by it, let alone under-appreciated.

Community

A different issue is whether these hundred thousand or so materials professionals enjoy any sense of community and mutual appreciation. The answer to that query is surprisingly affirmative, despite the historical balkanization and add-on character of the discipline. Materials people readily self-identify, share a great deal in common, and can readily communicate with each other in a well-understood vernacular. There exists, in short, a surprising sense of community — even among a collection of 6,000 conferees jumbled together at an MRS meeting. They have no trouble recognizing one another — and can talk thermodynamics, phase equilibria, and defects — but they experience obscurity within a greater population, remaining professionally anonymous, even to a neighbor next door. Incoming college freshpeople, even those exposed to intensive STEM curricula in their secondary schools, have never heard of the discipline nor recognize that there is a sophisticated science and cohort of accomplished practitioners behind all those 21st-century marvels that are so unthinkingly accepted, greeted with so little curiosity, by a majority of teenagers. Rocket scientists fare better! (though DARPA might well identify the real heroes of rocket science as the materials scientists).

The field of materials is far from insular: as should be readily evident in the foregoing accounts, the community is widespread domestically and wholly international. And, to coin a social-historian sort of word, *itinerancy* — both among national institutions but also across institutions internationally — is rampant, on a scale unheard of among journeymen in an older guild system, facilitated by near-saturation levels of mutual awareness through publication and conferencing. Personnel diffusion between university and university, between industry and academia, and between commercial laboratories and university research groups is so frequent as to be unremarkable. As chemical engineer Robert A. Brown (former MIT provost and later president of Boston University) once remarked, "technology transfers most effectively when wearing shoes." An astute themodynamicist might additionally observe that the interdiffusion also markedly increases the system entropy.

The same mobility can be said to characterize most professions today (not least among the dizzying array of high-tech start-ups), though only perhaps beginning in the1980s. (The classic example is the recent demise of the company-loyal "salaryman" in Japan, or even in the West the disappearance of those faithful engineers who once devoted their whole working careers to a single company — like Exxon, as it morphed from Humble Oil to Standard Oil to Esso to Exxon to ExxonMobil over the course of four decades.) But the diffusive itinerancy began earlier in materials science — likely because science is so universal — probably starting in the 1920s, when many younger materials chemists and physicists made pilgrimages to Germany (later reversed in the 1930s by the exodus of so many European scientific luminaries fleeing Nazi Germany and seeding Britain, North, and South America with their scientific acumen). The same journeyman dynamic is seen today with the influx of graduate students and postdocs from China populating the scientific graduate schools of the West, as any graduate-school dean will attest. The question might be posed whether the greater systemic permeability levels out the former delta functions of excellence.

In terms of inter-personal relationships, it must be stated that there has been remarkable inter-generational connectivity in mentoring chains, as is abundantly clear from the biographical accounts included in this review. And, should anyone doubt the enriching function of immigration, one need only invoke the mentor model and enumerate the corpus of inspired mentees of an immigrant Peter Hirsch or Rustum Roy or Arthur von Hippel or Norbert Kreidl or John Cahn. But, even beyond mentorship, there is remarkable connectivity among materials people; MSE folk are highly collaborative, and no individual is very far along in an Erdös-equivalent[95] chain, at least at present. What are more telling, from a sociological point of view, are probably the closer *kinship* relationships — *sibling*-like among all those supervised by or working for a particular mentor at some point, *first cousin*-like for those coming from the same department or division, *second cousin*-like for researching the same problem at competing institutions or in competing companies, etc. Such relationships are surprisingly real; whether they will prove of less (or more) import as the materials community continues to grow in number is uncertain. The same kinds of relationships can, of course, be fostered within a professional society, which remains an important societal function.

Socialization

There is little question that socialization constitutes a large component function in professional materials societies today, and was aimed at right from the start.[96] Norah Binns, an early Assistant secretary of the American Ceramic Society and daughter of one of the principal founders in 1899 (and later president) of ACerS, Charles Binns, communicated that social impetus in her 1925 history of the Society's first quarter century,[97] using the quaint but charming locutions of the time:

---

[95] Mathematicians often characterize their connectivity by the number of steps (degrees of removal) separating them from the remarkable Hungarian mathematician Paul Erdös (1913–1996): those co-authoring a paper with Erdös are assigned an *Erdös number* 1, those co-authoring a paper with an Erdös-1 individual are assigned *Erdös number 2*, etc. One could also track supervisory lineage, for example (in this author's case) he was supervised by P. B. Hirsch, who reported to W. L. Bragg, who was supervised by J. J. Thompson, etc. In this way, the connection to historical figures in a field can become more real, emphasizing how one's modest successes are always built upon the shoulders of talented predecessors.

[96] Scott (1912).

[97] Binns (1923).

> Intimacy naturally marked the relations between these men who were drawn together by no slight bond ... as if they were members of the same family (*ed. note*: the charter members were exclusively men, but the first woman member of the Society, (Miss) Mabel C. Farren, was in fact elected in 1915 and served as an officer of the Arts Division) ... The need for recreation was not satisfied by dancing and moving pictures. It was talk and more talk that these men craved ... in the evening, in a private room or around a table, refreshments of the good old variety were made available, cigars and pipes were lit ... Many of the members were working alone, so far as scientific sympathy or understanding was concerned. All of them had problems, all of them had a little knowledge. Put together, the bits of knowledge gradually began to penetrate the problems ... the earliest form of ceramic "cooperative research."

Ms. Binn's description could well describe a meeting of any materials society or its divisions or federations in the 21st-century (excepting the much larger scale and pace of such gatherings today, the absence of tobacco, and the closer gender equivalency); or, for that matter, the governing tenor of a Gordon or ECI conference in any materials-related field (See footnote 93). It regrettably sometimes describes less well the present interactions within a university between faculty members in the same field; or between professorial advisors and their graduate students — many of whom are becoming unionized and have injected an unfortunate adversarial impediment, sounding perhaps the first impending death knell of the academic guild system. Ms. Binn's account does capture well the sense of community engendered, which may still remain — even within a multidisciplinary MRS, MS&T, Euromat, or IUMRS meeting, not least for memberships who still enjoy surprisingly *scant* public recognition, given the revolutionary impact their efforts have made on 21st culture throughout the world. Generating that recognition remains an urgent unifying task facing all materials societies and their members, highlighting a pressing need for revelatory historical accounts like this series of volumes.

References

Artioli, G., Secco, M. and Addis, A. (2019). The Vitruvian legacy: Mortars and binders before and after the Roman world. *EMU Notes in Mineralogy* 20, 151–202.

Backer, S. (2002). *100 years of textiles at MIT* (MIT Press: Cambridge, MA).

Baker, W. O. and Cohen, M. (1975). *Report of the National Academy of Sciences Committee on the Survey of Materials Science and Engineering (COSMAT), Materials and Man's Needs: Materials Science and Engineering*, Vol. III (Washington, DC: National Academy Press).

Ball, P. and Pike, G. (November 2013). Four decades of materials development transform society. *MRS Bulletin* 38, 873–885.

Ballatto, J., Seddon, A., Claire, A., Petit, L., Hu, J. and Richardson, K. (July 1, 2022). Future of optical glass education. *Optica Materials Express* 10(7), 2626–2634.

Bell, J. L., Driemeyer, P. E., and Kriven, W. M (2009). Formation of ceramics from meta-kaolin-based geopolymers. Part I: Cs-based geopolymers. *Journal of the American Ceramic Society* 92(1), 1–8; Part II: K-based geopolymers. *Journal of the American Ceramic Society* 92(3), 607—615.

Bensaude-Vincent, B. (2001). The construction of a discipline: Materials Science in the United States. *Historical Studies in the Physical and Biological Sciences* 31(2), 223–248.

Bensaude-Vincent, B. and Hessenbruch, A. (2004). Materials science: A field about to explode. *Nature Materials* 3, 345–347.

Binns, N. W. (1923). Twenty-five years of the American Ceramic Society. *Journal of the American Ceramic Society* 6, 10–66.

Boomrungsiman, S., Gentleman, E., Carzaniga, R., Evans, N. D., McComb, D. W., Porter, A. E., and Stevens, M. M. (August 28, 2012). The role of intracellular calcium phosphate in osteoblast-mediated bone apatite formation. *Proceedings of the National Academy of Sciences of the United States of America* 109(35), 14170–14175. doi.org/10.1073/pnas.1028916109.

Boyet, P.–O. (2020). Eminent Structural engineer: Louis Vicat. *Structural Engineering International* 30, 128–30.

Brandon, C. J., Hohlfelder, R. L., Jackson, M. D., and Oleson, J. P. (2014). *Building for Eternity: The History and Technology of Roman Concrete Engineering in the Sea* (Oxford: Oxbow Books).

Brisard, S., Serdar, M., and Monteiro, P. J. M. (2020) Multiscale X-ray tomography on cementitious materials: A review," *Cement and Concrete Research* 128, 105824. doi.org/10.1016/j.cemconres.2019.105824.

Burns, R. M. (1977). *A History of the Electrochemical Society: 1901–1976* (Pennington, NJ: Electrochemical Society).

Chan, B. P. and Leong, K. W. (2008). Scaffolding in tissue engineering: General approaches and tissue-specific considerations. *European Spine Journal* 17(Suppl. 4) S467–79. doi.org/10.1007/s.00586-008-0745-3.

Chen, Z., Jiang, Y., Shao, Y.-T., Holtz, M. E., Odostrčil, M., Guizar-Sicairos, M., Hanke, I., Ganschow, S., Schlom, D. G., and Muller, D. A. (May 21, 2021). Electron ptychography achieves atom-resolution limits set by lattice vibrations. *Science* 372(6544), 826–831.

Choi, H. and Shields, B. (2015). A place for materials science: Laboratory buildings and interdisciplinary research at the University of Pennsylvania. *Minerva* 53, 21–42.

Corning, Inc. (2015). A day made of glass. https//www.corning.com/worldwide/en/innovation/a-day-made-of-glass.html.

Córdoba de la Llave, R. C. (2016). Guild authorities in late medieval Spain. In Jullien E. and Pauly, M. (eds), *Craftsmen and Guilds in the Medieval and Early Modern Periods* (Stuttgart: Franz Steiner Verlag) pp. 77–92.

Davidovits, J. (2020). *Geopolymer Chemistry and Applications* (5th edn.) (Saint-Quentin, France: Institut Geopolymère) p. 680.

Doremus, R. H. (2008). Ceramics. In G. Wnek and G. Bowlin (eds.), *Encyclopedia of Biomaterials and Biomedical Engineering,* 2nd edn. (Boca Raton, FL: CRC Press), pp. 585–593.

Dyer, N. A. (1980). The artificial eye. *Australian Journal of Ophthamology* 8, 325–327.

Fine, M. E. and Marcus, H. L. (1994) Materials science and engineering, an educational discipline. *Annual Review of Material Science* 24, 1–17.

Flemings, M. C. (1999). What next for Materials Science & Engineering departments. *Annual Revue of Materials Science* 29, 1–23.

Fleming, D. (1950). *John William Draper and the Religion of Science* (Philadelphia: University of Pennsylvania Press).

Fortes, M. A. and Nogueira, M. T. (1989). The Poisson effect in cork. *Materials Science and Engineering* A122, 227–232.

Francis, A. J. (1978). *The Cement Industry, 1796–1914: A History* (Exeter, Devon, UK).

Gatos, Harry (1993). How it all evolved. MRS Bulletin 18, 35–37 (20th Anniversary issue 1973–1993).

Gilligan, I. (2007). Clothing and farming origins: The Indo-Pacific evidence. *Bulletin of the Indo-Pacific Prehistory Association,* 27, 12–21.

Gilligan, I. (2010). The prehistoric development of clothing: Archaeological implications of a thermal model. *Journal of Archaeological Method and Theory* 17, 15–80.

Gupta, P. K. and Cooper, A. R. (1990). Topologically disordered networks of rigid polytopes. *J. Non-Crystalline Solids* 123, 14–25.

Gupta, P. K. and Heuer, A. H. (1993). "Alfred R. Cooper's contributions to glass science and technology," *Journal of the American Ceramic Society* 76 (4), 1077–1080.

Gupta, P. K. (2021). Topological constraint theory of inorganic glasses. In P. Richet, R. Conradt, A. Takada and J. Dyon (eds.), *Encyclopedia of Glass Science, Technology, History & Culture* (Ch. 2.7) (American Ceramic Society/Wiley On-Line Library), pp. 207–219.

Hargittal, I. (2022). Herman F. Mark — Pioneer of polymer chemistry and initiator of the gas-phase electron diffraction technique of molecular determination. *Structural Chemistry* 33, 1379–1384.

Hartley, E. N. (1957). *Ironworks on the Saugus* (Norman, OK: University of Oklahoma Press).

Hartman, P. L. (1992) (with a forward by Robert L. Sproull), *Cornell's Materials Science Center: The Early Years.* Also available as: Hartman, Paul L. (1992). MSC decade & a half: A brief history over the formative first decade and a half of the Materials Science Center at Cornell University. Available from Materials Science Center, Cornell University, Ithaca, NY.

Hench, L. L. and West, J. K (1996). Biological applications of bioactive glasses. *Life Chemistry Reports* 13, 187–241.

Hirsch, P. B., Howie, A., Nicholson, R. B., Pashley, D. W., and Whelan, M. J. (1965). *Electron Microscopy of Thin Crystals* (London: Butterworths).

Hobbs, L. W. and Revel, J.-P. (1989). The Invisible World. In Byron Preiss (ed.). *The Microverse* (New York, NY: Bantam Books), chapter 2, pp. 20–47.

Hobbs, L. W., Jesurum, C. E., Pulim, V. and Berger, B. (1998). Local topology of silica networks. *Philosophical Magazine* 178, 679–711.

Hobbs, L. W. (2007). Ceramics in biology and medicine. In Stephen Freiman (ed.), *Global Roadmap for Ceramic and Glass Technology* (New York: John Wiley & Sons), pp. 183–188.

Hobbs, L. W. (2008). Topological approaches to the structure of crystalline and amorphous atom assemblies. In Juan J. Novoa, Dario Braga and Lia Addadi (eds.), *Engineering of Crystalline Materials Properties*

(NATO Science for Peace and Security Series–B: Physics and Biophysics) (Dordrecht & London: Springer), pp. 193–230. ISBN: 1402068220.

Hobbs, L. W. and Siddall, R. (2011). Cementitious mortars of the ancient world. In Å. Ringbom and R.L. Hohlfelder (eds.). *Building Roma Aeterna, Finnish Society of Sciences and Letters, Commentaries on Human Literature* Vol. 128, pp. 35–60.

Hosler, D., Tarkanian, M. J. and Burkett, S. L. (1999). Prehistoric polymers: Rubber processing in ancient Mesoamerica. *Science* 284(5422), 1988–1991.

Hovestadt, H. (1900). *Jena Glass and its Applications to Science and Art* (Jena: Fischer). Abstract citation: R. T. G (1900). Jena glass and its applications to science and art. *Nature* 63, 173–174.

Kriegel, W. W. (1982). Keramos, a biographical history. In *Keramos Blue Books* (Vols. 1–8) (Columbus, OH: American Ceramic Society).

Langton, C. A. and Roy, D. M. (1984). Longevity of borehole and shaft sealing materials: Characterization of ancient cement-based building materials. *Materials Research Society Symposium Proceedings* 26, 543–549; Langton, C. A. and Roy, D. M. (September 1982). Longevity of borehole and shaft sealing materials: Characterization of cement-based ancient building materials. *Technical Report* ONWI-202, Distributions Category UC-70, National Technical Information Service, U.S. Department of Commerce.

Laudise, R. A. (1995). Federation of materials societies. *Advanced Materials* 7(10), 823–825.

Lechtman, H. and Hobbs, L. W. (1987). Roman concrete and the Roman architectural revolution. In W. D. Kingery (ed.), *Ceramics and Civilization* (Vol. 3) (American Ceramic Society, Columbus, OH), pp. 81–128.

Li, J., Geng, G., Myers, R., Yu, Y.-S., Shapiro, D., Carraro, C., Maboudian, R., and Monteiro, P. J. M. (2019). The chemistry and structure of calcium (alumino) silicate hydrate: A study by XANES, ptychographic imaging, and wide- and small-angle scattering. *Cement and Concrete Research* 115, 367–378. doi.org/10.1016/j.cemconres.2018.09.008.

Ling, J. C. and Hand, M. A. (1980). Federal funding in materials research. *Science* 209(4462), 1203–1207.

Marians, C. and Hobbs, L. W. (1988). The phase structure of aperiodic SiO_2 as a function of network topology. *Journal of Non-Crystalline Solids* 106, 309–312.

Mauro, J. C. and Zanotto, E. D. (2014). Two centuries of glass research: Historical trends, current status and grand challenges for the future. *Applied Glass Science* 5(3), 313–327.

Mazzoni, E., Iaquinta, M. R., Lanzillotti, C., Mazziotta, C., Maritati, M., Sprio, S., Tampieri, A., Tognon, M., and Martinin, F. (2021). Bioactive materials for soft tissue repair. *Frontiers in Bioeningeering and Biotechnology* 9, 0613787, 1–24. doi.org/10.3389/fbioe.2021.613787.

Mitchell, M. J., Billingsley, M. M., Haley, R. M., Wechsler, M. E., Peppas, N. A., and Langer, R. (2021). Engineering precision nanoparticles for drug delivery. *Nature Reviews, Drug Discovery* 20, 101–124. doi.org/10.1038/s41573-020-0090-8.

Musgraves, J. D., Hu, J. and Calvez, L. (eds.). (2019). *Springer Handbook of Glass* (Cham, Switzerland: Springer).

National Research Council (2013). *Optics and Photonics: Essential Technologies of our Nation* (Washington, DC: National Academies Press).

Neice, A. (2010). Methods and limitations of sub-wavelength imaging. *Advances in Imaging and Electron Physics* 163, 117–140, ISSN 1076-5670.

Oare, G. A. (November 2013). 40 years, one vision. *MRS Bulletin* 38, 958–962.

Ojovan, M. I. and Tournier, R. F. (2021). On structural rearrangements near the glass transition temperature in amorphous silica. *Materials* 14, 5235, 1–16.

Pappalardo, D., Mathisen, T., and Finne-Wistrand, A. (April 8, 2019). Biocompatibility of resorbable polymers: A historical perspective and cramework for the future. *Biomacromolecles* 20(4), 1465–1477. doi.org/10.1021/acs.biomac.9b00159.Epub 2019 Mar 20.

Picraux, S. T. (November 1993). MRS: The growth phase. *MRS Bulletin* (20th anniversary issue 1973–1993) 18, 77–83.

Porter, A., Hobbs, L. W., Rosen, V. B., and Spector, M. (2001). The ultrastructure of the plasma sprayed hydroxyapatite-bone interface predisposing to bone bonding. *Biomaterials* 23, 921–928.

Porter, A. E., Taak, P., Hobbs, L. W., Coathrup, M. J., Blunn, G. W. and Spector, M. (2004). Bone bonding to hydroxyapatite and titanium surfaces on femoral stems retrieved from human subjects at autopsy. *Biomaterials* 25, 5199–5208.

Ratner, B. D., Hoffman, A. S., Schoen, J., and Lemons, J. E. (eds.). (1997–2013). *Biomaterials Science: An Introduction to Materials in Medicine* (1st–3rd edns.) (London, UK: Academic Press).

Readey, D. W. (March 26, 2008). History of ceramic engineering by someone who has been there for almost half of it. In *2007 Faculty Senate Distinguished Lecture* (Golden, CO: Colorado School of Mines).

Richet, P., Conradt, R., Takada, A., and Dyon, J. (eds.). (2021). *Encyclopedia of Glass Science, Technology, History & Culture* (Westerville, OH: American Ceramic Society/Wiley On-Line Library).

Rosen, V. B., Hobbs, L. W., and Spector, M. (2001). The ultrastructure of anorganic bovine bone and selected synthetic hydroxyapatites used as bone graft substitute materials. *Biomaterials* 23, 725–733.

Roy, R. (ed.) (1970). *Materials Science and Engineering in the United States* (University Park, PA: Pennsylvania State University Press).

Roy, R. (September 1993). MRS: Conception gestation, birth and infancy. *MRS Bulletin* 18, 74–76 (20th Anniversary Issue 1973–1993); The Ideals and ideas that led to MRS, *ibid.* 84–85.

Sakdinawat, A. and Attwood, D. (November 30, 2010). Nanoscale X-ray imaging. *Nature Photonics* 4, 840–848. doi.org/10.1038/nphoton.2010.267.

Schwartz, L. H. (1987). The materials research laboratories: Reviewing the first twenty-five years. In *Advancing Materials Research* (Washington, DC, National Academies Press), pp. 35–48.

Scott, G. S. (September 24, 1912). The history and achievements of the Institute of Metals (British). In *Papers presented at the annual meeting of the American Institute of Metals*, Buffalo, NY (American Institute of Metals).

Sergi, C. and Trilio, J. (2022). The compression behavior and crashworthiness of cork: a review. *Polymers* 14, 134, pp. 1–21. doi.org/10.3390/polym14010134.

Seymour, L. M., Maragh, J., Sabatini, P., di Tommaso, M., Weaver, J. C., and Masic, A. (2023). Hot mixing: Mechanistic insights into the durability of ancient Roman Concrete. *Science Advances* 9(1), 1–13. doi.org/10.1126/sciadv.add1602.

Shah, A. M., Jung, H., and Skirrol, S. (2014). Materials used in cranioplasty: A history and analysis. *Neurosurgical Focus* 36(4), E19: 1–7. doi.org/10.3171/2014.2.focus13561.

Smith, C. S. (1960, 1988). *A History of Metallography: The Development of Ideas on the Structure of Metals before 1890* (Chicago: University of Chicago Press; Cambridge, MA: MIT Press).

Sproull, R. L. (1987). The early history of the Materials Research Laboratories. *Annual Review of Material Science* 17, 1–12.

Stefaniuk, D., Hajduczek, M., Ulm, F. J., and Masic, A. (March 28, 2023). Cementing CO_2 into C-S-H: A step towards concrete carbon neutrality. *PNAS Nexus* 2(3), 52. doi.org/10.1093/pnasnexus.pgad052.

Toups, M. A., Kitchen, A., Light, J. E., and Reed, D. L. (2022). Origin of clothing lice indicates early clothing used by anatomically modern humans in Africa. *Molecular Biology and Evolution* 28(1) 29–32.

Turnbull, D. (1983). A commentary on the emergence and evolution of "Materials Science." *Annual Review of Materials Science* 13, 1–7.

Turner, W. E. S. (1945). Glass technology at Sheffield: Prof. W. E. S. Turner OBE FRS. *Nature* 155, 420.

Vargason, A. M., Anselmo, A. C., and Mitragotri, S. (September 2021). The evolution of commercial drug delivery technologies. *Nature Biomedical Engineering* 5, 951–957.

Vicat, L.-J. (1818). *Recherches expérimentales sur les chaux de construction, les bétons et les mortiers ordinaires* (Paris: Goujon).

Vitruvius [Marcus(?) Vitruvius Pollio(?)] (1486). *De Architectura, libri decem.* First *printed* Latin editions, without illustrations (Roma: Fra Giovanni Sulpitus); in 1495/1496 (Venezia: Cristoforo Pensi). With reconstructed illustrations in 1511 (Venezia: Fra Giovanni Giocondo), in 1543 (Strasbourg: Georg Messerschmidt). Accessible in several English translations, e.g. Morgan, M. H. (translator) (1914). *Vitruvius: Ten Books on Architecture* (Cambridge, MA: Harvard University Press).

Wagner, W. R., Sakiyama-Ebert, S. E., Zhang, G. and Yaszemski, M. J. (eds.) (2020). *Biomaterials Science: An Introduction to Materials in Medicine* (4th edn.) (London, UK: Academic Press & Society for Biomaterials).

Yates, J. and Murphy, C. N. (2019). *Engineering Rules: Global Standards Setting* (Baltimore, MD: Johns Hopkins University Press).

Chapter 4.3

Materials Science Journals and Journals That Publish Materials Science

Daniel Ucko

Introduction

Materials science, like any academic field, generates publications, and like many of them, these publications appear in journals. Any academic field requires a mechanism for disseminating results, or making results public. This is done by a process called publication, which in itself is old. While previously most scientific publishing was done in books, because of a need for immediacy and a need for a forum for the exchange of ideas, in any academic field, journals arose in response to a community in need of communicating with itself as well as of reaching an interested audience.

The first scientific journals, *Philosophical Transactions* and *Comtes Rendus*, were created by the Royal Society of London and the *Académie Royale* of France respectively for the purpose of communication inside the societies (Biagioli, 2002). Later, specialized journals emerged, as more specialized societies were created. In addition, technological changes in the business of printing enabled the creation of periodicals to rival established society journals. These new publication outlets could sometimes have a quite limited scope. In general, such entities lived or died by the strength of their communities, and could as such be quite short lived, such as *London Mechanics' Register* (1824–1828),

Geologist (1842–1843), or *Entomological Magazine* (1832–1838) (Dawson *et al.*, 2020). However, the materials science community possesses some characteristics that differentiates it from other fields. To start with, it is highly interdisciplinary in nature. Depending on the perspective, materials science could be said to have its roots in metallurgy, crystallography, polymer science, or be an offshoot of engineering. Indeed, materials science departments today boast of their interdisciplinary credentials.

The matter of materials science journals thus forms an interesting case study of journals. With such varied constituent parts, how can one describe a cohesive community that can sustain a journal? Indeed, what we today think of as quintessential materials science journals were often spun off from existing journals, and often did not have the word "materials" in the title or in whatever editorial statement of scope that existed. The journals exist in a strange space of the phase diagram of journals. They are not multidisciplinary like the *Transactions*, *Nature*, or *Science*, journals that purport to represent all scientific fields and, often, to also reach an audience of interested lay people, but neither are they simply described as "specialized journals" Indeed, *Nature*, as a journal, explicitly defined itself as not being a specialized journal from the first solicitation of papers in 1869 (Baldwin, 2015). They are also not like a journal purporting to cover all areas of a single field such as *Physical Review Letters* or *Journal of the American Chemical Society*, both of which are wide-ranging but constrained in scope by disciplinarity. Spanning many (but not all) scientific disciplines, it is a challenge for materials science journals to determine and serve its audience.

In this chapter, I will sketch a history of materials science journals by considering the role such journals play. I will consider journals created *ab nihilo* as well as journals that grew out of other, more general journals. I will consider journals that "became" materials science journals even if their original mission appeared to be different. I will consider how technological developments have affected the field and thus also the journals. Lastly, I will attempt some prognostications of the future of materials science, reading the tea leaves through its journals.

What Is Materials Science?

In order to understand the community that would demand materials science journals, it is necessary to spend a little time considering the question: what is materials science? A material is something that differs from a mere substance, which could be solid, liquid, or even a gas, because it is considered in terms of its use. Studies of materials in materials science always have praxis in mind, even if the investigation is several steps removed from an actual device or application. One could make a study of the variation of viscosity in a substance without making the explicit connection to lubricants, their use and technological function, but even in the moment of conceptualizing experiments the end use looms like a specter over the investigation. Materials science investigations are characterized by Heidegger's *Zuhandenheit* (ready-to-handedness), the focus being on its use, even if the investigation can seem to be very abstracted and appear to be all about *Vorhandenheit* (present-at-handedness), or exist as an object of study in its own right.

The idea of examining substances in the world to elucidate their properties is old. Matthew Eddy, taking mineralogy as an example, traces the intellectual origins of what we now call materials science back to the Enlightenment, when there was an appetite for categorizing and systematizing nature, though he also notes that "the Latin words that were used to systematically classify stones were drawn from classical sources, particularly from Theoprastus' *De Lapidibus* or Pliny's *Historia Naturalis*," (Eddy, 2008) suggesting that even if the pursuit of investigating materials was newer, the claimed lineage was ancient. The term *materials science* itself, however, is relatively new, dating from the 1950s. Northwestern University in Illinois, USA, was the first university to start a dedicated materials science department, when it decided to change the name of the Graduate Department of Metallurgy to the "Graduate Department of Materials Science", in 1958. Other universities soon followed suit, though there were also universities that did not set up departments of materials science, but rather developed interdepartmental graduate programs with similar content spread over departments of physics, chemistry, biology, and engineering (Cahn, 2001). The arrival of the new discipline of materials science was not necessarily smooth. In a history of the Advanced Research Projects Agency (ARPA)-funded Laboratory for Research on the structure of Matter (LRSM) at the University of Pennsylvania, a part of the Interdisciplinary Laboratories program (IDL), Hyungsub Choi and Brit Shields reveal that "the new laboratory buildings spurred by ARPA became the object of intense negotiations among policymakers, university administrators, and scientists during the early years of the IDL program." (Choi and Shields, 2015) The debates concerned the role and nature of interdisciplinary research. Much like a research building like the LRSM, journals in materials science have a number of stakeholders whose interests have to be balanced. The creation of an interdisciplinary journal like those in materials science is very similar to the planning and construction of an interdisciplinary research center.

We certainly now refer to historical periods as the stone age, the bronze age, the iron age, but these are all retrospective labels, and classifying periods by materials, no matter how dominant, would not have made sense at the time (Sass, 2011). This is not to say there were no investigations or uses of materials before this time, but if journals are to be the lens through which we look at this field, then these distinctions matter. Making the change from metallurgy to materials indicates an increased openness and an expanded footprint. At the same time, it is important to remember that the meaning and scope of "Materials" has continued to evolve ever since, as have the techniques for studying them. As an example, the development of metamaterials — an engineered material that has properties not found in natural materials — has gone from a curiosity to a field with fundamental and technological implications. A special issue celebrating a century of metamaterials was published in the *Journal of Optics*, not a materials journal (Tretyakov *et al.*, 2017). Here, though the field itself is more than a century old, the term "metamaterial" was only coined in 2000, which coincides with proposals and realizations of materials with exciting properties such as negative refractive index materials and electromagnetic cloaking during the first decade of the 21st century. As the opening editorial in the *Journal of Optics* special issue states, "The research domain of metamaterials has now grown to a fully fledged discipline in its own right. It is a synthetic field of research with top talent from microwave and optical sciences, solid state physics, materials science, acoustics, mechanics, nanotechnology and high-performance computing all attracted by the subject." This example highlights the sometime complicated nesting of fields concerning materials science: metamaterials can be said to be a part of materials science, but at the same time, materials science is a constituent part of metamaterials as a research field.

As in the metamaterials example, the constituent disciplines that make up materials science often have long and storied histories, but it was only after World War II that the term materials science emerged. The fluidity of description of materials science comes with its challenges in studying materials science from the vantage point of publication cultures. In a bibliometrics study attempting to look at highly cited articles in materials science from 1900 to 2011, Yuh-Shan Ho (2014) drew data from eight separate Web of Science categories in the database: biomaterials (BMS; 25 journals); ceramic materials science (CRMS; 25 journals); characterization and testing materials science (CTMS; 32 journals); coatings and films materials science (CFMS; 18 journals); composites materials science (CPMS; 24 journals); multidisciplinary materials science (MMS: 232 journals); paper and wood materials (PWMS; 21 journals); textiles materials science (TMS; 21 journals). The search yielded 1,297,456 articles. Ho found that 70% of the highly cited articles were published in the last two decades of the time window, spread across 257 journals Ho, 2014). One has to correct for the growth in the field of materials science, since citations scale with the number of practitioners, but even keeping that in mind, it is a staggering growth. According to the United States Bureau of Labor Statistics, the number of "Chemists and Materials Scientists" is projected to grow by 6% in 2020–2030, which is about average for all professions, and which would not seem to account for the statistical imbalance in citations. However, the growth could also be due to the growth in the number of people who would cite research classified as materials science from other fields. As an example of this, Ho identified the most cited article in the sample as a review article in *Nature Materials* in 2007 by Andre Geim and Konstantin Novoselov, Nobel Prize winners for their work on graphene (Novoselov *et al.*, 2004). In terms of its impact and the citation patterns, they range far beyond materials. Is this a work of materials science, physics, nanotechnology, engineering, or all of the above? With fluid boundaries, identifying materials science publications is often a judgment call.

Is a materials science paper different from a "normal" scientific paper? As John Ziman puts it (Ziman, 2002), "At a mundane level, every scientific paper must contribute something new to the archive. It must suggest a new scientific problem, propose a new type of investigation, present new data, argue for a new theory, or offer a new explanation. Ideally, it should combine several such virtues." Certainly, materials science journals contribute something new, but the novelty is in a specific sense. Take *Nature Materials,* part of whose aim and scope reads "Nature Materials covers all applied and fundamental aspects of the synthesis/processing, structure/composition, properties and performance of materials." Compare this to *Physical Review B* (*PRB*), whose aims and scope statement asks that submitted manuscripts meet the following criteria:

- Present important and novel physics.
- Make a significant contribution in a specific research area and generate interest for PRB's readers.
- Represent an authoritative and substantive addition to the body of literature.
- Explore the subject matter comprehensively and thoroughly.

The novelty for *Nature Materials* is not necessarily a new material, even advances in synthesis and characterization are considered appropriate. *PRB*'s focus, however, is on new and important physics. It's not that *Nature Materials* is less selective than *PRB,* far from it, but the focus is different. A paper about a systematic search through the phase diagram of a material that does not contain a significant physics insight might not be appropriate for *PRB,* but may be for a materials journal.

The importance axis is of course significant in deciding on publication. Is the material one of unusual intrinsic or technological interest? Is the synthesis method presented easier, or does it give access to a different regime of behavior for the material? In a lot of cases an impactful materials paper might be one that sets up future work that may not even be published in a materials journal, and may even be work with much more significant consequences for industry than for fundamental science. The evaluation, by referees, of a submission to a materials science journal, or the materials section of a more broad-scope journal, necessitates adopting a different lens in order to properly judge the suitability of the paper for that journal. It is also particularly interesting in the case of a more broad-scope journal. Would an article on ceramic materials have to also be of interest to polymer scientists?

Types of Materials Science Journals

Early science journals came out of societies, and sought to reflect the full scope of the relevant society. However, as more specialized science periodicals emerged in the 19th century, some eventually established themselves as viable specialized journals addressing a dedicated readership. Such journals can be specialized in the way of addressing a specific field, like physics or chemistry, or some aspect of that field, for example, a journal of condensed matter physics like *PRB* or *Journal of Physics C: Condensed Matter.* There are also journals like *Cell*, which in spite of its apparently specialized name covers a broad swath of life science, and *Nano*, a journal whose scope is defined by the size of the system studied, covering as it does nanoscience and nanotechnology. Journals can also concentrate on a field of study inside of another, like *Journal of Magnetism and Magnetic Materials*, a specific aspect such as *Langmuir*, a journal that focuses on systems where the interface dominates structure and function.

Differently from the above examples of specialized journals that are more dedicated to a field, materials science journals seek to address an interdisciplinary audience. This is not the same as trying to address a discipline, for example, the whole of physics.

Physicists generally identify with their discipline, and condensed matter physicists feel they have more in common with particle physicists than with chemists, even though the methods and practices used by chemists are more directly comparable to those used by condensed matter physicists than those used by particle physicists. Materials science journals, however, need to address a readership community that contains the entire list of material science topics, from metals and alloys to ceramics, polymers, electrical, optical and magnetic materials, materials chemistry, plastics, surfaces and coatings, and even biomaterials. If this is a community, it is an uneasy alliance.

The Converts

Some of the journals that we now call materials science journals have long histories that predate the coining of the term "Materials Science". There are plenty of examples of journals that currently publish what we now think of as materials science but that either do not or did not originally trade on the term "materials". The *Journal of the American Chemical Society* (*JACS*) is an important and respected chemistry journal that was founded in 1879. Through its physical chemistry, polymer, and colloid sections, and others, it covers quite some ground in materials science. Another journal, from physics, that covers materials science is *PRB*, the strapline of which reads "Condensed Matter and Materials Physics", "Materials Physics" indicating the demarcation of its foothold in materials science. In contrast to the stance of *PRB*, the *Physical Review* has also added *Physical Review Materials* to its portfolio in 2017, whose aims and scope states that it "serves the multidisciplinary community working on the prediction, synthesis, processing, structure, properties, and modeling of a wide range of materials." Notable is the absence of the word "physics" in this sentence, indicating a greater openness to materials science from related fields.

In considering materials science journals, it is useful to separate journals that publish materials science as part of their output, like *JACS* and *PRB*, ones that have turned into materials science journals (e.g., *Acta Materialis*), and ones that were started as explicit materials science journals, such as *Nature Materials*. The publications landscape of material science journals leads to questions of disciplinarity. As previously mentioned, any journal that is founded explicitly as a materials science journal lays claim to represent the entirety of the field. Journals that unite these disparate fields face particular challenges, especially if they have more representation in some fields than others. According to R. W. Cahn, materials science journals can be broadly divided into two types, broad spectrum materials science journals and specialized materials science journals, that is, specialized journals within the category of already specialized materials science journals. Broad spectrum journals are such as *Journal of Materials Science, Materials Science and Engineering,* and *Journal of Materials Research.* Cahn points out, however, that some journals appear to be broad spectrum, but are on closer examination much more narrow than they appear. An example that he gives of this is *Materials Research Bulletin,* published by Elsevier on behalf of the Materials Research Society. When Cahn wrote about this journal in 2001, he could fairly categorize it as mostly covering "the synthesis of inorganic materials." Even now, two decades later, their aims and scope statement lays down a clear boundary, "Papers purely on thermodynamics or theoretical calculations (e.g., density functional theory) do not fall within the scope of the journal unless they also demonstrate a clear link to physical properties."

Specialized materials science journals focus on one or more aspects of materials science, but do not lay claim to representing the whole. There is a great number of such journals, but some outstanding examples are *Journal of Alloys and Compounds,* which Cahn describes as hovering "between broad and narrow scope", *Journal of Nuclear Materials,* and a great number of polymer journals such as *Polymer, Progress in Polymer Science,* and *New Polymeric Materials.* Other specialized-but-broad journals include *Carbon, Acta Crystallographica,* and *Small.*

As can be seen from the above incomplete list, many journals that publish materials science do not have "materials" in their name. Additionally, many journals that publish materials science do not consider themselves exclusively material science journals.

Another useful way of categorizing material science journals beyond the taxonomy offered by Cahn might be to follow an axis of degree of involvement in materials science, or by considering the history of the journals, whether they were created as materials journals or pivoted to materials from, for example, metallurgy, as *Acta Metallurgica* did.

Other journals that fit into this category include the *Journal of the American Ceramic Society*, founded in 1918, which is of interest as representing another important branch of materials science other than alloys and metals. It is published by Wiley on behalf of the American Ceramic Society, and is described, on the society's website (https://ceramics.org) as publishing "fundamental science reports on discovery, characterization, and analysis of new materials and processes." Another of these journals is *Acta Crystallographica*, published by the International Union of Crystallography (IUCr). This journal was founded in 1948 as a single journal, but is now split up into subjournals labeled Section A through F, with Section B, founded in 1968, currently subtitled *Structural Science, Crystal Engineering and Materials*. From 1968 to 1982, however, it was known as *Acta Crystallographica. Section B: Structural Crystallography and Crystal Chemistry*, between 1983 and 2013 it was *Acta Crystallographica Section B: Structural Science*. Since 2013, it has its current title, reflecting its current scope. Its mission statement (http://journals.iucr.org/b/services/about.html) reads "Knowledge of the arrangements of atoms, including their temporal variations and dependencies on temperature and pressure, is often the key to understanding physical and chemical phenomena and is crucial for the design of new materials and supramolecular devices." Crystallography thus becomes, in Section B, a means to the end of understanding and synthesizing materials, rather than a subject of study in its own right.

The Spawn

To paraphrase Shakespeare, some are born as materials science journals, some achieve it, and some have materials science thrust upon them. We have already mentioned *Acta Materialia*, which started life as *Acta Metalurgica* (1953), but changed its name in 1990. A materials science journal can also be spun off from a journal with broader scope, for instance *Advanced Materials* launched in 1988 as a supplement to the general chemistry journal *Angewante Chemie.* R. W. Cahn notes that there are a number of renowned societies that started off as dedicated to metals and alloys, such as the Metallurgical Society of the American Institute for Mining; Metallurgical and Petroleum Engineers; The American Society for Metals; the Institute of Metals in London; the Deutsche Gesellschaft für Metallkunde; the Société Française de Métallurgie; the Indian Institute of Metals; and the Japan Institute of Metals. Cahn observes that most of these have now changed their names replacing “metals” with “materials”, and some smaller societies have merged together to broaden scope, such as when the Institute of Metals in London turned into the Metals Society, “which merged with the Iron and Steel Institute to become the Institute of Metals once again, and eventually merged with other societies concerned with ceramics, polymers and rubber to become the Institute of Materials.” (Cahn, 2001, 508)

Cahn observes that the journals published by these societies have also changed names repeatedly for similar motivations, often swapping metal or metallurgy for materials or materials science. It is not sure, however, what effect these name changes have beyond irritating librarians (“These many name changes must be a librarian’s nightmare.”) As Cahn hypothesizes, “The underlying idea fueling the many changes of names of journals is that by changing the name, societies can bring about a broadening of content. By and large this has not happened and the journals have remained obstinately metallurgical in character, because when a journal is first published, it quickly acquires a firm identity in the minds of its readers and of those who submit papers to it, and a change of name does not modify this identity.” Nevertheless, sometimes a change in name does have an effect on the size of the footprint. A look at the table of contents of a recent issue of *Acta Mateialia* (Volume 234, published August 1, 2022) reveals that only 22 of 33 papers in this volume were straightforwardly about metals or alloys, and the rest on graphene, oxides, organics, semiconductors, ceramics, and other materials. Whether this balance is representative, or sufficient to warrant changing the name of a journal, is an editorial or publishing question.

Sometimes materials science journals are spun off from existing journals. An example of this is *Advanced Materials*, published by Wiley, which was launched in 1988 as a supplement to the general chemistry journal *Angewante Chemie*. Its launch editorial acknowledges a multidisciplinary motivation, "Modern science and technology rely on specialists. Nevertheless, the complexity of many problems is all too often an obstacle to gaining a deeper understanding of physical and chemical phenomena. In particular, the development of advanced materials and new technologies requires the joint efforts of many specialists from a variety of disciplines." (Gölitz, 1989) Even as journal is spun off from *Angewante Chemie*, the launch editorial is already laying claim to physical phenomena as well as technological applications that range beyond the journal's chemistry origins. As the volume of research in materials science increased significantly since the 1990s, several journals were spun off, including *Advanced Engineering Materials* (1999), *Advanced Functional Materials* (2001), *Small* (2005), and *Advanced Energy Materials* (2011). *Advanced Engineering Materials* claims, as its scope, a strong focus on new manufacturing techniques. *Advanced Functional Materials* meanwhile used to be *Advanced Materials for Optics and Electronics*. Its launch editorial set out its scope as mainly being about devices:

> "Covering such a wide range of research interests, not all papers will be of interest to all readers, but the Editorial Board hopes that by having papers on all aspects of this important area of research in the same journal, discussions will be initiated between researchers who might not otherwise meet and that this may lead on to new research strategies and insights. For example, device engineers will be able to be more familiar with the types of material available to them and of their potential, whilst precursor chemists will be better able to design compounds with desirable properties if they understand what the problems of materials scientists and device engineers are." (Cole, 1992)

This also lays claim to an interdisciplinary identity, hoping to bring together researchers who would not otherwise have met and exchanged ideas. *Small* similarly claimed, in its launch editorial, "We will endeavor to typify the multidisciplinary character of this burgeoning area by providing our readers with a wide range of articles that appeal to the chemist and the biologist, the materials scientist and the engineer, the physicist, the biomedical scientist, the theoretician, and beyond." (Gölitz *et al.*, 2004) It should come as no surprise that *Advanced Energy Materials*' launch editorial also touts the need for collaboration, but makes an interesting distinction, "At this stage we should clarify that traditional energy generation by means of fossil and nuclear fuels or the design and use of structural materials for wind or hydroelectric power generation are not covered by the special section. These engineering disciplines, though clearly energy-related, find coverage and different readerships elsewhere." So, the focus for this journal needs to be on energy materials, not materials for energy production. Even the interdisciplinary discipline of materials science has to have some boundaries.

Materials from the Beginning

Some journals were started explicitly as materials science journals. To the ranks of journals born to materials we count *Nature Materials* (2002) and *Functional Materials* (1994), but also a number of other, less high profile journals that were launched in response to demand from the materials science community.

Starting at probably the most prestigious of these, *Nature Materials* was the first specialized *Nature* journal in physical sciences. The launch editorial (Ziman, 2002), states that this is the Nature Publishing Group's first "research title in the physical sciences", and lays out the motivations for its creation, "Materials science represents one of the most exciting areas of interdisciplinary science and technology, blending physics, chemistry, biology and engineering into an innovative, if not always cohesive, whole" The editorial acknowledges the potential drawbacks of having an interdisciplinary nature, "This diversity and 'frontier' spirit can sometimes be problematic. With so many disciplines using different technologies and speaking different jargon, communication is not always easy — the prime example being the scientific cultural gap between scientists involved in research into 'soft' and 'hard' matter." These kinds of camp issues are a consistent theme of any materials science journal.

Another journal that was launched as fully formed is the *MRS Bulletin.* This journal, published by the Materials Research Society in partnership with Elsevier, was launched in 1974 as the *MRS Newsletter.* After a brief hiatus in 1981, it returned as *MRS Bulletin.* The Materials Research Society, established in 1973, purports to cover all aspects of materials science, but as we saw earlier the *MRS Bulletin* will not publish certain kinds of theoretical treatments such as density functional theory work. In addition to the *MRS Bulletin,* the Materials Research Society also publishes *Journal of Materials Research* (1986), *MRS Communications* (2011), *MRS Energy & Sustainability* (2014), and *MRS Advances* (2018).

The success of *Nature Materials* led the Nature Publishing Group to launch further research journals at a level subordinate to the original *Nature* journal. One of these, of interest to us, is *Nature Nanotechnology,* which was launched in 2006 and is an example of a materials journal that does not have "materials" in its name. Like *Nature Materials,* it was also created from nothing. The short editorial announcing the launch mentions materials twice, highlighting one of the motivations for starting this journal (Gerber and Lang, 2006). Indeed, nanoscience, including nanotechnology, is a fascinating subcategory of materials science journals. Other journals in this category include *ACS Nano* (published by the American Chemical Society, which also publishes *JACS*), *Nanoscale,* and *Small.* Here a scope demarcation is made in terms of the size of the system under consideration. Many of these journals contain papers more centered on devices, though of course the devices need to be made of a material, which is a cause for their inclusion into the category of materials science journals.

ACS Nano is a particularly interesting example of this type, since it represents the foray of the American Chemical Society into areas normally not viewed as being part of chemistry. The launch editorial

(Weiss, 2007) highlights the interdisciplinary motivation of the journal, stating "We see our role as critical in "going deep" in chemistry, biology, biomedicine, engineering, materials, physics, and related fields, all at once."

Nanoscience is of course enabled by technology, and in many ways technological advances have been a driving force for materials science. As another example of a materials science journal on the fringes, we have *Zeitschrift für Kristallographie* (1877). This journal was renamed *Zeitschrift für Kristallographie — Crystalline Materials* in 2010. In spite of its name, it publishes in English only since the last renaming. This journal would not have been launched but for the growth in the discipline of crystallography, and the acceleration of this field was only increased by the discovery of x-rays in 1895.

Indeed, a journal is not launched in the absence of a community that requires such a journal. There must be a perception of needs being unmet. Nor is one journal per community sufficient. A wide range of quality and importance of published papers requires a range of journals, to allow authors choice in where to publish. The other pillar that leads to the foundation of a journal is the financial interest that can support such a journal (Dawson and Topham, 2020). Here Materials Science has a structural advantage; because of links with industry, there is a large readership beyond practitioners that is interested in the consequences of research.

Conclusion: Future of Materials Science Journals

What is next for materials science journals? Increasingly, materials journals are interesting in publishing design and discovery studies as well as experimental characterization and synthesis studies. In this they are helped by emerging technologies such as "Machine Learning" algorithms, which provide another interface with theoretical modeling like neural network algorithms, large-scale molecular dynamics simulations, and first-principles calculations such as density functional theory.

Another emerging interface, beyond polymers, is the field of "biomaterials", which is another area that overlaps with materials science.

From the origins of materials for prostheses to current technologies like organic LED displays, this represents another front of materials science far from origin stories in areas like metallurgy and ceramics. And while on the subject of biomaterials, the Materials Genome Initiative, set up in 2011 but with the term "Materials Genome" first coined in 2002, takes its inspiration from the Human Genome Project. Its aim is to design, manufacture, and bring about materials with desirable properties through theoretical predictions (*Nature Reviews Materials*, 2021).

Journals are adapting and will continue to adapt to the changing field of materials science in order to serve an ever growing and ever more inclusive community of researchers. *Nature Reviews Materials,* launched only in 2016, currently has the seventh highest impact factor of all journals, according to Scopus preview (https://www.scopus.com/sources.uri). We certainly do not predict a slowing down in the growth of materials science journals.

To conclude our study, we find that while materials science journals are interdisciplinary, even broad scope ones can have a less than comprehensive coverage. The origin stories of materials science journals broadly fit into three categories: some materials science journals are created as such, others have become materials science journals, or been spun off from other, broader journals, and a third category is populated by journals that publish some materials science, but could not simply be described as materials science journals. Journals, of course, follow the course of the field that they represent. In the case of Materials Science, we observe that there is a turning point in the history of materials science journals which occurs at the same time as when materials science became a type of academic department, in spite of the fact that what we now call materials science was practiced before this watershed.

In spite of the above observations, it is very difficult to talk about a materials science journal as if it is only one thing, with a blend of outright materials journals that nevertheless differs in scope, often leaning toward or excluding some particular aspect of the materials science spectrum. This also holds for journals whose contents are only partially in materials science, in that they also have large no-go areas in which they do not publish. Furthermore, on the subject of disciplinarity, while materials science encompasses physics, chemistry, biology, engineering, etc., the readership of the journals can float in and out of materials science and may not engage equally with all parts of the field. A polymer researcher may not be all that inspired by metallurgy papers, and vice versa.

Nevertheless, it seems the future is bright for materials science and materials science journals. As technological advances enable us to not only understand, but also to make, more and better materials, it should follow that materials science journals will increase their scope and reach, keeping pace with advances and inspiring a new generation of readers.

References

Baldwin, M.C. *Making Nature: The History of a Scientific Journal.* Chicago, IL: University of Chicago Press; 2015. 309 p.

Biagioli M., "From Book Censorship to Academic Peer Review," *Emergences: Journal for the Study of Media & Composite Cultures* 12, no. 1 (May 2002): 11–45, https://doi.org/10.1080/1045722022000003435.

Brabec, C., Waidhas M. Advanced energy materials needed. *Adv Mater.* 2010 Feb 23;22(8):E4–E5.

Cahn RW. *The Coming of Materials Science,* 1st ed. [Internet]. Amsterdam; New York : Pergamon; 2001. 568 p. (Pergamon Materials Series).

Choi H, Shields B. A place for materials science: Laboratory buildings and interdisciplinary research at the University of Pennsylvania. *Minerva.* 2015 Mar;53(1):21–42.

Cole D. Editorial. *Adv Mater Opt Electron.* 1992 Feb;1(1):1.

Dawson G, Lightman BV, Shuttleworth S, Topham JR, editors. *Science Periodicals in Nineteenth-Century Britain: Constructing Scientific Communities.* Chicago, IL: University of Chicago Press; 2020.

Eddy M. *The Language of Mineralogy: John Walker, Chemistry and the Edinburgh Medical School, 1750–1800.* Farnham, England ; Burlington, VT: Ashgate Pub. Co.; 2008. 309 p. (Science, Technology, and Culture, 1700–1945).

Gerber C, Lang HP. How the doors to the nanoworld were opened. *Nat Nanotechnol.* 2006 Oct;1(1):3–5.

Gölitz P. Paving the way .... *Adv* Mater. 1989;1(1):3–4.

Gölitz P, Levy E, Horley GA. Think small! *Small.* 2004 Nov 3;1(1):3–4.

Ho YS. A bibliometric analysis of highly cited articles in materials science. *Curr Sci.* 2014;107(9):1565–1572.

Nature Reviews Materials. Rise of the machines. 2021 Aug;6(8):641.

Novoselov KS, Geim AK, Morozov SV, Jiang D, Zhang Y, Dubonos SV, *et al.* Electric field effect in atomically thin carbon films. *Science.* 2004 Oct 22;306(5696):666–669.

Sass SL. *The Substance of Civilization: Materials and Human History from the Stone Age to the Age of Silicon.* 2011.

Tretyakov S, Urbas A, Zheludev N. The century of metamaterials. *J Opt.* 2017 Aug 1;19(8):080404.

Weiss PS. Welcome to *ACS Nano*. *ACS Nano.* 2007 Aug;1(1):1.

Yuh-Shan Ho. A bibliometric analysis of highly cited articles in materials science. *Curr Sci.* 2014 Nov 10;107(9):1565–1572.

Ziman J. Our changing nature. *Nat Mater.* 2002 Sep;1(1):1.

Chapter 4.4

History of the Brazilian MRS Meetings

Guillermo Solòrzano

It took two years of dedicated planning, visiting different materials research groups all over Brazil and in continuous email discussions, for the Brazilian Materials Research Society (BMRS) (named in Portuguese as SBPMat for Sociedade Brasileira de Pesquisa em Materials) to be founded on June 26 in an assembly meeting in Rio de Janeiro and over the internet.

A year later, from July 7th to 10th 2002, almost 400 scientists, from 18 countries, gathered in Rio de Janeiro, at the Carlton Atlantica Hotel and Convention Center, for the inaugural meeting of the Brazil MRS, with English Rio as the official language, thus making the international environment of the meeting clear. In fact, the foreign participants included the presidents of MRS (Alex King) and the European MRS, E-MRS (Giovani Marletta), as well as the General Secretary of the International Union of MRS, IUMRS (Robert P Chang). Chaired by the Founding President Guillermo Solórzano (Catholic University of Rio de Janeiro), the program committee assembled five symposia ranging from nanostructured materials and systems to semiconductors and polymeric thin film materials. The program included plenary sessions and a workshop on the processing of powder materials. Slightly more than 400 technical presentations were made either orally or as posters, displayed in a neighboring area of a commercial exhibition, engaging the attendees in lively discussions.

The meeting emphasized international materials collaborations, which were the themes of three plenary sessions. RP Chang described his vision of a Materials World Net to complement the efforts of international materials research collaborations; Adrian the Graff of the US National Science Foundation, NSF, and Celso P de Mello of the Brazilian founding agency, CNPq, described the creation of mechanisms for joint funding toward establishing, with other governmental agencies, an Inter-American Materials Collaboration Program. At the closing session, the SBPMat governing body announced that, in view of the success of this conference, the society will make this meeting an annual event.

On October 26th–29th 2003, the second BMRS meeting took place also in Rio de Janeiro. The program, augmented to six symposia, covering a wider spectrum of subjects, including Nanostructured and Biomaterials II, Materials for Energy Conversion and Environmental Protection, Structural Alloys for Transport Systems, Composite and Supramolecular Materials, and a workshop on Semiconductors Group III nitrides. The highlights of this meeting were two plenary sessions on international collaboration: one session in the scope of the Inter-American Collaboration Program, CIAM, with authorities of the Brazilian funding agencies CNPq and FINEP, the NSF and MRS President (also representing the US National Academy of Sciences). The event also hosted a Pan American Advanced Study Institute, funded by NSF, with the participation of students from Brazil, the US and other Latin American countries. The meeting also provided a commercial exhibition of products and services of interest to the community of materials researchers. With close to 500 participants and held at the Pestana Rio Atlantica Hotel Convention Center, on Copacabana beach with Rio's other sights close by, the meeting provided a wide range of opportunities for extracurricular activities, but nevertheless held the attention of a well-engaged audience.

The proceedings of the first and second Brazil MRS Meetings were published in two journals: *Materials Characterization* and *Materials Research.* The former selects contributions mainly concerning materials microstructure and properties. The latter, a journal with a larger range of subjects, consider contributions dealing with theory, modeling, phenomenological studies and experimental papers on synthesis, processing, performance and applications of materials.

Brazil is a country of continental dimensions with several options for venues to host BMRS meetings. An important defining criterion, however, is to offer an appropriate relaxing environment with hotels and convenient accommodation for students, local tourist attractions together with ready access to an international airport. So, the third BMRS was held in world-famous Iguassu Falls, southwest of Brazil, from October 10th to13th 2004 at Hotel Bourbon Cataratas and Convention Center. The meeting consolidated the symposium model and the program committee successfully organized a set of topics enhancing the interdisciplinary scope of materials research in nine symposia, covering theory and frontiers on experimental research, from solid-state chemistry to environmental sciences applied to materials. One workshop on Spintronics in Semiconductors was also included. Joining efforts with sister scientific societies and led by Elson Longo (Federal University of Sao Carlos), BMRS president this year, other well-established materials-related events were incorporated in this BMRS edition: the IV Brazilian Symposium on Electroceramics and a Symposium on Nanoscale Structural Characterization of Materials, hosting the bi-annual MICROMAT meeting, which is a traditional event of the Brazilian Society for Microscopy and Microanalysis (SBMM). Three plenary lectures presented New Developments in Nanotechnologies (Cylon Gonçalves da Silva, General Secretary, Ministry of Science and Technology, MCT, Brazil), the Atomistic Structure of Internal Interfaces (Manfred Ruhle, Director Max Plank Institut, Germany) and International Cooperation on Nanotechnologies (Dietling Jering, The European Union). The number of registered participants at this third meeting was close to 1000.

The fourth BMRS Meeting moved to Recife, an important and historical capital in the northeast of Brazil, on October 16th–19th 2005. The program committee, chaired by Celso P de Mello and Anderson S.L. Gomes (Federal University of Pernambuco), successfully set up eight symposia, all on interdisciplinary topics, including 37 invited talks by national and international scientists, 157 oral and 599 posters presentations, providing a total of 793 technical contributions, involving over 800 participants in the meeting. The opening session had as the Keynote Speaker, Prof. Sérgio Rezende, the Brazilian Minister of Science and Technology, a world-renowned materials scientist himself. Four plenary lectures were also delivered by international experts on different themes of materials research: Atomic-scale Defect Analysis of Ceramic Materials (Barry Carter, Department of Chemical Engineering and Materials Science University of Minnesota, US), Construction of Molecular Nano-Structures and Electronic Properties (Takuji Ogawa, Center for Molecular-Scale Nanoscience, Institute of Molecular Science, IMS/Okazaki, Japan), Multifunctional Organic/Inorganic Hybrids (Luis A Ferreira M Dias Carlos, Department of Physics, Universidade de Aveiro. Portugal), and From High-Temperature Superconductivity to Complex Electron Systems Science (H. Takagi, Magnetic Materials Laboratory, RIKEN'S Discovery Research Institute). A well-attended roundtable discussion session was included in this year's program, aiming at establishing the perspectives of the materials science area in Brazil. As in previous years, the meeting also offered a commercial exhibition.

The efforts to provide the society with a professional administration structure, initiated in 2006 with Fernando Lazaro F Jr (Catholica University of Rio de Janeiro) as the President of BMRS, proved to be productive regarding the communication required for the organization of the following meetings. So the fifth BMRS Meeting was successfully held at Costao do Santinho Resort and Spa in Florianopolis, a city of spectacular natural scenarios in the south of Brazil, on October 8th–12th, 2006. Under the coordination of Aloisio N Klein (Federal University of Santa Catarina, UFSC), the program reached 13 symposia. In parallel, a well-attended commercial exhibition stimulated the attendee's interest in experimental materials research. This fifth edition was organized jointly with the MICROMAT meeting, contributing with four symposia to the event overall and enriching the scope of topics and the keynote lectures, such as the one delivered by the President of the International Federation of Microscopy Societies (IFSM) (Cristian Colliex, Orsay, France). This meeting also hosted the International Symposium on Hybrid Materials, chaired by Antony Cheetham (University of Santa Barbara, US), and the fifth Brazilian Symposium on Electroceramics, chaired by R Muccillo (IPEN, SP) and J.A. Varela (UNESP, SP) As a result, a record number of 1166 submitted abstracts and about 1000 participants proved this annual meeting to constitute the most important Brazilian conference of Materials Science and Engineering.

The following year, for the sixth edition, the meeting moved again to the North Easter region of Brazil, being held in the exuberant city of Natal at SERHS Grand Hotel and Convention Center, October 28th–November 1st, 2007. Coordinated by Dulce M de Araujo (Federal University of Rio Grande do Norte (UFRN)), the program consisted of plenary lectures and 10 symposia covering a wide range of topics, ranging from nanomaterials to superconductors, including organic semiconductors, multiferroic materials, catalysis, and computational modeling of materials. The meeting had 1057 submitted abstracts nearly reaching again one thousand participants, with significant numbers of delegates from abroad, namely, from Latin America and Europe. The commercial exhibition has continued to increase in number of stands and gained the name of VI Expomat.

The seventh Annual Meeting was held in Guaruja, a beach resort in the state of Sao Paulo, Southeast of Brazil, at Casa Grande Hotel Resort and Spa, Praia da Enseada, September 28th–October 2nd, 2008. Chaired by Aldo Craievich (University of Sao Paulo (USP)) and Reginaldo Muccillo (Institute for Nuclear Energy Research (IPEN)) the program offered eight plenary lectures delivered by outstanding scientists and successfully set up 14 symposia, including five symposia in partnership with MICROMAT meeting, and one microscopy pre-meeting course. With a total of 1417 submitted abstracts, the meeting reached a record of 1284 registered participants, including about 700 graduate and undergraduate students! Two *MRS* journalists covered this meeting *in loco* and reported it in a 2008 issue of the *MRS Bulletin.*

In 2009, Rio de Janeiro hosted the International Conference of Advanced Materials, ICAM 2009, for the first time in South America, with Guillermo Solórzano (Pontifical Catholic University of Rio de Janeiro, PUC-Rio) and Elisa B Saitovitch (Brazilian Center for Physics Research) as chairs. ICAM is the most prestigious conference of the International Union of Materials Research Societies (IUMRS) and in 2009 was organized together with the eighth Brazilian MRS Meeting from 20th to 25th September 2009. The program constituted 28 technical symposia, eight plenary lectures, two workshops, an energy forum and an exhibition. Each symposium featured invited talks and contributed oral and poster presentations. About 2000 scientists from 47 countries gathered at Windsor Convention Center in Rio during this event. The 28 symposia were assembled in seven major areas: Nanoscience and Nanotechnologies, Health and Biological Materials, Energy and Environment, Functional and Electronic Materials, Structural Materials, Theory and Phenomena, and General Interest. The meeting opened with an official ceremony which included address talks by the President of the Brazilin MRS, the Program Committee Chair, the Conference Chairman and Brazilian government authorities. The state Ministry for Science and Technology delivered the opening lecture. All the oral presentations were delivered in English, which emphasized the international flavor of the meeting. As in the previous BMRS meetings, a poster session was held each evening and there was also an equipment exhibition. More than 1800 technical presentations were made (485 orally and 1374 posters) and the strong interdisciplinary program provoked lively discussions both in the formal sessions and informally. The conference also featured two workshops, "Innovation in Materials Research" and "Global Nanotechnology Network," and an energy forum. The commercial exhibition has doubled in terms of space and number of stands.

With the BMRS entering its nine years of existence, the Brazilian materials research community was accustomed to the annual meeting. From 24th to 28th October, 2010, the ninth BMRS meeting took place in Minas Gerais, an important Brazilian central state, in Ouro Preto, a patrimonial city recognized by UNESCO, at UFOP Arts and

Convention Center. Chaired by Margareth Spangler Andrade (Center for Technology of Minas Gerais, CETEC — MG) and Evando Mirra de Paula e Silva (Federal University of Minas Gerais UFMG/CGEE), the event offered an interdisciplinary program of 12 Symposia and six plenary lectures delivered by distinguished invited speakers. With a new record of 1766 submitted abstracts, 195 presentations were delivered orally and 1456 as posters, the meeting reached 1562 registered participants, including graduate and undergraduate students!

The 10th edition of the BMRS meeting was held in Gramado, an emblematic mountain region city in the South of Brazil, at the Convention Center of the Federal University of Rio Grande do Sul, October 25th–29th, 2011. The program and meeting committee chaired by Paulo F Fichtner and Naira M Balzaretti (Federal University of Rio Grande do Sul) arranged 16 symposia in a well-balanced spectrum of subjects, nine plenary lectures delivered by the distinguished scientist from Brazil and abroad, and three pre-meeting courses on advanced X-ray diffraction techniques. Celebrating the occasion of the 10th meeting, BMRS President, Jose A Varela, inaugurated the "Joaquim da Costa Ribeiro Memorial Lecture" and presented an award in recognition for the accomplishments serving the society to Founding President Guillermo Solórzano and to former State Minister for Science and Technology, Sergio M Rezende. This year, numerical output continued to increase both in terms of submitted abstracts (1907) and presentations (266 oral and 1546 posters) with 1840 registered participants, establishing a new record for a national meeting. The commercial exhibitions have become a well-established parallel event.

The BMRS meeting has become a consolidated event in Brazil and is known abroad. The symposium–based model has been successfully accepted by the local scientific community and gained increased attention. In the following years, the meeting experienced a steady growth in the number of participants, as summarized in the following paragraphs, with no less than 20 and up to 40 different countries represented, mainly from Latin America and Europe. The meeting sites have moved year to year across the country, selected by the local arrangement committees. The highlight of the meetings was the continuous participation of renowned scientists as plenary and keynote lecturers on various state-of-the-art topics in Materials Science and Technologies. No less important has been the involvement of young students both in oral presentations and, more prominently, in the ebullient poster sessions. In fact, aiming at promoting and better organizing student participation, the Society successfully established university chapters. On the other hand, the commercial exhibition has become a well-established parallel event.

The 11th BMRS meeting was held again in Florianopolis, from September 23th to 27th, 2012, at the same venue of 2006. The meeting was chaired by Aloisio N Klein and Adre A Pasa (Federal University of Santa Catarina) and the program consisted of 16 Symposia, seven plenary lectures and a round table discussion session on science, technology and innovation in Brazil. BMRS President Roberto M Faria (University of São Paulo USP) established the Bernhard Gross Award in this meeting. From the 1863 abstracts submitted to this meeting, 230 were presented orally and 1460 as posters with 1708 registered participants.

The 12th BMR meeting moved to Campos to Jordão, a touristic mountain region city in Sao Paulo state, at the Convention Center, from September 29th to October 3rd 2013. The conference was chaired by José Alberto Giacometti (University of Sao Paulo, USP) and Julio R Sambrano (Faculty of Sciences, UNESP), and the program featured 16 symposia, an opening lecture, seven plenary lectures, a round table discussion session on Science, Engineering of Industrial, Electronic and Biomedical Devices, a Humboldt Kolleg 2013 — Science & Technology in Contemporary Life: Impacts and Horizons and a "Bernhard Gross Award" closing ceremony. The number of submitted abstracts was 1835, of which 215 were presented orally and 1547 as posters with 1679 registered participants.

From September 28th to October 2nd 2014, the 13th BMRS meeting moved back to Northeast Brazil, being held in the historical city of João Pessoa and the capital of Paraiba state, at the

Convention Center. The meeting was chaired by Ieda M Garcia dos Santos (Department of Chemistry, Federal University of Paraiba, UFPB) and Severino J Guedes de Lima (Department of Mechanical Engineering, UFPB) and the program featured 19 symposia, including International Symposium Cementitious Materials (ICSM-2014), an opening lecture, seven plenary lectures and the **Bernhard Gross Award** Closing ceremony. The submitted posters reached the record number of 2528, enabling 294 oral presentations and 1800 as posters with 2148 registered participants, including 1652 effective attendees.

The 14th BMRS meeting was held again in Rio de Janeiro, at the Sul America Convention Center, from September 27th to October 1st 2015. The meeting chairs Marco Cremona (Department of Physics, PUC-Rio) and Fernando Lazaro F Jr (Brazilian Center for Physics Research, CBPF) assembled a program of 27 symposia, an opening session/memorial lecture, seven plenary lectures and two workshops, one on Nanofabrication and Innovating Manufacturing Technologies and the other on Organic/Thin films Electronics in Industry and Processes, and a closing ceremony awards presentation session. The 27 symposia were organized into seven major groups: Nanomaterials and Synthesis, Theory Characterization and Modelling, Electronics and Photonics, Energy and Sustainability (which included the eighth International Summit on Organic and Hybrid Solar Cell Stability (ISOS-8), Biomaterials and Soft Materials, and New Horizons in Materials Science. A new record was established in terms of submitted abstracts, 1607, and in terms of presentations, 373 orally and 1880 as posters, with 1987 registered participants including 1400 attendees.

In spite of the economic situation affecting Brazil since the last decade, BMRS President Oswaldo Nde Oliveira Jr (University of Sao Paulo USP) and his team were successful in maintaining the impetus and quality of the society's well-established activities. For the 15th edition of the BMRS annual meeting stayed in the South East of Brazil, in Campinas, a technological hub city in the state of Sao Paulo, 25th–29th September, 2016, at the Expo Dom Pedro

Convention Center. Chaired by Ana Flavia Nogueira (Institute of Chemistry, Unicamp) and Monica A Cotta (Institute of Physics, Unicamp), the meeting program features 23 symposia, an opening session/memorial lecture, five plenary lectures, three meeting panels, one workshop (school of scientists and hands-on tutorial), and a closing ceremony. From the 2133 abstracts submitted, 276 were presented orally and 1567 as posters with 1768 registered participants and 1416 effective attendees.

The 16th BMRS meeting moved again to Gramado, at FAURGS, September 10th–14th, 2017, chaired by Daniel E Weibel (Federal University of Rio Grande do Sul- UFRGS). Keeping up with the previous meetings, this edition programmed 23 symposia, one workshop (Young Researches School: How to Produce and Publish High Impact papers), an opening ceremony/memorial lecture and a closing ceremony. From the 2029 abstracts submitted, 345 were presented orally and 1487 as posters with 1507 registered participants and 1275 attendees from about 20 countries.

As a venue for the 17th BMRS meeting, the city of Natal in northeast Brazil was chosen for the second time. The event took place at the Praiamar Natal Hotel and Convention Center, September 10th–14th, 2018. Chaired by Antonio E Martinelli (Federal University of Rio Grande do Norte), the meeting featured 21 symposia, one workshop (re-edition of the successful Young Researchers School), an opening ceremony/memorial lecture, eight plenary lectures and the closing ceremony. From the 1743 abstracts submitted, 343 were presented orally and 1256 as posters with 1861 registered participants and 1071 effective attendees.

The 18th Annual Meeting was held in Camburiu, a beach resort city in the southern state of Santa Catarina, September 22nd–28th, 2019. The meeting, chaired by Ivan H Bechtold (Department of Physics, Federal University of Santa Catarina, UFSC) and Hugo Gallardo (Department of Chemistry, UFSC), assembled a program constituting 23 symposia, an opening ceremony/memorial lecture, eight plenary lectures, a pre-meeting workshop and the closing ceremony. The 23 Symposia were organized into five major groups: Bio-related Materials and Applications, Materials Properties and

Emerging Technologies, Materials Characterization and Degradation, Materials Synthesis and Processing, and Nanostructured and Functional Materials. The number of submitted abstracts established a new record of 2616 as well as the 467 oral presentations and 1868 posters, with 1987 registered participants including 1686 attendees.

The 19th BMRS meeting, planned to be held in Foz de Iguazu in 2020, did not take place because of the pandemic year. So, it was postponed and held virtually from August 30th to September 3rd 2021 together with the 2021 International Conference on Electronic Materials — IUMRS ICEM 2021. The meeting was chaired by Gustavo M Dalpian (Federal University of ABC) and Carlos CB Bufon (National Nanotechnology Laboratory) were successful in setting up a full program consisting of 23 symposia, an opening ceremony/memorial lecture, nine plenary lectures and a closing ceremony awards presentation, besides a pre-meeting school (Young researchers), a workshop (Women in Science) and a parallel virtual exhibition. In addition, a discussion on CBAN, Brazil–Argentina Nanotechnology Center, was established. From the 1003 abstracts submitted, 304 were presented orally and 706 as posters with 1307 registered participants including 1038 effective attendees.

In summary, over the past two decades, the Brazilian MRS meeting has grown steadily and consolidated as the major interdisciplinary meeting of Materials Science, Technology and Engineering in Brazil. It has reached a reputation that transcends country borders, is continuously attracting international participation, mainly from Latin America, USA and European countries, consolidated as the second Materials Research Meeting in the Americas and perhaps as number 3 in the western hemisphere. The merit of this success is due to the enthusiastic involvement of volunteer society members led by competent and dedicated Brazil MRS executives, with the effective assistance of a small and committed BMRS staff and the electronic and data processing Aptor system over the years. The continuous support of (older) sister societies MRS and E-MRS as well as of IUMRS have provided additional incentives.

Recognition is noted for the financial support received periodically from the federal funding agencies, MCT, CNPq, FINEP and CAPES, as well as from the state funding agencies, namely, FAPESP and FAPERJ, for financing part of the infrastructure and granting the students participation. Other state agencies such as FAPEMIG, FAPERGS, and FAPESC have also contributed to local arrangements in their states. On the other hand, the partnership established with the commercial exhibitors, enriching the success of the meetings, is also acknowledged.

Part V
Regional Networks

Part V shows non-Western countries evolving their materials science institutions. In each case, an entire network of institutions — sites, funding agencies, journals, and so forth — was required to make the materials science happen, each needing to be flexibly and efficiently managed.

Hiroto Kono's chapter, "Building the Blocks of Sciences for Materials: Materials Science Institutions in Japan from a Historical Perspective," concerns a national network that emerged apart from the Western model but became entangled with it. The chapter begins shortly after the arrival of Commodore Perry in 1853 and Japan's subsequent opening to the West. The Meiji Restoration, which followed 15 years later, brought enormous changes, including the introduction of Western technologies, and the creation of steel, chemical, and electric industries. The two World Wars brought an enormous demand for materials research due to the unavailability of many materials previously imported. New institutions had to be created, many patterned after Western models, such as the American Society for Testing Materials, ARPA, IDLs, and so on. Here, we witness the formation and evolution of a national materials science network under the impact of a combination of internal needs and external pressures, and eventually international models. The chapter cleverly charts the transformations by changes in the Japanese words used in connection with materials science: *fugen*, *shigen*, *genryô*, *zairyô*, *sozai*, and *busshitsu*.

Wei Qian's chapter on China is exceptional in being about the most ancient and longest institutional network in the world. China surely had perhaps the most ancient traditional technologies: pottery has been unearthed from 20,000 years ago, and later porcelain, iron, and steel. Modern materials science began in 1895 with the Department of Mining and Metallurgy at the founding of the earliest university in China; the first society engaged in materials research in 1927; and a materials test department in the defense ministry in 1938. A major shift came in 1949 after the establishment of the People's Republic of China in 1949 in the wake of which came the founding of the Chinese Academy of Sciences. In 1956, the "12-year Plan" laid out more progress in materials science and engineering, and more research institutes were established and majors in various materials fields created at numerous universities. Yet another major shift came in 1978 after the reform, when materials research began to become interdisciplinary, with international cooperation, and based on Western models, led to the creation of the "Shenyang National (Joing) Laboratory of Materials Science," the first national research laboratory in China.

The chapter by Wole Soboyejo, Samuel Chigome, and Iwnetim Abate, "The Development of Materials Science Institutions in Africa," reviews the story of materials science research in Africa from early efforts in South Africa to the springing up of materials research elsewhere mainly within traditional departments such as physics, chemistry, and engineering, to the establishment of the African MRS in 2002, a major step. This chapter then traces the development of this society, with the aid of partners in Europe, Asia, and the United States, into an international network of scientists and engineers.

Armando Fernández Guillermet's chapter, "The Roots of Materials Science in Argentina: Jorge A. Sabato's Multifaceted Metallurgy Project," is the remarkable story of how a single person, the physicist, and metallurgist Jorge Sabato, pursued an ambitious vision to develop modern metallurgy in Argentina, to have it recognized in that country as a specific field within solid-state physics, and to develop metallurgy specifically for nuclear applications. Guillermet shows how, in order to realize this vision, Sabato had to design and develop an entire network of materials science institutions to educate, fund, and connect researchers in Argentina and throughout Latin America.

Chapter 5.1

Building the Blocks of Sciences for Materials: Materials Science Institutions in Japan from a Historical Perspective

Hiroto Kono

Introduction

Discussing the emergence and evolution of Materials Science and Engineering (MSE) in the United States, historian Bernadette Bensaude-Vincent suggests characterizing related endeavors as pointed responses to challenges. In addition to the challenge from the Soviet Union's launch of Sputnik, which helped forge the generic notion of materials science, she writes of the "Japanese challenge": "Competing with Japan and regaining the leadership ... became the prime concern or obsession of U.S. materials projects in the 1980s" (Bensaude-Vincent, 2001, p. 230) and as a result "the Japanese challenge prompted a re-orientation toward processing and industrial competition" (Bensaude-Vincent, 2001, p. 238; Forester, 1988). Indeed, Japan, in the role of challenger to American industrial preeminence, had made remarkable achievements in discovering and engineering materials, and Japan's competitiveness manifested in industrial applications.

Such achievements did not grow out of the framework of today's understanding of MSE; rather, they originated from scientific and technological practices related to materials dating to the beginning of (or even before) Japan's modernization. From then on, materials have been a central concern of Japan's industrial development, where each sector, constantly motivated by its distinctive challenges and crises, came up with solutions by building its own science of materials. These largely separate blocks were able to persist due to the success of their industrial applications, and the consequent "Japanese challenge" rendered it unnecessary to forge a single scientific framework that encompasses every kind of material while bridging basic and applied research — the very ideal of material science as was taking shape in the United States. It took a long time for the US model to replace the Japanese framework of various and somewhat independent blocks for the sciences of materials.

This chapter focuses on the development of Japan's MSE institutions and explores several specific aspects of the Japanese history of MSE, which still wants for comprehensive historical investigation. As some predecessors of MSE institutions in Japan were established at the beginning of the country's modernization (the Meiji Restoration in the late 1860s), the chapter begins with the Meiji Era (1868–1912) (see the first section). It then discusses the Taisho Era (1912–1926) (see the second section), when the experience of World War I restructured the research scheme; then it turns to the prewar and wartime period, the early Showa Era (1926–1989), when the mobilization of science and technology toward the total war loomed large (see the third section); finally, it addresses the reconstructing and building a high-tech nation in the postwar period, the late Showa Era, and onward up to the turn of the century (see the fourth section). The conclusion (see the fifth section) discusses the outline and some aspects of the history of MSE in Japan and suggests several points to be addressed in future research.

The Project of Modernizing Japan

In 1853, Commodore Matthew Perry sailed into the harbor at Edo bay (later Tokyo bay) with a message from the US President, Millard Fillmore. The United States sought to open the Japanese economy to the Western world. The event is symbolic of the growing external pressure Japan faced in the mid-18th century to engage globally as well as to protect itself amid the storm of colonialism. The newly formed Meiji State rushed to establish a modern social structure and to foster industry with the famous slogan "*Hukoku Kyôhei*" ("Enrich the Country, Strengthen the Armed Forces"). Regarding the introduction of Western science and technology, the state mainly pursued the strategy already initiated since the preceding Tokugawa period: inviting foreigners and hiring them as teachers or engineers, while sending Japanese students overseas to learn Western knowledge (Hirosige, 2002, Chapter 1; Bartholomew, 1989, Chapter 3; Morris-Suzuki, 1994, Chapter 3).

Mining and metallurgy, the forerunners of MSE, were typical examples of how Japan introduced Western technology. Although Japan enjoyed considerable indigenous and empirical knowledge of metalworking, as symbolized by traditional Japanese swords (Smith, 1988, pp. 40–62; Notis, 2000), the import of Western mining and metallurgy techniques was an urgent need as massive steel production was demanded by the project of modernizing and industrializing Japan. Thus, departments for educating experts in this field were included in modern educational institutions at their inception: a Department of Mining in Kôgakuryô (the Imperial College of Engineering, later known as the Kôbu Daigakkô) established under Kôbushô (variously translated as follows: Ministry of Public Works; Industry; Engineering) in 1873, and in Kaisei Gakkô (renamed Tokyo University in 1877), growing from the Tokugawa Shogunate's Bansho Shirabesho (Institute for the Study of Barbarian Books) and set up under the jurisdiction of Monbushô (Ministry of Education) in the same year. When these two institutions were merged into Imperial University in 1885, the mission of educating mining experts and metallurgists was assumed by the university's Department of Mining and Metallurgy in the engineering faculty (*Tokyo Daigaku Hyakunen-shi Henshû Iinkai*, 1984; 1987, pp. 5–11).

As a section of the engineering faculty, the department functioned as a strong and then virtually unique institution for educating professional mining experts and metallurgists. The German influence was clear and strong. The first teacher of modern metallurgy (at Tokyo University) was the German metallurgist Curt A. Netto (1847–1909), who had worked at the mine in Kosaka as an engineer. His lectures were documented by his students and published as *Ne-shi Yakingaku* (Netto's Metallurgy) (Netto, 1884), in which he emphasized the scientific method in metallurgy rather than the empirical one (Netto, 1884, p. 16). The book was the earliest textbook of modern metallurgical technology in Japan. While learning the German metallurgy in Japan, many students were also sent to Germany, especially Netto's alma mater Freiberg Bergakademie: 41 students had been enrolled by the outbreak of World War I (Yajima, 2010, pp. 62–63). Among the teachers there was Adolph Ledebur (1837–1906), whose *Handbuch der Eisenhüttenkunde* (Ledebur, 1884) was then one of the most used textbooks of modern ferrous metallurgy in Japan (Notis, 2000, p. 255; Yajima, 2010, Chapter 2; Iida, 1987, pp. 63–74).

While many graduates actively worked as mining engineers at private enterprises, the department gradually came to function also as a research institution, where Japanese metallurgists researched and even formed a research tradition or school. Tawara Kuni-ichi (1872–1958),[1] a graduate of the department, studied under Ledebur and John Oliver Arnold (1858–1930) at the Bergakademie and became a Professor at his alma mater. He bought the microscope from Germany in 1903 and established the new tradition of metallographic study there (Notis, 2000, pp. 255–256; Iida, 1987, pp. 171–174).

As the steel industry developed and the demand for experts rapidly increased, departments of mining and metallurgy were also set up in the newly established technical schools and universities. Among them was the Department of Mining and Metallurgy at Kyoto Imperial University, founded in 1898, where the first three professors had studied at the Bergakademie (*Kyoto Teikoku Daigaku*, 1943, pp. 493–531). Thus, metallurgy cultivated institutional roots at an early stage in the state's modernization strategy, namely the import of Western knowledge, especially from German in this case.

[1] In the main text of this paper, Japanese names are written in the traditional order, where family names come first and given names second.

Some university metallurgy departments established in this era, including Tokyo's and Kyoto's, would become MSE departments a century later (see the fourth section).

Inorganic chemistry, another forerunner of MSE, was also among the industries targeted for modernization, but the state's approach was different. The emphasis here was on fostering indigenous industries as well as recasting their knowledge in scientific form. The main institution that took charge of this mission was Nôshômushô (Ministry of Agriculture and Commerce). In 1882, the ministry set up Chishitsu Chôsajo (Geological Survey Institute) to investigate *fugen* (natural resources) as well as to conduct testing and analysis for the promotion of industry. Its analysis division, which was transferred to the ministry in 1886 and retransferred to the institute in 1890, worked on (raw) materials in industrial use such as glass, lacquer, clay, coal, ore, and cement in response to the requirements of the state or private enterprise, or as their original study (Sugiyama, 2010, pp. 23–27; Kamatani, 1988, Part II).

While university departments of chemistry or applied chemistry fostered experts in the field, the Tokyo Shokkô Gakkô (Tokyo Vocational School), established in 1881 by the Ministry of Education, offered more practical, technical education to cultivate engineers. In 1884, the department dedicated to ceramics and glass was set up, where traditional crafts such as pottery, glass, and lacquerware were forged into modern industries with large-scale production under the leadership of Gottfried Wagener (1831–1892). The school would be now the Tokyo Institute of Technology, and its educational and research scheme for MSE partly originated in this department.

Based on the experience and achievements of the Chishitsu Chôsajo's analysis division, the long-outstanding idea of establishing the testing and research institution dedicated to the industry at large was finally realized as Kôgyô Shikenjô (Industrial Research Laboratory) in 1900. Its mission was to serve in a "technical advisory" role to industry. Starting with two divisions of "general analysis" and "general chemical industry," it gradually expanded its mission with the divisions of "ceramics industry" and "color dyeing industry." While working on the testing and improvement of products important for industry or export, the institute began to function as a modern research institution producing scientific knowledge on materials in the chemical industry (Sugiyama, 2010, pp. 28–29; Kamatani, 1988, Part III, Chapters 1–2).

Testing and improving materials were also essential in the electric industry. In 1891, Denki Shikenjo (Electrotechnical Laboratory, ETL) was set up under the auspices of Teishinshô (Ministry of Communications and Telegraphs) as the institution charged with testing and validating the performance of electrical equipment. It set up the section for *zairyô* (materials) in 1898 to work on the properties of materials used in telegraphs and telephones (*Denki Shikenjo Gojûnenshi*, 1944, p. 10). As it expanded its mission toward research on electric power and its application in 1903, materials with better performance naturally came under its scope (Hirosige, 2002, p. 23).

Toward the construction of a modern industrialized nation, the fields that would be the breeding grounds of MSE were nurtured, earning dedicated institutions in each context. Metallurgy epitomized the large-scale import of Western knowledge, with its dedicated institutions in universities. Inorganic chemistry was among the industries developed through recasting indigenous empirical knowledge and practice into the language of science based on the state's strategy for the promotion of local industries. Materials were of interest also in emerging industries like the electrical industry, earning dedicated sections in the context of testing and later research. All these were linked to industrial contexts, but the type of institutions differed. These research activities on materials would gather further attention when another external pressure came: the serious shortage of materials due to World War I.

The Crisis of Materials Shortage

World War I restructured Japan's economy and industry through the enormous wartime demand for both materials and technologies made from them. When the war ended, Japan developed into a creditor country and an industrialized nation with more than 1 million workers in the industry sector. However, the war also caused serious shortages of certain materials that had been imported from

Europe, Germany especially, which were now scarce, underlining Japan's self-recognition as a natural resource-poor nation. The state took the strategy of protecting and fostering the industries they regarded as key to a self-sufficient economy, leading to a period marked by burgeoning research laboratories both in academia and in private enterprise (Hirosige, 2002, Chapter 3).

The chemical industry was among the industries to be promoted, as it was seriously damaged by wartime shortages and its vulnerability was vividly demonstrated. A wide range of industries that had used the imports of this field, like the medical industry, were damaged, causing significant confusion as well as attracting public attention. In response to this crisis, the state established Kagaku Kôgyô Chôsa Kai (the Chemical Industry Study Commission, CISC) to investigate the situation in the chemical industry and to propose actions to nurture it (Kamatani, 1988, Part IV).

Among the proposals the CISC made was the setting up of an institution dedicated to research on physics and chemistry, the revival of a prewar idea. The result was the establishment of Rikagaku Kenkyûjo (also known as Riken; the Institute of Physical and Chemical Research), which would become one of the institutions with a strength in MSE. Although initiated with strong state support, it was set up as a private institution. Under the strong leadership of the third Director Okôchi Masatoshi (1878–1952), the links between its basic research and industry were promoted and the network of enterprises that industrialized the inventions of Riken, the so-called Riken Konzern, was formed (Bartholomew, 1989, Chapter 7; Morris-Suzuki, 1994, Chapter 5; Kamatani, 1988, Part IV).

The CISC also made proposals on specific issues in the chemical industry, some of which resulted in institutional reactions. Rinji Chisso Kenkyûjo (Provisional Nitrogen Research Institute, later merged into Tokyo Kôgyô Shikenjo in 1928) was one example: its establishment was based on the CISC's suggestion that an institution be dedicated to industrializing the Haber process (Hirosige, 2002, p. 107).

Meanwhile, the CISC attempted to strengthen local research industrial institutions and the network linking them, along the lines of the strategy the Ministry of Agriculture and Commerce had pursued since the Meiji era. Local institutions such as Osaka Kyôgyô Shikenjô (Osaka Industrial Research Laboratory) were established in 1918, and Kôgyô Shikenjo (see the first section) was renamed Tokyo Kôgyô Shikenjo and positioned at the apex of the network of these local institutions.

With several acts for protection, chemical industries were developed rapidly and earned institutions in both state universities and private enterprises. For examples of the former, Rinji Rikagaku Kenkyûjo (Provisional Institute of Physical and Chemical Research) was established temporarily at Tohoku Imperial University to research organic synthetic industrial materials such as Bakelite and nonflammable celluloid in 1915, and temporal Kagaku Tokubetsu Kenkyûjo (later renamed Kagaku Kenkyûjo, Institute for Chemical Research, and made non-temporal in 1926) was set up at Kyoto Imperial University to pursue especially the production of salvarsan used in the medical industry (The Physical Society of Japan, 1978, p. 255).

Steel was another example of wartime need. Since the steel industry could not fully meet a great deal of wartime demands, the state took an approach similar to its approach to the chemical industry: Seitetsugyô Chôsakai (the iron industry study commission) was set up to make proposals; some acts were enacted to protect and promote the industry (Hirosige, 2002, pp. 107–108). Educational institutions in universities (mentioned in the first section) continued to cultivate mining experts and metallurgists. Nihon Tekko Kyôkai (the Iron and Steel Institute of Japan) was formed in 1915 as a professional society.

Metallurgy was then being transformed into physical metallurgy, an emerging field that employed physical techniques like X-ray diffraction to address the relation between structure and properties (Bensaude-Vincent, 2001, pp. 224–225). In Japan, Kinzoku Zairyô Kenkyûjo (Research Institute for Iron, Steel, and Other Metals; RIISOM; but the literal translation of the Japanese name is "Research Institute for Metal Materials") was established at Tohoku Imperial University in 1922 and became one of the centers for this new approach to metals. The institution began as the second division of the Rinji Rikagaku Kenkyûjo at the university in 1915 when the

Sumitomo family offered to subsidize the research on metals of Honda Kôtarô (1870–1954).

Honda worked on metallurgy but came from a context different from the mining and metallurgy department of engineering faculty at Tokyo Imperial University (see the first section): Honda graduated from the physics department at the university, where the tradition of research on magnetism had remained prominent due to the influence of the British teacher's Sir James Alfred Ewing (1855–1935) and Cargill Gilston Knott (1856–1922). Honda entered the study of metals through the work on magnetostriction of his mentor Nagaoka Hantarô (1865–1950), who had studied under these British teachers. Honda then went abroad and became a postdoctoral student of Gustav Tammann (1861–1938) at Göttingen from 1907 to 1911 and developed his interest and knowledge on physical metallurgy.

Honda continued his studies of physical metallurgy after he returned to Japan and was appointed to Tohoku Imperial University's physics department, where he further developed his study on metal in the new institute as its chief researcher. Collaborating with researchers in other institutions, like Tawara at Tokyo (see the first section) as well as engineers in private enterprise, he and his disciples began to make remarkable achievements, including KS steel, which was named after the sponsor Sumitomo Kichizaemon. Based on further subsidy from the family's business, the institute was made non-temporal and renamed Tekkô Kenkyûjo (Iron and Steel Research Institute, ISRI) in 1919 and then RIISOM in 1922. Honda's strong leadership resulted in the formation of the so-called "Honda School" tradition of physical metallurgy, which explored a variety of functional alloys such as new KS magnet steel, Sendust, Superinvar, and Coelinvar (Iida, 1987, Chapter 11; Tôhoku Daigaku Hyakunenshi Hensyû Iinkai, 2006, Chapter 1).

The university set up the Department of Kinzoku Kôgaku (metal engineering) at the engineering faculty in 1924 to further strengthen research in this direction (Tôhoku Daigaku Hyakunen-shi Hensyû Iinkai, 2006, p. 7). Together with RIISOM (which changed its English name to the Institute for Materials Research in 1987), the department would be the origin of the university's now mighty research scheme for MSE (see the fourth section). It is important to attend to the gradual shift of naming from the traditional usage of metallurgy to metal engineering, and from iron and steel to metal materials, which underlines the expanding range of objects as well as applications. Nippon Kinzoku Gakkai (the Japan Institute of Metals) was established as an academic society established in 1937 and headed by Honda.

A Resource-Poor Nation Toward the Total War

While it demanded that the crisis of shortages be dealt with immediately, World War I also gave rise to the fear of total war and forced Japan to face the problem of *shigen* (resources). It was often said then that the power of science enabled Germany to endure more than four years of war instead of the shortage of resources (Hirosige, 2002, pp. 130–131).

Thus, in the gradual process of the mobilization of science during the 1920s, resource shortages remained the central concern or motivation. *Shigen*, a then-recent term (Sato, 2011, Chapter 1), designated the mobilization and use of both human and material resources. In 1927, Shigen Shingikai (Resource Council) was formed as an advisory board on human and material resources and their control and operation. Shigen Kyoku (Resource Bureau; later reconstructed into Kikaku-In, Planning Board, in 1937) was established in the same year as the state's organization dedicated to the same problem and became the central organization for promoting measures for mobilizing science and technology. The list published by the Bureau in 1933 enumerates 40 research themes that were "recognized as urgent from the viewpoint of preparation of control and operation of resource." Among the topics were research on materials such as metals, alloys, insulating materials, electric heating materials, and rubber. In the context of efficient use of resources, research on materials was recognized as urgent, although at this point the large-scale expenditure following this list was not realized (Hirosige, 2002, Sections 4.1–4.3; The Physical Society of Japan, 1978, Section 4.1; The History of Science Society of Japan, 1966, Chapter 1, pp. 50–53).

The Manchurian Incident in 1931 deepened militarism, which magnified military expenditure and resulted in inflation. In this context, a large budget for the promotion of science and technology came into existence. Established in 1932, Nihon Gakujutsu Shinkôkai (Japan Society for the Promotion of Science, known as Gakushin), began subsidizing research funds for the following year, which was about three times the annual amount of research funding that had been granted to scientific research by the Ministry of Education up to then (Hirosige, 2002, Sections 4.4–4.6).

The method of distribution was also characteristic: its emphasis was on the category of Sôgô Kenkyû (comprehensive research) rather than that of Kojin Kenkyû (individual research). This comprehensive research was conducted by the subcommittees designed for each research theme (the special subcommittee for themes that cover two or more fields). The committees gathered members widely from universities, government research institutions, the military, and companies. In conducting research, they further involved a large number of researchers in addition to the members. This style of organizing was a response to the growing demand for rationalization of the research system. Research efficiency was pursued through the mutual exchange in subcommittees beyond the barriers between research institutions, and cooperation between fields was promoted by introducing special subcommittees.

Gakushin's initiative played a decisive role in organizing scientific research in line with industrial and military demands, which in part sought to resolve resource problems and thus required research on materials. The top 10 committees to which the largest amount of research funds were allocated included research themes such as "steel for special purposes" (by a special subcommittee), and "organic synthesis" and "electrical materials" (by subcommittees). Research on materials thus began to be pursued beyond existing institutions and academic fields based on the military and industrial demands of resource problems (Hirosige, 2002, Section 4.6).

After the Sino-Japanese war broke out in 1937, the promotion of science and technology was accelerated in the wake of growing military demands. The budget for scientific research was enormously expanded when the Ministry of Education enlarged its funding in 1939, and the number of research and educational institutions increased rapidly. The research labs in private enterprises grew to 695 in 1943 (Hirosige, 2002, Section 6.1; Kagaku Dôin Kyôkai, 1943, Part IV Chapter 5).

As the tide of the war turned against Japan — in particular, the Pacific War in 1941 — scientists were integrated more fully into the context of wartime mobilization. Some pursued research in the army and navy; others were mobilized via a system similar to Gakushin's comprehensive research: Kenkyû-han (Research Group) and Kenkyû-Tonarigumi (Research Neighborhood Group). The former was organized by Gakujutsu Kenkyû Kaigi (the National Research Council of Japan, NRCJ), which had been established in 1920 and charged with allocating the research budget of the Ministry of Education; the latter was formed by the Cabinet Technology Agency and carried out by its extra-governmental organization. In both systems, budgets were allocated to specific research themes, in which a wide range of researchers were involved. Among the topics addressed were materials such as electrical weapons materials, heat-resistant insulating materials, photoelectric materials, and light metal materials (in the former) and fireproof materials for exterior walls, fluorescent materials, and high permittivity materials (in the latter). Some materials-related scientific achievements were reported in these projects: for example, in the research neighborhood group for high permittivity materials, a collaboration between electrochemists and theoretical physicists resulted in the discovery of barium titanate (Hirosige, 2003, Chapter 8; The History of Science Society of Japan, 1966, Chapter 8; Takagi, 1990; Aoki and Hiramoto, 2003; Aoki, 2006).

While researchers were thus mobilized from all kinds of institutions, including existing organizations mentioned in the first and second sections, and some of them researched materials around specific issues of military demands, in this era, an important field of science for materials emerged — polymer science (Furubayashi, 2004). The field grew out of the textile industry, which was among the major export industries. The production of rayon was at a high level from the end of the Taisho era to the mid-1930s. The departments of applied or industrial chemistry at the Imperial Universities at Tokyo and Kyoto fostered textile researchers and several private companies in the industry sprung up. In 1935, Sen'i Kagaku Kenkyûjo (research institute for textile science) was established at Osaka Imperial University, and Nihon Kagaku Sen'i Kenkyûjo (research institute for chemical textiles) was established at Kyoto Imperial University the following year, both of which were based on donations from the textile industry. The response to the arrival of nylon in October 1938, later known as the "nylon shock," was the establishment of Nihon Gôsei Sen'i Kenkyû Kyôkai (Japanese Research Association of Synthetic Fibers, JASF). Funded by the public and private sectors, JASF set up and supervised several laboratories at universities and organized a network of researchers. It brought together a wide range of researchers in related fields including organic chemistry and rubber research, and its purpose shifted to the production of materials using synthetic polymers in military use as the tide of the war worsened (renamed Kôbunshi Kagaku Kyôkai, Japanese Association of Polymer Chemistry, in 1944), which later resulted in widening the research field to include all polymer materials, not just fibers. The association would later become Kôbunshi Gakkai (The Society of Polymer Science) in 1951. In 1959, the first department for research on polymer was set up at Osaka University, based on the assets of Sen'i Kagaku Kenkyûjo.

While the sciences for each category of materials, including polymer science, developed in close relation to both industrial and military needs, the movement for all kinds of *zairyô* (materials) was born in this era also in the industrial context. Zairyo Kenkyûkai (materials research society) was formed in the fall of 1932 from the community of industrial and academic researchers with applied interests, mainly in Tokyo, to share and cultivate the knowledge of *zairyô* (Zairyô Kenkyûkai, 1934b). It held colloquia on the specific themes and published annually *Zairyô Bunkenshû* (collections of literature on materials), in which all the articles related to materials published in the year were introduced. It enumerated more than 500 works of literature (for example, 556 in the fiscal year 1932), most of which were accompanied by the authors' summaries (Zairyô Kenkyûkai, 1934a). In 1934, it earned the branch in the Kansai region and its membership exceeded 500 (Zairyô Kenkyûkai, 1932; 1934c; 1934d). The society also published *Zairyô Binran* (handbook of materials) for metals and non-metals, in which knowledge of properties, theories, and processing of materials was introduced (Zairyô Kenkyûkai, 1936; 1937), the textbook series of *Saishin Kôgyô Zairyô Sôsho* (new industrial materials series), and since 1943 its own journals, all of which was terminated due to the war (Nagai, 1952).

Society was motivated and planned in the industrial context. The International Association of Testing Materials (IATM) had requested some Japanese researchers to form an association for materials in Japan. Nagai Shôichirô (1894–1970), a chemist at Tokyo Imperial University who was among the founding members of the society, had attended the IATM's conference while he was in Germany from 1930 to 1932. After returning to Japan, he and other researchers received inquiries from IATM but could not respond to them as there was no corresponding organization in Japan, which motivated them to form Zairyô Kenkyûkai (Zairyô Kenkyûkai, 1934c; 1934d; Nagai, 1952). The early version of the constitution of the association explicitly stated that the liaison with IATM was among the society's mission (Zairyô Kenkyûkai, 1932).

This context shaped the composition of the society. At the first general meeting in 1933, the society set up four divisions following IATM: metal materials; non-metal materials; organic materials; and general problems (Zairyô Kenkyûkai, 1934b; 1934c). Subsequent activities and publications of the association followed this framework of adding up the subcategories. This categorization, forged in the industrial context, fit well with the existing structure of institutions, which itself had been formed in the industrial context, and this industrial framework for materials would be taken over through the postwar period.

Materials Sciences or Stacking the Sciences for Materials

As Japan embarked on the pathway from wartime devastation toward becoming a high-growth nation, knowledge of materials was

Table 1. Several major material science and engineering institutions (excluding university departments) in postwar Japan.

1952	Japan Society for Testing Materials (JSTM)
1956	National Research Institute for Metals (NRIM)
1957	Institute for Solid State Physics (ISSP), The University of Tokyo
1958	Research Laboratory of Engineering Materials (RLEM), Tokyo Institute of Technology (Merger of Research Laboratory of Building Materials and Research Laboratory of Ceramic Industry) (Renamed Materials and Structures Laboratory in 1996 and Laboratory for Materials and Structures in 2016)
1963	Society of Materials Science, Japan (JSMS) (Renamed from JSTM)
1963	Materials Science Society of Japan (MSSJ)
1966	National Institute for Research in Inorganic Materials (NIRIM)
1969	Research Institute for Polymers and Textiles (Renamed from Textile Research Institute)
1993	National Institute of Materials and Chemical Research (NIMCR) (Merger of three research institutes including National Chemical Laboratory for Industry (former Tokyo Kogyô Shikenjo) and Research Institute for Polymers and Textiles)
2001	National Institute of Advanced Industrial Science and Technology (Merger of the Agency of Industrial Science and Technology's 15 research institutes, including NIMCR, Osaka Kogyô Gijutsu Kenkyûjo, and ETL, and the Weights and Measures Training Institute)
2001	National Institute for Materials Science (NIMS) (Merger of NRIM and NIRIM)

eagerly needed, again in the industrial context. Nihon Zairyô Shiken Kyôkai (the Japan Society for Testing Materials, JSTM) was established in 1952 to share the knowledge of the materials in industrial use and testing methods. Although it was not a direct successor of the Zairyô Kenkyûkai, Nagai wrote to the society of his delight in seeing the new society as the descendant of his previous endeavor (Nagai, 1952).

Indeed, the motivations and directions of the two were similar. The influence of the American Society for Testing Materials (ASTM), which had evolved from the American Section of IATM and already exerted a strong influence on the industry through standards, was clear: information about ASTM's activities and Japanese standards was constantly introduced in its own journal *Zairyô Shiken*. The journal's "review of overseas literature" section reveals the continuity of the industrial framework of materials: it was categorized into metal materials, electrical materials, civil engineering, and architectural materials, inorganic materials, and organic materials, along with testing methods and machines.

In the meantime, the state pursued the need to establish comprehensive research institutes for materials by developing the aircraft, nuclear power, and electronics industries. The result was the establishment of Kinzoku Zairyô Gijutsu Kenkyûjo (National Research Institute for Metals, NRIM) in 1956, which came from a budget request by the Ministry of International Trade and Industry but was placed under the charge and jurisdiction of the newly established Science and Technology Agency (Tsukahara, 1995, p. 155). The scope of the institute was limited to metals, but the prospectus included a plan to expand it into a large research institute for all kinds of *zairyô* (materials) in the future (Tsukahara, 1995, pp. 55–56). An example of one of the few institutions established during this period that focused on materials in general is Tokyo Institute of Technology's Kôgyô Zairyô Kenkyûjo (Research Laboratory of Engineering Materials, RLEM), which was a merger of Research Laboratory of Building Materials and Research Laboratory of Ceramic Industry. It started with five sections: chemical metallurgy; inorganic melting materials; inorganic forming materials; solid state physics; and basic measurements, which exhibit the industrial framework of materials.

These movements developed mainly in the industrial context and remained somewhat apart from the academic context, where another development of the science of matter proceeded. In wartime, a network of physicists and chemists interested in the properties of matter developed their own scientific framework for research on materials called *Busseiron* (Kono, 2022). The research institution dedicated to this framework, *Bussei Kenkyûjo* (Institute for Solid State Physics), was established as a joint research center at the University of Tokyo in 1957. Research in the institution focused on fields of statistical and condensed matter physics with an academic rather than industrial focus.

In 1963, the two societies with MS in their names appeared: Nihon Zairyô Kagaku Kai (the Materials Science Society of Japan, MSSJ) and Nihon Zairyô Gakkai (the Society of Materials Science, Japan, JSMS), which was renamed from JSTM. When ASTM put "and" in between "T (Testing)" and "M (Materials)" in 1961, JSTM followed immediately in the English name of the journal, but the discussion on naming resulted in rebranding the society into JSMS, and the journal into *Zairyô* (materials) (*Zairyô* 12, 1963, p. 66, 115, 191, pp. 258–259).

Meanwhile in the United States, the framework of Materials Science was emerging. Supported by the Advanced Research Projects Agency (ARPA), a series of university-hosted interdisciplinary laboratories (IDLs) opened and created the research field of materials science (Bensaude-Vincent, 2001; Martin, 2015; Hentschel, 2011; Mody and Martin, 2020). Japanese researchers gradually began to accept this burgeoning framework. Lawrence Van Vlack's *Elements of Materials Science*, the first major textbook for materials scientists and engineers, which appeared in 1959, was translated into Japanese as *Zairyo Kagaku Yôron* (Essences in Materials Science) in 1964 and the Japanese authors' textbooks with *zairyô kagaku* (materials science) in their titles also appeared in the 1960s. Among them were the *Lectures on Materials Science* (1968–1969), consisting of seven volumes for (1) fundamentals of Materials Science, (2) crystal lattice defects, (3) strength of crystals, (4) electrical properties of materials, (5) magnetic properties of materials, (6) thin film and surface phenomena, and (7) exotic materials, and *Materials Science Series* (1969–1972), consisting of four volumes for (1) sound waves and properties of matters, (2) recrystallization, (3) carbon materials, and (4) elasticity and inelasticity. These categorizations exhibit a striking contrast to those in the traditional, industrial context: another series for industrial materials in the same period, *Atarashî Kôgyô Zairyô no Kagaku* (1963–1969), still consisted of subcategories of materials such as electrical materials and organic materials.

The establishment of MSSJ was clearly influenced by this emerging framework of MS in the United States. The society's prospectus explicitly mentioned, while noting that there had been some activity in this direction for more than 10 years, that departments for MS were set up in several major universities and the research in the framework of MS was quite active in the US (*Zairyô Kagaku*, 1964). In its journal *Zairyô Kagaku*, many Japanese advocates of this new field wrote their experiences, analysis, and ideas, referring to the United States' MS. One of the boarding members wrote of his encounter with MS in his journey to the US, from the west coast to the east coast, and documented how enthusiastic scientists were in pursuing the new interdisciplinary framework of MS in universities (Sakui, 1964). Sôma Junkichi (1923–1994), a co-translator of Van Vlack's textbook, reported the educational and institutional situation of MS in the United States and raised the field's four characteristics; (1) MS seeks the understanding of the properties and behavior of materials in relation to their structure and surroundings and researches their interactions; (2) MS is an attitude that relies on — instead of taking a trial and error approach — the theoretically grounded system of knowledge on materials in general for the development of materials and that seeks to apply this knowledge strategically and tactically for the application; (3) MS is essentially at the interface between science and engineering; and (4) MS is a field of cooperation among specialists, and also an integrated or comprehensive discipline for materials in general (Sôma, 1965).

The society pursued these ideals of American MS. The editorial board defined MS as "a comprehensive science that seeks to understand, develop, and utilize, in a unified manner, all the properties of all materials under all conditions," and declared that this would be the future direction of the journal (The Editorial Board of *Zairyô Kagaku*, 1964).

The society continued to dedicate itself to the framework of MS through meetings and publications. Although there were tensions with JSMS as a similar society dealing with materials in general, there were certain differences in various aspects such as approach, interest, and community — all of the first board members (except for the auditors) of MSSJ were university faculty members (*Zairyô Kagaku*, 1964), whereas that of JSMS contained the member from the industry (*Zairyô*, 1963) — and the ideal image of the science of materials. Upon the establishment of MSSJ, the two societies had "reached a complete point" that "understand and respect each other's positions and characteristics, and work hand in hand with each other" through a preliminary meeting (*Zairyô Kagaku*, 1964).

This twofold approach to materials, however, could not find a place in the existing educational and research institutions. One of the few examples of departments for MSE is one set up in 1962 at the newly established School of Engineering Science at Osaka University. Physicist Nagamiya Takeo (1910–2006), who had supervised the translation of *The Structure and Properties of Materials in Four Volumes* edited by John Wulff into Japanese as *Zairyô Kagaku Nyûmon* (An Introduction to Materials Science), played a definitive role in this establishment. Based on the requests from researchers, including university researchers with industrial interests, the National Institute for Research in Inorganic Materials (NIRIM) was set up in 1966, but this meant a setback to the idea of expanding NRIM into a large research institute dedicated to materials at large (Tsukahara, 1995, pp. 55–56).

Nakagawa Yûzô (c. 1899-1992–1992), a Professor at Kyoto University and one of the first board members of JSTM, wrote in 1954 that "the necessary condition for the realization of university departments of *zairyô* (materials) is that the content of this department should not be a combination that of existing departments, but should have an independent subject and perspective. Only when these two are established can it have any meaning" (Nakagawa, 1954). However, the "independent subject and perspective," although they were pursued also in the movement for importing MS in the US like MSSJ, were not found, or even if they were found, they were not compelling enough to make the establishment of new educational or research institutions of materials feasible. A decade later, upon the renewal of the society's journal to *Zairyô*, Nakagawa wrote to recommend, although he did not give up hope of establishing a department of materials, to have a "second specialty" in *zairyô*, and called on members to become active in the society in order to develop this specialty (Nakagawa, 1963). However, these two societies themselves remained small compared to other societies dedicated to each category of materials in the traditional framework.

Thus, Japanese materials achievements that formed the background of the Japanese challenge were realized within the existing institutions and framework — the tradition of research on each kind of materials in the industrial context, as well as the close links with the private industrial enterprises. Japan's overall research and development had been led by the private sector since the 1950s, but in the early 1960s, the establishment of corporate central laboratories proliferated, taking charge of industrialization from basic research and producing achievements, some of them related to materials (Nakayama, 1995).

One of the most important examples in this context is the development and industrialization of carbon fiber, which can be used for Carbon Fiber Reinforced Plastics (CFRP) and in which Japan has a large share. At Osaka Kôgyô Gijutsu Shikenjo (renamed from Osaka Kogyô Shikenjo, see the second section, in 1952), where research on carbon had been conducted since the production of carbon brushes for electricity in 1922, Shindo Akio (1926–2016) developed a carbon fiber composites based on polyacrylonitrile (PAN) and private corporates sought to industrialize it. Toray, established as Tôyô Rêyon in 1926 among the startups in the field at that era (mentioned in the third section), began commercializing carbon fiber in 1971, when the demand for carbon fiber was still low, and opened up the market. Toray's decision to develop carbon fiber was based on the research at its basic research laboratory established in 1962, as well as on the Shindo's patent rights it had obtained in 1970 (Takamatsu, 2002).

These commercial successes obscured the need to institutionalize MS until the prestige of the Japanese economy began to eclipse, as Kuroda Kotaro observes: "at the time when the US was shifting [to MS], Japan was in the midst of the period of high economic growth, and neither universities nor industry recognized the need to transform the departments of metallurgy and metals into those of a new concept of materials science and engineering." Indeed, it was only after or on the occasion of the deregulation of the university act in 1991 or Central Government Reform in 2001 that MSE widely earned dedicated research and educational institutions. Whereas in the US many metallurgy departments were renamed MS departments in the 1960s, the same phenomenon occurred in Japan at the end of the century: for example, MSE (or MS, ME) departments were established in the 1990s at universities in Tokyo, Kyoto, and Tohoku, all of which were based on the departments of metallurgy or metals engineering established in the Meiji and Taisho periods (see the first and second sections). Kuroda notes: "This change of department name was initiated by [the Ministry of Education, or] the Ministry of Education, Culture, Sports, Science, and Technology, first in technical colleges, and then spread to universities" (Kuroda, 2014). In 1993, three research institutes, including Kagaku Gijutsu Shikenjo (National Chemical Laboratory for Industry; former Tokyo Kogyô Shikenjo mentioned in the first and second sections) and Research Institute for Polymers and Textiles, were merged into Busshitsu Kôgaku Kôgyô Gijutsu Shikenjo (National Institute of Materials and Chemical Research, NIMCR). Upon the Central Government Reform in 2001, 16 research institutes, including NIMCR, Osaka Kogyô Gijutsu Kenkyûjo, and ETL (see the first section) into Sangyô Gijutsu Sôgô Kenkyûjo (National Institute of Advanced Industrial Science and Technology, AIST), and NRIM and NIRIM was merged into Busshitsu Zairyô Kenkyû Kikô (National Institute for Materials Science, NIMS): almost half a century had passed since the establishment of the NRIM and its future-plan of the research institute of *zairyô* at large.

Conclusion

Putting Japanese MSE institutions into historical perspective helps us to understand how each industry forged its own science of materials. While building these blocks, the industrial context also gave birth to the perspective to deal with *zairyô* (materials) in industrial use at large, which, quite naturally, resulted simply in stacking the blocks. Meanwhile, MS imported from the United States aimed to become a single framework that encompasses every kind of material while bridging basic and applied sciences, but it was not compelling enough to replace the existing blocks — or their assembly — that had already been implemented in Japan. This, in turn, articulates the uniqueness of how the US framework of MS was forged — as symbolized by the responses to challenges (Bensaude-Vincent, 2001) and the center model (Mody and Choi, 2013), for example, and even recast a light on the fact that the generic entity of materials itself emerged as a social construction (Mody and Martin, 2020; Bensaude-Vincent, 2011). Only after the Japanese challenge was over and Japan's economic prestige waned did serious consideration or commitment to the framework of MSE in Japan begin. The investigation of the subsequent development of this Japanese-style MSE, along with further research on the development in the period this paper has addressed and comparisons with the cases in other countries (Hentschel, 2011; Mody and Martin, 2020), would be a future challenge in exploring the Japanese history of MSE.

Based on Japan's self-recognition as a resource-poor nation, especially after World War I, the effective use of limited material resources has long been of primary concern to Japan, thereby giving a favorable ground for scientific research on materials. The approach or attitude to materials, however, varied in each era and context, as the variation of the terms around materials might articulate: *fugen* (natural resources), *shigen* (resources), *genryô* (raw materials), *zairyô* (materials for specific use), *sozai* (raw materials; materials) and *busshitsu* (substances). The development of these words and concepts (Sato, 2011, Chapter 1) in relation to the sciences for materials should be further investigated, as Bensaude-Vincent did for the case of "materials" (Bensaude-Vincent, 2011).

Acknowledgments

This work was partially supported by JSPS KAKENHI Grant Numbers JP19J22731 and JP23K00270.

References

Aoki, H. (2006). The mobilization of science and the 'Research Groups' under the National Research Council of Japan during World War II [in Japanese]. *Socio-economic History* 72, 331–353.

Aoki, H. and Hiramoto, A. (2003). The mobilization of science and technology and 'the Research Neighborhood Groups': Research collaboration in Japan during World War II [in Japanese]. *Socio-Economic History*, 68, 501–522.

Bartholomew, J. R. (1989). *The formation of science in Japan: Building a research tradition.* (New Haven: Yale University Press).

Bensaude-Vincent, B. (2001). The construction of a discipline: Materials science in the United States. *Historical Studies in the Physical and Biological Sciences* 31, 223–248.

Bensaude-Vincent, B. (2011). Das Konzept von Werkstoffen in historischer Perspektive. *NTM Zeitschrift für Geschichte der Wissenschaften, Technik und Medizin* 19, 107–123.

Denki Shikenjo. (1944). *Denki Shikenjo Gojûnenshi.* (1944). (Tokyo: Denki Nihonsha).

Forester, T. (1988). *The Materials Revolution: Superconductors, New Materials and the Japanese Challenge.* (Oxford: Blackwell).

Furubayashi, Y. (2004). *The establishment of Macromolecular Chemistry in Japan and the Japanese Research Association of Synthetic Fibers During World War II* [in Japanese; Unpublished master's thesis] Tokyo Institute of Technology.

Hentschel, K. (2011). Von der Werkstoffforschung zur. *Materials Science. NTM Zeitschrift für Geschichte der Wissenschaften, Technik und Medizin* 19, 5–40.

Hirosige, T. (2002). *Kagaku no Shakaishi (Jô).* (Tokyo: Iwanami Shoten).

Hirosige, T. (2003). *Kagaku no Shakaishi (Ge).* (Tokyo: Iwanami Shoten).

Iida, K. (1987). *Jinbutsu Tekkô Gijutsu-shi.* (Tokyo: Nikkan Kôgyô Shinbunsha).

Kagaku Dôin Kyôkai. (Ed.). (1943). *Kagaku Gijutsu Nenkan Shôwa 18 nendo.* (Tokyo: Dai Nippon Insatsu)

Zairyô Kagaku. (1964). *Kaihô. Zairyô Kagaku* 1, 54.

Kamatani C. (1988). *Gijutsu taikoku hyakunen no kei: Nihon no kindaika to kokuritsu kenkyu kikan.* (Tokyo: Heibonsha).

Zairyô Kenkyûkai. (Ed.). (1936). *Kôgyô Zairyô Binran (Hi Kinzoku).* (Tokyo: Maruzen).

Zairyô Kenkyûkai. (Ed.). (1937). *Kôgyô Zairyô Binran (Kinzoku).* (Tokyo: Tokiwa Shobô).

Kono, H. (2022). The Emergence of "*Busseiron*" and its Historiographical Position [in Japanese]. *The Journal of the Japanese Society for the History of Chemistry* 49, 157–185.

Kuroda, K. (2014). Issues of materials education in 21st Century [in Japanese]. *Trends in the Sciences* 19, 17–25.

Kyoto Teikoku Daigaku. (Ed.). (1943). *Kyoto Teikoku Daigaku-shi.* (Kyoto: Kyoto Teikoku Daigaku).

Ledebur, A. (1884). *Handbuch der Eisenhüttenkunde* (Felix: Leipzig).

Martin, J. D. (2015). What's in a name change? *Physics in Perspective* 17, 3–32.

Mody, C. C. M. and Choi, H. (2013). From materials science to nanotechnology: Interdisciplinary center programs at Cornell university, 1960–2000. *Historical Studies in the Natural Sciences* 43, 121–161.

Mody, C. C. M. and Martin, J. D. (2020). Materials science. In *Encyclopedia of the History of Science.* Retrieved June 11, 2021, from https://doi.org/10.34758/6afy-w006.

Morris-Suzuki, T. (1994). *The Technological Transformation of Japan: From the Seventeenth to the Twenty-First Century.* (Cambridge: Cambridge University Press).

Nagai, S. (1952). Nihon Zairyô Shiken Kyôkai no Hakkai ni saishi Zairyô no Seizô Kagaku Gijutsu no Jûyôsei wo Kyôchô shitai. *Zairyô Shiken* 1, 4.

Nakagawa, Y. (1954). Zairyô Gakka to iu Senmon ga Kanô ka. *Zairyô Shiken* 3, 71.

Nakagawa, Y. (1963). A Specialty named Materials is also Everyman's Course [in Japanese]. *Zairyô* 12, 1.

Nakayama, S. (1995). The rise of corporate R&D activities: The central laboratory boom [in Japanese]. In S. Nakayama, K. Goto, and H. Yoshioka (Eds.), *Tsûshi Nihon no Kagaku Gijutsu* (Vol. 3) (Tokyo: Gakuyô Shobô), pp. 44–50.

Netto, C. (1884). *Ne-shi Yakingaku Jô-satsu.* (Tokyo: Monbushô Henshûkyoku).

The History of Science Society of Japan. (Ed.) (1966). *Nihon Kagaku Gijutsushi Taikei* (Vol. 4) (Tokyo: Dai'ichi Hôki Shuppan).

The Physical Society of Japan. (Ed.). (1978). *Nihon no Butsurigakushi* (vol. 1). (Tokai University Press).

Zairyô Kagaku. (1964). Nihon Zairyô Gakkai Setsuritsu Shuisho. *Zairyô Kagaku* 1, 1–2.

Zairyô. (1963). Nihon Zairyô Gakkai Teikan. *Zairyô* 12, 256–8.

Zairyô Kagaku. (1964). Nihon Zairyô Kagaku Kai Shôwa 38, 39 nendo Yakuin. *Zairyô Kagaku* 1, 59.

Notis, M. R. (2000). The history of the metallographic study of the Japanese Sword. *Materials Characterization* 45, 253–258.

Sakui, S. (1964). Zairyô Kagaku tono deai. *Zairyô Kagaku* 1, 4–5.

Sato J. (2011). *'Motazaru Kuni' no Shigenron.* (Tokyo: The University of Tokyo Press).

Smith, C. S. (1988). *A history of metallography: The development of ideas on the structure of metals before 1890.* (Cambridge, Mass.: MIT Press).

Sôma, J. (1965). Amerika Zairyô Kagaku no Inshô: Kyôiku wo shu to shite. *Zairyô Kagaku* 2, 1–8.

Sugiyama S. (2010). *Nihon no Kindai Kagakushi.* (Tokyo: Asakura Shoten).

Takagi, Y. (1990). Kyôyûdentai to shiteno Chitabari. In Murata Manufacturing (ed.), *Kyoui no Chitabari.* (Tokyo: Maruzen), pp. 41–48.

Takamatsu T. (2002). PAN kei Tanso Sen'i no Kaihatsu. In T. Nakaoka (Ed.), *Sengo Nihon no Gijutsu Keisei: Mohô ka Sôzô ka* (Tokyo: Nihon Keizai Hyôronsha), pp. 59–93.

The Editorial Board of *Zairyô Kagaku.* (1964). Henshû no Kihon Hôshin ni tsuite. *Zairyô Kagaku* 1, 61.

Tôhoku Daigaku Hyakunen-shi Hensyû Iinkai. (Ed.). (2006). *Tôhoku Daigaku Hyakunen-shi: Bukyoku-shi* (vol. 4). (Sendai: Tôhoku Daigaku Kenkyû Kyôiku Shinkô Zaidan).

Tokyo Daigaku Hyakunen-shi Henshû Iinkai. (Ed.). (1987). *Tokyo Daigaku Hyakunen-shi: Bukyoku-shi* (Vol. 3). (Tokyo: University of Tokyo Press).

Tokyo Daigaku Hyakunen-shi Henshû Iinkai. (Ed.). (1984). *Tokyo Daigaku Hyakunen-shi: Tsû-shi* (Vol. 1). (Tokyo: University of Tokyo Press).

Tsukahara, S. (1995). The expansion of government research activity into the forefront of research [in Japanese]. In S. Nakayama, K. Goto, and H. Yoshioka (Eds.), *Tsûshi Nihon no Kagaku Gijutsu* (Vol. 3) (Tokyo: Gakuyô Shobô), pp. 51–60.

Tsukahara, S. (1995). The strengthening of government research institutes [in Japanese]. In S. Nakayama, K. Goto, and H. Yoshioka (Eds.), *Tsûshi Nihon no Kagaku Gijutsu* (Vol. 2) (Tokyo: Gakuyô Shobô), pp. 147–157.

Yajima, T. (2010). *Kan'ei seitetsujo kara Tôhoku Teikoku Daigaku Kinzoku Kôgakka e.* (Sendai: Tohoku University Press).

Zairyô Kenkyûkai (Ed.). (1934a). *Zairyô Bunkensyû: Shôwa 7 nendo.* (Tokyo: Tokiwa Shobô).

Zairyô Kenkyûkai. (1934b). Zairyô Kenkyûkai Dôsei. In Zairyô Kenkyûkai (Ed.), *Zairyô Bunken Shû: Showa 7 nendo* (Tokyo: Tokiwa Shobô), p. 60.
Zairyô Kenkyûkai. (1934c). Zairyô Kenkyûkai Kiji. *Engineer* 13, 408–11.
Zairyô Kenkyûkai. (1934d). Zairyô Kenkyûkai no Hatten. *Kikai* 7, 17.
Zairyô Kenkyûkai. (1932). Zairyô Kenkyûkai. *Semento Kai Ihô* 295, 29–30.

Chapter 5.2

The Rise of Materials Science in China: Historical Aspects of Educational and Research Institutions

Wei Qian

In the Foreword of the book *The Coming of Materials Science,* Robert W. Cahn told a Chinese story: when learning from the Soviet Union in the 1950s, some Chinese universities set up majors in "metal physics," which was actually a misunderstanding of the word *metalovedenie* (similar to metallography or physical metallurgy). However, due to the lack of training required by engineers in metal heat treatment, the majors had to stop teaching and only a few of them changed to materials science in the 1980s.[1] It is interesting to review the history of the discipline of materials science and engineering (MSE) in China by tracing the role played by various institutions. Although a single article does not suffice to meet that goal, it is possible to explore some microcosms of China's scientific development under the multiple transforms between the Western general education model and the Soviet specialized education model.

This chapter is divided into the following five sections in chronological order. Before the founding of the People's Republic of China in 1949, young students returning from the West established a few institutions to start scientific research on metallurgy and ceramics (see the first section). At the beginning of the Red China, the

[1] Cahn (2002).

institutions were reorganized learning from the Soviet Union, attached importance to the study of applied technology, and laid the industrial foundation of the new state (see the second section). From the beginning of the Reform and Open in 1978, the discipline of MSE was reorganized according to the model of the United States (see the third section). In the new century, in line with the development trend of science and technology, the institutions were appropriately adjusted and gradually formed a national strategic research force (see the fourth section). Finally, and in conclusion, some prospects will be put forward that point to a bright future for MSE in China in the new round of adjustment of the institutions (see the fifth section).

From Traditional to Scientific Material Technology (Before 1949)

An Old Country With Traditional Ceramics and Iron Technology

China is one of the great ancient civilizations with many excellent traditional technologies. The earliest pottery, unearthed from a cave in Jiangxi province, dated to around 20,000 years ago, rewriting the history of humans using natural materials to produce artificial materials.[2] In the 13th century BC, proto-porcelain appeared, and then various types of porcelain products were developed. Since ancient times, preferring the clean and opaque art of jade, Chinese people led a different porcelain development path from western glassware. After the 14th century, Jingdezhen became the center of porcelain manufacturing. As early as the 6th century BC, liquid iron smelting technology was invented in China and was developed into a variety of steel-making technologies, forming a unique traditional iron and steel technology system. In the 14th–16th centuries, Zunhua Iron Workshop became the largest iron and steel manufacturing base in the Ming Dynasty and even in the world. Fu Jun, a chief inspector of Zunhua Iron Workshop, wrote *Tieye Zhi* (*Iron Metallurgy Annals),* which summarized the technology, operation, and organization at

---

[2] Wu *et al.* (2012).

that time. *Tiangong Kaiwu*, written by Song Yingxing in 1637, records many traditional processes, such as metallurgy, casting, and ceramics production; it can be called the Encyclopedia of Chinese Crafts in the 17th century. In the mid-18th century, Zhu Yan's *Tao Shuo*, focusing on Jingdezhen porcelain, described in detail the characteristics of porcelain kilns and porcelain in various periods of China and the development history of ceramic technology.

However, China's traditional technology was mainly based on the craftsmen's experience and had not summarized more advanced physical and chemical knowledge from practice, resulting in a rapid lag behind a series of material process improvements and development after the European Industrial Revolution. When Josiah Wedgwood (1730–1795) discovered the secrets of porcelain and Henry Bessemer (1813–1898) invented the converter steelmaking in England, China's advantages in traditional materials technology became non-existent. The dumping of modern industrial production of foreign iron and porcelain into the Chinese market has brought about a huge impact, forcing Chinese craftsmen to turn to Western modern science and technology. The new metallurgical and ceramic technology and the scientific knowledge were just like the surging river into this old country with good material manufacturing traditions.

A typical example is the Jiangnan Manufacturing Bureau, which was established in 1865. Many scientific translations were published. Among them, *Yejin Lu* (*Metallurgy Record*) translated by John Fryer (1839–1928) is a special book focusing on metallurgy and casting, which played a great role in the development of China's metallurgy industry at the end of the 19th century.

Metallurgy and Ceramics Education in Early Chinese Universities

The establishment of materials science in modern Chinese universities can be traced back to the mining and metallurgy discipline of Peiyang University, the earliest university in China. When Peiyang University was founded in Tianjin in 1895, the Department of Mining and Metallurgy was set up, creating a precedent for modern materials education in China. Charles D. Tenney (1857–1930) served as the Provost of Peiyang University from 1895 to 1906.

In 1898, he established the teaching system of the University by imitating Yale University and hired geologists, mineralogists, and metallurgists such as Noah F. Drake, George I. Adams, Walter H. Adams, Franklin L. Barker, and Ronald A. Mills. Among them, Edwin A. Sperry (1857–1935) stayed the longest and had the greatest influence. He graduated from the Colorado School of Mining and was a Professor of Metallurgy at Peiyang University from 1911 to 1935. He was buried in Tianjin in 1935. In the same year, Peiyang University set up an engineering research institute with a mining and metallurgy branch, becoming the earliest MSE institute in Chinese universities.

Most of the students trained by Peiyang University in the early years had been abroad to further study. After returning, they became the backbone of national industrial construction and scientific research. One of the important figures is Chungyu Wang (1879–1958) who graduated from Peiyang University majoring in mining and metallurgy in 1899. From 1901 to 1904, he studied at the University of California at Berkeley, Columbia University, and obtained a master's degree. After returning to China, he established China's first antimony refinery using modern methods for Huachang Company and served as the Chief Engineer. He obtained several patents on antimony and tungsten metallurgy in the United States and had written *Antimony*, *Tungsten*, etc. He is a pioneer of modern antimony smelting technology in China and has made important contributions to the development of China's modern steel and nonferrous metal industry. His English monograph *Antimony*, which was published by Charles Griffin Publishing Co, Ltd. in 1909,[3] is the first monograph on antimony in the world. and comprehensively discusses the history, nature, geology, metallurgy, analysis, application, and economic evaluation of antimony. *Tungsten*, an English monograph jointly written with Dr. Guoqin Li (1887–1961), Manager and industrialist of Huachang Company, is the 94th chemical monograph series organized by the American Chemical Society in 1943.[4]

[3] Wang (1909).

[4] Wang and Li (1943).

Another famous figure is Shoukun Wei (1907–2014), who graduated from Peiyang University in 1929 with a bachelor's degree in mining and metallurgy. After graduation, he began to study in Germany. In 1935, he obtained a doctor's degree in engineering from the Department of Chemistry at Technische Universität Dresden, Germany. Then he returned to China and became a Professor in the Department of Mining and Metallurgy at Peiyang University. In addition to teaching, he researched the iron and alloy steel, aluminum, and molybdenum smelting methods, and refractory production methods based on the mineral resources in China.

Additionally, Tangshan Institute of Technology (1905), Peking University (1910), Shanxi University (1910), Hunan University (1911), Northeast University (1912), Wuhan University (1913), Chongqing University (1935), and other universities successively set up the departments of mining and metallurgy. At the beginning of the 20th century, Jiaotong University, and Tsinghua University established material laboratories, setting a precedent for modern material research in China. Tsinghua University began to prepare the Institute of Metals in 1936. It was formally established in Kunming in the summer of 1938. Youxun Wu (1897–1977), the Director of the Department of Physics of the National Southwest Associated University, also served as the Director of the Institute, but it was soon suspended.

Facing the backwardness of China's ceramic technology, many scholars with lofty ideals in the late Qing Dynasty thought of starting with education and introducing advanced western ceramic science. Hunan Ceramics School was established in 1906, and China Ceramics School was founded (initially located in Poyang, Jiangxi Province, and later moved to Jingdezhen) in 1910. With the efforts of a group of modern ceramic science communicators represented by Hao Zhang (1876–1954), the ceramic school education based on Jingdezhen became an important position for the dissemination of modern ceramic science and technology, and truly transformed the traditional ceramic technology into science. Fan Wang (1905–1990), who graduated from the Tokyo Institute of Technology (TIT) in Japan, returned to China and served as the president of Jiangxi

Ceramics School. In 1936, he translated into Chinese the book *Ceramic Industry* by Keiji Kondo, Director of the kiln department of TIT, which was published by the Commercial Press. It was the earliest ceramic translation in China. He is the first person in China to do systematic scientific research on electric porcelain, chemical porcelain, and color glaze, and pioneered the localization of electric porcelain and chemical porcelain.[5]

Meanwhile, there were only a few specialized vocational schools for ceramic education, such as Jingdezhen in Jiangxi, Yixing, and Nanjing in Jiangsu, Jiangjin in Sichuan, and Gaopo in Guangdong The higher education was almost a blank except for the kiln group in the Department of Chemistry at Yenching University. The lack of professional talents, especially senior talents, was also an important factor affecting the development of this major. In 1946, after Tsinghua University returned to Beijing from Kunming, it added the Department of Chemical Engineering to the School of Engineering, expanding the education of nonmetallic materials. In 1947, Zhongwei Wu (1918–2000) established the first concrete research laboratory at National Central University (later named Nanjing University).

Before 1949, China's MSE education was mainly to cultivate talents in mining and metallurgy. There were only a small number of professional ceramic talents and few in polymer materials. The outstanding feature of MSE education in this period is that there is no division of majors, and the teaching includes mining, mineral processing, metallurgy, materials, etc. It is a wide-field training mode. Many senior experts in the material industry came out of the school under this wide caliber discipline background, and have made significant contributions to the development of China's material science, technology, education, and industry in the long-term practice.

Early Material Research Institutions in China

In February 1927, as the first MSE society, the Chinese Institute of Mining and Metallurgy was established in Beijing. The purpose of

[5]Zheng *et al.* (2013).

the Institute is to contact comrades, study academia, and develop China's mining and metallurgical industry. Chungyu Wang served as the Chair of the academic committee of the section of iron and steel industry. The Institute founded the journal of *Kuangye* (*Mining and Metallurgy*), which became an academic field for MSE colleagues.[6]

The study of modern metallurgy in China can be divided into three stages from 1927. At first, because geological research paid attention to the analysis of various minerals, each mining and smelting plant wanted to determine its economic value and also engaged in ore analysis, which was the preparation stage of metallurgical research. Later, scholars paid attention to the problem of how to extract metals from ores. In 1928, the Yellow Sea Chemical Industry Research Society began to study how to extract aluminum oxide from Boshan Aluminum Mine in Shandong Province, and Peking University went on to study how to extract aluminum, which was the beginning of metallurgical research. However, all experiments were carried out in the laboratory, less attention was paid to industrial technical problems, and no contact was made with the actual factory. In the early days of the Sino-Japanese War (1937–1945), there was a lack of metal materials. In order to solve local self-sufficiency, many practical metallurgical problems arose at the historic moment. The research goal was to take the laboratory as the starting point and gradually promote its application in factories, which can be described as the third stage.[7]

In 1928, the Academia Sinica was established in Nanjing, with the Institute of Engineering under it. Ren Zhou (1892–1973) who graduated from Cornell University was appointed as the Director. He advocated that the purpose of engineering research was to "focus on the improvement of old industries and the creation of new industries in China". At the beginning of the establishment of the Institute, Ren Zhou set up two laboratories for metallurgy and ceramics. When preparing to build the National Iron and Steel Laboratory in Shanghai, he purchased three-phase electric

---

[6] Lei *et al.* (2018).

[7] Wei (1947).

arc furnaces from the United States and introduced electric furnace smelting technology to solve the smelting difficulties of high-quality steel. Thanks to the efforts of Ren Zhou, Xingjian Zhou (1895–1986), and Enyu Yan (1886–?), various steel products such as stainless steel, manganese steel and high-speed steel had been successfully trial produced, and a variety of equipment components were generated, such as train crankshafts and couplers, which opened the way for China's steel research. During the Sino-Japanese War (1937–1945), Ren Zhou was responsible for establishing the China Electric Steel Works in Kunming, and served as General Manager and Chief Engineer to produce wire ropes and other alloy steels for brine extraction from flowing wells.[8]

The National Ceramic Laboratory was built in Nanjing in 1928 when special industrial ceramics and refractory materials were developed. It was in the exciting trend of the times of "saving the country through industry" that Zhou Ren used the professional knowledge he learned in the United States to transform and improve the ceramic industry. He led to the establishment of the National Ceramic Laboratory, which is the first national research institution named after "ceramics" in modern China. The Laboratory had the following missions: (1) to research and imitate ancient porcelain, including excavating famous ancient kilns, investigating and testing porcelain clay, and studying the formula of body, mud, and glaze; (2) to revitalize household porcelain, such as research on porcelain printing paper, new molding mechanism; (3) to manufacture industrial porcelain, mainly electric porcelain, chemical porcelain, and refractory materials; (4) to develop the machinery and equipment needed for ceramic production. Ren Zhou believed that chemistry and electricity were the basis for the prosperity of science, while there were many places where chemical and electrical instruments needed ceramics, and these instruments were all imported at that time, so it was incumbent on him to engage in the development of electric and chemical porcelain. The research results of the

[8] Zhou (1987).

ceramic proving ground were quickly applied and had a wide impact on society.[9]

In order to utilize the mining and smelting resources during the Sino-Japanese War (1937–1945), the Ministry of Economic Affairs merged the mining section with the metallurgical section of the Raw Resources Committee in 1938, established the National Bureau of Mining and Metallurgical Research, appointed Yulun Zhu (1901–1989), the Director of the former mining section, as the Director of the new institute, and transferred Politicu Chinu (1902–1971), the Director of the former metallurgical section, to Chongqing to establish an electrochemical smelting plant. The National Bureau of Mining and Metallurgical Research is the first national institution to specialize in mining and metallurgical academic research in China. Five research laboratories were set up, respectively, for mining, mineral processing, metallurgy, coke making, and materials testing. Later, antimony, light metal, refractory, and metallography research laboratories were successively added. According to the urgent need of the War, the scientific research of it paid special attention to the following four aspects: coal and iron industry as the industrial basis; as the main materials to enrich the treasury fund and exchange foreign exchange, such as gold, tungsten, antimony, and tin; relevant urgently needed metals for the military industry and civil use, such as copper, zinc and aluminum; and research on technology policy.[10]

In 1938, The Ministry of Defense established the Material Test Department at the Bureau of Military Industry, which became one of the few institutions in China that used physical metallurgy to develop materials. In 1939, Zhihong Zhou (1897–1991), who obtained his doctor's degree in physical metallurgy from Harvard University, and Yuchi Qiu (1907–1962), who obtained his doctor's degree from RWTH Aachen in Germany, set out to research the crucible steelmaking, cooperate with the 28th Factory of the Bureau of Military Industry, to develop and produce alloy steels

---

[9] Zheng (2012).

[10] Lei *et al.* (2016).

and ferroalloys, such as high-speed steels and hubbing steels, which were urgently needed for military and civil use, and made great contributions to national defense. Since then, Chinese scientists began to pay attention to physical metallurgy. They used scientific methods to study various alloy steels, heat treatment, mechanical treatment, performance test, and microstructure. For example, the metallographic study of magnetic steel showed that the dissolution of carbide was different, resulting in different microstructures and properties. In 1948, Shoukun Wei wrote about physical metallurgy: "For more than a decade, a considerable number of domestic students in chemistry, chemical engineering, and even in physics, mechanical engineering, and aviation engineering switched to metallurgy abroad. Because of the rapid development of physical metallurgy abroad in the past decade, it has almost separated from metallurgical engineering and become a pure science, and many scholars in foreign countries also study physical metallurgy."[11]

In 1930, the National Bureau of Industrial Research of the Ministry of Economic Affairs was established in Nanjing, headed by Shanxiang Xu (1882–1969), to test industrial raw materials, improve manufacturing methods and identify industrial products. The laboratory initially set up two groups: the chemical group was responsible for the research on the analysis and application methods of industrial raw materials, and the mechanical group was on the test and improvement of mechanical materials and engineering materials. During the Sino-Japanese War (1037–1945), the Institute expanded its business to chemical and light industry production, involving wood, fiber materials, textile printing and dyeing, leather tanning, etc., and became an important organic material research institution in China.[12]

In 1934, Ruwei Shi (1901–1983), who got his doctoral degree at Yale University and returned from the US, started to establish China's first magnetic research laboratory at the Institute of Physics of Academia Sinica. He carried out research work on metal

---

[11] Wei (1947).

[12] Zhang (1990).

magnetism, such as preparing magnetic alloy samples, conducting phase and structure analysis, and observing magnetic domain pattern and magnetic measurement. Before 1936, many magnetic researches were completed or carried out, such as the magnetic anisotropy of nickel cobalt single crystal, the magnetism of isotropic ferromagnetic materials, and the domain powder pattern of the magnetite crystal and polycrystalline ferromagnet.[13]

In addition, the crystallographic research in the Institute of Physics of the National Academy of Peiping was also the fundamental research of materials. In 1936, Xueshan Lu (1905–1981) returned from the United Kingdom to continue his work of researching crystal structure with X-rays at Manchester University. During the Sino-Japanese War (1937–1945), the Institute of Physics successively provided more than 1000 high-quality crystal oscillators of various thicknesses to the military telecommunications and broadcasting departments and also solved several badly needed crystal oscillators for the US military at Kunming and the British Royal Air Force in India.[14] Others, such as the Yellow Sea Chemical Industry Research Society, the Institute of Chemistry of Academia Sinica, and other chemical research institutions, have participated in some material research work.

In 1945, the Chinese Ceramic Society was established in Chongqing, which was the beginning of academic discussion and exchange organized by Chinese scientists and technicians on ceramics, glass, cement, and refractory. The first Chairman was Qifang Lai (1899–1985).[15]

Since China's industrial foundation was very weak before 1949, a group of returned students from Europe, America, and Japan, with the ambition of saving the country through industry, gradually replaced foreign engineers and teachers, established schools and research institutes, and applied scientific means to develop metallurgy and ceramic technology, aiming to improve the country's basic

---

[13] Pan and Wu (1984).

[14] Hu (1997).

[15] Lai (1985).

materials industry. However, due to the frequent wars and financial problems, the effect was very little. Their work was mainly based on material testing and development according to actual needs. They utilized local resources to develop products. As a whole, they were still at a relatively preliminary stage. There were few original scientific research achievements with world influence at that time.

Foundation of China's Materials Science (1949–1977)

Overview

On October 1, 1949, the People's Republic of China was founded, while everything was waiting to be revived. In the 1950s, on the premise of learning from the Soviet Union in an all-round way, China's educational and research institutions were reorganized following the Soviet model. This practice has completely broken the original wide caliber talent training method of learning from Europe and the United States and turned to the path of Soviet professional training. A number of new professional engineering schools and research institutions have been established.

In 1956, the Long-Term Plan for the Development of Science and Technology from 1956 to 1967 (commonly known as the "Twelve Year Plan") was drawn up, proposed 57 important scientific and technological tasks from 13 fields, and grasped 12 critical research priorities. These included establishing an alloy system and seeking new metallurgical processes in combination with national resources, and developing heavy organic synthesis by comprehensive utilization of fuels. At the same time, "four emergency measures," such as computers, radio electronics, semiconductors, automation, were planned to develop. Those would play a key role in the development of modern science and technology, so that they can change the status quo in a short time and approach the international level.

During this period, the research on MSE was carried out rapidly around the development of heavy industries such as metallurgy, building materials, petroleum, and chemical industry. At the same time, close attention was also paid to the research and production of materials related to the military industry, and made outstanding achievements in the field of materials used in atomic bombs, missiles, artificial satellites, semiconductors, etc. Scientific research institutions were reorganized as a new pattern of "Five Routes Armies" composed of the Chinese Academy of Sciences, colleges, and universities, industrial institutions, military institutions, and local institutions.

MSE in Colleges and Universities

After the founding of the People's Republic of China, due to the need for economic construction, higher education inherited the teaching system of the Soviet Union, and changed from the wide-field training to the narrow specialty training. In 1951, the Department of Metallurgy of Dalian University of Technology was transferred to the Northeast Institute of Technology (now Northeastern University) to strengthen the teaching scale of its metal materials discipline. Harbin Institute of Technology was a typical engineering college learning from the Soviet Union. In 1952, it first established the majors of metal materials and heat treatment, casting, forging and welding in China, and later established the Department of Metal Materials Technology. In April 1952, the Beijing University of Iron and Steel Technology (now the University of Science and Technology Beijing) was founded on the basis of the mining and metallurgy departments of six universities, including Peiyang University (now Tianjin University) and Tsinghua University. After the establishment, it added materials majors such as metallography, steel rolling, metal material heat treatment, corrosion and protection. In the same year, the Departments of Mining and Metallurgy of Wuhan University, Hunan University, and Guangxi University was adjusted to the Central South Institute of Mining and Metallurgy (now Central South University) in Changsha, and materials majors such as nonferrous metal metallurgy and heat treatment, nonferrous metal and its alloy pressure processing, powder metallurgy physics and chemistry were successively added. The newly established Beijing University of Aeronautics and Astronautics (now Beihang University) added polymer materials including composite materials and other majors after the adjustment of the departments.[16]

In order to match the basic theoretical needs of engineer training, Beijing University of Iron and Steel Technology established the first metal physics major in China in 1956 under the leadership of Tsun Ko (1917–2017), who had just returned from the University of Birmingham in the UK. Having ignored the opposition of Soviet experts during the teaching process, he introduced from the West the new theory of physical metallurgy, strengthened the guidance of theoretical basis, opened experimental courses such as X-ray and electronic microscope, increased the proportion of physics courses, and formed a characteristic major combining science and engineering.[17] More universities such as Peking University also set up metal physics majors in the departments of physics subsequently, which have become the basic force of later materials science.

The discipline of ceramic materials in colleges and universities was established only after the adjustment of national colleges and departments in 1952. At that time, the major of silicate engineering was first established in the chemical engineering departments of Tianjin University, East China Institute of Chemical Technology (now East China University of Science and Technology), South China University of Science and Technology, and Nanjing Institute of Technology (now Southeast University). Subsequently, the education ceramic materials went through a period of steady development, and specialized colleges such as Beijing Institute of Building Materials (later moved to Wuhan, now Wuhan University of Technology) were established. In the 1950s, they focused on process and engineering. In addition to general basic courses and chemical engineering, they also offered the following courses: silicate mineralogy and petrography, silicate physical chemistry, silicate technology, silicate industrial thermal equipment and kilns, and silicate industrial machinery and equipment. Around 1960, advised by

---

[16] Zuo (2009).

[17] Han and Shi (2012).

Dongsheng Yan (1918–2016) who returned from UIUC and became the Deputy Director of the Shanghai Institute of Ceramics in CAS, Shanghai University of Science and Technology (now Shanghai University) first began to adjust the teaching plan of ceramic engineering to the direction of material science, reducing engineering courses and adding basic courses of modern chemistry and physics, such as introduction to solid physics, structural chemistry, physical properties of ceramic materials, modern experimental technology, and other courses. At the same time, many other universities still maintained their original training objectives, and developed their professional direction to a narrow technological direction, such as cement engineering, glass engineering, ceramic engineering, refractory materials, optical materials, daily glass, daily ceramics, concrete products, etc.[18]

In 1956, the first semiconductor major in China was prepared at Peking University, which was composed of the teachers from Peking University, Fudan University, Xiamen University, Jilin University, and Nanjing University. Kun Huang (1919–2005) and Xide Xie (1921–2020) became the directors, who edited *Semiconductor Physics*, the first professional textbook in China's semiconductor field. Then, a few universities have established semiconductor teaching and research sections and increased the enrollment scale. For example, the semiconductor section of Tsinghua University was established at the end of 1957. At the same time, the country has set up some new engineering colleges, such as Xidian University and Chengdu Institute of Telecommunication Engineering (now China University of Electronic Science and Technology), both of which have established semiconductor-related majors since then.[19]

The establishment of polymer majors in China started since 1953, which was formed and separated from the majors of chemistry and chemical engineering. A polymer chemistry section was established in the Department of Chemistry at Peking University, while the plastic engineering section was established in the Department of

[18] Ding and Yan (1985).
[19] Wu and Qian (2021).

Chemical Engineering at Chengdu Institute of Technology (now Sichuan University). The earliest Department of Polymer Chemistry and Physics was founded in the University of Science and Technology of China. Beijing University of Chemical Technology was the first university in China to set up a comprehensive department of polymer science. Since the 1950s, polymer majors in Chinese universities successively included polymer chemistry, polymer physics, polymer chemical engineering, plastic engineering, rubber and plastic engineering, polymer materials, composite materials, synthetic rubber, chemical fiber and others.[20]

According to the training mode and teaching system of the Soviet Union, Chinese MSE talents were divided into more than ten majors from metallurgy, machinery, chemical engineering, physics, chemistry, etc. Metal materials major alone was subdivided into metallurgical physical chemistry, metal materials and heat treatment, casting, welding, pressure processing, metal corrosion and protection, powder metallurgy, superalloys, precision alloys, etc. Because of absorbing the scientific and technological knowledge of the Soviet Union at the international level at that time, most of the students were able to work in the corresponding industry and quickly adapt to their posts. The education has made great contributions to China's economic construction and development. During that period, the level of MSE education in China has been greatly improved and the scale has been greatly developed. However, this kind of education mode has brought about a very detailed and unreasonable division of disciplines and majors, and the shortcomings of narrow knowledge of students have gradually been exposed.

While science and technology are developing rapidly, the first level discipline field of MSE has been formed, and the important period when European and American countries have carried out the MSE education reform in succession, China's MSE education has been at a standstill due to the impact of Cultural Revolution (1966–1976), which has widened the distance from MSE education in the West.

---

[20] Zhao and Gu (2008).

Chinese Academy of Sciences

In 1949, the Chinese Academy of Sciences (CAS) was established as a national scientific research center, mainly inheriting the personnel and facilities of the former Academia Sinica (AS) in Nanjing and the National Academy of Peiping (NAP) in Beijing. In 1950, the first batch of institutes of CAS were established. The Institute of Applied Physics in Beijing was a combination of the metal and optical parts of the Institute of Physics of AS and NAP. Its main research fields were optics, spectroscopy, crystallization, and metals. Its Director was Jici Yan (1901–1996), and Xueshan Lu (1905–1981) as the Deputy Director. The Engineering Laboratory in Shanghai was rebuilt by the Institute of Engineering of AS, with the main work focused on metallurgy, glass, and ceramics. Its Director is Ren Zhou and Xingjian Zhou as the Deputy Director. In 1953, the Laboratory was renamed as Shanghai Institute of Metallurgy and Ceramics, and later was divided into two: Shanghai Institute of Metallurgy and Shanghai Institute of Ceramics.

In 1951, the Engineering Laboratory successfully developed spheroidal graphite cast iron. After many experiments, Ren Zhou and his colleagues finally found a suitable graphite nodulizing additive, mastered the production and heat treatment process of nodular cast iron, whose performance reached the international level at that time, and promoted its application to automobile and diesel engine crankshafts. This work opened the way for the promotion of casting instead of the forging process. In 1953, in order to restore the Chinese ancient porcelain craft, Ren Zhou began to tackle the problem of ancient porcelain. He used modern scientific means to conduct chemical analysis and scientific identification of historical masterpieces, investigated and analyzed raw materials in various places. After repeated tests, batches of high-grade porcelain with traditional characteristics were successfully trial produced, representing the level of ancient ceramics in China. In 1958, the publication of *Research on Jingdezhen Porcelain* by Ren Zhou was a milestone in the history of Chinese ceramics.[21]

[21] Zhou (1987).

In 1953, in order to support the construction of the heavy industry base in Northeast China, the CAS established the Institute of Metals in Shenyang. Its research fields included metal physics, metallurgical physical chemistry, metal processing, mineral processing, refractory, metallographic technology, and chemical analysis. Xun Li (1913–1983), who returned from the University of Sheffield in England, became the Director of the Institute of Metals. He led the Institute in expanding the fields of metal physics, metallurgical physical chemistry, metal processing, mineral processing, refractory, metallograph, chemical analysis, and the application of rare earth in steel, and made many important achievements.[22] The Institute was a research center in the field of metal materials, so a large number of returned overseas talents successively joined. Tingsui Kê (1913–2000), a former research assistant of Cyril S. Smith at the Institute of Metals at the University of Chicago, was once the deputy director of the Institute of Metals, CAS. He was mainly engaged in the research of metal internal friction, grain boundary defects, and metal mechanical properties. The first "Kê's torsion pendulum" was created, and the peak of grain boundary internal friction (Kê's peak) was found for the first time.[23]

According to the "Twelve Year Plan," CAS actively carried out research on semiconductors. In 1956, Shouwu Wang (1919–2014) carried out research on semiconductor germanium in the Semiconductor Research Section of the Institute of Applied Physics. He researched the purification and doping process of germanium single crystals and developed germanium transistors. By the beginning of 1958, it was possible to control the conductivity type, resistivity, and minority carrier life of germanium single crystals, meeting the requirements of device production for materials. In August of the same year, the first batch of germanium high-frequency transistors in China was successfully developed by the alloy diffusion method. At the same time, under the leadership of Lanying Lin (1918–2003), research on silicon materials was carried out, and the

---

[22] Shi and Tan (1985).

[23] Hou (2006).

first silicon single crystal in China was grown in 1958. On this basis, dislocation-free silicon single crystals were prepared by the Czochralski method in 1962. Then, the relationship between dislocation density and growth conditions was studied, and the optimal conditions for the growth of low dislocation density zone fused silicon single crystals were found. Under the international embargo of equipment and technical blockade, China continued to make research achievements in semiconductor silicon materials by relying on its own efforts, creating a prerequisite for China to comprehensively develop semiconductor silicon devices in the 1960s.[24]

In December 1959, the State Science and Technology Commission and the Planning Commission established the "Central New Technology Materials Group," which was responsible for proposing requirements for new materials, arranging scientific research, pilot scale test, and industrial production. In February 1960, the Group decided that the CAS would be responsible for establishing a national test base. By 1962, there had been more than 5000 testing technical teams, arranged nearly 600 research projects, recommended 299 testing methods (for trial implementation) as the national unified standard to the country, and developed several new testing equipment, which made important contributions to the development of China's high-tech.

In 1959, the Soviet Union tore up the technical cooperation agreement with China and withdrew the experts in China. In the process of nuclear bomb research, the key technology of uranium separation was facing great difficulties. Ziliang Wu (1917–2008) of the Shanghai Institute of Metallurgy undertook the development of Type A separation membrane, and the Institute of Metals also was involved in the entrusted tasks of nuclear fuel and related materials analysis, chemical corrosion of nuclear fuel, raw materials chemical research, and nuclear fuel element development. Even more, the Institute of Metals undertook the research and development of nuclear submarine power reactor components, including component cores, component welding processes, and related chemical

---

[24]Wu and Qian (2021).

analysis and physical testing. In 1966, most of the personnel involved in this work were transferred to the newly established professional research institute of the Second Ministry of Machinery (later the Ministry of Nuclear Industry). For example, metallurgist Peilin Zhang (1917–2005) was transferred to the Nuclear Fuel Bureau of the Second Ministry of Machinery.

Since 1957, Changxu Shi (1918–2014) and his colleagues in the Institute of Metals have studied the materials and processes of turbine blades and discs of turbojet engines. Since the Soviet Union withdrew the experts and stopped the supply in 1959, the aircraft engine urgently needed high-temperature alloys. They immediately cooperated with Fushun Steel Plant to overcome the production technology of high-temperature alloys for aircraft engines. At that time, the supply of nickel in China was tight. In order to save nickel, they developed their own iron base superalloy GH135 in 1958 and put it into production. In March 1965, the Aviation Research Institute of the National Defense Science and Technology Commission proposed the task of hollow turbine blades. They soon developed a hollow turbine blade with porous casting in the laboratory, and the cooling effect reached 120°C. In November 1965, it passed the first ground test run, and soon it successfully flew again. China adopted this technology only five years later than the United States and became the second country in the world to use cast alloy turbine blades, which was earlier than the Soviet Union and the United Kingdom.[25]

Since 1958, Changchun Institute of Applied Chemistry has begun to research the rare earth catalytic system for directional polymerization and has successively explored titanium, cobalt, and nickel-based cis polybutadiene rubber suitable for industrial production. In the early 1960s, natural rubber and synthetic rubber, as important strategic materials of the country, were listed as embargoes by the West, and the making technology of synthetic rubber was strictly blocked. In order to independently establish the synthetic rubber technology, the Lanzhou Institute of Petroleum (now Lanzhou

[25] Fan (2001).

Institute of Chemical Physics) has developed a series of molybdenum ternary mixed oxide catalysts since June 1961, which was the first in the world to realize the new reaction of oxidative dehydrogenation of butene to butadiene, and has obtained the main monomer butadiene for the production of cis polybutadiene rubber. The Ministry of Petroleum Industry, the Ministry of Chemical Industry, the Ministry of Higher Education, the First Machinery Industry, and the CAS jointly formed a technical battle for the whole process pilot test of cis- polybutadiene rubber, which finally opened up the industrial production process in 1975 to achieve long-term stable operation.[26]

Research Institutions in Industrial and Military Departments

After 1949, a number of research institutions directly under the industrial ministries have been established, forming central research centers for industrial technological innovation. The industrial institutions are an important part of the material science research and technology development. They belong to several industrial raw material industry departments such as the Ministry of Metallurgical Industry, the Ministry of Chemical Industry, China Nonferrous Metals Corporation, the State Building Materials Industry Bureau, and the aviation and aerospace departments, and others that research and utilize the new materials.[27]

The General Institute of Iron and Steel Research was founded in Beijing in 1951, formerly under the Ministry of Heavy Industry, and later under the Ministry of Metallurgical Industry. It mainly engaged in basic research and development of new varieties of precision alloys, powder metallurgy, superalloys, alloy steels, corrosion and protection, welding of materials, amorphous metals, and alloys.

Beijing Nonferrous Metals Research Institute was founded in Beijing in 1952 and was subordinate to the China Nonferrous Metals Corporation. It was the largest comprehensive R&D unit of

---

[26] Fang and Sun (2019).
[27] Shi (1995).

nonferrous metal materials science and technology in China. It mainly engaged in nonferrous, rare, and precious metal ore dressing and metallurgical processes, new material development and processing technology, semiconductor materials, rare earth materials development, and physical testing and analysis inspection.

China Academy of Building Materials Science was founded in Beijing in 1954 and became subordinate to the State Building Materials Industry Bureau. It was mainly engaged in the R&D of cement, concrete additives, new building materials, aviation glass, secondary processing glass, special functional glass, cast stone, glass crystal materials, special glass fibers and products, quartz glass, optical fiber panels, functional ceramics, structural ceramics, refractories for cement kilns and glass furnaces, building ceramics, new thermal insulation materials, etc.

Shanghai Materials Research Institute was founded in Shanghai in 1956 and was a research institute directly under the Ministry of Machinery Industry. It was mainly engaged in material strength and its testing technology, nondestructive testing technology of materials and parts, testing and analysis of material composition and microstructure, as well as development and application technology of new materials such as metal materials, powder metallurgy materials, engineering plastics, and engineering ceramics to adapt to the development and quality improvement of new products in the mechanical industry.

Beijing Research Institute of Chemical Industry was founded in Beijing in 1958 and was a key research unit directly under the Ministry of Chemical Industry. With petrochemical research and deep processing as the main research direction, it conducted comprehensive chemical development and application research, mainly including the cracking, separation, and comprehensive utilization of petroleum fractions, the synthesis process of organic raw materials, the process of synthetic resin and rubber, the processing and modification technology of plastics and rubber, and the development of fine chemical products.

Luoyang Refractory Research Institute, established in 1963, was subordinate to the Ministry of Metallurgical Industry and was the largest comprehensive research institute of refractory science and technology in China. It was mainly engaged in the research and development of refractory raw materials, new refractory materials, engineering ceramics, high-temperature energy-saving insulation materials, and responsible for the quality inspection and arbitration of refractory materials, as well as the research and formulation of technical standards and testing methods for refractory materials.

Nanjing Institute of Glass Fiber Research and Design was founded in Nanjing in 1964 and was the key research unit of the State Building Materials Industry Bureau. It was mainly engaged in the research of new varieties, new processes, new equipment and application technologies of glass fiber, the research and development of new product technologies of glass wool, mineral wool, and special inorganic fiber materials, special-shaped woven fabric technology, and product trial production, and the development of optical fiber molding technology and product application.

Chenguang Research Institute of Chemical Industry, founded in Chengdu in 1964, was subordinate to the Ministry of Chemical Industry. It was mainly engaged in the research and development of fine chemical products such as organic silicon, organic fluorine, new engineering plastics and new structural materials, special adhesives, functional polymer materials, electronic chemicals, leather, and textile auxiliaries.

Northwest Institute of Nonferrous Metals Research, founded in 1965 in Baoji, Shaanxi, is a key research unit directly under China Nonferrous Metals Corporation. It is mainly engaged in research and development of processing technology, powder metallurgy, explosive processing and welding of titanium and titanium alloys, refractory metals and materials for the nuclear industry, superconducting materials, precious metals, and rare metals.

In addition, the military department has also set up relevant institutions for material research. The Ministry of Weapon Industry set up institutions specializing in the research of ordnance materials such as the 52th Research Institute. In the process of nuclear bomb and missile research, many materials scientists have been dispatched from all around the country to the Chinese Academy of Atomic

Energy (now the Chinese Academy of Engineering Physics) and the Institute of Aeronautics in the Ministry of Aerospace Industry.

Beijing Institute of Aeronautical Materials was founded in Beijing in 1956 and was subordinate to the Ministry of Aerospace Industry. It was the center of comprehensive application of aeronautical materials science and technology, mainly engaged in the research and development of metal and nonmetallic materials, material processing technology, nonmetallic component forming technology, physical and chemical performance testing, etc., in the aviation field.

Beijing Institute of Materials and Technology, founded in 1957 in Beijing, was the materials and technology research center of the China Aerospace Industry Corporation. It was mainly engaged in the development and research of new materials and processes for missiles, satellites, launch vehicles, and other spacecraft. It included metal materials, non-metallic materials and application research, parts design and processing, composite material design and processing technology, as well as supporting metal materials casting, heat treatment, spinning forming, chemical milling, electroplating, surface treatment, various welding and brazing technology, etc.

The 52nd Research Institute, founded in 1959 in Baotou, was a research unit directly under China National Weapon Corporation. It was mainly engaged in the research and development of ferrous and non-ferrous materials, composite materials, ceramic materials, special functional materials, powder metallurgy materials, rare earth materials, etc. it carried out the supporting detection technology, process technology, engineering mechanics, and theoretical research on an application basis.

In the 1950s, China Institute for Mining and Metallurgical Engineering moved to Taiwan to continue its work, while two important academic organizations were developed. One is the Chinese Society of Metals, which was founded in 1956. Ren Zhou was the first Chairman of the Society. It is mainly a mass academic group of Chinese metallurgical science and technology scientists and engineers. The other is the Chinese Ceramic Society, which was reestablished in 1959, inheriting the preparation committee of the

Society in 1945 (renamed as the Chinese Society of Kiln Engineering in 1951), and is a mass academic group of Chinese scientists and engineers in ceramics research.

Reform of the Discipline of Materials Science and Engineering (1978–2000)

MSE in Colleges and Universities

After the Reform and Open policy starting in 1978, while accelerating the development of materials science and technology, with the understanding of MSE education reform in Europe and the United States, China has also started the process of the reform. In the face of the development of international materials science and technology, the change in the demand for training MSE talents due to the development of high-tech, and the drawbacks of China's MSE education mode and content are gradually recognized. With the development of economy, society, and science, the boundary between material science and material engineering began to blur, and there were more internal links and commonalities between several major materials. The emergence and wide application of new materials such as composite materials, ceramic materials, and functional materials, and the rapid promotion of advanced technologies like computers made innovation in all aspects more emphasis on the foundation and the connection between horizontal and vertical. In fact, various disciplines are increasingly crossing, infiltrating, drawing on, and transplanting each other. In terms of application, more and more large-scale mutual substitution and combination had become an objective fact.

In this context, Zhejiang University took the lead in setting up the Department of Materials Science and Engineering. In 1978, Qidong Wang (1921–2019) took the opportunity of an American metallurgical and materials science delegation to visit China for exchange and consulted visiting professors about the experience of some American universities in setting up MSE departments. After that, he immediately started to transfer the whole group or part of the faculty engaged in metal materials (casting, metallograph and heat treatment), semiconductor materials, magnetic materials, inorganic non-metallic materials, and polymer materials from the Department of Machinery, the Department of Radio, the Department of Physics, and the Department of Chemical Engineering, respectively, and merged to form the Department of MSE of Zhejiang University. This is the first department of MSE established by Chinese universities. Qidong Wang was appointed the first Chair of the department. In 1983, Zhejiang University also established the academic journal *Journal of Materials Science and Engineering*, which was the pioneer.[28] Tsun Ko once said that they discussed the importance of setting up a department of MSE, and thought Zhejiang University had a relatively balanced orientation in deferent materials, which made it the most suitable place to become the first Chinese department of MSE.

Since then, the University of Science and Technology Beijing, Fudan University, Tsinghua University, and other key universities have successively set up the departments of MSE (or a similar name). Over the next decade, more and more universities participated in the reform. During that period, the main contents of the reform were as follows: expanding the contents on the basis of the original majors such as adding non-metallic engineering materials and functional materials in the teaching of metal materials and heat treatment majors. The trial establishment of new majors, such as material science, material engineering, and material physics, generally reflects the reform trend of gradually breaking the boundaries of the original majors and strengthening the infiltration and connection between majors.

In 1997, the Ministry of Education issued a new version of the academic degree and subject catalog, setting up the first-level discipline of MSE, and three second-level disciplines as materials physics and chemistry, materials science, and materials processing engineering, which recruit master's and doctoral students. In 1998, the Ministry of Education revised the existing majors and reformed the phenomena: the division of majors in colleges and universities was

---

[28] Chen (2015).

too detailed and too narrow, some of the names of majors were not scientific and standardized, and majors were set repeatedly among categories. Based on the principles of standardization and broadening, combined with the national conditions of the country, and drawing on the successful experience of foreign higher education major setting, the transition from profession division to discipline division was made. The new professional directory has been reduced. Mineral and rock materials have been merged into material physics. Iron and steel metallurgy, nonferrous metal metallurgy, metallurgical physical chemistry have been merged into metallurgical engineering. Metal pressure processing, powder metallurgy, corrosion and protection, welding process and equipment, some metal materials and heat treatment, composite materials, plastic or plastic processing processes, and equipment have been merged into metal material engineering. Inorganic non-metallic materials, silicate engineering, and some composite materials are merged into inorganic non-metallic material engineering. Some composite materials, polymer materials, and chemistry are merged into polymer material science and engineering. In addition, the merged metallurgical engineering, metal material engineering, inorganic non-metallic material engineering, and polymer material science and engineering are also included in the guiding directory as the first-level discipline MSE. These measures have further promoted the reform and development of MSE education in China and created a new situation for exploring and forming an MSE education training model with Chinese characteristics.[29]

In the late 1990s, under the guidance of the policy of major adjustment toward the 21st century and the reform of curriculum system and teaching content, the Ministry of Education integrated the setting of undergraduate material majors in colleges and universities. In the 1990s, Tsun Ko promoted the reform of national engineering education and established a pilot class of "big materials" at the University of Science and Technology Beijing, integrating the original metallurgy and materials majors, aiming to break through

[29] Zuo (2009).

the original discipline boundaries and cultivate generalists in MSE research, which had a wide impact throughout the country.[30]

MSE in Chinese colleges and universities generally developed from two types. One was the discipline established in engineering schools through the three major categories of materials in metallurgy and machinery: metal, inorganic nonmetal, and polymer, as well as the disciplines on which their composite materials depended. This type of discipline focused on exploring the performance evaluation and use of new materials from the perspective of specific applications. On the other hand, some comprehensive universities, on the basis of tracking the frontier of science and technology, developed and differentiated from physics and chemistry to form new disciplines of material physics and material chemistry, and established departments of material sciences, which was characterized by the intersection of materials science, physics, chemistry, and other disciplines. These two kinds of MSE disciplines with different starting points are approaching each other through self-improvement, combining science with engineering, and gradually developing toward the direction of combining basic research with applied research.

MSE in Chinese Academy of Sciences

In the spring of 1978, a National Science Conference was held in Beijing, ushering in the "Spring of Science". The CAS has gradually turned its attention from military service to civil service, and many research institutes have made important progress in MSE.

In 1984, the Institute of Physics successfully developed NdFeB permanent magnetic materials and established the Three Ring New Materials Company in 1985. China became the third country in the world to industrialize magnetic materials.[31]

In 1983, Fujian Institute of Material Structure, under the guidance of Jiaxi Lu (1915–2001), cultivated a nonlinear optical

---

[30] Han and Shi (2012).

[31] Wang and Hu (2002).

crystal — low-temperature phase barium metaborate crystal (b-BaB2O4, abbreviated as BBO) through a large number of experimental studies such as structure selection, material synthesis, physicochemical and phase diagram analysis, crystal cultivation and performance testing. In 1987, Chuangtian Chen (1937–2018) successfully developed a new nonlinear optical crystal, lithium boronate (LBO), which was listed as one of the world's top ten products in the field of laser and optoelectronics in 1989 by the American magazine *Laser and Electronic Optics*.[32]

In the autumn of 1984, American scientists first found quasiperiodic crystals ("quasicrystals" for short) in the aluminum manganese quenched alloy. In the spring of 1985, the research team led by Ke Hsin Kuo (1923–2006) of the Institute of Metals found icosahedral quasicrystals with quintic symmetry in Ti–V–Ni quenched alloys. This was another new quasicrystal reported after the aluminum manganese quasicrystal, known as the "Chinese phase".[33]

In 1987, the high critical temperature superconductor research group led by Zhongxian Zhao (1941–) and Liquan Chen (1940–) of the Institute of Physics found that the initial transition temperature of YBaCuO superconducting material was above 100K. This achievement marks that China's superconducting research has leaped into the world's advanced ranks.[34]

In order to further develop the industrial production of special polypropylene spinning materials, the Institute of Chemistry cooperated with Liaoyang Petroleum Chemical Fiber Company to carry out more than ten difficult technical transformations on the polypropylene plant and production process, solved the technical problems of polymerization, purification, refining, drying, granulation, packaging, measuring instruments, etc., and developed a series of mobile polypropylene spinning materials created by chemical degradation.

The research team led by Sishen Xie (1942–) of the Institute of Physics has successfully prepared an ultra-long aligned carbon

[32] Ye (2019).

[33] Zhang (1997).

[34] Wu and Liu (2012).

nanotube array with a length of 2–3 mm and has realized the use of conventional experimental methods to test the physical properties of carbon nanotubes. This research result attracted the attention of the scientific and technological circles after it was published in *Nature* magazine in 1998. It is believed that this is the first time to increase the length of carbon nanotubes to the millimeter scale.[35]

Societies for MSE

During this period, with the development of MSE, in addition to the existing Chinese Society of Metals and the Chinese Ceramic Society, a number of new MSE societies emerged. Besides that, there are also some divisions of MSE in the Chemical Industry and Engineering Society of China, China Textile Engineering Society, etc.[36]

Founded in Beijing in May 1991, the Chinese Materials Research Society is an academic group of Chinese MSE scholars. It is a member of the International Union of Materials Research Societies (IUMRS). Hengde Li (1921–2019) was appointed as the first Chairperson of the Society. The key job of the Society is to promote the research and development of metal materials, inorganic nonmetallic materials, organic polymer materials, composite materials, etc., especially the research of material science frontier and the intersection of disciplines, and at the same time, strive to promote the practical application of new materials, new processes, and new technologies in industries, so as to play a positive role in national economic construction. The Society is one of the sponsors of the *Chinese Journal of Materials Research.*

The Nonferrous Metals Society of China was established on November 1984. It is an academic social organization formed voluntarily by scientific and technological scholars and relevant units in nonferrous metals and related disciplines. It is an important social force to develop China's nonferrous metals science and technology, and a scientific and technological organization in China's nonferrous

[35] Yang (2001).
[36] Shi (1995).

metals industry. The journals sponsored by the society include *Transactions of Nonferrous Metals Society of China*, etc.

The Chinese Society of Rare Earths (CSRE), founded in November 1979, is a national academic social organization voluntarily formed by rare earth science and technology scholars and relevant units. It is an important social force for the development of rare earth science and technology, and a scientific organization in China's rare earth industry. It sponsors the *Journal of Rare Earths.*

The Chinese Society for Corrosion and Protection (CSCP) was established in Hangzhou in November 1979. The Society aims to carry out academic exchanges and cooperation at home and abroad, disseminate popular science knowledge and advanced production technology, edit and publish academic journals, carry out continuing education, serve the construction of corrosion and protection disciplines, and serve scientific and technological scholars, enterprises and institutions in the corrosion and protection industry. *Journals of Chinese Society for Corrosion and Protection* are published. The participating international organizations are as follows: International Corrosion Council, Asia Pacific & Corrosion Association and World Corrosion Organization.

In January 1989, the China Composite Society was initiated and established. It is a social organization formed voluntarily by scientific, technical, and management workers and relevant units engaged in composite materials and related disciplines. It sponsors academic journals, such as *Acta Materiae Compositee Sinica* and *Composites Communications.*

The Chinese Society of Crystals, formerly known as the Chinese Committee of the International Union of Crystallography (IUCr), joined the Union under that name in 1978. It was approved to be established independently in 1993. The Society is a voluntary academic and public welfare social organization formed by Chinese crystallographic scholars and an important social force for the development of crystal science and technology in China. Youqi Tang (1921–2022) was the first Chairman of the Society.

At this time, the concept of MSE has gradually become popular. In 1995, Changxu Shi wrote as follows in the Preface of *Encyclopedia of Materials Science and Technology*: "MSE is to study the composition, structure, performance, process and use efficiency of materials as well as their relationship. MSE is characterized by the intersection and combination of physics, chemistry, metallurgy, ceramics, polymer science, computational mathematics, and other disciplines. The materials science and technology proposed in recent years is not fundamentally different from MSE."[37]

Material research plays a fundamental role in the development of social economy and high-tech. Since the beginning of the period of the Reform and Open, governments at all levels have paid more and more attention to it. A series of scientific research plans and development plans launched, such as the "863 Plan", the "973 Plan", and the National Science and Technology Research Plan (later renamed the National Science and Technology Support Plan), give priority support to the MSE field. During this period, the state implemented the strategy of revitalizing the country through science and education, under the guide of the action plan for revitalizing education in the 21st century, the "211 Project" and the "985 Project" by universities, and the "Knowledge Innovation Project" by the CAS. The state made great efforts in the introduction of advanced equipment for MSE and the introduction and training of high-level teachers. A large number of national, provincial, and ministerial key laboratories, engineering test centers construction projects, were launched to make the conventional and advanced instruments and equipment required for materials teaching and research step by step.

Materials Science and Engineering in the New Century

Institutional Reform and New Pattern of MSE

China formally joined the World Trade Organization in 2001, which promoted the large-scale institutional reform of the State Council and enterprise system reform to adapt to the market economy system. Almost all specialized industrial economic departments were

---

[37] Shi (1995).

abolished. In this way, a number of ministries and commissions have been transformed into enterprises. Some of the central research institutes have been transformed into science and technology companies affiliated to central enterprises. For example, the China Academy of Building Materials Sciences has become a subsidiary of China Building Materials Group Corporation, and the Beijing Institute of Aeronautical Materials of the Ministry of Aviation has been transformed into a subsidiary of China Aeroengine Group. Some directly became central enterprises, such as Iron and Steel Research Institute changed into China Iron and Steel Research Group Corporation, and China Nonferrous Metals Research Institute changed into General Research Institute of Non-ferrous Metals (GRINM) Group Corporation. So far, the central research institutes have lost their original status as the national technology center, and have also transformed to develop their own key technologies. For example, GRINM Group carried out excellent silicon material research and formed a joint-stock company with its silicon material semiconductor research center as the core, and was listed as *Youyan Guigu* on the Shanghai Stock Exchange on March 1999. This is the first material science and technology company Stock Exchange in China. At the same time, some pure production enterprises, such as Baosteel Group, have begun to strengthen their research on iron and steel technology. Its subordinate Baosteel Research Institute has become the leader in domestic steel material research and development.

The national innovation system united government, industry, universities and research institutions attaching great importance. Due to the important role of materials in the national economy, the state has given top priority to the development of materials research. "The National Medium and Long-Term Science and Technology Development Plan" was launched in 2006. Advanced materials such as "superstructural materials, new generation functional materials and devices, environment-friendly materials, biomedical and bionic materials" were listed as key developing fields. The rapid improvement of materials research conditions and achievements provided a good opportunity for the development of MSE education and industrial utilization.

National Laboratory and National Research Center

China has been exploring the model of the US National Laboratory. On October 27, 2000, the Ministry of Science and Technology officially approved the establishment of the Shenyang National (Joint) Laboratory of Materials Science, which is the first national research laboratory in China, relying on the Institute of Metals of CAS. It was officially put into operation in June 2001. It is mainly engaged in basic research of material science and application, covering the preparation and processing of materials, structural characterization of materials, material theory and computational simulation, performance evaluation and service behavior of materials.

On November 21, 2017, with the approval of the Ministry of Science and Technology, Shenyang National (Joint) Laboratory of Materials Science was officially reorganized into Shenyang National Research Center of Materials Science. On April 14, 2018, the Center was founded with an organized expert demonstration meeting on the implementation scheme. The competent department of the Center is the CAS, and the first Director of the Center is material scientist Ke Lu (1965–).

The National Center for Materials Service Safety (NCMS) was approved and established by the State Development and Reform Commission in December 2008. It is one of the major scientific infrastructures planned to be constructed "Materials Service Safety Assessment Facilities" (MSAF), which is under the responsibility of University of Science and Technology Beijing. The Institute of Metals, CAS, etc. participated in the construction. NCMS is the first national science center contracted by universities. It focuses on four key fields of service safety of major engineering structural materials such as scale domain, environment domain, time domain and safety evaluation method. The construction includes eight sets of large-scale scientific research and test devices for service safety of engineering structural materials, including the multi-factor coupling environmental material test device. They comprehensively improved the test and research capabilities of large-scale/full-size components and materials and the overall strength of safety evaluation technology and established China's own standards and specifications for the safe service of engineering materials.

State Key Laboratory and National Engineering Research Center

In addition, there are a number of high-level state key laboratories distributed in universities and research institutes of the CAS. The State Key Laboratory is an important innovation base where the state organizes basic research and applied basic research, gathers and trains outstanding scientific talents, and conducts high-level academic exchanges, with advanced scientific research equipment. It is an important part of the national innovation system. Since the 1980s, there are about 21 State Key Laboratories in the MSE field approved by the Ministry of Science and Technology as shown in Table 1.

Table 1. State key laboratories in MSE field.

No.	Name	Supporting institutions	Approval year
1	State Key Laboratory of Optoelectronic Materials and Technologies	Sun Yat-sen University	1984
2	State Key Laboratory of Crystal Materials	Shandong University	1984
3	State Key Laboratory of Silicon Materials	Zhejiang University	1985
4	State Key Laboratory of Advanced Technology for Material Synthesis and Processing	Wuhan University of Technology	1987
5	The State Key Lab of High Performance Ceramics and Superfine Microstructure	Shanghai Institute of Ceramics, CAS	1988
6	State Key Lab of Metal Matrix Composites	Shanghai Jiaotong University	1988
7	State Key Laboratory of Solidification Processing	Northwestern Polytechnical University	1989
8	State Key Laboratory of Polymer Materials Engineering	Sichuan University	1989
9	State Key Laboratory of Powder Metallurgy	Central South University	1989

(*Continued*)

Table 1. (*Continued*)

No.	Name	Supporting institutions	Approval year
10	State Key Laboratory of Superhard Materials	Jilin University	1989
11	State Key Laboratory of Pulp and Paper Engineering	South China University Of Technology	1989
12	State Key Laboratory for Mechanical Behavior of Materials	Xi'an Jiaotong University	1990
13	State Key Laboratory for Advanced Metals and Materials	University of Science and Technology Beijing	1990
14	The State Key Laboratory of Functional Materials for Informatics	Shanghai Institute of Microsystems and Information Technology, CAS	1991
15	State Key Laboratory of New Ceramics and Fine Processing	Tsinghua University	1991
16	State Key Laboratory for Modification of Chemical Fibers and Polymer Materials	Donghua University	1992
17	State Key Laboratory of Solid Lubrication	Lanzhou Institute of Chemical Physics, CAS	1999
18	State Key Laboratory of Metastable Materials Science and Technology	Yanshan University	2006
19	State Key Laboratory of Luminescent Materials and Devices	South China University of Technology	2011
20	State Key Laboratory of Organic–Inorganic Composites	Beijing University of Chemical Technology	2011
21	State Key Laboratory of Silicate Materials for Architectures	Wuhan University of Technology	2011

Table 2. National research center for MSE.

No.	Name	Supporting institutions	Approval year
1	National Engineering Technology Research Center for Nonferrous Metal Composites	GRINM Group Co. Ltd	1992
2	National Catalysis Engineering Technology Research Center	Dalian Institute of Chemical Physics, CAS	1992
3	National Research Center for Reaction Injection Molding Engineering Technology	Liming Research Institute of Chemical Industry	1993
4	National Fiberglass and Products Engineering Research Center	Sino-Materials Science & Technology Co. Ltd	1993
5	National Engineering Research Center for Superhard Materials and Products	Zhengzhou Institute of Abrasive Grinding	1993
6	National Engineering and Technology Research Center for Magnetic Materials	BGRIMM Materials & Technology Co. Ltd.	1993
7	National Engineering Technology Research Center for Amorphous Microcrystalline Alloys	Central Iron & Steel Research Institute	1995
8	National Glass Deep Processing Engineering Technology Research Center	China Building Materials Academy	1999
9	National Research Center for Photosensitive Materials Engineering Technology	China Lucky Film Corporation	1999
10	National Fluorine Material Engineering Technology Research Center	Juhua Group Corporation	2002
11	National Engineering Technology Research Center for Composite Modified Polymer Materials	Guizhou Material Technology Innovation Base	2004

The National Engineering Research Center, mainly relying on key scientific research institutions, enterprises or universities with strong power in the industry and scientific field, has a domestic first-class professional talent team for engineering technology research and development, design and test. It has relatively complete engineering supporting test conditions, can provide a variety of comprehensive services, and is closely connected with relevant enterprises. At the same time, it is a scientific research and development entity with a self-cycle mechanism. The Ministry of Science and Technology announced 11 national engineering research centers for MSE since 1992.

Summary

The MSE major has trained a large number of high-quality talents who serve the preparation, processing, production, application, research, development and management of materials for the society, and has made important contributions to the national economy and scientific and technological construction. The core of undergraduate talent training for MSE is to focus on the composition and structure, characterization, preparation and performance of materials. Accordingly, major backbone courses such as fundamentals of materials science, methods of material analysis and testing, physical properties of materials, and materials technology have been set up. From the perspective of where the students will go after graduation, the students trained in this major are not only engaged in traditional material industry-related work, such as steelmaking, automobile, cement, glass, ceramics, chemical industry and other industries, but also engaged in technology and management related work in the new material industry. In addition, a considerable proportion of students choose to continue their studies and engage in material science research. Nowadays, in the process of cultivating talents in MSE, the teaching proportion of traditional materials has declined. More and more colleges and universities have opened such new materials courses as new energy materials, biomaterials, and computational materials, which reflects the importance of the development trend of materials frontier fields.

The formation of MSE has experienced the development, differentiation, reorganization and integration of a series of different disciplines. It has been nearly 40 years since it officially became an independent discipline. At present, MSE at China and abroad is growing, and its connotation and related fields are constantly enriched and expanded, and gradually become mature. With the world's attention to the research and application of new materials, MSE will usher in a good opportunity for development. From the development stage of MSE education in China, we can see that the formation and development of MSE education follow two basic principles: one is to adapt to social and economic needs (demand traction), and the other is to drive the development of science and technology (discipline promotion). From the level of discipline and professional knowledge, it reflects the scientific development law from broad to subdivided and from subdivided to comprehensive.

Some new research fields of MSE are increasingly valued. For example, the CAS has set up a nano science research center, and universities have also set up a series of nano material research institutions, showing a thriving trend. Another example is computational materials science, which also has a new upsurge. In 2015, the Ministry of Science and Technology set up a key technologies and support platform for material genetic engineering to build three types of demonstration platforms, including efficient computing, high-throughput experiments and databases, that support material genetic engineering research and collaborative innovation and development. In addition, Beijing, Shenzhen, Shanghai, Yunnan, and some large enterprises related to materials have also set up relevant research projects to build material genetic engineering research facilities and promote material design and development.

China's influence in the field of materials science is growing day by day. China has become a leader in the field of materials science, such as nanomaterials, condensed matter physics and structural materials. According to statistics, from 2010 to 2019, the number of materials science papers published in China reached 35.41% of the world's, and from 2010 to 2020, the number of citations of materials science papers published in China ranked first in the world. It is believed that more Chinese scholars and research institutions will be seen in the field of MSE in the future.

At present, China has made it clear that the national strategic scientific and technological force is composed of national laboratories, national scientific institutions, high-level research universities and leading hi-tech enterprises. Shenyang National Laboratory of Materials Science (United) has been built into a national research center on material science, and a number of national research centers have been established or planned. Some provinces and cities have also started to plan national laboratories for materials. For example, Gusu Laboratory for Materials Science is a new research and development institution aiming at national laboratory standards and international first-class standards. The state key laboratories of MSE are being reorganized, called "national key laboratories", to focus more on doing important things. The MSE discipline of universities has achieved unprecedented development. In the second round of "Double First-Class" construction list released, 32 MSE disciplines in universities have been approved as first-class disciplines, accounting for 8.3% of the 410 in China, and ranking first in the number of disciplines. A number of high-tech enterprises characterized by MSE are also rising.

Faced with the development, demand and promotion of scientific and technological progress in the 21st century, the reform of MSE education and research in China is reforming and innovating in a wider range and deeper level, with a view to forming a great development of MSE research that is in line not only with the international advanced MSE but also with Chinese characteristics.

References

Cahn, R. W. (2002). *The Coming of Materials Science* (Oxford: Pergamon), pp. 13–14.

Chen, Lixin (2015). Qidong Wang [edited by Yong Gan in Chinese]. *Overview of the Academic Achievements of China's Famous Scientists in the*

20th Century — Materials Science and Engineering (Beijing: Science Press), pp. 152–160.

Ding, Zishang and Yan, Dongsheng (1985). Education of inorganic non-metallic materials in New China [in Chinese]. *Guisuanyan Tongbao* 5(3), 59–64.

Fan, Guialn (2001). Pioneer of China's superalloy materials Academician Shi Changxu, a famous material scientist [in Chinese]. *Zhongguo Kexue Jijin* 1, 12–14.

Fang, Yibing and Sun, Song (2019). The role of CAS in the combined tackling of the key problem of butadiene rubber (1958–1976): The case of the R&D Work of the Lanzhou Institute of Chemical Physics [in Chinese]. *Studies in the History of Natural Sciences* 38(3), 290–307.

Han, Rubin and Shi, Xinmin (2012). *Ke Jun Zhuan (Biography of Tsun Ko)* [in Chinese]. (Beijing: Science Press), pp. 185–200.

Han, Rubin and Shi, Xinmin (2012). *Ke Jun Zhuan (Biography of Tsun Ko)* [in Chinese]. (Beijing: Science Press), pp. 213–220.

Hou, Yude (2006). The outstanding representative of Chinese good physicist TS Kê [in Chinese]. *College Physics*, 26(1), 44–48.

Hu, Shenghua (1997). A review on the research at the Institute of Physics of National Academy of Peiping (1929～1949) [in Chinese]. *Wuli* (10), 57–62.

Lai, Qifang (1985). In memory of the Chinese Ceramic Society 40th anniversary [in Chinese]. *Guisuanyan Tongbao* (2), 62–63.

Lei, Lifang, Fang, Yibing and Qian, Wei (2016). Research on metallurgical fuels in the rear area during the Anti-Japanese War: A case study of the National Bureau of Mining and Metallurgical Research [in Chinese]. *The Chinese Journal for the History of Science and Technology* 37(3), 271–283.

Lei, Lifang, Fang, Yibing and Qian, Wei (2018). The Chinese institute of mining & metallurgy in the decade before the war of resistance against Japan [in Chinese]. *Ziran Bianzhengfa Tongxun* 40(04), 74–80.

Pan, Xiaoshuo and Wu, Qianzhang (1984). In memory of Mr. Ruwei Shi [in Chinese]. *Chinese Historical Materials of Science and Technology* 5(2), 73–77.

Shi, Changxu (1995). *Encyclopedia of Materials Science and Technology* [in Chinese] (Beijing: China Encyclopedia Press), pp. 1251–1253.

Shi, Changxu (1995). *Encyclopedia of Materials Science and Technology* [in Chinese] (Beijing: China Encyclopedia Press). pp. 1247–1248, 1253–1257.

Shi, Changxu (1995). *Encyclopedia of Materials Science and Technology* [in Chinese] (Beijing: China Encyclopedia Press), p. 5.

Shi, Changxu and Tan, Bingyu (1985). Deeply missing Comrade Xun Li [in Chinese]. *Jinshu Xuebao* 3, 2–7.

Wang, Chungyu (1909). *Antimony* (London: Charles Griffin Publishing).

Wang, Chungyu and Li, Guoqin (1943). *Tungsten* (New York: Reinhold)

Wang, Zhenxi and Hu, Boping (2002). The development of San Huan Ltd and magnetics research [in Chinese]. *Wuli* 7, 422–429.

Wei, Shoukun (1947). Metallurgical research in China in recent 20 years [in Chinese]. *Kexue* 31(3), 67–72.

Wu, Xi and Qian, Wei (2021). The start and growth of China's semiconductor industry technology (1949–1978) [in Chinese]. *Journal of Engineering Studies* 13(6), 589–600.

Wu, Xiaohong, Zhang, Chi, Goldberg, Paul, Cohen, David, Pan, Yan, Arpin, Trina and Bar-Yosef, Ofer (2012). Early pottery at 20,000 years ago in Xianrendong Cave, China. *Science* 6089, 1696–1700.

Wu, Yan and Liu, Bing (2012). Study on the method of the breakthrough which Zhao Zhongxian and his team obtained during 1986～1987 based on written and oral sources. *Chinese Journal for the History of Science and Technology* 33(2), 143–155.

Yang, Liuchun (2001). Standing at the forefront of nanomaterial research: Interpret Sishen Xie and his research team [in Chinese]. *Zhongguo Kexueyuan Yuankan* 1, 70–71.

Ye, Hengqiang (2019). In memory of Chuangtian Chen [in Chinese]. *Journal of Synthetic Crystals* 10, 1771–1772.

Zhang, Bichun (1990). Machine testing, design and manufacture of National Bureau of Industrial Research [in Chinese]. *Chinese Historical Materials of Science and Technology* 11(2), 66–72.

Zhang, Ze (1997). Progress in quasicrystals research [in Chinese]. *Keji Daobao* 7, 52–55.

Zhao, Changsheng and Gu, Yi (2008). Development and present situation of polymer materials and engineering [in Chinese]. *China Plastics Industry* 36(1), 70–71.

Zheng, Kangni (2012). A primary study on National Ceramics Laboratory and its Working Reports [in Chinese]. *Zhongguo Taoci* 48(3), 74–78.

Zheng, Naizhang, Li, Yuanye, Zheng, Kangni, Mo, Yunjie and Yao, Lifeng (2013). The Transmission of the modern ceramic science and technology in Jindezhen [in Chinese]. *Guisuanyan Tongbao* 32(6), 1092–1096.

Zhou, Peide (1987). A brief biography of Ren Zhou [in Chinese]. *Kexue* (1), 58–64.

Zuo, Tieyong (2009). Major achievements and future prospects of the development and reform of MSE education in China [in Chinese]. *Xincailiao Chanye* 10, 39–47.

Chapter 5.3

The Development of Materials Research Institutions in Africa

Wole Soboyejo, Samuel Chigome and Iwnetim Abate

Introduction: The Materials and Energy Resources of Africa

The materials resources of Africa are very significant, as shown in Fig. 1, in which some of Africa's materials resources are presented schematically in the map of Africa's resources. These include large deposits of gold, diamond, coal, copper, and mineral resources that are present in quantities that are sufficient to motivate their industrial development. However, with the exception of South Africa, Egypt, and Morocco, the transition from raw materials economies to value-added economies has been very limited. This has been attributed to the lack of human resources that can extract, process, and add value to these resources in ways that can stimulate economic growth and sustainable development (Soboyejo, 2005, 2007, 2015, Soboyejo and Madsen, 2018 Soboyejo, 2019).

Similarly, the energy resources of Africa are very significant (Fig. 2). These include the following: the solar energy resource of the Sahara desert, which can produce one-third of the continent's energy needs; the hydropower energy resource of the Congo River, which can also produce about one-third of the continent's energy needs, and the geothermal energy resources (which extend from

Fig. 1. The materials resources of Africa.

Egypt in the North down to Tanzania) can also provide one-third of the continent's energy needs. Thus, the continent could be energy independent, even without exploiting the oil and gas in the Gulf of Guinea (that runs across the coast of West Africa to the Western Coast of Central Africa all the way to Angola), the newly discovered oil and gas in Tanzania and South Sudan, the coal reserves in South Africa and Nigeria, and the potential for wind, hydro, biomass and solar energy that exists across the African continent.

For Africa to develop, there is a need to develop the human capacity to extract, process, and add value to its materials and energy resources (Soboyejo, 2005). Within this context, materials science and engineering is an important field for the development of Africa. This is a field that provides the knowledge that is needed

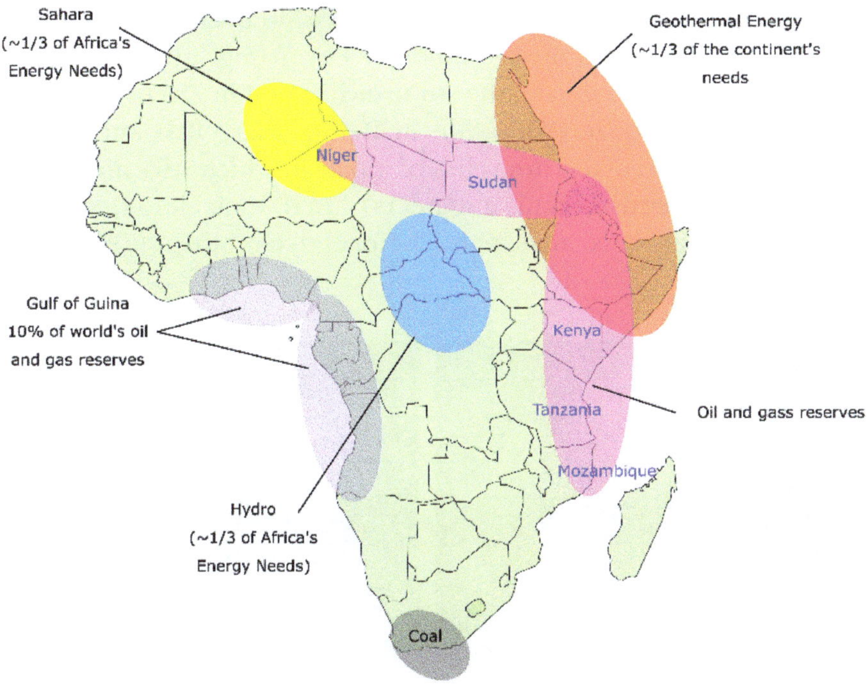

Fig. 2. The energy resources of Africa.

to extract, process, and package raw materials into value-added materials that could transform the lives of Africans from poverty into wealth.

However, in most African countries, the raw materials and waste materials are often extracted and exported to foreign countries, where they are processed into value-added materials. They are then resold back to Africa at much higher costs. There is, therefore, a need for Africa to use materials science and engineering as a field as it can enable Africa to realize the full value of its materials and energy resources in ways that will improve the quality of life.

The Origins of Materials Research in Africa

Ironically, early Africans had a very different approach to materials. There is clear historical and archaeological evidence of the development and use of materials by early Africans. These include the products of "early African materials research" in the relics of ancient African civilizations in the pyramids in Egypt (Fig. 3(a)), the obelisks in Ethiopia (Fig. 3(b)), the stone civilization in Zimbabwe (Fig. 3(c)), and the bronze civilization in Benin (Figs. 3(d) and 3(e)). These historical artifacts are relics of a past in which African tinkerers (artists, scientists, and engineers) used their knowledge to extract and process materials, before assembling them into

Fig. 3. Selected Examples of Materials from Ancient Africa: (a) Earth-Based Materials in the Pyramids; (b) Ethiopian Obelisks; (c) Stone Culture from Zimbabwe, and (d, e) Benin Bronze. Images reproduced with permission from: https://www.history.com/topics/ancient-egypt/the-egyptian-pyramids; https://www.gavinpublishers.com/article/view/Ethiopia's Material Culture: Some Notes on the Obelisks of Axum and the Rock-Hewn Churches of Lalibela (gavinpublishers.com); https://www.britannica.com/place/Great-Zimbabwe; https://www.bing.com/search?q=images+of+benin+bronzes&cvid=f8f6689d01024e6794fe14c401a5db77&aqs=edge.0.0j69i57j0l7j69i11004.8887j0j4&FORM=ANAB01&PC=U531.

impressive structures that have survived many centuries (Kete-Asante, 2019; Harms, 2018; Ade Ajayi and Crowder, 1985).

In the case of Benin, this led to an empire that spanned over much of West Africa, in ways that were strengthened by their wealth of knowledge in the processing of bronze into statues and weapons (Figs. 3(a)–(e)) (Kete-Asante, 2019; Harms, 2018; Ade Ajayi and Crowder, 1985). The Egyptians also developed sustainable building materials for homes that could use dome or desert architectures and well-controlled heat diffusion characteristics to develop building materials and structures that enabled them to manage the thermal comfort of their homes (Thomas *et al.*, 2006). These ideas have been documented in the work of Hassan Faty in his classic work entitled "Architecture for the Poor" (Fathy, 1973). The underlying heat diffusion that controls the temperatures in such buildings has been modeled in work by Thomas *et al.* (2006).

The Modern Development of Materials Research in Africa

During the early part of the 20th century, South Africa developed strong institutions for materials research. These included universities such as the University of Pretoria, the University of the Witwatersrand, and the University of Cape Town. The materials research at these universities was highly connected to the needs of the mining and materials processing industries. South African universities also trained competent materials engineering graduates that were able to support the development of an industrialized economy.

The Council for Scientific and Industrial Research (CSIR) and Mintek are two notable institutions that also played a major role in materials research and the development of South Africa. The CSIR was established in 1945, while Mintek was started in 1934. Both organizations have engaged in applied research and development that provided South Africa with the technologies that were needed to transform its mineral deposits into wealth.

The activities of the CSIR have spanned the full range from cutting-edge research on titanium alloys to nanotechnology and laser processing, while those of Mintek have focused on extractive metallurgy, mineral processing, foundry technology, wrought processing of ingot and powder metallurgy alloys, nanotechnology and computational materials science. Both organizations have developed an international reputation for their work on extractive metallurgy and metallurgical engineering.

Beyond South Africa, some African countries established departments that focused on metallurgical engineering or metallurgy in the post-colonial era. These include the following: the University of Zambia, the Obafemi Awolowo University (OAU) in Ile-Ife in Nigeria, the Kwame Nkrumah University of Science and Technology (KNUST) in Ghana, and the Federal University of Technology in Akure (FUTA), and the University of Zimbabwe. These universities played a major role in training a new generation of African metallurgists that are now emerging as senior researchers and metallurgists/metallurgical engineers.

The department at OAU was started by Prof. Olusegun Adewoye and Prof. Adeniyi Afonja in the 1980s. The department was initiated as part of an engineering physics program in the 1970s before it evolved into a department in the 1980s. During the early stages of engineering physics, the program was well recognized for the work of Adewoye on ceramics (Adewoye and Page, 1976). Prof. Adeniyi Afonja also worked on the strain rate sensitivity and the activation energy of micro-duplex stainless steels in the early 1980s (Afonja, 1982). Prof. Benjamin Imasogie (Yang *et al.*, 2008), from the OAU materials program, has also done important work on the fatigue and fracture of LIGA Ni MEMs thin films.

The materials research program at KNUST in Ghana was initiated within the Department of Mineral Processing and Extractive Metallurgy of the Institute of Mining and Mineral Engineering at KNUST. Prof. Francis Warings Yao Momade (Momade and Momade, 1999) led this program in the early 2000s before Prof. Sam Kwofie (2001) took over the department in the late 2000s. The department has trained several materials engineers that are now teaching and engaging in materials research at KNUST (Arthur *et al.*, 2015) and abroad (Ampaw *et al.*, 2019; Klenam *et al.*, 2015).

Prof. Elifas Bisanda initiated materials research at the Kigali Institute of Science and Technology (KIST), which is now part of the University of Rwanda, in the early 2000s. His work focused on polymer composite materials that were fabricated from tannin-based resins reinforced with rice husks (Ndazi *et al.*, 2006). He is currently the Vice Chancellor of the Open University of Tanzania. In Zambia, Prof. Chris Chama has been engaged in metallurgical research in the Department of Metallurgy and Material Processing at the University of Zambia, since the late 1970s (Opoku and Chama, 1979). He has also collaborated with Prof. Joost Vlassak (Harvard) and Prof. Wole Soboyejo (then at Princeton) on the mechanical properties of copper thin films (Chama *et al.*, 2012).

However, a significant amount of African materials research has occurred within departments of chemistry, physics, electrical engineering, mechanical engineering, chemical engineering, and civil engineering, where the research efforts have focused on solving problems in the areas of passive/active solar energy and electrical properties (mostly departments of chemistry, physics, and electrical engineering) (Koech *et al.*, 2022; Abdalla *et al.*, 2004; Ergete *et al.*, 2015; Tong *et al.*, 2009; Manyala *et al.*, 2008; Fru *et al.*, 2021; Ahmed *et al.*, 2021; Orisekeh *et al.*, 2021; Chen *et al.*, 2011), biomaterials (mostly biomedical engineering) (Marei and El Backly, 2018; Meng *et al.*, 2009; Hampp *et al.*, 2012; Obayemi *et al.*, 2014; Cornish, 2017) and construction materials (mostly civil engineering) (Ballim and Reid, 2003; Kenai, *et al.*, 2004; Savastano, *et al.*, 2009; Tan *et al.*, 2011; Yakub and Soboyejo, 2013; Ismaiel *et al.*, 2009).

Materials programs that developed within departments of chemistry (University of Addis Ababa, Stellenbosch University, and the University of Benin) and departments of physics (University of Pretoria, Makerere University, University of Nairobi, University of Dar-es-Salaam, Witwatersrand University, University of South Africa (UNISA), University of Pretoria, Cheikh Anta Diop University, University of Ghana, University of Nigeria-Nsukka (UNN), University of Lagos, and the African University of Science and Technology in Abuja) have tended to focus on thin films and materials for energy. These include research groups led by the following: Prof. Banti

Workalemahu, Prof. Teketel Yohannes, and Prof. Shimelis Admassie at the University of Addis-Ababa (Abdalla *et al.*, 2004; Ergete *et al.*, 2015); Dr. Babaniyi Babatope (Tong *et al.*, 2009) at OAU, Prof. Bernard Aduda at the University of Nairobi (Aduda *et al.*, 2003); Prof. Ncholu Manyala (Manyala *et al.*, 2008) and Prof. Mmantsae Diale (Fru *et al.*, 2021) at the University of Pretoria, Prof. Malik Maaza at UNISA and iThemba Labs (Matinise *et al.*, 2017); Prof. Aboubaker Beye (Dione *et al.*, 2017), Prof. Papa Douta Tall (Tal *et al.*, 2007), and Prof. Balla Diop Ngom (Ngom *et al.*, 2009) at the Cheikh Anta Diop University, led by; Prof. Abdulhakeem Bello (Ahmed *et al.*, 2021) and Prof. Vitalis Anye Chioh (Du *et al.*, 2014) at AUST; Prof. Benjamin Agyei-Tuffour (Asare *et al.*, 2016) and Prof. Joseph Asare (Agyei-Tuffour *et al.*, 2016) at the University of Ghana; and Prof. Rose Osuji and Prof. Fabian Ezema at the University of Nigeria-Nsukka (UNN) (Nwanya *et al.*, 2014).

The Origins of the African Materials Research Society (AMRS)

The African Materials Research Society (AMRS) is a Pan-African Society that was formed in 2002, following a US/Africa Materials Workshop that was held at the Farm Inn in Pretoria, South Africa, in 2000 (Savastano *et al.*, 2006). The meeting, which was organized jointly by Dr. Adrian De Graaf of the US National Science Foundation (NSF) and Jill Sawyer of the National Research Foundation (NRF) of South Africa, brought together 70 materials researchers from the US and Africa to discuss areas of potential for research collaborations between materials scientists and engineers in Africa and the United States.

The idea of forming an African Materials Research Society was suggested at the US/Africa Materials Workshop in 2000. Potential areas for future US/Africa collaborations were also identified. These included materials research to address problems in energy and the environment, water, agriculture, building materials, transportation, and health.[1] These ideas were further developed at

[1] https://www.africanmrs.net.

subsequent meetings that were held in San Juan, Puerto Rico, and San Diego, CA. The workshop in Puerto Rico focused on materials for infrastructure, while the workshop in San Diego explored a wider range of opportunities for materials research and education (Savastano *et al.*, 2006; Soboyejo, 2007). These ideas were later integrated into the mission and vision of the US/Africa Materials Institute (USAMI) (Soboyejo, 2004), which was initiated in 2003 at Princeton University, with support from the Division of Materials Research of the National Science Foundation.

USAMI brought together researchers from several African countries to work with their US counterparts at US universities such as Harvard, Princeton, Yale, University of Illinois-Urbana-Champaign, Rutgers, Penn State, University of Pennsylvania, Arizona State University, and Lehigh University. Between 2003 and 2009, USAMI supported the visits of 100 Visiting African Fellows that visited US universities to work with collaborators with research programs on materials for health, energy, affordable building materials, and water purification (Soboyejo, 2004).

USAMI Visiting Researchers included Prof. Fawzy Hammad of Egypt, Prof. Olusegun Adewoye of Nigeria, Prof. Aboubaker Beye of Senegal, Dr. Joe Zimba of Zimbabwe, Prof. Benny Imasogie of Nigeria, Prof. Banti Workalemahu of Ethiopia, Prof. Shimelis Admassie of Ethiopia, Prof. Papa Douta-Tall of Senegal, Prof. Said Kenai of Algeria, Prof. Tom Otiti of Uganda, Prof. Francis Nyongesa (University of Nairobi), Dr. Sharif Al-Ghohary of the Egyptian Atomic Energy Authority, Dr. Mohammed al Khodhary (University of Alexandra), Prof. Modupe Oyatogun (Obafemi Awolowo University), Architect Mbake Niang of Senegal, Prof. Sam Kwofie of KNUST, Prof. John Amazigo of the University of Nigeria-Nsukka, and Prof. John Chama of the University of Zambia. A total of 99 Visiting Fellowships were supported by USAMI.

Following their visits, the African researchers returned to Africa to continue with their respective institutions to continue with their research institutions. Two of them (now Dr. Modupe Oyatogun and Dr. Mohammed al Khodhary) also received their PhDs, based on research work that they did during their visits to Princeton

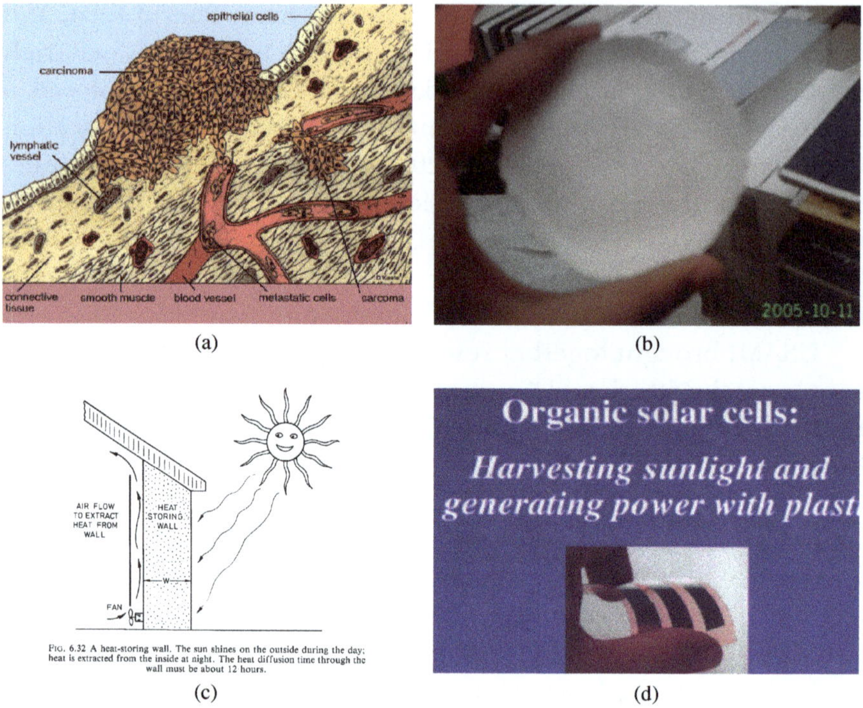

Fig. 4. Four Thematic Areas of USAMI: (a) biomaterials; (b) materials for water purification; (c) materials for thermal comfort and affordable housing.

University. They also established successful research groups in Africa, following their research visits to the US.

Several US undergraduate and graduate students also visited Africa to engage in research with African collaborators. These include the following: Dr. Tiffany Tong, who worked with Prof. Babaniyi Babatope on solar organic photovoltaics at the Engineering Materials Development Institute (EMDI) in Akure, Nigeria; Prof. Daniel Cohen and Prof. Randy Bly, who worked as undergraduates with Prof. Mona Marei on biomaterials at Alexandria University) (Soboyejo, 2004; Bly *et al.*, 2007), and other students that participated in Africa/African MRS Meetings. At least 40 US undergraduate and graduate students were supported by USAMI to participate in US/Africa workshops, AMRS conferences, and research and innovation projects across Africa between 2003 and 2009.

The Formation of the Africa/African Materials Research Society (AMRS)

The AMRS[2] was formed in Dakar, Senegal, in 2002, after two years of effort to develop an African Materials Network that was well connected to the Materials Research Society in the US[3] and the International Union of Materials Research Societies (IUMRS).[4] The first meeting of the IUMRS in Dakar was hosted by Prof. Aboubaker Beye of the Cheikh Anta Diop University, after an earlier planning meeting that was also held in Dakar, Senegal. The planning meeting brought together many of Africa's leading materials scientists and representatives of African governments to discuss the structure of the new society.

Figures 5(a)–(h) present some photographs of people that were present at the planning meeting. The keynote was presented by Prof. Wole Soboyejo, who presented a vision for the context and the future of the AMRS (Fig. 5(a)). Two of Africa's first PhDs in engineering from Stanford (Alfred Soboyejo from Ohio State University and Bert Marsh from the NSF Engineering Directorate) were also present (Fig. 5(b)), along with Prof. Olusegun Adewoye (left) and Dr. Adeoye Soyemi (middle) (Fig. 5(c)). Professor Chris Chama (right), Dr. Joe Zimba (second from right), and Prof. Banti Workalemahu (third from right) also participated actively in the program (Fig. 5(d)), along with Prof. Papa Douta Tall, Prof. Aboubaker Beye, Dr. Dele Popoola and Prof. Holmer Savastano (Fig. 5(e)).

Other participants at the meeting were the following: Prof. Elifas Bisanda (right), Prof. Mohammed Sassi (second from right), Prof. Banti Workalemahu (third from right), and Prof. Said Kenai (fourth from right) (Fig. 5(f)); Prof. Mohammed Sassi (right) and Prof. Abdel Azeez Jazouli (second from right) (Fig. 5(g)); Prof. Dan Nosenchuck (right) and Dr. Dele Popoola (second from right).

---

[2] Ibid.

[3] https://www.mrs.org.

[4] https://iumrs.org.

(a)

(b)

(c)

(d)

(e)

(f)

(g)

(h)

Several delegates wanted the emerging society to be closely connected to the bureaucratic structures of the governments of all the countries in Africa. However, in the end, a more conventional structure and approach were proposed for the fledgling society, without the bureaucratic connections to the governments of the different countries. It was also agreed that the Africa Materials Research Society, as it was then known, would be registered in Senegal.

The first meeting of the AMRS was attended by Prof. Alex King (President of the MRS in the US, Prof. Robert Chang (General Secretary of the IUMRS), and Ms. Anati Canca (South African NRF); Prof. Oumar Sock (Director, Polytechnic School, Cheikh Anta Diop University, Dakar, and Stanley Israel (Member of the Board of the American Ceramic Society) (Fig. 6).

After the first meeting of the AMRS in Dakar, Senegal, in 2002, the second meeting of the AMRS was held in Johannesburg, South Africa, at the University of Witwatersrand in 2003. The meeting was organized by Prof. Arthur Every and Prof. Yunus Ballim of the University of Witwatersrand. At the end of this meeting, Prof. Yunus Ballim was elected as the President of the AMRS, while Prof. Abdelaziz El Jazouly was selected to organize the third meeting of the AMRS in Marrakech, Morocco, in 2005. This was followed by the

Fig. 5. Images from the 2002 Meeting of the Africa MRS in Dakar, Senegal: (a) Prof. Wole Soboyejo (Princeton) keynote at 1st AMRS Meeting; (b) Dr. Elbert Marsh (NSF) and Prof. Alfred Soboyejo (The Ohio State University); (c) Prof. Olusegun Adewoye (EMDI) and Dr. Kunle Soyemi (Middle); (d) Dr. Joe Zimba, Prof. Chris Chama (University of Zambia), Prof. Banti Workalemahu (University of Addis-Ababa); (e) Prof. Holmer Savastano (University of Sao Paulo), Dr. Dele Popoola (Ford Motor Company), Prof. Aboubaker Beye (Cheikh Anta Diop University), Dr. Papa Douta-Tall (Cheikh Anta Diop University); (f) Prof. Said Kenai (University of Blida), Prof. Mohammed Sassi (National Engineering School of Monastir), Prof. Elifas Bisanda (Kigali Institute of Science and Technology), (g) Prof. Wole Soboyejo (Princeton), Prof. Abdelaziz El Jazouli (Mohammed Hassan II), Prof. Mohammed Sassi (National Engineering School of Monsastir), and (h) Dr. Dele Popoola (Ford Motor Company) and Prof. Dan Nosenchuck (Princeton).

Plenary Speakers at the Opening Session of the Africa Materials Research Society (MRS-Africa) inaugural meeting included (left to right) Aboubaker Chédikh Beye (University Cheikh Anta Diop [UCAD] Dakar and chair of the organizing committee), Anati Canca (National Research Foundation of South Africa), Alexander King (2002 President of the Materials Research Society), Robert Chang (General Secretary of the International Union of Materials Research Societies), Oumar Sock (Director, Polytechnic School, UCAD Dakar), and Stanley Israel (Member of the Board of the American Ceramic Society).

Fig. 6. Plenary Speakers at the first AMRS Meeting in Dakar, Senegal (Adapted from Conference Reports of MRS Bulletin, www.mrs.org/publications/bulletin.

fourth meeting of the AMRS, organized by Prof. Joseph Tesha at the University of Dar-es-Salaam in Tanzania in 2007.

Prof. Olusegun Adewoye was then elected to organize the fifth meeting of the AMRS, which was held in Abuja, Nigeria, in 2009, where Prof. Mona Marei of Alexandria University was elected as the first female President of the AMRS. Dr. Joseph Zimba was also selected to organize the sixth meeting of the AMRS meeting in Victoria Falls, Zimbabwe in 2011. Prof. Yohannes Teketel of the University of Addis Ababa was then selected to organize the seventh meeting of the AMRS in Addis Ababa in 2013 when Prof. Wole Soboyejo was elected to serve as President of the AMRS. This was followed by the eighth meeting of the AMRS that was organized by Prof. Dodoo-Arhin at the University of Ghana in Legon, Ghana, in 2015.

Dr. Samuel Chigome of the Botswana Institute for Technology Research and Innovation (BITRI) was selected to organize the ninth meeting of the AMRS in Gaborone, Botswana, in 2017, when the Society voted to change its name from the "Africa Materials Research Society" to the "African Materials Research Society." The logo of the AMRS was also changed (Fig. 7) to reflect the new name of the society in 2018. Due to the significant growth of African materials research-related activities coordinated from the secretariat office at BITRI in Botswana, Dr. Samuel Chigome was tasked by the International Union of Materials Research Societies (IUMRS) to set up the first-ever African Regional Office of the IUMRS at BITRI in Botswana. The functions of the Regional Office take advantage of the existing African MRS secretariat office at BITRI. The logo of the African Regional Office of the IUMRS is presented in Fig. 8.

Some of the special guests that attended the ninth AMRS conference in Gaborone, Botswana, are shown in Fig. 9. They include the former Vice President of Botswana, Dr. Mokgweetsi Eric Keabetswe

Fig. 7. Logo change that occurred after the AMRS name change from "Africa Materials Research Society" to "African Materials Research Society".

Fig. 8. Logo of the African Regional Office of the International Union of Materials Research Societies (IUMRS).

Fig. 9. Conference Photograph from the ninth International Conference of the AfricanMRS.

Masisi, who is now the President of Botswana. The guests included delegates from Europe, Asia, North America, and Africa.

Prof. Hulda Swai was also elected as President of AMRS, and chosen to organize the 10th meeting of the AMRS in Arusha, Tanzania, in 2019 (Fig. 10), while Dr. Samuel Chigome of BITRI was elected as the first-ever Vice President of the AMRS in 2017 tasked with setting up an AfricanMRS secretariat office at BITRI. Dr. Samuel Chigome was later elected as the President of the AMRS in 2019 and remained in this position for three years until December 2022, when Prof. Balla Diop Ngom was elected as the President of the AMRS after co-organizing the 11th meeting of the AMRS along with Prof. Aboubaker Beye of the Cheikh Anta Diop University in Dakar, Senegal. Prof Omololu-Akin Ojo of the East African Institute of Fundamental Research (EAIFR) at the University of Rwanda was also elected as the Vice President of the AfricanMRS.

Much of this US engagement in the AMRS has been financially supported by the US National Science Foundation (NSF). This includes the initial US/Africa Workshops, which were supported by Dr. Adrian De Graaf, and the subsequent MRS engagements that

Fig. 10. Conference Photograph from the 10th African MRS in Arusha, Tanzania. During the AMRS meeting in Dar-es-Salaam in 2009, Prof. Eric Garfunkel of Rutgers University volunteered to help with the organization of AMRS Meetings. Since then, Prof. Garfunkel has been involved in the organization of US interactions with the co-organizers of AMRS Meetings in Nigeria, Zimbabwe, Ethiopia, Ghana, Botswana, Tanzania, and Senegal. At each of these meetings, Prof. Garfunkel and Dr. Johanna Bernstein have supported teams of US researchers and students, and US/African co-organizers to participate in the meetings of the AMRS.

have been supported by Dr. Guebre Tessema and Ms. Margaret Anne Wampamba (now retired from the NSF) who have been helpful in organizing US/Africa collaborations, workshops and conferences since the inception of US/Africa meetings in 2000. Dr. Mamadou Diallo has also sponsored a recent US/Africa Meeting in South Africa in 2019. He also participated actively in the 11th AMRS Meeting that was held in Dakar, Senegal.

In any case, over the past 20 years, the African MRS, as it is now called, has developed into an international network of scientists and engineers with a passion for the development of materials science and engineering in Africa. The meetings have typically attracted between 400 and 500 delegates from all over the world. The meeting agenda typically includes educational workshops for graduate students; plenary talks that are often given by Nobel Laureates, and a range of keynote talks by leading materials scientists and engineers. The topics of interest cover the full range of materials for energy to biomaterials, structural materials, construction materials, materials for water purification, and materials for agriculture. The proceedings of the meetings have also been published in journals such as Materials Advances.[5]

Finally in this section, it is of interest to highlight the Nobel Laureates that have participated in the AMRS Meetings. They include the following: Prof. Robert Grubbs of Caltech (2005 Nobel Laureate in Chemistry for his work on olefin metathesis); Prof. Gerard Mourou (2018 Nobel Prize in Physics for his work on ultra-fast lasers), and Prof. Jean-Marie Lehn (1987 Nobel Prize in Chemistry for work on cryptands and supramolecular chemistry).

Prof. Grubbs was the plenary speaker at the AMRS meeting that was held in Addis Ababa in 2013. He gave an exceptional talk on catalysis and sustainability. Prof. Jean-Marie Lean was the keynote speaker at the African MRS meeting that was held in Gaborone, Botswana, in December 2017, while Prof. Gerard Mourou was a plenary speaker at the AMRS meeting that was held in Dakar, Senegal, in December 2022.

The Gaborone AMRS meeting was also attended by His Excellency Dr. Mokgweetsi Eric Keabetswe Masisi, who was then the Vice President of Botswana. Dr. Masisi presented a keynote talk at the 2017 AMRS conference. He has been the President of Botswana since 2018. It is also through his strong commitment to Materials Science that a state-of-the-art Center for Materials Science was set up at BITRI under the leadership of the founding CEO of BITRI, Prof Nelson Torto.

---

[5] https://www.cambridge.org/core/journals/mrs-advances/.

Collaborations of the AMRS with the IUMRS, ASESMA, and JUAMI

The African MRS has also worked closely with partners in Africa, Europe, and Asia.[6,7,8,9] These include interactions with the International Union of Materials Research Society (IUMRS)[10] and the Joint Undertaking for an Africa Materials Institute (JUAMI)[11] that was established in 2012 by Prof. Sossina Haile of Northwestern University, Prof. Simon Billinge (Columbia University) and Prof. Peter Green of the University of Michigan, to build collaborations between the United States and Africa in materials research and education.

Since 2012, JUAMI has organized Schools in East Africa that have introduced students from Africa and the US to materials science, electrochemistry, sustainable energy, clean water, and solar fuel production. These schools have featured the world's leading materials scientists and experiences that have provided the students with exposure to grant writing.[12] They have been funded by the US National Science Foundation.

Prof. Richard Martin (Stanford University and the University of Illinois at Urbana-Champaign), Dr. Nicola Seriani (International Center for Theoretical Physics, Trieste), and Dr. Omololu Akin-Ojo (East African Institute for Fundamental Research) have also worked with the African School for Electronic Structure Methods and Applications (ASESMA) to develop a US/Africa initiative on Electronic Structure. Their initiative has been partly funded by the American Physical Society.

---

[6] https://www.africanmrs.net.

[7] https://www.mrs.org.

[8] https://iumrs.org.

[9] https://www.juami.org.

[10] https://iumrs.org.

[11] https://www.juami.org.

[12] Ibid.

The IUMRS[13] has also worked closely with the AMRS since its inception in 2002. Their involvement ranged from the participation of Robert Chang (one of the founders of the IUMRS) in the original US/Africa Workshop in Pretoria in 2000 to meetings that led to the establishment of the AMRS, and the support of African materials scientists and graduate students to attend IUMRS meetings. As a result of the establishment of the African Regional Office of the IUMRS, AMRS through a team from Ghana led by Prof. Dodoo-Arhin of the University of Ghana is organizing the first-ever International Union of Materials Research Societies–International Conference on Advanced Materials (IUMRS–ICAM2025) in 2025.

Finally in this section, it is important to note that independent materials societies have been established in Egypt and Nigeria. These include the Egyptian Materials Research Society[14] and the Materials Society of Nigeria (MSN)[15] which have members that meet annually to discuss progress in materials research in these countries. Other country societies include the Materials Research Society of Kenya, the Ethiopian Materials Research Society, and the Materials Research Society of Ghana. Regional MRS meetings have also been organized in Monastir, Tunisia, and Accra, Ghana, by the AMRS. However, most of the conferences of the AMRS have occurred biennially over the past 20 years.

Interactions with the African Union (AU), the United Nations Economic Commission for Africa (UNECA), and United Nations Educational Scientific and Cultural Organization (UNESCO)

Since 2013, the AMRS has interacted with the African Union (AU). The interactions enabled the AMRS to host its seventh Meeting at the Headquarters of the African Union in Addis Ababa, Ethiopia, in 2013. The meeting also enabled some interactions between the

---

[13] https://iumrs.org.

[14] https://egyptmrs.wixsite.com/eg-mrs.

[15] https://msn.ng.

Fig. 11. Some of the speakers at the UNECA — AfricanMRS Nanotechnology Research and Innovation Bootcamp.

Board members of the AMRS (Prof. Eric Garfunkel and Prof. Wole Soboyejo) and the scientific leaders of the AU. More recently, Dr. Samuel Chigome (the most recent prior President of the AMRS) has developed an ongoing collaboration between the UN through two UN agencies UNECA and UNESCO, and the AMRS. A program cooperation agreement between the AfricanMRS and the UNECA was signed in 2020 with the aim to promote nanotechnology in Africa. This has led to a series of nanotechnology workshops that have brought together leading scientists to interact with African graduate students in cutting-edge areas of materials science and engineering (Fig. 11). Additionally, it made a significant contribution toward promoting a holistic view of materials science covering policy, research, education, entrepreneurship, and innovation in line with the objectives that form the foundation of the AfricanMRS.

Materials Research by Africans

The work at the University of Addis Ababa resulted in some major contributions by Workalemahu *et al.* (2004), Yohannes *et al.* (2015), and Admassie *et al.* (2009) in the area of organic electronics for solar cells. Similarly, the work at the University of Nairobi on solar cells was led by Prof. Aduda (2003), who also trained Prof. Francis Nyongesa, Dr. Robinson Juma Musembi, and Dr. Justus Simiyu, who are now members of the physics faculty at the University of Nairobi. The group at Makerere University in Uganda, which is in the physics department, has also been active with contributions from Prof. Tom

Otiti to the development of solar cells and nanoparticles for cancer imaging (Meng *et al.*, 2009; Otiti and Soboyejo, 2006). Rwenyagila in the Department of Physics at the University of Dar-es-Salaam in Tanzania has conducted computational studies of the effects of surface texture on the optical properties of silicon structures (Rwenyagila *et al.*, 2014; Rwenyagila *et al.*, 2015).

In West Africa, the groups in Nigeria, Senegal, and Ghana have made significant contributions. These include contributions to ceramic research by Adewoye and Page (1976) of OAU and thin film research by Beye *et al.* (2017), Tall *et al.* (2007), Ngom *et al.* (2009). Prof. Archana Bhaw-Luximon and Prof. Dhanjay Jhurry have also made very significant contributions to the field of biomaterials (2017).

Significant efforts have also been made to develop materials research programs in departments with a focus on engineering physics, mechanical engineering (University of Cape Town, University of Pretoria, University of Lagos, and the University of Nairobi), chemical and petroleum engineering (University of Port Harcourt, Stellenbosch University and African University of Science and Technology), biomedical engineering and biomaterials (University of Ghana, Alexandria University and the University of Cape Town), and construction materials (University of Blida, University of Ibadan, the African University of Science and Technology, and the Federal University of Science and Technology in Akure).

The outcomes of the research in engineering departments have been quite significant. In the case of the engineering physics groups at OAU, Prof. Oladele Ajayi and Prof. A. M. Salau established a fabrication facility for the study of thin films and microelectronic structures in the 1970s. The group at the OAU also produced notable graduates, such as Prof. Akintayo Akinwande, who went on to become a professor of electrical and computer engineering at MIT. Prof. John Amazigo of the University of Nigeria Nsukka (UNN) has also made outstanding contributions to the field of mechanics and materials (Budiansky *et al.*, 1988a, 1988b, 1989), while Prof. Alex Animalu of UNN is a pioneer in materials physics. The

Volker-Heine–Animalu potential (Animalu and Heine, 1965) is named after him and his Cambridge PhD advisor, Volker Heine.

Similarly, the groups in South Africa have made outstanding contributions to materials science. These include the notable contributions of Prof. Frank Nabarro to dislocation theory Nabarro (1987) and the work of Prof. Lesley Cornish in physical metallurgy (Cornish, 2017), and the contributions of Prof. Malik Maaza to nanoscience (Matinise *et al.*, 2017). The contributions from South Africa also include electronic materials and thin films research by Prof. Mmantsae Diale (Fru *et al.*, 2021), and work on electronic materials by Prof. Ncholu Manyala at the University of Pretoria (Manyala *et al.*, 2008). The work of Dr. Esau Masuku of BITRI on low-cost construction materials has also resulted in the construction of sustainable buildings in Botswana (Masuku and Maedza, 2006).

In North Africa, there have been major contributions from Egypt, Algeria, and Morocco. These include outstanding work by the Egyptian materials scientist, Prof. Mona Marei of Alexandria University (Chen *et al.*, 2011; Marei and El Backly, 2018). Prof. Mona Marei, who was also the first female President of the AMRS, is a pioneer in the field of tissue engineering (Chen *et al.*, 2011; Marei and El Backly, 2018). She has also contributed significantly to the development of Alexandra University as a Center of Excellence in Tissue Engineering.

Beyond Egypt, and within North Africa, significant contributions have been made by Kenai and co-workers (Kenai *et al.*, 2004) to the development of green concrete. These include limestone concrete (Kenai *et al.*, 2004) and other novel approaches to the development of concrete chemistries that reduce the carbon dioxide emissions associated with cement production. The work of Prof. Abdelaziz El Jazouli (El Jazouli *et al.*, 1988) of Mohammed Hassan II University is also acknowledged, especially in glass formation, and the structure and thermodynamic properties of phosphate glasses.

In Central Africa, it is important to acknowledge the contributions of Dr. J. Foba-Tendo and Dr. Tido Stanislas, who have worked with Prof. Holmer Savastano of the University of Sao Paulo on the use of nano-fibrillated cellulose as a reinforcement in earth-based composite materials for construction (Stanislas *et al.*, 2020).

The Emergence of African Centers of Excellence in Materials Research and Education

The development of the African Materials Network has enabled African Centers of Excellence to emerge over the past few decades. The initial centers of excellence were in South Africa. They include Centers at the University of Witwatersrand (Center for Strong Materials[16] and AMSEN[17]); the Pan African Materials Institute (PAMI)[18] at AUST; the Center for Research, Agricultural Advancement, Teaching Excellence and Sustainability (CREATES)[19] in Food and Nutritional Security in Africa.

The Center of Excellence in Strong Materials (CoE-SM)[20] focused on new ways of increasing South Africa's strength in exploring hard materials beyond diamonds. It was established in 2004 with funding from the NRF and the Department of Science and Innovation (DSI) in South Africa. It is hosted by the University of Witwatersrand. However, it also includes collaborations with Nelson Mandela University, Rhodes University, the Universities of Johannesburg, KwaZulu-Natal, Limpopo, Cape Town, and Stellenbosch, in addition to the Science Councils and the South African Energy Corporation. The areas of research include hard materials for cutting tools (carbides and cermets), carbon nanotubes, strong composites, ceramics, diamonds, thin films, and strong alloys.

The African Materials Science and Engineering Network (AMSEN)[21] is another African Center of Excellence that was established at Wits in 2008. Led by Prof. Lesley Cornish of the University of the Witwatersrand, AMSEN was funded originally by the Carnegie

---

[16] https://www.wits.ac.za/strongmaterials.

[17] https://www.wits.ac.za/strongmaterials/amsen.

[18] https://www/aust.ng/pami.

[19] https://www.creates-nmaist.ac.tz.

[20] https://www.wits.ac.za/strongmaterials.

[21] https://www.wits.ac.za/strongmaterials/amsen.

Foundation, and active between 2008 and 2016. It was one of five networks that were funded by the Carnegie Institute for Advanced Studies Regional Initiative for Science and Engineering (RISE) Program. The goal was to prepare future faculty members through research and collaborations between the nodes that were located at Wits in South Africa, the University of Botswana, the University of Ghana, the University of Nairobi, the University of Namibia, and the Federal University of Technology, Akure (FUTA). Their work focused largely on metallic materials and their alloys. AMSEN produced high-quality PhDs that have gone on to successful careers in academia across Africa.

The Pan African Materials Institute (PAMI)[22] was established at AUST, with funding from the World Bank African Centers of Excellence (ACE) program, the program brought together collaborators from Ghana, Nigeria, and Cameroon to work on materials for health, energy, water and the affordable building materials. Research at PAMI resulted in targeted drugs and drug delivery systems for breast cancer treatment, novel ceramic water filters for the filtration of microbes and chemical pathogens from contaminated water, earth-based materials for sustainable buildings, and improved devices (solar cells, light emitting devices, batteries, and supercapacitors), for energy harvesting and energy storage.

The PAMI program has trained more than 50 PhDs in materials science and engineering that are now working mostly as faculty members in Africa and other parts of the world. They include graduates that are professors at MIT, AUST, Nile University, Penn State University, Buffalo State University, Baze University, University of Dar-es-Salaam, 2–E — Institut Internationa' d'Ingénierie d' l'Eau et de l'Environnement, Kwara State University (KWASU), the University of Buea, the University of Ghana, the Kwame Nkrumah University of Science and Technology (KNUST), the University of Liberia, the University of Abuja, Ashesi University, Sunyani Technical University, and the Sheda Science and Technology Complex (SHESTCO) in Abuja, Nigeria.

---

[22] https://www/aust.ng/pami.

In addition to many students in materials science benefitting from the capacity-building bootcamps in nanotechnology innovation, entrepreneurship, and research, the collaboration with UNECA resulted in the development of a Pan-African Curriculum in Materials Science and Nanotechnology. The curriculum aims to ensure that any materials science student in Africa has access to research and learning facilities without the limitation of funding. Furthermore, the collaboration has resulted in AfricanMRS contributing to the African Science Technology (STI) and Innovation landscape through the annual Africa Science Technology and Innovation Forum.

A notable initiative is the ORIGIN research and innovation lab, a concept that creates a multidisciplinary platform for the best brains in Africa to come together and solve African problems. Through this initiative where the first regional lab has been set up in Kenya at the Dedan Kimathi University of Technology and led by Prof Kioni, materials scientists in Africa will benefit to ensure that they develop solutions that make a significant impact in society. Overall, the dream of AfricanMRS seeing the realization of a triple helix model that will result in sustainable relationships between industry, governments, and academia in materials research is slowly becoming a reality. This could lead ultimately to the sustainable funding of materials research across Africa.

Finally, it is important to appreciate the work of the Nanoscience African Network (Nanoafnet)[23]. This was established in 2005 by Prof. Malik Maaza of iThemba Labs and the University of South Africa (UNISA). Nanoafnet, which was established with support from the International Center for Theoretical Physics (ICTP), has trained African scientists to make important contributions to nanoscience and nanotechnology. It has also produced Dr. Samuel Chigome and Prof. Balla Ngom Diop, who have emerged as young leaders of the African MRS in recent years. These are all signs of a new era in which a new generation of African materials researchers is well poised to use materials science and engineering as an engine for sustainable African development.

[23] https://nanoafnet.tlabs.ac.za.

The Role of the African Diaspora

It is of interest to comment on Africans in the Diaspora who have made important contributions to materials research and efforts to develop materials research and education in Africa. These include the following: Prof. Kwadwo Osseo-Asare of Penn State University, who has made important contributions to materials synthesis and processing, hydrometallurgy and electrochemistry, and the teaching of Africans in these areas; Prof. Sossina Haile of Northwestern University, who has made important contributions to the development of fuel cells along with teaching at JUAMI and AUST, and Prof. Ange Nzihou of the University of Albi-Toulouse in France, who has been recognized by the President of France on the materials science of carbon-based materials.

Other Africans that have made major contributions to materials research include the following: Prof. Deji Akinwande of the University of Texas at Austin, for his work on carbon nanotubes and device physics, and the development of technology start-ups; Prof. Tayo Akinwande of MIT for his research on Micro-Electro-Mechanical Systems (MEMS) and nanotechnology, and his work on training African STEM academics at MIT; Prof. Ilesanmi Adesida of Nazarbayev University in Kazakhstan, for his research on nanoscale devices and optoelectronics, and Prof. Wole Soboyejo for his research on multifunctional materials for health, energy, and the environment, and STEM outreach to Africa.

Emerging African materials researchers in the Diaspora include the following: Prof. Iwnetim Abate of MIT for his work on electrochemistry and materials for energy storage and computing and his work on STEM outreach to Ethiopia (SCIFRO INC); Prof. Loza Tadesse of MIT for her work on medical diagnostics, cell type monitoring and detection, and her work on STEM outreach to Ethiopia (SCIFRO INC); Prof. Aristide Gumyusenge of MIT for his work on polymers for computing and sensing, Prof. Chibueze Amanchukwu of the University of Chicago for his work on batteries and STEM outreach in Nigeria, Prof. David Kwabi of the University of Michigan for his work on batteries and carbon-based materials, Prof. Kwabena

Bediako of the University of California Berkeley for his work on electrochemistry and two-dimensional materials, Prof. Boubacar Kanté of the University of California Berkeley for his work on nanophotonics and quantum optics and Prof. Ismaila Dabo of Penn State University for his work on computational catalysis and energy storage, and Prof. Nelson Dzade of Penn State University for his work on energy materials and computational modeling. These are just some of the Africans that are leveraging their experience in the Diaspora, while also making important contributions to materials research and education in Africa. Some of these faculty are active participants of the African MRS conference by presenting their research and offering workshop lectures. Fruitful collaborations between the scientists in the diaspora and those in the continent have a strong potential to train the next generation and develop impactful STEM solutions to local African problems.

Implications

Africa also has the potential to leverage materials research in the development of a strategy for the utilization of Africa's rich array of human, materials, and energy resources for the development of new industries that will add value to Africa's mineral deposits and "waste" materials. Such industries, as they have in many parts of the world that have engaged in materials processing, are likely to create wealth, while solving many of the local problems associated with Africa's challenges in health, energy and the environment, agriculture, transportation, and infrastructure.

Furthermore, as Africa attempts to leap-frog its way into the 21st century, the fourth Industrial Revolution offers Africa the opportunity to use digital knowledge and innovation to leverage fields such as data science in the development of new materials, novel green building materials, autonomous electric cars that are powered by batteries produced from African raw materials and recycled waste materials, and several new cutting edge technologies that will enable the transformation of Africa through research and development in materials science and engineering.

Summary and Concluding Remarks

This paper presents an overview of the history of materials research in Africa and the role that different stakeholders have played in this history. Following a brief description of the early history of materials in Africa, we focus mostly on the evolution of African materials research over the past century. This includes the early efforts in South Africa, which includes the research at Mintek, CSIR, Wits, University of Pretoria, the University of Cape Town, Stellenbosch University, University of KwaZulu Natal, and University of Johannesburg, and the development of materials programs in Egypt, Ethiopia, Algeria, Morocco, Kenya, Uganda, Tanzania, Zambia, Senegal, Ghana, Nigeria, Namibia, Cameroon, Zimbabwe, and Zambia.

In all these places, materials research efforts have emerged over the past century. AMRS has also evolved as an organization that has woven these Pan-African efforts into an international network of collaborations with researchers from Europe, Asia, North and South America, and the Caribbean. The collaborations have also engaged the African Diaspora as partners in the development of African capacity in materials research and education.

The major contributions of Africa to materials research include the following: bronzes (the Benin Empire); earth-based materials (the Egyptians); dislocation theory (Frank Nabarro); the Volker-Heine–Animalu Potential (Alex Animalu); mining and extractive metallurgy (Mintek); titanium metallurgy (CSIR); nanomaterials (Malik Maaza); fuel cells (Sossina Haile); biomaterials and tissue engineering (Mona Marei); mechanics and materials (John Amazigo); carbo-nitriding (Olusegun Adewoye), batteries and carbon-based materials (Ncholu Manyala), Lesley Cornish (metallurgy); Ange Nzihou (carbon-based materials), MEMS and nanotechnology (Akintayo Akinwande), optoelectronics (Ilesanmi Adesida); carbon-based devices (Deji Akinwande); cement-based materials (Yunus Ballim); green concrete (Said Kenai), Esau Masuku (low-cost construction materials), Archana Bhaw-Luximon and Dhanjay Jhurry (biomaterials), and multifunctional materials (Wole Soboyejo).

However, there is a need to develop stronger linkages between African materials researchers and industry that could further enable the translation of research into products that create value that can transform the minerals and "waste" materials of Africa into wealth. There is also the opportunity for African governments and the African Union to leverage the African researchers and the African Diaspora/International Collaborators in the transformation of Africa's minerals and waste materials via a manufacturing initiative that could lead to the development of a new generation of "African Lions," just like the "Asian Tigers" before them. These are some of the opportunities for the future of materials research for the sustainable development of Africa.

Acknowledgments

The authors are grateful to Prof. Robert Crease for his assistance with the preparation of this paper on the history of materials research in Africa. His encouragement and support are much appreciated. The authors would also like to thank the Late Dr. Arne Hessenbruch (Lecturer at MIT) for his enthusiastic support of our efforts to prepare this paper.

References

Agyei-Tuffour, B., Rwenyagila, E. R., Asare, J., Zebaze Kana, M. G. and Soboyejo, W. O. (2016). Effects of pressure on the nano- and micron-scale morphological changes in conjugated polymer photovoltaic cells. *Journal of Materials Research* 31, 3187–3195.

Abdalla, Taha H., Wendimagegn, Mammo and Bantikassegn, Workalemahu (2004). Electronic and photovoltaic properties of a single layer poly[3-(2″,5″-diheptyloxyphenyl)-2,2′-bithiophene] devices. *Synthetic Metals* 144, 213–219.

Ade Ajayi, J. F. and Crowder, M. (January 1, 1985). *West Africa in History, Vol. 1: Before 1800* (3rd edn.). (London: Longman Sc & Tech).

Adewoye, O. and Page, T. (1976). *Journal of Materials Science* 11, 981–984.

Aduda, B. O., Ravirajan, P., Choy, K. L. and Nelson, J. (2003). Anisotropic behaviour of etched hardness indentations. *International Journal of Photoenergy* 6(3), 141–147.

Afonja, A. (1982). Grain size dependence of the strain rate sensitivity and activation energy of high temperature deformation of a microduplex stainless steel. *Materials Science and Engineering* 54, 257–263.

Ahmed, R. A., Ebechidi, N., Ichwani, Re., Orisekeh, K., Huda, A., Bello, A., Oyewole, O. and Soboyejo, W. O. (2021). Pressure-induced interfacial contacts and the deformation in all solid-state Li-ion batteries. *Journal of Power Sources* 521 230939.

Ampaw, E. K., Arthur, E. K., Badmos, A., Obayemi, J. D., Adetunji, A. R., Olusunle, S. O. O., and Soboyejo, W. O. (2019). *Journal of Materials Engineering and Performance* 28. doi.org/10.1007/s11665-019-04471-8.

Animalu, A. O. E. and Heine, V. (1965). The screened model potential for 25 elements. *Philosophical Magazine* 12(120), 1249–1270.

Arthur, E. K., Ampaw, E., Kana, M. G. Zebaze, Adetunji, A. R., Olusunle, S. O. O., Adewoye, O. O. and Soboyejo, W. O. (2015). Nano- and Macro-wear of Bio-carbo-nitrided AISI 8620 Steel Surfaces. *Metallurgical and Materials Transactions* 46, 5810–5829.

Asare, J., Adetunji, S. A., Oyewole, O. K., Agyei-Tuffour, B., Du, J., Arthur, E., Anye, V. C., Zebaze Kana, M. G. and Soboyejo, W. O. (2016). Cold welding of organic light emitting diodes: Interfacial and contact models. *Physics Advances* 6(6), 1–12.

Ashby, M. F. (2016). *Materials Selection in Mechanical Design* (5th edn.). (Butterworth-Heinemann, Oxford: Elsevier).

Ballim, Y. and Reid, J. C. (2003). Reinforcement Corrosion and the Deflection of RC Beams: An Experimental Critique of Current Test Methods. *Cement and Concrete Composites*, 25, 625–632.

Bhaw-Luximon, A. and Jhurry, D. (2017). Artemisinin and its derivatives in cancer therapy: status of progress, mechanism of action, and future perspectives. *Cancer Chemother Pharmacol* 49, 451–466.

Bisanda, E. T. N., Ogola, W. O. and Tesha, J. V. (2003). *Cement and Concrete Composites* 25, 593–598.

Bly, R. A., Cao, Y., Moore, W., Gao, Z., Cuitino, A. and Soboyejo, W. O. (2007). Investigation of the effects of alkane phosphonic acid/RGD coatings on cell spreading and the interfacial strength between human osteosarcoma cells and Ti-6Al-4V. *Materials Science and Engineering C* 27, 83–89.

Budiansky, B. and Amazigo, J. C. (1989). Toughening by aligned, frictionally constrained fibers. *Journal of the Mechanics and Physics of Solids* 37, 93–109.

Budiansky, B., Amazigo, J. C. and Evans, A. G. (1988). Small-scale crack bridging and the fracture toughness of particulate-reinforced ceramics. *Journal of the Mechanics and Physics of Solids* 36, 167–187.

Budiansky, B., Amazigo, J. C. and Evans, A. G. (1988). Interaction of particulate and transformation toughening. *Journal of the Mechanics and Physics of Solids* 36, 581–595.

Chama, C. C., Vlassak, J. and Soboyejo, W. O. (2012). Stress evolution in copper-silver thin films during thermal cycling. *Materials Science and Engineering A* 532, 414–419.

Chen, J., Bly, R. A., Saad, M. M., Al Khodary, M. A., El-Backly, R. M., Cohen, D. J., Kattamis, N., Fatta, M. M., Moore, W. A., Arnold, C. B., Marei, M. K., Soboyejo, W. O. (2011). In-vivo study of adhesion and bone growth around implanted laser groove/RGD-functionalized Ti-6Al-4V pins in rabbit femurs. *Materials Science and Engineering* 31, 826–832.

Cornish, L. (2017). PGMs: A cornucopia of possible applications. *Journal of the South African Institute of Mining and Metallurgy* 117, 969–974.

Dione, A., Ndaiye, S., Gueye, El H., Dioum, A., Diouf, A., Abderman, M., Ngom, B. D., Beye, A. (2017). *Nanomaterials* 5, 68–71.

Du, J., Anye, V., Yu, D., Vodah, E., Kana, Z. and Soboyejo, W. O. (2014). *Journal of Applied Physics* 115, 233703.

El Jazouli, A., Viala, J. C., Parent, C., Le Flem, G. and Hagenmuller, P. (1988). Structural investigation of glasses belonging to the Na2O Nb2O5 P2O5 system. *Journal of Solid State Chemistry* 73(2), 433–439.

Ergete, Assegid, Kedir, F., Li, Yonfang and Anshebo, T. Y. (2015). Effect of solvent additives and P3HT on PDTSTTz/PCBM-based bulk heterojunction solar cells. *Journal of Photonics for Energy* 5, 057209.

Fathy, H. (1973). *Architecture for the Poor: An Experiment in Rural Egypt* (1427 E. 60th Street Chicago, IL 60637 USA: University of Chicago Press).

Fru, J., Nombona, N. and Dialo, M. (May 2021). Characterization of Thin MAPb(I1–xBrx)3 Alloy Halide Perovskite Films Prepared by Sequential Physical Vapor Deposition. *Frontiers in Energy Research* 9.

Hampp, E., Botah, R., Odusanya, O. S., Anuku, N., Malatesta, K. A. and Soboyejo, W. O. (2012). Biosynthesis and adhesion of gold nanoparticles for cancer detection and treatment. *Journal of Materials Research* 27, 2891–2901.

Harms, R. (2018). *Africa in Global History with Sources* (1st edn.) (New York, USA: W. W. Norton & Company), Print ISBN 9780393927573, 0393927571.

Ismaiel, Y., Plappaly, A., Leftwich, M., Malatesta, K., Friedman, K. C., Rivera, R., Soboyejo, A. B. O. and Soboyejo, W. O. (2009). Porosity, flow and filtration characteristics of frustrum-shaped ceramic water filters. *Journal of Environmental Engineering* 139, 986–994.

Kenai, S., Soboyejo, W. and Soboyejo, A. B. O. (2004). Some engineering properties of limestone concrete. *Materials and Manufacturing Processes* 19, 949–962.

Kete-Asante, M. (2019). *The History of Africa the Quest for Eternal Harmony* (3rd edn.) (London, United Kingdom: Routledge), Print ISBN 9781138710382, 1138710385.

Klenam, D., Polese, C., Chown, L. H. and Kwofie, S., Cornish L. A. (2015). Mechanical behaviour of pack carburized AISI 316L austenitic stainless steel. *Journal of the South African Institute of Mining and Metallurgy* 115, 1183–1191.

Koech, R., Olanrewaju, Y. A., Ichwani, R., Kigozi, M., Oyewole, D., Oyelade, O. V., Sanni, D. M., Adeniji, S. A., Collin-Ulloa, E., Titova, L., Martin, J., Grimm, R., Bello, A., Oyewole, O., Ntsoenzok, E. and Soboyejo, W. O. (2022). Thermally induced failure mechanisms in double and triple cations perovskite solar cells. *Scientific Reports* 12, 12860.

Kwofie, S. (2001). An Exponential Stress Function for Predicting Fatigue Strength and Life Due to Mean Stresses. *International Journal of Fatigue* 23, 829–836.

Manyala, N., DiTusa, J. F., Aeppli, G. and Ramirez, A. P. (2008). Doping a semiconductor to create an unconventional metal. *Nature* 454, 976–980.

Marei, M. K. and El Backly, R. (2018). *Frontiers in Bioengineering and Biotechnology* 6. doi:10.3389/fbioe.2018.00049.

Masuku, E. U. and Maedza, Botshelo H. (2006). The development of an affordable Kgalagadi Sand Building Block (KSBB) — a position paper. *Journal of Materials Science* 41, 6949–6955.

Matinise, N., Fuku, X. G., Kaviyarasu, K., Mayedwa, N., Maaza, M. (2017). *Applied Surface Science* 406, 339–347.

Meng, J., Fan, J., Galiana, G., Branca, R. T., Clasen, P. L., Ma, S., Zhou, J., Leuschner, C., Kumar, C. S. S. R., Hormes, J., Otiti, T., Beye, A. C., Harmer, M. C., Kiely, C. J., Warren, W., Haataja, M., and Soboyejo, W. O. (May 5, 2009). LHRH-functionalized superparamagnetic iron oxide nanoparticles for breast cancer targeting and contrast enhancement in MRI. *Materials Science and Engineering C* 29(4), 1467–1479.

Momade, F. Y. and Momade, Zs. G. (1999). Reductive Leaching of Manganese Oxide Ore in Aqueous Methanol- Sulphuric Acid Medium. *Hydrometallurgy* 51, 103–113.

Nabarro, F. R. N. (1987). *Theory of Crystal Dislocations* (New York, USA: Dover Publications), pp. 1–821.

Ndazi, B., Tesha, J. V., Karlsson, S. and Bisanda, E. (2006). Production of rice husks composites with Acacia mimosa tannin-based resin. *Journal of Materials Science* 41, 6978–6983.

Ngom, B. D., Mpahane, T., Manyala, N., Nemraoui, O., Buttner, U., Kana, J. B., Fasasi, A. Y., Maaza, M. and Beye, A. C. (2009). 255, 4153–4158.

Nwanya, A. C., Jafta, Charles J., Ejikeme, Paul M., Ugwuoke, Paulinus E., Reddy, M. V., Osuji, Rose U., Ozoemena, Kenneth I. and Ezema, Fabian I. (2014). Electrochromic and electrochemical capacitive properties of tungsten oxide and its polyaniline nanocomposite films obtained by chemical bath deposition method. *Electrochimica Acta* 128, 218–225.

Obayemi, J. D., Odusanya, S., Malatesta, K., Anuku, N. and Soboyejo, W. O. (2014). Biosynthesis and the conjugation of magnetite nanoparticles with luteinizing hormone releasing hormone. *Materials Science and Engineering C* 43, 482–496.

Opoku, J. and Chama, C. C. (1979). An Investigation of Environment-Assisted Cracking in Mild Steel. *Zambia Journal of Science and Technology* 4, 45–52.

Orisekeh, K., Singh, B., Olanrewajua, Y., Kigozi, M., Ihekweme, G., Umare, S., Anye, V., Bello, A., Parida, S., Soboyejo, W. O. (2021). Processing of α-Fe_2O_3, nanoparticles on activated carbon cloth as binder-free electrode material for supercapacitor energy storage. *Journal of Energy Storage* 33, 102042.

Otiti, T. and Soboyejo, W. O. (2006). Limited contribution of photovoltaic energy to the economic development of sub-Saharan Africa. *Perspectives on Global Development and Technology* 5, 69–80.

Rwenyagila, E. R., Akin Ojo, O., Onogu, K. O., Zebaze-Kana, M. G., Agyei-Tuffour, B., Alford, T. L. and Soboyejo, W. O.(2015). Computational modeling of transparent electrodes for multilayered ZnO/Al/ZnO optical structures. *Optical Letters* 40, 3914–3919.

Rwenyagila, E. R., Akin Ojo, O., Zebaze-Kana, M. G. and Soboyejo, W. O. (2014). Optical properties of ZnO/Al/ZnO multilayer films for large area transparent electrodes. *Journal of Materials Research* 29, 2912–2920.

Savastano, H., Ghavami, K., Beye, A. C. and Soboyejo, W. O.(2006). Materials for affordable housing. Special Edition on Materials for Affordable Housing and Infrastructure. *Journal of Materials Science* 41, 6895–6964.

Savastano, H., Santos, S. F., Radonjic, M. and Soboyejo, W. O. (2009). Fracture and fatigue of natural fiber-reinforced composites. *Cement and Concrete Composites* 31, 232–243.

Soboyejo, W. (March 22, 2007). Technological skills essential for continent's growth. *Business Daily*, Nation Media Group, Nairobi, pp. 22–23.

Soboyejo, W. O. (2005). The Role of Science and Technology in National Development, Nigerian Economic Summit Group, Public Lecture Series, Invited Lecturer, 17th of March 2005, Published By the Nigerian Economic Summit Group Ltd., ISBN No. 978-37849-2-7.

Soboyejo, W. O. (2007). Materials issues in bioengineering. *The Journal of Materials Science: Materials in Medicine* 18, 1–179.

Soboyejo, W. O. (2015). *Can Energy Access Help to Improve Health? Smart Villages: New Thinking for Off-Grid Communities Worldwide* (Banson, Cambridge, UK: Lavenham Press).

Soboyejo, W. O. (2019). History of the African materials research society. *Bulletin of the African Materials Research Society* 1(1).

Soboyejo, W. O. (January 2004). The US/Africa materials institute. *IUMRS Facets*, 3(1), 11–14.

Soboyejo, W. O. and Madsen, L. D. (July 2018). Strengthening diversity and cooperation through international collaborations: A focus on Africa, South America and the Caribbean. *Materials Research Bulletin* 43, 541–547.

Stanislas, Tido, Foba-Tendo, J., Ojo, Emeso, Ngasoh, Odette Fayen, Onwualu, Peter Azikiwe, Njeugna, Ebenezer and Savastano, Holmer (2020). Production and characterization of pulp and nanofibrillated cellulose from selected tropical plants. *Journal of Natural Fibers* 19, 1–17.

Tall, P. D., Ndaiye, S., Beye, A. C., Zong, Z., Soboyejo, W., Lee, H. J., Ramirez, A. G. and Rajan, K. (2007). *Materials and Manufacturing Processes* 22, 175–179.

Tan, T., Meng, J., Allameh, S. M., Kwofie, S., Dissmore, D., Ghavami, K. and Soboyejo, W. O. (2011). Mechanical properties of functionally graded hierarchical bamboo structures. *Acta Biomaterialia* 7, 3796–3803.

Thomas, J. M., Algohary, S., Hammad, F. and Soboyejo, W. O. (2006). Materials selection for thermal comfort in passive solar buildings. *Journal of Materials Science* 41, 6897–6907.

Tong, T., Akogwu, O., Meng, J., Eleruja, M., Admassie, S., Babatope, B., Woodson, T., Zong, Z., Rahbar, N., Watola, C. and Soboyejo, W. O.

(2009). Adhesion in organic electronic structures. *Journal of Applied Physics* 083708-8.

Yakub, I. and Soboyejo, W. (2013). Adsorption of fluoride from water using sintered clay-hydroxyapatite composites. *Journal of Environmental Engineering* 139, 995–1003.

Yang, Y., Yao, N., Imasogie, B. and Soboyejo, W. O. (2008). Nanoscale and submicron fatigue crack growth in nickel microbeams. *Acta Materialia* 55, 4305–4315.

Chapter 5.4

The Roots of Materials Science in Argentina: Jorge A. Sabato's Multifaceted Metallurgy Project

Armando Fernández Guillermet

Introduction

In Chapter 14 of his comprehensive book *The Coming of Materials Science*, Robert W. Cahn (Cahn, 2003) discussed the development of materials science in particular places. His approach was to select "particular institutions" and paint "brief portraits of the people that brought them into being." In the case of Argentina, he selected two centers linked to the development of atomic energy in the country, one of them in Buenos Aires, and the other in a city in Patagonia: San Carlos de Bariloche (hereinafter: Bariloche). Briefly analyzing the development that metallurgy and materials science reached in the 20th century in these places, Cahn highlighted the initiatives of the Argentine physics professor and metallurgist Jorge Alberto Sabato.

Advancing in the direction chosen by Cahn, this paper aims to analyze the activity that Jorge A. Sabato developed in the almost 15 years in which he acted in the Atomic Energy Commission of Argentina (CNEA), starting in 1955. Specifically, the purpose of the current work is, in the first place, to expand the portrait of Sabato presented by Cahn (2003), namely, "an Argentinian metallurgist who had recently joined the [...] laboratory from the local industry," and "a dynamo of energy and originality." To this end, a synoptic picture of the family, cultural and academic environment in which he developed is offered. Second, the present study aims to conceptualize what later in the work will be called Sabato's project. For this purpose, an analysis is reported of its political, institutional, and academic contexts, as well as the strategy chosen by him to (a) build up in Argentina research capability and metallurgical expertise relevant to nuclear technology as well as to the local industry and (b) strengthen the higher education in metallurgy and the training of metallurgical engineers in Argentina and other countries in Latin America.

Since the objective of the current work is to obtain a synoptic image of Sabato's interrelated initiatives, the main sources of information are as follows: (i) the articles in which Sabato himself analyzes the work at the CNEA in chronological order: (Sabato, 1968), (Sabato, 1970), (Sabato, 1972), (Sabato, 1973) and (Sabato, 1981); (ii) the institutional report (CNEA, 1972) on the activities carried out by CNEA's Metallurgy Department in the period 1955–1972; (iii) the annual reports ("Memorias Anuales") and informative bulletins of the CNEA corresponding to the same period: (CNEA, 1955), (CNEA, 1958), (CNEA, 1961), (CNEA, 1962a), (CNEA 1962,b), (CNEA 1962-1963), (CNEA, 1964), (CNEA, 1965), (CNEA, 1966), (CNEA 1967), (CNEA, 1969), (CNEA, 1970), (CNEA, 1971) and (CNEA, 1973); (iv) the works published by Sabato's collaborators (Martínez Vidal, 1993 and CNEA-UNSAM, 1998); and, (v) other works related to particular aspects of Sabato's activity (Alegría *et al.*, 1972; Bonfiglioli *et al.*, 1972; Galvele, 2001; Iurman, 2009; Enríquez, 2011, 2012 and Ovejero García, 2018). This specific information was complemented with that offered by more general historical studies, which are indicated at the beginning of the corresponding sections, as well as other works in which Sabato discusses the relations between science, technology and development in Latin America (Sabato and Botana, 1968).

Finally, it should be emphasized that the current work does not attempt, as Cahn (2003) put it, "the hopeless task of painting a complete portrait" of the development of materials science in Argentina.

Certainly, the analysis of the advances in this field of knowledge in the half century that has elapsed since 1970 requires a study that is far beyond the scope of the present investigation.

Jorge Sabato's Personal Journey to Metallurgy

Family History

Jorge Alberto Sabato Condron, as was his full surname, was born on June 4, 1924, in Rojas, a city located 240 km from Buenos Aires. He was the second son of Vicente Esteban Sabato (1895–1972) and Brígida Condron (1898–1926), who descended from Italian and Irish immigrants, respectively. Brígida was one of the six daughters of James Condron Cormick (1856–1909) and Sara Tobin Fergus (1869–1937), and, like two of her sisters Isabel (1900–1971) and Margarita (1902–1997) she is mentioned in the records as "teacher." Vicente Esteban was the son of Francesco Sabato (1869–1949), a Calabrian immigrant who settled in Rojas at the beginning of the 20th century with his wife Giovanna Ferrari (1874–1963). Vicente Esteban Sabato dedicated himself to agricultural activity, but three of his brothers had a significant performance in the 20th-century Argentine university, cultural and political life (Labrada, 2010). Juan Sabato (1904–1988), an engineer, professor, and university rector, was a member of the energy cabinet of President Arturo Illia (1963–1966). Arturo Sabato (1913–1998) received his doctorate in chemistry and during the government of President Arturo Frondizi (1958–1962), he played a key role in national oil policy. Ernesto Roque (1911–2011), who wrote his last name with an accent on the first syllable, i.e., "Sábato," received his doctorate in physics. He subsequently devoted himself to literature, becoming a renowned novelist and essayist.

Jorge Sabato — Teacher and Physics Professor

When his mother, Brígida, died in 1926, Jorge Alberto Sabato was left in Rojas in charge of his Irish relatives. There he completed his primary school studies, and in the mid-1930s he moved to the city of

Buenos Aires where his father had settled. In the city of Quilmes, located 17 km from Buenos Aires, he graduated in 1942 as a primary school teacher ("Maestro Normal"). Later he entered the "Instituto Superior del Profesorado Secundario" (ISPS), a higher education institute for secondary school teachers in Buenos Aires, to study physics.

Founded in 1904, the ISPS had in its early years hired foreign teachers, mostly German. One of them, the physicist Georg Berndt (1880–1972) contributed to the equipping of the Department of Physics. In 1912, the ISPS was directed by a former collaborator of Albert Einstein, Johann Jakob Laub (1882–1962), who also carried out some research work there. In the following decades, the research activity declined, but in 1940, the ISPS had among its professors prominent scientists and humanists, such as the Spanish mathematician Julio Rey Pastor (1888–1962), the Dominican critic and writer Pedro Henríquez Ureña (1884–1946), the Argentine/Spanish philologist Amado Alonso (1896–1952), the Argentine philosophers Francisco Romero (1891–1962) and Ángel Vasallo (1902–1978) and the Argentine historian Abraham Rosenvasser (1896–1983). Physicist Ernesto Sábato, the aforementioned Jorge Alberto's uncle, also taught at the ISPS around 1940, when he returned to Argentina after his stays at the Joliot-Curie laboratory (1938) and at the Massachusetts Institute of Technology (1939).

In 1947, Jorge Sabato completed his studies at the ISPS, graduating as a secondary school teacher ("Profesor de Física"). Later he developed various activities, including teaching and journalism. In 1949, he began writing a secondary school — physics textbook in collaboration with Alberto P. Maiztegui (1920–2018), another ISPS-trained teacher, who developed an outstanding academic career promoting physics education in Argentina. Maiztegui and Sabato's textbook was published in the early 1950s and through several reissues it became a classic.

A Self-Taught Metallurgist

In 1952, Jorge Sabato was hired as director of the research laboratory of the metallurgical company Guillermo Decker S.A., which had two plants in Buenos Aires dedicated to the production of brass and copper tubes and semi-finished products. The passage from physics and journalism to industrial metallurgy required Sabato a great self-study effort. Consequences of his work were the publication in 1954 of his first scientific articles and the creation, together with chemist Luis A. Boschi, of "Investigaciones Metalúrgicas" (IMET) a metallurgical research laboratory aimed to provide technical support to the local industry.

In summary, the industrial and self-study activity carried out by Sabato in the period 1952–1954 allowed him to start as a metallurgist and become familiar with the challenge posed by research in metallurgy, an area with little academic tradition in the country. Both aspects, which reflect Sabato's ability to, as he put it, "learn by doing," were relevant to the subsequent stages of his career.

Jorge Sabato's Multifaceted Project for Modern Metallurgy

The Political and Institutional Context

This section is partly based on the studies by López Dávalos and Badino, (1999), Marzorati, (2000), García and Reising (2002, 2003), Marzorati (2003), Marzorati (2006), Marzorati (2012) and Hurtado de Mendoza (2005).

At the beginning of the 1950s, the Argentine government of Juan D. Perón (1895–1974) made several decisions that shaped the institutional setting in which Jorge Sabato conceived and carried out his project. First, in May 1950 the government created the "Comisión Nacional de la Energía Atómica" (hereinafter: the Commission), an administrative body chaired by Perón himself, responsible for coordinating and controlling activities in the area. In practice, the Commission made it possible to formalize and give administrative support to the activities carried out at Huemul Island on Lake Nahuel Huapi (Bariloche) by the Austrian physicist Ronald Richter (1909–1991), to produce energy through a controlled nuclear fusion reaction. Second, in 1951 the government created, within the framework of the Commission, the "Dirección Nacional de la Energía Atómica National" (DNEA), an executive body based in

Buenos Aires that would be in charge of training the technical personnel necessary for the institutional projects. The DNEA was left in charge of the Secretary of the Commission, a man from the Army: Colonel Enrique P. González (1896–1969).

In 1952, several committees of scientists, one of which was the physicist José Antonio Balseiro (1919–1962), were convened to evaluate Richter's project. Since the activities were considered devoid of scientific bases, Richter was removed from his position and the project was canceled in November 1952. In addition, the control of atomic energy passed from the hands of the Army to those of the Navy. These facts had two relevant consequences for the subject of this work. First, after the closure of the Huemul Island Project, the Commission's activities in Bariloche shifted to an establishment called "Planta Experimental de Altas Temperaturas" (PEAT). There, a few years later, an institute dedicated to the higher education of physicists began to function, which is analyzed later in this work.

Second, within the DNEA, led since April 1952 by Captain of the Navy: Pedro E. Iraolagoitía (1912–1990), the organization of research groups in nuclear physics, radiochemistry and other areas was consolidated, through the incorporation of scientists and the acquisition of equipment. In particular, in May 1953 a Cockroft–Walton accelerator was put into operation in Buenos Aires and in September 1954 a synchrocyclotron. The new equipment favored the development of the nuclear physics area, as well as the radiochemistry group organized by Walter Seelmann-Eggebert (1915–1988), a German scientist who had trained with Otto Hahn (1879–1968) and collaborated with Fritz Strassmann (1902–1980).

In addition to continuing with the prospecting and extraction of uranium ore, the Commission carried out other activities related to nuclear energy, in particular: (i) the construction in Ezeiza, some 30 km from the city of Buenos Aires, of a pilot plant for obtaining metallic uranium, and, (ii) the realization, as of September 1953, of the first course on nuclear reactors, which continued in Bariloche between January and March 1954. On the basis of the various scientific advances achieved within the Commission, Argentina presented

37 papers on various topics at the First International Conference "Atoms for Peace" held in Geneva in August 1955. In particular, the radiochemists of the Buenos Aires group reported the discovery of a dozen new radioisotopes.

In September 1956, a military coup overthrew President Perón, and in December 1956 the atomic energy area was reorganized. The Commission created in 1950 ceased to exist, and on the basis of the DNEA's activity, the "Comisión Nacional de Energía Atómica" (CNEA) was structured. This institution is referred to in the rest of this work using the acronym "CNEA." Captain Oscar Quihillalt (1913–2001) was appointed president of the CNEA, continuing in the position almost without interruption until 1973. This fact allowed considerable continuity in the pursuit of institutional objectives, despite the political changes that experienced Argentina in the same period.

The Origin of Jorge Sabato's Multifaceted Project

On December 15, 1954, Jorge Sabato entered the Commission on probation for a term of six months, as personal advisor to Iraolagoitía and representative of the IMET. At the beginning of 1955, he was in charge of the organization of a metallurgy laboratory, aimed to solve problems related to nuclear fuels for research or power reactors. Faced with this challenge, one possible strategy was to hire experts in nuclear metallurgy and send fellows to some of the excellent laboratories that existed in Europe or the USA. However, in Sabato's view, this strategy required that there be a good general metallurgy laboratory in the country capable of offering scientific support for work in nuclear metallurgy. But such a laboratory did not exist in 1955. Certainly, in several places in Argentina, there were engineers who carried out pioneering work in the study and teaching of metallurgy, among others, Gregorio Cusminsky and Antonio E. Sturla at the National University of La Plata (UNLP) and Eduardo R. Abril (1915–2009) in Córdoba. Also at the University of Buenos Aires (UBA), there were activities in the field. However, in general: (i) there were no degrees in metallurgical engineering and the universities did not produce doctors or graduates in the science of metallurgy; (ii) no specific subjects in physical metallurgy or physical-chemistry of metals were taught in engineering, physics or chemistry courses; and, (iii) there were no scientists dedicated full-time to metallurgical research, and there were no physicists working on the theory of metals.

In view of the panorama described, Jorge Sabato proposed to the authorities of the Commission (and of the CNEA) to create a laboratory not exclusively dedicated to nuclear metallurgy but capable of solving general metallurgical problems. For him, there were several reasons for this choice. In the first place, to develop nuclear energy in a country, nuclear metallurgy in the narrow sense of the metallurgy of uranium and nuclear fuels is not enough: in the construction of a nuclear power plant, for example, problems related to steels, welding or fracture, appear. Second, a laboratory with a broad base of knowledge could also promote and support the development of metallurgy in other places, helping to create an organized academic activity in Argentina, with courses, scientific papers, conferences, etc. Third, for Argentina to be able to acquire nuclear capacity it would be necessary for the local electromechanical and metallurgical industry to also have the required quality levels. To achieve that it would be useful to have a laboratory capable of offering to the industry the necessary scientific and technical support.

On these conceptual bases, Jorge A. Sabato began in 1955 a long-term project to develop in Argentina: (a) "modern metallurgy," that is, metallurgy as it had been transformed by recent scientific developments; (b) metals physics, as a specific area of solid-state physics; and, (c) metals technology for nuclear applications. To these ends, the project required, first of all, the training of scientists and engineers and the dissemination of metallurgical knowledge among the metallurgical community. But also, to achieve nuclear capacity, it was necessary for the local metallurgical and metalworking industry to incorporate knowledge into industrial practices. In the rest of this work, the initiatives carried out by Jorge A. Sabato between 1955 and 1970 within this framework of ideas, values and institutional objectives are synoptically analyzed.

Early Academic Initiatives on Metallurgy and Metals Physics

Physical Metallurgy in Buenos Aires

Lacking trained metallurgists, Sabato's strategy to create his work team involved three major institutional decisions. The first one was to incorporate, as of April 1955, a group of young graduates, as follows: (i) initially three engineers, one aeronautical (Heraldo Biloni); one electromechanical (Carlos A. Martínez Vidal); one civilian (Arnoldo Leyt) and two chemists (Luis A. Boschi and Antonio J. Carrea); and, (ii) later, four chemical engineers (Jorge A. Mazza, César M. Libanati, Nelly H. Ambrosis and Jorge Kittl); a mechanical engineer (Ignacio Silbert); an industrial engineer (Oscar Wortman) and three chemists (Jorge A. Coll, Juan C. Di Primio and Carlos Aráoz).

The second decision was to organize, as of July 1955, a series of courses and seminars for the metallurgical training of the new team and other university graduates. These initiatives counted as professors, in addition to Sabato himself: (a) Argentine scientists such as Heberto A. Puente, from the Commission's Physical-Chemistry Laboratory, who contributed to the training in physical-chemistry of metals and alloys, and Ernesto Galloni (1906–1987), from the UBA, in crystallography and X-rays; and, (b) invited foreign experts such as Robert W. Cahn (1924–2007) (UK), Erich Gebhardt (Germany) and Paul Lacombe (1911–1997) (France), who visited the laboratory between 1955 and 1957.

The Metallurgy Course for University Graduates was taught for the first time between July 1955 and October 1956, it was repeated between March 1959 and June 1960, and, in collaboration with the UBA, between August and November 1960. In the 1960s, the Pan-American Courses and a large number of other courses were organized in Buenos Aires for professors and industry personnel. Such courses are discussed later in this work. Following the orientation proposed by Sabato, all the courses organized since 1955 focused on the basic principles and problems of physical metallurgy and the basic processes of industrial metallurgy.

The third decision was to initiate lines of research on problems chosen for their general metallurgical interest and for the possibility of addressing them with the available equipment. To this end, it was decided that members of the group would carry out training stays in leading scientific centers abroad. In October 1955, the first scholarship recipients went to Germany (Biloni and Kittl) and England (Carrea and Mazza). Between 1956 and 1958, other members of the group went to Germany (Di Primio and Martínez Vidal) and France (Ambrosis and Libanati). Sabato himself made research stays in Birmingham (1957–1958), invited by Robert W. Cahn, and at Stanford (1963–1964), invited by Oleg D. Sherby (1925–2016).

The Metallurgy Laboratory was transformed in 1957 into the Metallurgy Division (hereinafter MDiv), and in 1958 into the Metallurgy Department (MDep). In 1960 the CNEA created 5 "Gerencias," one of which was the Gerencia de Tecnología (GT), in charge of Sabato. The GT carried out the following three programs: (a) nuclear metallurgy, which involved the study and production of fuel elements; (b) general metallurgy, which included basic research on the subject; and, (c) training and assistance to the industry, which took shape in 1961 with the creation of the "Servicio de Asistencia Técnica a la Industria" (SATI), an industry helpdesk that is discussed later in this chapter. Around 1963, a fourth program, devoted to Development, was added. A few years later in the 1960s, the GT comprised four Departments, three of which (Metallurgy, General Metallurgy, and Nuclear Metallurgy) dealt with research and development. The fourth comprised the SATI.

Sabato led the GT until 1969. In 1971, the GT comprised three departments: Nuclear Fuels, MDep, and SATI. In 1972, shortly after Jorge Sabato left his positions at the CNEA, both the GT and the MDep were led by his first collaborators: Carlos Aráoz and Jorge Kittl, respectively. The MDep had 127 research and development professionals, 37 research fellows and 62 technicians. Of the professionals, 27% were physicists, 17% were chemists, and the remaining

56% were engineers from various specialties, mostly mechanics. Some of the members of the first group had presented their doctoral theses abroad: Ambrosis and Libanati (France, 1959) and others had done so in Argentina in the 1960s: Aráoz (1960), Carrea (1961), Coll (1963), Boschi (1963) and Di Primio (1965). Also, more than 20 doctoral theses had been carried out by members of the MDep or researchers associated with it since 1963.

In 1972, the research activity involved the following: (i) several working groups developed since the late 1950s: Solidification and casting (1957); Diffusion in metals (1960); Gases in metals (1960); X-ray diffraction (1960); Radiation damage (1965); Corrosion (1966); radiation damage in electron microscopy (1969); (ii) a division that dealt with Mechanical Transformations; (iii) a more recent research group on Steels and another on Phase transformations and Heat treatments. The members of the MDep working on these subjects had published more than 180 articles and 220 presentations to congresses, internal reports and other publications. Such activity received national financial support from the National Council for Scientific and Technical Research (CONICET) of Argentina and the Commission for Scientific Research of the Province of Buenos Aires (CIC), among others, and from international organizations.

Metals Physics at the Bariloche Institute

This section is partly based on the studies by López Dávalos and Badino (1999), García and Reising (2002, 2003) and Marzorati (2003).

Created in 1955 by an agreement between the Commission and the National University of Cuyo (UNCUYO), a regional university based in the Province of Mendoza, the "Instituto de Física" (IF) established in the PEAT of Bariloche, began its activities in August 1955, under the direction of the aforementioned José Antonio Balseiro. The purpose of the IF was to train physicists with a capacity for pure and technological research, in particular, to face problems inherent to the following: (i) scientific and technological metallurgy; and (ii) nuclear physics, such as those related to research reactors and as sources of energy. As of January 1957 the PEAT became the "Centro Atómico Bariloche" (CAB).

In his book, Cahn (2003) relates that Jorge Sabato promoted metals physics as one of the orientations for the training of the Institute's physicists. For this reason, in a brief visit to the IF in 1955, and at the request of Sabato, Cahn gave Balseiro some guidance to build the metallurgy laboratories. In 1957, Sabato offered the collaboration of his team to the following: (a) teach an introductory course on physical metallurgy for PhD students; (b) advise on the laboratory work that could be developed with the available equipment. As part of the cooperation, Heraldo Biloni and Antonio J. Carrea, two members of the MDiv who had returned from their stays abroad, taught a two-month course on phase diagrams in early 1958.

At the beginning of 1959, some of the first graduates of the IF were awarded scholarships to study metallurgy in the USA (Edgardo Bisogni), UK (Aldo Vidoz) and Germany (Teodoro Halpern). In addition, that year, courses in physical-chemistry of metals and alloys (Heberto A. Puente) and crystallography (Robert W. Cahn) were taught. Moreover, two European scientists linked to metals physics arrived. One of them, the Austrian physicist Gunther Schöck (1928–2015), who had been contacted by Jorge Sabato, remained in Bariloche until September 1961. During this period, Schöck directed the work of several students (Manuel A. Mondino, José Saura and Conrado Hoffmann) and participated in local research projects. The other visitor, Glyn Davies (UK) came to stay for two years, contributing to the activities in experimental metal physics.

At the beginning of 1960, with the support of UNESCO and the International Atomic Energy Agency (IAEA), and the participation of numerous foreign professors, a Summer School was held in Bariloche, directed by Jorge Sabato, which included activities in metallurgy, nuclear and theoretical physics.

Between 1960 and 1961, those who had been abroad began to return: Teodoro Halpern and Aldo Vidoz joined the metal physics group organized by Schöck, while Edgardo Bisogni joined Sabato's. On March 26, 1962, the first director of the IF died, and it was renamed "Instituto Balseiro" (IB) in his honor. The metal physics area, a result of the early cooperation between Jorge A. Sabato and

José A. Balseiro, continued to develop thanks to the institutional support of the CNEA and the UNCUYO. In particular, during the 1960s and the first years of the 1970s, several physicists presented their doctoral theses at the IB and joined the MDep and GT: Bisogni (1965), Bonfiglioli (1965), Ipohorski (1967), Pampillo (1967), Cabo (1969), Fainstein (1970) and Audero (1972).

As Cahn also highlights in his book, the academic center formed by the close relation between the IB and the CAB has continued to expand its activity in the area of metals physics and materials science. However, the detailed study of the progress made in Bariloche since the early years exceeds the scope of this synoptic study and should be the subject of a specific investigation.

A Key International Colloquium

Among the academic initiatives of Sabato to consolidate research in physical metallurgy in Buenos Aires and metals physics in Bariloche, there was the International Colloquium on the Impact of Physical Metallurgy on Technology, held in April 1962, which was inaugurated in Buenos Aires and continued in Bariloche. Organized with the collaboration of Robert W. Cahn and the specialist in nuclear science and engineering John P. Howe (1910–1999), the Colloquium brought together various metallurgists and academics to discuss the following: (i) the link between scientific research and industrial technology; and, (ii) the relationship between research topics and practical problems. Fifteen international academics from the USA (8), England (3), Germany (1), Australia (1), the Soviet Union (1) and Brazil (1) were invited to the Colloquium, among which were those who contributed significantly to 20th-century metallurgy: Bruce Chalmers (1907–1990), Werner Köster (1896–1989) and Cyril S. Smith (1903–1992).

Promoting Metallurgy in Other Argentinian Institutions

This section is partly based on the work by Bonfiglioli *et al.* (1972).

At the beginning of the 1960s, the GT had established cooperative relationships with the Technological Institute of Buenos Aires

(ITBA) in the area of Metallurgical Engineering, with the Faculty of Physico-Mathematical Sciences of the UNLP and with the Institute of Mathematics, Astronomy and Physics (IMAF) from the National University of Córdoba (UNC). The IMAF was created in 1956 at the initiative of the Argentine physicist and astronomer Ramón Enrique Gaviola (1900–1989). By agreement of 1963, the GT dictated a semester course in physical metallurgy at the IMAF that included laboratory work. In 1966, these activities were expanded to offer physics students: (i) courses in crystallography, X-rays, phase transformations, crystal defects, and solidification, among others; and, (ii) work stays in the Buenos Aires laboratories. By the early 1970s more than 20 people, a significant number for the IMAF in those days, had participated in this program through courses, research papers, doctoral theses, or as members of local working groups. In this institution, which is currently the Faculty of Mathematics, Astronomy, Physics and Computing of the UNC, the study of materials continues to be one of the lines of research in Physics. Toward the end of the Sabato period in charge of the GT, cooperative relations had expanded to other science or engineering faculties located in Bahía Blanca, Rosario, Tucumán and Córdoba.

Further Initiatives for Advanced Training in Metallurgy and Materials

Between 1969 and 1972, 18 courses were held in Buenos Aires for researchers and university professors on specific metallurgical topics, lasting 1 or 2 weeks. These courses were often taught by visiting professors from the USA, UK or France. The total participation in the indicated period was 248 people.

A National Forum for Local Metallurgists

The "Sociedad Argentina de Metales" (SAM) was founded on December 20, 1955, in an assembly that brought together about 70 professionals and technicians, called by Jorge A. Sabato along with the aforementioned Eduardo R. Abril, Antonio E. Sturla and other metallurgists. The objectives of the SAM were, among others, the following: (i) promote knowledge of scientific and technological metallurgy; (ii) promote, coordinate and guide their research and teaching; (iii) link people who dedicate themselves to it or are interested in its development; and, (iv) form a bibliographic consultation center, linking or associating with similar foreign or international organizations. Some of the early initiatives inspired by these goals are listed below, while some of the most recent developments are mentioned at the end of this chapter.

In July 1956, the association initiated a series of postgraduate metallurgy courses, and in May 1957, at the UNLP, a series of metallurgical outreach conferences, which included a lecture by Paul Lacombe. In 1958 the association offered postgraduate courses in metallurgy and optical metallography. In March 1959, it began to publish the journal "Metalurgia Moderna." In 1975, due to financial difficulties, and when 36 numbers had been published, the publication was discontinued. Some numbers were published between 1986 and 1992.

In November 1959, the first Argentine congress on metallurgy organized by the CNEA and the SAM under the name of "Primeras Jornadas Metalúrgicas" took place. The congress featured contributions from foreign metallurgists such as R.W. Cahn and P. Lacombe. During the rest of the period in which Jorge Sabato held executive positions, the GT collaborated with the SAM in the organization of the congresses on metallurgy that were held in 1962, 1964, 1970 and 1972.

Nuclear Metallurgy at the Beginning of the Reactor Program

In July 1955, Argentina joined the Atoms for Peace program promoted by President Dwight Eisenhower. The cooperation agreement established that the USA would supply the enriched uranium for the research reactors. In April 1957, the CNEA decided to install its first reactor and that, unlike what Brazil and other countries did, the reactor would not be acquired abroad. The construction of the reactor, carried out in Buenos Aires following the Argonaut

("Argonne Nuclear Assembly for University Training") design, developed at Argonne National Laboratory in Chicago, was entirely national. The MDep was in charge of manufacturing the fuel elements and developed an original technique, despite the limitations in equipment and the lack of experience in the field. The reactor, named RA-1, was officially inaugurated on January 20, 1958 (CNEA; 1998). In August of the same year, the know-how for manufacturing the fuel elements, which involved the co-extrusion of aluminum and uranium oxide, was sold to Degussa-Leybold AG (Frankfort, Germany). In September 1958, the work on RA-1 was presented at the Second International Conference "Atoms for Peace" held in Geneva.

In 1959, the CNEA decided to increase the power of the RA-1 reactor, for which a critical testing facility called RA-0 was built. Subsequently, it was decided to build a reactor for the commercial production of radioisotopes for medical uses: the RA-3 as well as a facility for critical tests called RA-2. The RA-3 was inaugurated in December 1967. The MDep was in charge of developing the fuel elements for these new reactors.

In 1965, the Argentine government commissioned the CNEA to carry out a feasibility study for the construction of a nuclear power plant. In 1966, a price offer was called for the construction of what is known today as the "Central Nuclear Atucha I" (CNA I), located about 100 km from the city of Buenos Aires, and in 1968 the contract with Siemens AG (Germany) was signed for a 319 MW plant with heavy water cooled and moderated natural uranium reactor. The plant began operating in 1974. Based on the experience acquired, the CNEA initiated in 1976 the construction of a pilot plant to produce the fuel elements for the CNA I. It reached an industrial scale in 1982, within the framework of a new company created by the association of the CNEA with local industrialists.

The Interaction with the Local Metallurgical Industry

Metallurgy Training Courses

Between 1960 and 1972, the GT gave 32 short courses, lasting 1–2 weeks, for technical personnel on metallography, welding, heat treatments and other topics of industrial interest. In some cases the courses were taught by foreign professors from the USA, UK or France, who were visiting Buenos Aires, and sometimes in industrial plants. In the period under study, the total participation in these courses was 608 people.

Technical Assistance

In March 1961, the CNEA and the "Asociación de Industriales Metalúrgicos de la República Argentina" (ADIMRA) created a non-profit organization: the "Servicio de Asistencia Técnica a la Industria" (SATI). The activities of SATI were developed following two groups of objectives. The first group included the following: (i) bringing modern knowledge and new techniques closer to the industry; (ii) constituting an agile consultative body to help the industry solve its problems; (iii) serving as a core for metallurgical engineering training. SATI's activities began with advisory tasks on various topics of transformation metallurgy, later dealing with developments that represented a technological advance for the industry.

Promoting Participation in CNEA's Projects

Starting in the mid-1960s, SATI made use of its contacts with the productive sector for a second set of objectives: (a) to obtain detailed technical information on the manufacture of components for nuclear reactors; (b) to know the possibilities of the local industry to participate in the CNEA projects; and, (c) to favor such participation, due to its economic importance and the technological development that it could stimulate. In this regard, it should be noted that 32 companies had participated in the construction of RA-1, and in that of RA-3 the local participation had reached 90%. In the case of CNA I, SATI's participation was relevant for the full application of the so-called "opening of the technological package," a criterion introduced by Sabato to achieve the greatest possible participation of the national industry. In the case of CNA I, contracted with Siemens AG, this participation reached 40%.

Other Activities in Metallurgy and Metals Technology

Early Latin American Meetings

The MDep participated in the organization of the following: (i) the First Latin American Colloquium on Powder Metallurgy, in November 1961; and (ii) the First Latin American Conference of Experts in Metallurgy of Transformation, organized in August 1964, with the participation of some 50 specialists from Argentina, Brazil, Colombia, Chile, Mexico, Peru, Uruguay and the USA. In this Conference the following issues were debated: (a) the impact of metallurgical research on the economy of developing countries; (b) the possibility of establishing priorities in relation to national needs and possibilities; (c) the objectives of regional collaboration on technological research; and, (d) the organization of meetings and technical courses in the region.

Pan-American Courses in Metallurgy

The First Pan-American Course on Nuclear Metallurgy, aimed at graduates in metallurgy, physics, chemistry, or engineering from any of the countries in the region, was organized by the CNEA with the support of the Organization of American States (OAS) and was held in Buenos Aires between March and December 1962. Directed by Jorge Sabato and taught by researchers from the CNEA and other Argentine and foreign university professors, the Course was intensive and divided into two cycles with theoretical and practical classes: a basic one of 597 hours, dictated between March and July, and a specialization cycle of 650 hours, between July and December. This Course summoned 13 Latin American students. The Second Pan-American Course on Nuclear Metallurgy was held in 1965. As of the third, held in 1967, the course was called Pan-American Course on Metallurgy and was held annually.

These courses had other impacts on academic activity in Argentina. In particular, the National University of the South (UNS) in Bahía Blanca, which had created the Doctorate in Engineering in 1970, considered that to qualify for the doctorate degree, in addition to the thesis work, the Pan-American Metallurgy Course was valid as a postgraduate course.

The Multinational Metallurgy Program (PMM)

In 1967, within the framework of the OAS Scientific and Technological Development Program, priority was assigned to the Multinational Metallurgy Project. The CNEA MDep was evaluated as a center of excellence for the region and was responsible for designing it. Thus arose the OAS-CNEA "Programa Multinacional de Metalurgia" (PMM), which began in 1969 with the following objectives: (i) contribute to the development of Latin American Centers trained for postgraduate training and research of regional interest; (ii) conduct research on metallurgical problems of special interest to the region, in processes and materials; and (iii) provide technical assistance for the creation and development of new metallurgical centers.

The PMM included activities in three areas: (a) training; (b) research and development in lines of common interest to Latin American countries: solidification and casting, physical-chemistry of metals, phase transformations and heat-treatments, plastic deformation, and mechanical working, metallurgy of ferrous metals and corrosion of metals and alloys; and, (c) general assistance activities: to universities and institutes, in particular, for access to high-cost equipment and technical assistance to the industry. Centers in Brazil, Colombia, Chile, Ecuador, Mexico, and Peru were part of the PMM. Within this framework, between 1969 and 1972, 21 associate researchers from Bolivia, Brazil, Colombia, Chile, Honduras, Mexico, and Uruguay carried out advanced training or postgraduate thesis.

Summary

Between 1962 and 1974, 10 Pan-American courses were organized in Argentina, attracting 190 students: 40% of whom came from Argentina and the rest came from Bolivia, Brazil, Colombia, Costa Rica, Chile, Ecuador, Guatemala, El Salvador, Spain, Honduras, Mexico, Nicaragua, Peru, Dominican Republic, USA and Venezuela.

From July 1974–1981 the Pan-American Metallurgy Course was held in Mexico.

Between 1976 and 1980, five "Cursos de Entrenamiento Avanzado en Metalurgia" were held in Buenos Aires within the framework of the "Programa de Apoyo Científico y Tecnológico" (PACITE). Between 1981 and 1993, 13 annual courses on "Metalurgia y Tecnología de Materiales" were organized. Around 340 people attended the complete cycles. Furthermore, 10 two-month Latin American "Postdoctoral Seminars" were held with a total of 261 participants.

Finally, within the framework of the Multinational Program, 98 undergraduate theses and 112 doctoral theses were finalized, and about 450 papers were published.

Concluding Remarks: From Metallurgy to Materials Science

As a contribution to the study of materials science in Argentina, this work set out to offer a synoptic image of a project designed and promoted by Jorge A. Sabato in the period 1955–1970, in which he held executive positions at the CNEA. Due to the diversity of the project's objectives and strategies, the work focused on Sabato's early initiatives and their evolution during the 1960s.

However, there are two developments of an institutional nature that are directly linked to Sabato's initiatives, which took place in Argentina a decade after his death, which occurred in Buenos Aires on November 16, 1983.

One of them occurred on March 15, 1993, in the city of Rosario, Province of Santa Fe, when the "Sociedad Argentina de Metales," whose creation was promoted by Sabato in 1955, was renamed as "SAM-Asociación Argentina de Materiales," preserving the original purposes. The association has regularly organized an annual congress, which in recent decades has been held in conjunction with the Chilean Society of Metallurgy and Materials (SOCHIM). Furthermore, since 2004, the association has been publishing "Revista SAM" in digital format.

The other development took place on November 16, 1993, when the "Instituto de Tecnología Profesor Jorge A. Sabato" (ITJS) a university institution was created by agreement between the CNEA and the National University of San Martín. On the basis of the experience gained since the days of the Pan-American Course and the multinational programs, the ITJS began in 1994 to offer a Master's Degree program in Materials Science and Technology. Since 1996 a Materials Engineering Degree program is offered. Subsequently, the Doctorate in Science and Technology was implemented, with an orientation to materials (in 1997) and physics (in 1998). Other specialized careers in topics of industrial and nuclear interest have also been created.

Suggestions for Future Research

In addition to the evolution of the institutions considered in this work, subsequent studies should address the generation in Argentina of a diversity of academic institutions of science and engineering dedicated to training and/or research in materials science, which have been established, among other places, in Bahía Blanca, Córdoba, La Plata, Mar del Plata, Neuquén, Río Cuarto, Rosario, Santa Fe, Tandil and Tucumán.

Acknowledgments

The author would like to thank Gastón Burlot and Dalía Bertoldi for their valuable assistance with the bibliography. The financial support from Universidad Nacional de Cuyo through Proyecto Bienal 06/M016-T1 is gratefully acknowledged.

References

Alegría, J. L., Coll, J. A. and Suter, T. (1972). *Una breve reseña histórica de la CNEA* (Buenos Aires: Argentina: CNEA).

Bonfiglioli, A., Craievich, A. and Maiztegui, A. (1972). Ciencia de los materiales en una Escuela de Física. In *III Inter-American Conference on Materials Technology*, Centro Regional de Ayuda Técnica, Buenos Aires, pp. 701–704.

Cahn, R. W. (2003). *The Coming of Materials Science* (Oxford, UK: Elsevier Science).

CNEA (1955). Comisión Nacional de la Energía Atómica. *Memoria Año 1955* (Buenos Aires, Argentina)

CNEA (1958). Regreso del Dr. Heberto Puente. Comisión Nacional de Energía Atómica. Boletín Informativo. *Año II. Junio de 1958.* (Buenos Aires, Argentina), pp.12–13

CNEA (1961) Curso Panamericano de Metalurgia Nuclear. Comisión Nacional de Energía Atómica. *Boletín Informativo.* Año V, No 1. Junio 14, 1961. (Buenos Aires, Argentina), pp.18–20

CNEA (1962a) Coloquios. El Impacto de la Metalurgia Física en la Tecnología. Comisión Nacional de Energía Atómica. *Boletin Informativo.* Año VI, No 2. Mayo 1, 1962. (Buenos Aires, Argentina), pp. 9–13

CNEA (1962b) Curso Panamericano de Metalurgia Nuclear. Comisión Nacional de Energía Atómica. *Boletin Informativo.* Año VI, No 2. Mayo 1, 1962. (Buenos Aires, Argentina), pp. 14–20

CNEA (1962-1963). Comisión Nacional de Energía Atómica. *Memoria Anual Correspondiente al Ejercicio 01-11-1962 a 30-10-1963.* (Buenos Aires, Argentina).

CNEA (1964). Comisión Nacional de Energía Atómica. *Memoria Anual 1964.* (Buenos Aires, Argentina).

CNEA (1965). Comisión Nacional de Energía Atómica. *Memoria Anual 1965.* (Buenos Aires, Argentina).

CNEA (1966). Comisión Nacional de Energía Atómica. *Memoria Anual 1966.* (Buenos Aires, Argentina).

CNEA (1967). Comisión Nacional de Energía Atómica. *Memoria Anual 1967.* (Buenos Aires, Argentina).

CNEA (1969). *Reseña General de las Actividades que realiza la Comisión Nacional de Energía Atómica.* (Buenos Aires, Argentina), 1969

CNEA (1970) Comisión Nacional de Energía Atómica. *Memoria Anual 1970.* (Buenos Aires, Argentina).

CNEA (1971) Comisión Nacional de Energía Atómica. *Memoria Anual 1971.* (Buenos Aires, Argentina).

(CNEA, 1973). *Comisión Nacional de Energía Atómica. Actividades de la Gerencia de Tecnología 1955–1972.* (Buenos Aires, Argentina).

CNEA-UNSAM (1998). *Sabato en CNEA* (Buenos Aires, Argentina: Instituto de Tecnología "Profesor Jorge A. Sabato").

CNEA (1998). Comisión Nacional de Energía Atómica. CAC-RA1 1958-1998. Recopilación y Edición Dra. E. Forlerer - Ing. T. Palacios. Centro Atómico Constituyentes (General San Martín, Argentina), 1998.

Enríquez, S. N. (2011): A 50 años de la creación del Servicio de Asistencia Técnica a la Industria (SATI): homenaje y legado. *Revista SAM* 8(3), 25–31.

Enríquez, S. N. (2012). Significación de la Transferencia de Tecnología en la Comisión Nacional de Energía Atómica (CNEA): un análisis histórico-comparativo. *Fundamentos en humanidades* Año XIII, II (26), pp. 215–229.

Fernández, J.R. (2011) El surgimiento de las comisiones de energía atómica en Argentina y Brasil (1945–1956). *Revista eä,* Vol.2, No 3, (Abril 2011). (ISSN 1852-4680)".

Galvele, Jose R. (2001). Characteristics of the Institute of Technology "Prof. Jorge A. Sabato." In *International seminar nuclear energy, education, employment and environment, Buenos Aires, Argentina, 24–26 April 2001, CNEA.*

García, M. C. and Reising, A. M. (2002). La consolidación del Centro Atómico Bariloche: Una aproximación desde el desarrollo de la física experimental. *Saber y Tiempo* 4(14), 33–54.

García, M. C. and Reising, A. M. (2003). Los inicios del Instituto Balseiro: sus ideales de ciencia y formación científica. *Epistemología e Historia de la Ciencia* 9(9), 167–173.

Hurtado de Mendoza, D. (2005). Autonomy, even Regional Hegemony: Argentina and the "Hard Way" towards Its First Research Reactor (1945–1958). *Science in Context* 18(2), 285–308.

Iurman, L. (2009). Evolución de la enseñanza de los materiales, *Revista SAM,* Vol.6, No2, pp. 4–12.

Labrada, A. (2010) La familia Sabato, en Historias de Rojas (Enero de 2010). Accessed September 29th, 2023. {https://www.historiasderojas.com.ar/index1.php?id=id00009}.

López Dávalos, A. and Badino, N. (1999). *J.A.Balseiro: crónica de una ilusión. Una historia de la física en la Argentina* (Argentina, Buenos Aires: Fondo de Cultura Económica de).

Martínez Vidal, C. E. (1993). Sabato en el pensamiento sobre el desarrollo científico-tecnológico latinoamericano. *Arbor CXLVI,* 575, 45–76.

Marzorati, Z. (2000). La participación argentina en la I Conferencia "Átomos para la Paz" de Ginebra (1955). *Epistemología e Historia de la Ciencia* 8(6), 275–282.

Marzorati, Z. (2003). Plantear utopías. La formación de la comunidad científica: CNEA (1950–1955). *Cuadernos de Antropología Social* 18, 123–140.

Marzorati, Z. (2006). Un desarrollo científico-tecnológico autónomo: la construcción del RA-1. *Cuadernos de Antropología Social* 23, 105–116.

Marzorati, Z. (2012). *Plantear utopías: la conformación del campo científico-tecnológico nuclear en Argentina (1950–1955)* (Buenos Aires, Argentina: Fundación CICCUS).

Ovejero García, J. (September 2018). El nacimiento de la metalurgia moderna en Argentina. *Revista SAM* 2, 28–44.

Sabato, J. A. (1960). Metallurgy in the development of atomic power. *American Journal of Physics* 28, 228–234.

Sabato, J. A. (1968). Energía atómica en Argentina. *Revista Estudios Internacionales* 2(3), 332–357.

Sabato, J. A. (1970). Para el prontuario del Plan Nuclear Argentino. *Ciencia Nueva* 1, 32–46.

Sabato, J. A. (1972). Quince años de Metalurgia en la CNEA. *Ciencia Nueva* 15, 7–15.

Sabato, J. A. (1973). Atomic energy in Argentina: A case history. *World Development* 1(8), 23–38.

Sabato, J. A. (1981). Nuclear metallurgy in Argentina. *Journal of Nuclear Materials* 100, 137–138.

Sabato, J. and Botana, N. (November 1968). La ciencia y la tecnología en el desarrollo futuro de América Latina. *Revista de la Integración* 1(3), 15–36.

Part VI
Institutions of Institutions

Some institutions engaged in materials science themselves consist of networks of institutions. As Terry Quinn points out in "The Origin and Early Days of Material Research in Metrology Laboratories," metrology is an example. The International Bureau of Weights and Measures, or BIPM after the French acronym, was established in 1872 and set out to manufacture metric standards. But the standards failed to meet the desired requirements, and the BIPM wound up making several different contributions to materials science: (1) study of materials used in the standards in the days when these were material objects; (2) testing materials and standards; (3) coordination with national metrological institutions and their standards, including the German National Metrology Laboratory (PTB), the UK National Physical Laboratory (NPL), the US National Bureau of Standards (NBS), and later the National Institute of Standards and Technology (NIST), and a host of other international agencies. From the first, the BIPM was the center of a global network and institution of institutions.

Another international institution created from national institutions, CERN, is known for building and operating the world's largest scientific instrument, the Large Hadron Collider. But as the five authors of "CERN: A Global Hub for Materials Science" — Panagiotis Charitos, Paul Lecoq, Michael Moll, Cristoforo Benvenutti, and Luca Bottura — show, to do so required the expertise in materials science that the laboratory had accumulated over decades. Developing the LHC, its predecessors, and also its detectors, required developing and testing materials and coating processes that would stand up in extreme conditions. The authors describe how this was done for scintillating crystals, for the huge demands placed on materials to be used in the LHC's detectors, for coating technology in high vacuum beam pipes, and for superconducting material. Taken together, this work makes CERN, whose reputation is as a high-energy physics laboratory, also a world-class materials science institution.

Sometimes, networks of institutions spring up, concerned with entirely new materials. Patrick McCray's chapter, "Will Small be Beautiful? Making Policies for Our Nanotech Future," is about one such network of institutions, in which the National Nanotechnology Initiative (NNI, 2000) resulted in the creation of Nanoscale Science Research Centers (NSRC). McCray shows how, in the wake of the NNI and the institutions it spawned, these networks are now both "an active area of real-life research and development" and a "futuristic vision."

Chapter 6.1

The Origin and Early Days of Material Research in Metrology Laboratories

Terry Quinn

Introduction

One can distinguish two quite distinct origins of materials research related to measurement standards. The first was the evident need to study the materials used in the manufacture of the standards themselves in the days when the units of measurement were defined in terms of material objects. This activity long predates the creation of what we call today the metrology laboratories or institutes. The second, which came much later, was when the metrology institutes began what became a major activity, namely, the testing of materials and standards.

The English unit of length, the yard, originally called the Ulna, was defined by Edward 1 in 1305 as equal to 3 feet, each foot equal to 12 inches and each inch equal to the length of three grains of barley, dry and round. It was presumably preceded by studies as to the uniformity of lengths of grains of barley. The unit is linked to specific natural objects so that if the original representation of the unit is lost, it can be reproduced.[1]

The *toise*, the pre-metric unit of length in France, has its origin in antiquity but was represented in modern times by an iron bar which was sealed in the stone on the steps of the Grande Chatelet in Paris in 1668. It was the reference taken for the manufacture of two famous copies of the toise made of iron in 1735. They were made by the artisan Langlois whose official title was *ingenieur du roi et de MM de l'Academie des Sciences pour les instuments d'Astronomie,* He worked under the direction of an Academician named Godet. These two objects came to be known as the *Toise du Perou* and *Toise du Nord* after they had been taken, respectively, to Peru and the North of Lapland to determine the length on the ground of one degree of latitude. The result confirmed that the Earth was slightly flattened at the Poles. Both *Toises* still exist and are conserved at the Paris Observatory in their original wooden cases. Figure 1 shows one end of the *Toise du Perou.* It can be seen that the material is a uniform fine-grained

Fig. 1 The view from one end of the *Toise du Perou.*

[1]This is the same principle adopted in 2018 by the 26th General Conference on Weights and Measures when it redefined the Base Units of the SI by linking them to fixed numerical values of the fundamental constants. See *Metrologia,* Vol. 51, No. 2, 2014 and Vol. 53, No. 5, 2016 for review papers on this subject.

structure. The whole *Toise* remains in perfect condition and the quality of both the material and manufacture are excellent.[2]

Toward the end of the 18th century, at the time of the French Revolution when the Metric System was created, new standards of length and mass were required. It had been decided that the new unit of length would be the Meter, defined as one ten millionth of the quadrant of the Earth passing through Paris and the new unit of mass would be the Kilogram, defined as the mass of one cubic decimeter of distilled water at the temperature of melting ice. By that time, knowledge of materials and technology had advanced and it was decided that the practical representations of the new definitions would be made of a material considered to be more permanent than iron, namely platinum. After the enormous work of measuring a fraction of the quadrant of the Earth passing through Paris and then determining the weight of a cubic decimeter of water, the new material standards to represent these abstract definitions were made. The most experienced artisans of the time carried out this work: Etienne Lenoir for the Meter and Nicolas Fortin for the Kilogram. These two objects, the Meter and Kilogram of the Archives, were locked in the "Armoire de Fer" in the Archives of France in 1799 where they remain today (Figure 2). By the middle of the 19th century, many countries had adopted the metric system, but the fact that the basic reference standards were locked in the French Archives and thus essentially inaccessible started to become a problem.

The Meter Convention of 1875 and the Creation of New Metric Standards

In geodesy, the unit of length had long been the toise with results being traceable to the Toise de Perou. Many European institutes had copies of the Toise de Peru and regular comparisons among

[2]A toise is divided into 6 feet, each foot into 12 inches and each inch into 12 lines. The Metre, one ten millionth of the quadrant of the Earth, was determined to be 443.296 lines of the Toise de Perou, that is slightly more than half the length of a toise.

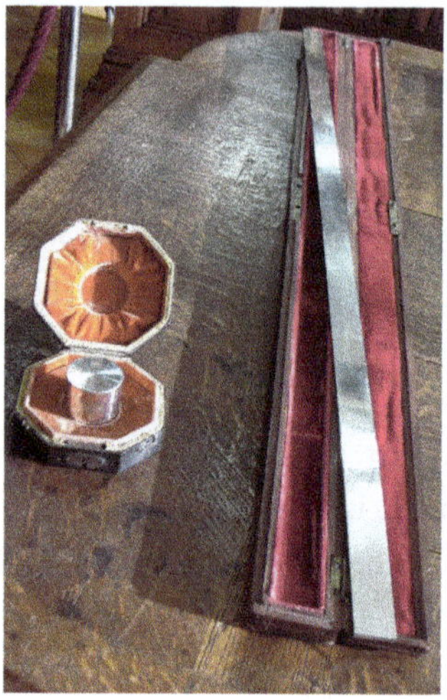

Fig. 2 The Kilogram and Meter of the Archives.

them took place. It was in the field of geodesy that there were the most demanding requirements for uniformity. During the period 1821–1824, the great Carl Friedrick Gauss carried out a survey of the Kingdom of Hannover. During the calculation of the results, he invented the least squares method of adjusting experimental data. This led to much-improved uniformity all over Europe, but the absence of traceability to the Meter in Paris began to be a real problem. At a Conference in Berlin in 1867, calls were made for the construction of a new European prototype of the Meter, to be copied from the Meter of the Archives but to be deposited in a European International Bureau of Weights and Measures and thus accessible to all.

The outcome, 2 years later, was the creation by the French Government of an International Meter Commission. After meetings in Paris in 1870 and 1872, a proposal was indeed made for the creation of new international prototypes of the Meter and Kilogram to be deposited in an International Bureau of Weights and Measures under the authority of an international Convention of the Meter. The French Section of the Meter Commission was charged with the construction of the new prototypes under the guidance of the Commission. The work of the International Metre Commission was extensively documented. It undertook a serious study of the physical and chemical properties of the proposed alloy of platinum and iridium, the processes of manufacture and construction of the proposed new metric standards.

It was decided that the new prototypes should be made of Pt-10% Ir to achieve a hard, easily machineable alloy. The work on the alloy was led by one of the most famous French chemists, Sainte-Clare Deville, who was a good friend and long-time collaborator of George Matthey of London. Together in 1874, they cast in Paris a 250-kg ingot, from which 40 m bars and 40 kilograms were to be made.[3] It was soon found, however, that there were problems with samples cut from this large ingot. The density was found to be significantly lower than expected, 21.115 g cm^{-3} instead of 21.4 g cm^{-3}, the known value for an alloy made of Pt-10% Ir. Chemical analysis showed the presence of about 3% of foreign metal. Chemical analysis showed small amounts of iron and ruthenium. This led to much argument and dissension in the Committee, but nevertheless, the French Section of the Commission decided to accept the ingot and proceed.

Meanwhile, in parallel, diplomatic discussions continued which led to the signing of the Metre Convention in 1875 creating the International Bureau of Weights under the authority of an International Committee for Weights and Measures itself under the authority of a General Conference of Weights and Measures made up of delegates of signatory States to the Convention. The new International Committee immediately took charge of the fabrication of the new metric standards and decided that the 1874 alloy was unsuitable. Instead, it decided to place an order with the Johnson Matthey Company in London for the fabrication of the new prototypes. Among the first activities in the laboratories of the new International Bureau were studies related to the annealing of the

[3]This was the number sufficient to supply the States that had participated in the Metre Commission.

initial specimens of the London alloy for the new meter bars and the polishing and engraving of the surface. This was been completed in 1882 and the order for 40 ingots of 1 kg and the same number of meter bars were ordered from the Johnson Matthey in August 1882. The alloy was specified to be of 90% Pt and 10% Iridium with limits of 0.25% and impurities should not include more than 0.1% ruthenium, 0.1% iron, 0.015% of rhodium or palladium or 0.2% silver, copper or other metals. The density must be at least 21.5 g cm^{-3}. It took some time to produce all these and it was only in 1888 that sufficient had been made for the International Committee to convoke the first General Conference on Weights and Measures to approve all this work and formally adopt the new metric standards. The conference took place in September 1889.

The Conservatoire National des Arts et Métiers in Paris (CNAM)

All of the work carried out by the French Section of the Metre Commission took place at the Conservatoire National des Arts et Métiers in Paris. This was (and remains) one of the institutions created at the time of the Revolution. It was at the instigation of Abbe Gregoire with the aim of helping the French industry by "putting at the disposal of the economic players the designs, tools, and machines useful in the arts and trades." It was thus essentially a teaching institute. It still exists and is today a major teaching institute and one of the "Grandes Écoles." In 1848, through the transfer of the official depot of the metric standards and prototypes, it had also been given responsibility for the French metric standards and metrology. At the time of the creation of the International Metre Commission, the Director was General Arthur Morin who along with his deputy Henri Tresca took a key role in all of what happened at that time. It was Henri Tresca who designed the famous X cross-section of the new international prototype of the Metre with the defining lines being engraved on the top surface of the central crossbar. In 1901, a small testing laboratory was installed in the Conservatoire with the name Laboratoire National d'Essais (LNE), but the increasingly important electrical standards remained with the Laboratoire Centrale d'Electricite (LCIE). This was created after the 1881 International Conference on Electricity held in Paris. No single institute was given overall responsibility for measurement standards in France as happened later in Germany, the UK, and the USA.

The French Section of the Metre Commission met regularly from 1869 until its final meeting, the 132nd, in 1886. During this long period, the immense work carried out by the French Section was published in a long series of reports. The plan to produce, at the end, a volume containing all of this never came to fruition. This is a sad story because as the years went by, the members who had been so influential in its work had progressively died and at the end, only Fizeau and Cornu were left and they did not have the energy! Sadly, no record exists of the details of the many optical and mechanical instruments they constructed as these were described in words in the many reports of meetings, but without drawings, it is impossible to reconstruct what they actually did. All of this is described in more detail in Quinn (2011). The successive reports of the meetings and activities of the French Section are listed in References section at the end.

The International Bureau of Weights and Measures, BIPM, 1875–1920 and the Discovery of Invar and Elinvar

While the creation of the three major metrology institutes, the PTR in 1887, the NPL in 2000, and the NBS in 2001, marked a turning point in international metrology, by far the most important work on materials related to metrology took place at the BIPM during the first two decades of the 20th century. It was undertaken by Charles Eduard Guillaume, Director of the BIPM from 1915 to 1936. It was the study of alloys of nickel that led to the discovery of what came to be known as Invar and Elinvar. These were two different alloys of nickel and iron that showed almost zero thermal expansion at room temperature (Invar) and almost zero thermal coefficients of elasticity (Elinvar) at room temperature. The immediate impact on horology and geodesy was enormous and led to him being awarded the Nobel Prize for Physics in 1920. The whole story began in 1895 and is recounted in Guillaume's Nobel lecture. He begins by recounting the story of the creation of the BIPM and the manufacture of the new International prototypes of the meter and the kilogram. Then he goes on to the search for a material less expensive than Pt–Ir from which working standards of the meter could be made. The properties of these alloys were the result of many years of work on the properties of alloys of nickel that began by the chance observation in a study, conducted at the International Bureau by J. R. Benoit, of a meter standard made from an iron alloy containing 22% nickel and 3% chromium, where its expansibility was close to that of brass. It was a non-magnetic alloy and thus was doubly anomalous. He goes on:

> "I first examined this problem in 1891 and soon discovered the really excellent properties of pure nickel and still today this is the metal used to make a non-oxidizable standard, unaffected by the passage of time, rigid and of average expansibility. However, one difficulty prevented me generalizing its use. A geodesic standard 4 metres long was required, and no nickel-producing factory would undertake to supply, perfectly sound and crack-free, a suitable bar. Subsequent studies were guided by a few strokes of good luck. In 1895, at the request of the Ordnance Technical Department (Section technique de l'artillerie) in Paris, J. R. Benoit had undertaken the study of a standard made from an iron alloy containing 22% nickel and 3% chromium and found its expansibility to be close to that of brass. It was a non-magnetic alloy and thus was doubly anomalous. The very curious phenomena discovered by John Hopkinson had already been known for a few years, notably that after forging certain alloys of iron and nickel containing about 25% of the latter are non-magnetic and not very hard; however, once they have been cooled in, say, solid carbon dioxide, they are hard and strongly magnetic; furthermore, during this transformation, their volume has increased by about 2%.The phenomena discovered by Hopkinson and Benoit were clearly related. But

however interesting they might be for the physical chemist, they were strictly taboo for the metrologist. An alloy that is changeable, and another of high expansibility, are unsuitable for designing length standards. The matter took on a different aspect when in 1896 I was put on the trail of a new and quite unexpected fact closely connected with those just reported. A bar of steel containing 30% nickel arrived at the International Bureau and I found its expansibility to be about one third less than that of platinum. The continuation of a study so begun augured well and I pursued it with stubborn obstinacy."

In the 30 pages that follow, Guillaume gives a detailed account of his study of the phase equilibrium diagrams of a large range of nickel steels together with measurements of their physical properties. All this was carried out in the laboratories of the BIPM in close collaboration with the manufacturer of the alloys Commentry-Fourchambault Company in France. His final paragraphs, headed "Conclusion," are the following:

"Let us now glance back to the origins of the studies of which the results have just been reported. The point at issue was essentially to look for a metal having qualities which would enable it to be used to make standards of length that are inexpensive while satisfying metrological requirements. An initial solution had been found which required only slight refinement and it was during the search for this slight refinement that an unsuspected anomaly came to light which was a direct manifestation of the struggle between the various states of iron to which the association of nickel gives an unexpected shape, creating a class of alloys whose properties, unique so far, satisfy in a quite unhoped-for manner the wishes which metrologists would have formulated had it not been assumed that such properties could not exist in metals. Then the problem broadened: from the measurement of length, which was the only aim of the original programme, the measurement of time arose of its own accord as a direct and to some extent obvious consequence, so true is it that in metrology everything is related and that a carefully executed measurement invariably entails generalizations.

The whole of the research programme required the constant collaboration of a factory and a laboratory. The great advantage of cooperation of this type is nowadays commonly appreciated and it would be superfluous to insist on it except to highlight a particular feature. When in the spring of 1896, the Société de Commentry-Fourchambault& Decazeville undertook to assist a programme of research which, failing such assistance, would not have advanced beyond its initial stages, they did so without the directors imagining that they would be rewarded with anything but the satisfaction of having lent effective assistance to a task of scientific exploration. But the meticulous examination to which the alloys were subjected, and the increasingly stringent requirements imposed on the Company as regards the compositions as well as the treatments, led, in the jointly explored field, to the creation of a kind of precision metallurgy, the result of which has been the manufacture of products considered impossible a few years ago. That has been a new and unexpected outcome of the scope of precision measurements. Demanding unflagging and occasionally laborious effort of those who apply themselves to it, precision measurement research could not long be pursued without the guidance and illumination of an ideal of perfection. For the research worker himself, this ideal stimulated by a belief in the utility — even afar off — of every unselfish effort is his mainstay at all times and the counsellor of every initiative. The highest approval which you have just conferred on my work highlights a task stretching back a quarter of a century and for that reason I feel obliged right here to express my deepest gratitude to the illustrious members of the Nobel Committee and of the Swedish Academy of Sciences, as well as to the memory of Alfred Nobel. From generation to generation my descendants will look with pride on the magnificent diploma and the Nobel medal, treasured testimony of your kindness and esteem. But also I wish to express my gratitude to you on behalf of the International Bureau which I have served for thirty-seven years and whose existence reflects at this moment the dramatic changes occurring in the world. Next I am grateful to you on behalf of metrology on which the conferment of a Nobel Prize casts a bright lustre well-fitted to commend it as a vocation. Lastly my gratitude is on behalf of the two countries with which my life is associated: Switzerland, of which I am a citizen and which I left equipped for life, and France, where I have known so many esteemed friendships."

I strongly recommend the reading of the whole of Guillaume's Nobel lecture not only for the importance of its content but also for the fluidity and clarity of his text.

Creation of the three "Grand Laboratoires": The Physikalisch Technische Reichenstalt (PTR) in Berlin in 1887, the National Physical Laboratory (NPL) Teddington UK in 1900 and the National Bureau of Standards (NBS) Washington DC in 1901

The creation of the three major scientific institutes, PTR, NPL, and NBS, at the turn of the 20th century reflected the rapidly growing need accepted by their governments that industry needed access to the results of advanced scientific research and advice on how this could be used in manufacturing. In addition, it was also clear that high-level calibrations of measuring instruments had also become essential. International competition also played an important part in persuading governments to act. The first was the PTR, which was created at the behest of two major figures in German science and industry, Helmholtz and Siemens. Following the first Electrical Conference held in France in 1881, Siemens was convinced that the then recently formed electrotechnical laboratory in France would give an unfair advantage to the French electrical industry. Helmholtz agreed and added that scientific research at a high level would become increasingly important for both science and industry.

The foundation of the PTR was however achieved only after much difficult discussion and argument in Germany. The universities were jealous of the pre-eminence in scientific research and industry was frightened of particular industries being given unfair advantages from work carried out in this proposed government institute. Nevertheless, the institute was created in 1887, but it initially had two distinct sections: the Scientific Section and the Technical Section. It was only after the First World War that they were brought together, much to the advantage of German industry. Materials research as such did not appear as part of the original remit of the institute, as it did a little later in the original role of the National Physical Laboratory in the UK. It did appear, however, indirectly as part of the large activity of testing materials and instruments.

The evident success of the PTR in giving support and assistance to German industry was, however, one of the arguments put to the British Government for the foundation of the National Physical Laboratory. Right from the beginning of 1900, in addition to metrology and standards, there were plans for an Engineering Department in which wide-ranging physical and chemical studies of the properties of new alloys would be undertaken. This Engineering Department was well equipped by 1905 with a rolling mill for the preparation of a sample for study, and in 1906, a new electrotecnics laboratory was opened. By 1907, there existed metallurgy and metallurgical Chemistry Departments, thus considerably widening the area of activities. A large rolling mill was installed. The range of work was even more extended when in 1910 work began in aeronautics and a wind tunnel was constructed. This was followed by the building of a 150 m ship tank. The NPL by the time of the outbreak of the First World War was established as a wide-ranging institute carrying out not just the development and maintenance of measurement standards but also research in many areas of engineering and what we would now call "Materials Science".

In the USA, the creation of a National Standards Institute followed soon after the creation of the NPL. It was established however despite the overriding principles of free and unregulated trade. The urgent need for uniformity of standards in the electrical and lighting industries led to a National Congress of Electricians in September 1884, which in turn led to a Conference at the Franklin Institute supported by the US Government, the outcome of which was a recommendation for the establishment of a Bureau of Physical Standards. It was not until some 16 years later that Congress finally decided to create the Bureau of Standards. The word "National" was added to its name some years later. In so doing however, Congress gave it wide-ranging responsibilities not only for the establishment and maintenance of physical standards and testing but the resolving of problems that arise in connection with these standards, the determination of physical constants and the properties of materials when such data are of great importance to scientific and manufacturing interests. This was a very broad remit and the National Bureau of Standards became an important center for, among other things, the study of materials.

Thus, by 1914, there existed the three principal national standards institutes, all of which were deeply involved in the study of materials, not just those directly used for measurement standards but also, particularly at NPL and NBS, materials of importance to the manufacturing industry and science. Much later, the National Bureau of Standards began a major program for the preparation and supply of standard reference materials in the areas of physics, chemistry, engineering, and, more recently, in medicine.

Conclusions

While the term "National Metrology Institute (NMI)" is very recent,[4] from the very beginning these institutes were deeply involved in materials research. This was not only during the period when national and international standards were material objects but also as a result of the expertise in measurement which allowed them to produce reliable data on material properties. This continues today

[4] The acronym NMI, standing for National Metrology Institute, was invented only in the 1990s during the drawing up of the CIPM Mutual Recognition Arrangement for National Measurement Standards (CIPM MRA).

and is an important part of their activities. A prime example of modern materials research is that which underpinned the determination of the Avogadro constant, N_A, through the manufacture of a nearly perfect sphere of mono-isotopic silicon. Extensive studies were carried out by a consortium of some 15 national metrology institutes on the manufacture of nearly perfect spheres of silicon, on its lattice constant, and on the properties of the final surface and its oxide layers. The value of N_A thus obtained confirmed the value of the Planck constant, obtained directly using a Kibble balance, used in the new definition of the Kilogram adopted by the 26th General Conference on Weights and Measures in 2018.

References

Guillaume, C. E. Nobel Lecture 1920.

Quinn, T. J. (2011). *From Artefacts to Atoms, The BIPM and the Search for Ultimate Measurement Standards* (New York: Oxford University Press).

Meetings of the International Metre Commission

Comité des recherches préparatoires avril 1872 (1872) (Paris: Imprimerie Viéville et Capiomont), p. 90.

Comité permanent, Procès-verbaux des Séances 1872–1873 (1873) (Paris: Imprimerie Nationale), p. 40.

Comité permanent, Procès-verbaux des Séances 1874 (1874) (Paris: Imprimerie Nationale), p. 36.

Organisation (1 September 1870 Session de 1870, 1872) (Paris: Imprimerie Nationale), p. 8.

Réunions générales de1872 (1872) (Paris: Imprimerie Nationale), p. 231.

Session de 1870, Procès-verbaux des Séances (1871) (Paris: Librairie Polytechnique de J. Baudry, Editeur), p. 51.

Tables de correction, 1875 (1875) (Paris: Imprimerie Nationale), p. 24.

Section Française de la Commission Internationale du Metre

Annexes des procès-verbaux des Séances de 1869 et 1870 (1872) (Paris: Librairie Polytechnique de J. Baudry, Editeur), p. 24.

Exposé de la situation des travaux au 1 october 1873 (1873) (Paris: Imprimerie Nationale), p. 60.

Procès-verbaux des Séances de 1870 (1871b) (Paris: Librairie Polytechnique de J. Baudry, Editeur), p. 51.

Procès-verbaux des Séances de 1871–1872 (1872) (Paris: Imprimerie Nationale), p. 69.

Procès-verbaux des Séances de 1872–1873 (1873) (Paris: Imprimerie Nationale), p. 117.

Procès-verbaux des Séances de 1873–1874 (1874) (Paris: Imprimerie Nationale), p. 174.

Procès-verbaux des Séances de 1874–1876 (1876) (Paris: Imprimerie Nationale), p. 215.

Procès-verbaux des Séances de 1877 (1878) (Paris, Imprimerie Nationale), p. 158.

Procès-verbaux des Séances de 1878–1882 (1883) (Paris: Imprimerie Nationale), p. 94.

Procès-verbaux des Séances de 1886–1890 (1892) (Paris: Imprimerie Nationale), p. 116.

Réunions des Membres Français*Procès-verbaux des Séances de 1869 et 1870* (1871a) (Paris: Librairie Polytechnique de J. Baudry, Editeur), p. 70.

Similar reports were made in 1874, p. 154; 1876, p. 86; 1877, p. 25; 1878, p. 2 and 1879, p. 13.

Creation of the PTR, NPL and NBS

Cahan, D. (1989). *An Institute for an Empire, the Physikalisch-Technische Reichenstalt 1871–1918* (Cambridge, UK: Cambridge University Press).

Cochrane, R. C. (1966). *Measures for Progress, A History of the National Bureau of Standards* (Washington, DC, USA: US Department of Commerce).

Magnello, E. (2000). *A Century of Measurement, an Illustrated History of the National Physical Laboratory* (London: HMSO).

Pyatt, E. (1983). *The National Physical Laboratory — A History* (London: Adam Hilger Ltd.).

Chapter 6.2

CERN: A Global Hub for Materials Science*

Panagiotis Charitos, Paul Lecoq, Michael Moll, Cristoforo Benvenuti and Luca Bottura

Introduction

The European Organization for Nuclear Research (CERN), home to the world's largest scientific tool, the Large Hadron Collider (LHC), carries decades of accumulated expertise in advanced material characterization and analysis. The challenges of building and operating at large scale the components of the LHC offer CERN a unique position in the design, development, and assessment of materials and coating processes to meet demanding requirements, often at the extremes of temperature, radiation, vacuum, and magnetism. CERN experts, together with researchers from around the world and industrial partners, are working to translate these requirements into new methods and recipes that also take into account the constraints of large-scale industrialization required for the demands of Big Science.

The following sections describe some of the challenges in the design, development, and building of new detector and accelerator

---

* *Editor's Note*: This chapter is jointly authored by the listed authors, but each author wrote a section of the chapter. Each authored section is again labelled with the respective author's name to give proper credit. The Introduction was contributed by Panagiotis Charitos.

technologies and reflect on CERN's history as a hub for collaborations between academia and the industry as well as on the training of thousands of students and fellows in international collaborations since the early 1990s. The first two parts focus on material-related challenges driven by detector developments. CERN personnel together with its global collaborators develop, prototype, construct, maintain, and operate some of the largest and most complex detectors in the world. This begins at the micro scale, with the production of PCBs, thick-film and thin-film hybrids, flex-rigid boards, fine-line flex circuits, large-size boards, and sequential buildup multilayers for mechanical support and cooling.

A new phase in the history of scintillating crystals started with the construction of the LHC and the design of detectors for the four LHC experiments involving academic and industrial partners from all over the world. Paul Lecoq reflects on the development of scintillating crystals and the history of the Crystal Clear Collaboration that was formed in 1990 and extensively worked on identifying the most adequate scintillator crystals for the LHC. CCC made an extensive investigation for the most adequate scintillator crystals for the LHC. Lead tungstate was chosen by CMS and ALICE as the most cost-effective crystal compliant with LHC conditions, and today, PWO crystals are installed in two of the LHC experiments, namely CMS (75848 PWO crystals) and ALICE (17920 crystals). Furthermore, the new crystals developed by the CCC are used in medical imaging applications and specifically in PET devices. Today, the CCC remains active in academic and applied researches on scintillating materials and novel ionizing radiation detecting devices for particle physics and medical imaging, as well as on the development of medical imaging prototypes. A second significant effort, concerning the development of new materials for the LHC experiments, was undertaken by CERN's RD50 collaboration. RD50 was launched to address the challenges in the harsh environment of the High Luminosity LHC and develop semi-conductor detectors that could cope with the radiation environment of HL-LHC. The collaboration's activities lie in the field of material engineering to suppress microscopic defects with a detrimental effect on the macroscopic detector parameters, and device engineering, to produce more radiation-tolerant detector geometries.

The other two essays focus on accelerator components and more specifically on coating technology for maintaining the high vacuum on the beam pipe as well as the development of superconducting material for high-field magnets. CERN components must meet demanding expectations in performance and longevity. To achieve this, CERN experts have accumulated over 50 years of experience in the development and application of novel thin-film coatings for functional and optical use, including in ultra-high vacuum (UHV) systems and high-energy detectors. CERN uses a wide range of superconducting (SC) materials in the construction of its accelerators and experiments. Niobium titanium has been the workhorse for the LHC. Nb3Sn is required to produce the high-field magnets needed for the high luminosity upgrade. A wide range of materials are being explored to enable the cables, magnets, and devices required for the future. This also entails the construction of comprehensive test facilities and expertise for characterizing and benchmarking wires and cables and an extensive database of superconducting material and conductor characteristics.

HEP and the Development of Scintillating Crystals

Paul Lecoq

Scintillators are among the most popular ionizing radiation detectors. They have played and are still playing a major role in the development of modern physics. The visual observation of scintillation on a Zinc Sulfide screen has allowed E. Rutherford to observe alpha-particles, an event which can be considered as the starting point of modern nuclear physics. Until the end of the Second World War, Zinc Sulfide and Calcium Tungstate were among the most popular particle detectors found in nuclear physics laboratories. The intensive development of atomic projects in the postwar period stimulated the development of new ionizing radiation detecting techniques, including scintillation counters. With the development of experimental physics, and in particular, with the occurrence of photoelectric multipliers, it became clear that scintillating materials are ideal devices to detect elementary particles, identify them and measure their energy with great precision (Broser and Kallmann, 1947; Marshall and Coltman, 1947; Coltman and Marshall, 1947). In a rather short time (1947–1951), scintillation was observed in various organic and inorganic crystalline media (Moon, 1948; Kallmann, 1949; Collins and Hoyt, 1948; Bell, 1948) as well as fluids (Kallmann, 1950; Kallmann and Furst, 1950, 1951a, 1951b), gases (Reynolds, 1950; Swank, 1954), and polymeric compounds (Schorr and Torney, 1950). At the same time, the still most widely used scintillating crystalline material NaI (Tl) (Hofstadter, 1949) has been discovered.

Scintillators are used in a large variety of applications:

- particle detectors for high energy and nuclear physics;
- spectrometry of low energy *g*-quanta;
- space missions and astrophysics;
- medical imaging and nuclear medicine;
- homeland security;
- nuclear waste monitoring;
- well-oil logging;
- non-destructive inspection methods.

When Hofstadter (1949) introduced in 1948 Thallium-doped Sodium Iodide, NaI(Tl), he probably did not realize that it was going to be the most popular scintillator for the next 35 years. The best example of the high discovery potential of scintillators in high energy physics (HEP) was first given by the NaI(Tl) Crystal Ball experiment at SLAC (Oreglia, 1982) which allowed reconstructing of the precise spectroscopy of charmonium states (Fig. 1).

This success has motivated a number of physicists to propose in the following decades several experiments using scintillator-based electromagnetic calorimeters and HEP has become a driving force, together with medical imaging, for the development of new scintillation materials, due to the very challenging requirements as well as to the large volume of scintillators needed in modern experiments.

This trend started with Bismuth Germanate ($Bi_4Ge_3O_{12}$ or BGO) for the L3 experiment at the large electron-positron collider (LEP)

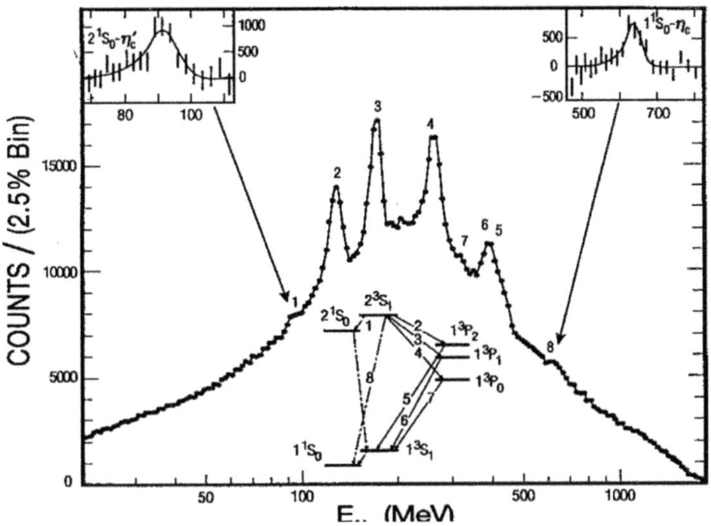

Fig. 1. Charmonium spectroscopy with the Crystal Ball at SLAC (Oreglia, 1982).

at CERN (Adeva *et al.*, 1990) and Cesium Iodide (CsI either Thallium doped or pure) for Cleo II experiment at Cornell's Electron Storage Ring (Kubota *et al.*, 1991), Crystal Barrel at CERN's LEAR (Amsler, 1998), KTeV at the University of Chicago, US (Ray, 1994), Belle at KEK[1] and BaBar at SLAC[2]; experiments based both in Japan and the USA that were already known but in small sizes and small quantities only. It became even more evident with Barium fluoride (BaF_2) for TAPS (Novotny *et al.*, 1987) and GEM,[3] Cerium Fluoride (CeF3) as a candidate for CMS,[4] and L3P[5] and finally Lead Tungstate (PbWO4 or PWO) for CMS (see footnote 3) and ALICE[6] at the CERN LHC, which were essentially developed for HEP experiments.

The following section describes how the difficult physics constraints and harsh experimental conditions at high luminosity colliders impose very tight specifications on scintillators. Following this

[1] The Belle Collaboration. Technical Design Report, KEK Report 95-1 (1995).

[2] The BaBar Collaboration. BaBar Technical Design Report, SLAC-R-95-457.

[3] GEM Letter of Intent, SSCL SR-1184, November 1991.

[4] CMS Technical Proposal, CERN/LHCC 94-38, December 1994.

[5] L3P Letter of Intent, CERN/LHCC 92-5 (1992).

[6] ALICE Collaboration Technical Proposal, CERN/LHCC/95-71 (1995).

brief technical discussion, we will offer examples of the development of BGO and PWO, respectively, for the L3 and the CMS experiments at CERN and conclude this essay with perspectives for future HEP experiments will be given.

Requirements for Scintillators Used in HEP Experiments

Because of their excellent energy resolution, scintillators are used in HEP for compact, high-precision, homogeneous electromagnetic calorimetry. The purpose is to measure with the highest achievable precision the energy of electrons and photons, directly produced or through the decay of unstable heavier particles, over the widest possible energy range. The criteria for the choice of the scintillator are based on the density, the scintillation properties, and the radiation hardness. The cost is also an important issue taking into consideration the very large volumes of several cubic meters considered for such detectors.

Density

The compactness of the calorimeter is essential in order to reduce the detector volume and cost. This is achieved by using high stopping power materials resulting in a small radiation length X_0. A high-density material ($\rho > 5$ g/cm^3) is therefore needed, but not necessarily with very high Z components as often mentioned. The density should be preferably based on the high compactness of the crystal lattice (a large number of atoms per unit volume), keeping the atomic number of the components not necessarily too large in order to reduce the lateral shower size defined by the so-called Moliere radius (RM $\approx X_0$ $(Z+1.2)/37.74$). A small Moliere radius will limit the contamination of the energy measurement by other particles from the same or other events (pile-up) and help for the position reconstruction as well as for the $\pi0$ rejection, which is the dominant background in the electromagnetic calorimeter at the LHC. Finally, a compact material helps reduce the lateral spread of the shower in a high magnetic field.

Scintillation properties

Pile-up imposes severe constraints on the detector readout, particularly at high-luminosity colliders. It is therefore essential to collect as much signal as possible in a short time, to keep a good signal-to-noise ratio in the electronics chain. Decay times of the order of the bunch crossing time or even less are necessary. Only optically allowed (inter-configuration) transitions (like the transition 5d → 4f for Ce3+), cross-luminescence, which is intrinsically fast and temperature independent, and strongly quenched intrinsic and activator's luminescence can give rise to a fast and bright scintillation, allowing acquisition rates at several tens of MHz.

As HEP electromagnetic calorimeters are usually installed in magnets, readout by photodiodes or avalanche photodiodes is desired. These photo-detectors have a gain, which is either 1 for PIN diodes or a few hundreds in the case of avalanche photodiodes, which is significantly lower than photomultipliers. This implies a sufficient light yield (a few 100 pe-/MeV of deposited energy), and an emission wavelength above 250–300 nm, where the quantum efficiency of the photodiodes becomes high. Light emission in the visible spectrum will also ease the problems of light collection in long crystals. The development of SiPMs with a gain comparable to vacuum photomultipliers tubes restricts these limitations and allows considering calorimeter designs with much finer granularity, with energy deposits in each cell at the MIP (minimum ionizing particle) level.

The energy resolution of the calorimeter will strongly depend on all possible sources of non-uniformity. The light collection in a pointing geometry will introduce non-uniformity due to the focusing effect, which depends on the refractive index of the crystal. Fluoride crystals and glasses, with a refractive index around 1.5 will limit this effect to a much smaller value (and therefore make it much easier to correct) than for the BGO (index 2.15) or PWO (index 2.3). The material can be intrinsically luminescent or doped with a scintillating impurity. Intrinsic scintillators are generally preferred, as it is easier to control the light yield uniformity in long crystals. On the other hand, a controlled distribution of the doping ions could help correct for the non-uniformity caused by the light collection in a pointing geometry. In addition, the scintillation yield should be as independent as possible from the temperature. Large temperature coefficients increase the complexity of the detector and of the software corrections, and temperature gradients between the front and back faces of the crystals introduce non-uniformities affecting the constant term of the resolution.

Radiation hardness

It is now well established that the most significant damage in the majority of inorganic scintillators results from the formation of color centers in the bulk of the material, which absorbs part of the scintillation light on its path to the photo-detector (Lecoq *et al.*, 1991). A short radiation length will therefore reduce the total attenuation for a given damage.

In addition, as the majority of color centers absorb mostly in the UV (more precisely they involve traps that are in the vicinity of the fundamental absorption edge), crystals emitting light in the visible part of the spectrum are likely to be less severely damaged. Some surface effects may also occur but it is generally very small and saturates quickly with the dose.

The formation of color centers results from the trapping of electric charges by crystal structural defects or impurities and is therefore directly related to the quality of the raw material. A large effort should be made to purify the raw materials to the best quality. However, in some cases, a specific doping of the crystal has proved to be an efficient and economical way to significantly improve the radiation hardness (Annenkov *et al.*, 1999).

Cost considerations

The factors building-up the price of crystals must be identified and carefully analyzed. In some cases, the raw material of the desired purity can represent a substantial fraction of the cost. For this reason, all crystals based on rare and expensive lutetium are generally discarded for applications were tons of crystals are needed, although some of them are very fast cross-luminescent materials ($BaLu_2F_8$) or very efficient scintillators (LSO:Ce). Cerium is more favorable, as it is the most abundant rare-earth in the earth's crust. The crystal growth technique and the crucible material are dominant parameters and crystals, which can be grown for instance by using the relatively cheap Bridgeman method in graphite crucibles, have obvious advantages. The cost of the energy plays also an important role, directly impacted by a combination of the melting temperature and the pulling rate. From this point of view, the high melting point of GSO or LSO (gadolinium and lutetium orthosilicates) (>1900°C) and the high cleavage susceptibility of their crystallographic matrix implying a low pulling rate (1–2 mm/h) may put some limits to the minimum obtainable price for such crystals. On the other hand, the low melting point of PbF_2 (lead fluoride, 822°C) and PbWO4 (lead tungstate, 1123°C) and the abundance of the raw material make them low-cost materials similar to NaI(Tl). A high crystal density directly limits the total volume for a calorimeter, but also the crystal dimensions and consequently the furnace and crucible sizes will be a major parameter of the total cost for a crystal-based detector. Finally, the very good mechanical properties of PbWO4 for instance, allow a high production yield of the mechanical processing, lowering consequently the final cost.

Examples of Crystals Developed for Large Collider Experiments at CERN

BGO and the L3 experiment at LEP

At the time the LEP program was launched the top quark still had to be discovered and its mass was expected to be in the energy range of the LEP collider. The electromagnetic calorimeter of the L3 experiment was designed to achieve the best possible energy resolution down to 100 MeV in order to explore the detailed spectroscopy of the toponium system, the putative bound state of the top-antitop quarks. A similar situation was encountered a few years earlier after the discovery of the J/Ψ, the charm-anticharm bound system, by B. Richter and S. Ting (the spokesperson of the L3 experiment) with the unprecedented performance of the NaI:Tl Crystal Ball at PEP to study with high precision the charmonium states. This success convinced the community that a crystal-based homogeneous electromagnetic calorimeter was the best approach to reach a very good energy resolution, particularly in the 100 MeV–10 GeV energy range.

The L3 collaboration decided therefore to build a homogeneous electromagnetic calorimeter based on a recently discovered crystal, bismuth germanate (BGO). BGO had been known as a scintillator for less than 10 years in 1973 (Weber and Monchamp, 1973). The reason for this choice was motivated by the much higher stopping power of BGO for photons and electrons compared to NaI:Tl. This allowed for building a compact, highly granular, and high-performance calorimeter with a minimum impact on the dimensions and cost of the external detectors (hadron calorimeter and muon chambers) as well as on the magnet. Furthermore, BGO has good scintillation properties and is non-hygroscopic in contrast to NaI:Tl.

This calorimeter consisted of nearly 12,000 BGO crystals readout by photodiodes insensitive to the 0.5 T magnetic field. It was composed of a cylindrical barrel with 7680 crystals and two endcaps of 1536 crystals each in a pointing geometry with a small azimuth angular offset of 10 mrad to avoid photon leakage in the gaps between crystals. Each crystal, 24 cm long, was a truncated pyramid of about 2×2 cm^2 at the entrance face and 3×3 cm^2 at the outer end.

Transforming the BGO crystal from a laboratory curiosity to a mass-producible technology meeting the stringent L3 experimental specifications required a new style of R&D. A large-scale and multidisciplinary R&D effort was organized, a novelty for detector R&D in HEP. This effort involved a number of institutes of the L3 collaboration and also experts in crystallography, solid-state physics, and luminescence, including the BGO "father" M.J. Weber. This collaboration would become the seed of the future Crystal Clear

Fig. 2. The L3 BGO-based electromagnetic calorimeter being assembled at CERN.

collaboration, which made major contributions to the development and construction of the Lead Tungstate-based calorimeter of the CSM experiment at the LHC one decade later. As a result of this effort, a BGO mass production technology could be developed at the Shanghai Institute of Ceramics in Shanghai, where 12,000 BGO crystals (10 tons) of excellent and reproducible quality were produced (Fig. 2).

The mass production of BGO became a mature technology so that the excellent properties of this crystal could be exploited in other domains, such as nuclear physics and space experiments, as well as in medical imaging devices, where BGO became the crystal of choice for the second generation of PET scanners for about two decades.

PWO and the CMS experiment at the LHC

When the LHC program was launched at CERN it was realized that an ambitious and aggressive R&D effort had to be organized to develop the detector technologies for the harsh operating conditions of the LHC. For this purpose, the DRDC (Detector Research and Development Committee) was created and offered an ideal framework for launching a large multidisciplinary effort gathering a number of world experts in different aspects of material sciences (crystallography, solid state physics, luminescence, defects in solids) working together for the development of new inorganic scintillators. Created in 1990, and still very active today, he Crystal Clear Collaboration, registered as RD-18 by the DRDC,[7] organized a worldwide R&D effort allowing both the CMS and ALICE collaborations to decide in 1994 to select lead tungstate ($PbWO_4$) scintillator for their electromagnetic calorimeters. Indeed, providing that a suitable crystal could be developed for this purpose, a homogeneous crystal-based electromagnetic calorimeter was considered to be an ideal detector to meet the very challenging requirement of detecting narrow states, like the Higgs boson decaying into two γ-rays over a large background for Higgs bosons in the intermediate mass range, for which the natural width is completely dominated by the instrumental resolution. This was the choice made by the CMS experiment for its electromagnetic calorimeter.

For about 10 years, a systematic approach has been followed by all the actors involved in this multidisciplinary endeavor for optimizing the performances of Lead Tungstate crystals to cope with the stringent operating conditions of the LHC, in particular in terms of scintillation dynamics and radiation hardness. A large effort has also been spent to develop mass production technologies allowing to grow 80,000 large-size crystals for a total weight of 100 tons in a few years' time at an affordable cost.

The first objective was to grow ingots large enough to extract crystals with the required dimensions: 23 cm in length, $2.05 \times 2.05\ cm^2$ front section, and $2.38 \times 2.38\ cm^2$ back section. It took several months to obtain crystals of these dimensions with good mechanical stability and therefore a high cutting and polishing yield.

---

[7] Crystal Clear Collaboration, RD18, CERN/DRDC/P27/91-15.

A second problem was the suppression of slow components in the scintillation of $PbWO_4$. It was observed in 1995 that the optimization for a higher light yield had very often the consequence of producing slow components at a few percent levels in the crystals. A thorough investigation was made which led to the conclusion that a combination of molybdenum contamination and bad annealing conditions could introduce such components.

A very critical parameter for Lead Tungstate crystals is their ability to survive in the high radiation environment of the LHC in the CMS experiment, with the minimum short-term variations of collected light, as well as during the lifetime of the experiment. This problem was the most difficult to solve, as it implies a perfect modelization of the radiation conditions in the LHC machine and a very good understanding of the chemistry of defects in this crystal. All these aspects have been systematically investigated and impressive progress has been made.

Another important concern was to set up a production and quality control infrastructure at an industrial scale for the production of 100 tons of scintillators. The possibility to make use of the large production facilities installed during the cold war in the former Soviet Union has been a key to the success. This has motivated the collaboration with the International Science and Technology Centre (ISTC)[8] for the conversion of the Bogoroditsk Techno-Chemical Plant in Russia for the production of a large fraction of the 80'000 Lead Tungstate crystals for the CMS experiment at CERN, with up to 238 crystal growth ovens involved in the production of up to 1800 crystals per month (Fig. 3). Around 5% of the crystals were grown at the Shanghai Institute of Ceramics, previously involved in

[8] The International Science and Technology Center (ISTC) is an intergovernmental organization established in 1992 by an agreement between the European Union, Japan, the Russian Federation, and the United States. It was signed in Moscow in 1992 and provisionally entered into force on 2 March 1994, pending ratification by the Duma. The ISTC serves as a clearinghouse for developing, approving, financing, and monitoring projects aimed at engaging weapons scientists, technicians, and engineers from the Commonwealth of Independent States (CIS) in peaceful, civilian science and technology activities.

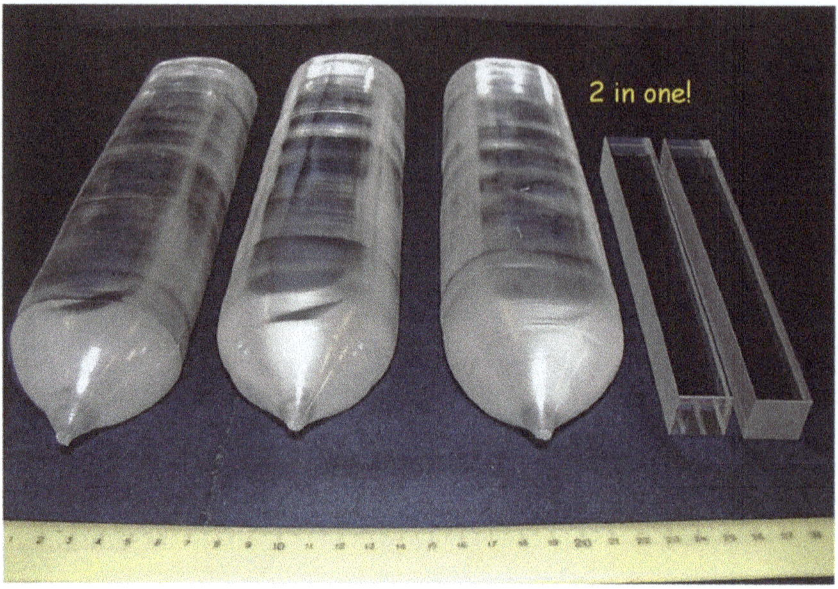

Fig. 3. 65 mm ingots with 2 CMS barrel crystals cut per ingot.

the development and production of the BGO crystals for the L3 experiment at LEP.

Future Perspectives: Addressing the Increased Demand for Fast Timing

The search and development of scintillators in the last decades have been mainly oriented toward higher light yield and better proportionality in order to improve the energy resolution, at high energy to detect narrow states like the Higgs boson over a large background and at low energy for precise spectroscopy in applications like homeland security.

Recent years have seen the emergence of fast timing capability as a new requirement, mainly driven by HEP to cope with higher event rates while minimizing pile-up and time-of-flight positron emission tomography (TOF-PET) applications to improve the image signal-to-noise ratio and effective sensitivity. Timing resolutions in the 10ps range are required in both cases, which boots the research for scintillators with a high light yield, a short rise time, and decay time, as well as for ultrafast scintillation mechanisms to produce prompt photons as a result of transient phenomena and/or quantum confinement in nanocrystals.

The quest for rare events in HEP requires the construction of colliders with high luminosity, which is achieved by densely spaced particle bunches at the High Luminosity LHC, the Compact Linear Collider CLIC or the Future Circular Collider (FCC). Precise timing capability will be a key aspect of particle detectors at these high luminosity colliders, as the time information will be an essential component of the event reconstruction. Indeed, the high track density and event pile-up pose serious challenges for physics event reconstruction and analysis. For energy measurements, one important source of degradation is represented by the contamination of neutral particles originating from secondary vertexes. A precise timing of both calorimeter deposits and vertexes can aid in the reconstruction, allowing the rejection of spurious energy deposits that are not consistent with the primary vertex time.

The development of ultrafast radiation hard scintillators is therefore mandatory to prepare highly performant detectors for this new generation of accelerators and physics. We can fortunately benefit from the large organized R&D effort launched at the time of the LHC, in which the Crystal Clear collaboration is particularly active, to prepare for this future. For instance, through detailed modeling of the band structure of several host scintillators, it has been possible to define the density of states in the conduction and valence bands of several scintillators and to better understand the mechanisms of charge and energy transfer in these materials in the presence of different doping ions. This has led to the development of co-doping strategies in several scintillators, which are particularly effective in improving their radiation hardness (Fasoli *et al.*, 2011).

Similarly, a large effort has been launched recently to improve the timing performance of scintillators, with a target objective of 10ps resolution for low energy deposits as small as 511 keV, in the context of the 10ps challenge (Lecoq *et al.*, 2020). If the initial motivation is driven by a high demand to increase the effective sensitivity of PET scanners in medical imaging, this large-scale multidisciplinary effort, which has always prevailed between HEP and medical physics in a spirit of cross-fertilization, will necessarily benefit a new generation of particle physics detectors (Lecoq, 2017). The possibility to exploit ultrafast scintillation from transient phenomena or quantum confinement processes in nanocrystals is, from this point of view, particularly attractive.

Conclusions

Fundamental physicists have from the very beginning recognized the importance of scintillators for the discovery of new particles, and both disciplines have developed and cross-fertilized each other over more than a century, until now. This has not only benefited nuclear physics and HEP, but a number of other scientific technical, and industrial domains.

Dark matter searches, in particular through the direct detection of weakly interacting massive particles (WIMPS) as well as neutrinoless double beta processes, and more generally neutrino physics are using scintillating crystals with a specific additional requirement for a very high radio-purity of their constituents.

Astrophysics and space missions are also using scintillators for detecting galactic and extragalactic gamma-ray bursts or analyzing the composition of the planet's crust.

Along the same line, scintillating crystals are used for geophysical studies, in particular for well oil logging. Nuclear waste monitoring, homeland security, and border control are also large scintillator consumers.

A number of industrial processes are using scintillators, for non-destructive tests of tanks, welds, automatic thickness control of steel plates, a.o. But a very important domain for scintillator applications is nuclear medical imaging with single photon emission computed tomography (SPECT) and positron emission tomography (PET), with a production of several tens of tons of scintillators every year. In a way similar to particle physics, this is also a very dynamic area, with new medical challenges requiring ever-better scintillator performances.

These application domains are pushing for maintaining and even reinforcing the coordinated multidisciplinary R&D effort, which has been boosted 30 years ago by the LHC endeavor and thanks to the global collaboration of a vast network of research centers, universities, and representative industries motivated by the ambitious goals of this program and coordinated by CERN.

Development of Radiation-Tolerant Silicon Detectors for HEP

Michael Moll

Introduction

The importance and potential of semiconductor-based detectors for particle detection in nuclear and high energy physics were already anticipated in about 1960 (Snell, 1960; Yuan, 1961), but took two further decades to develop into reliable full detector concepts. First, silicon-based tracking detectors appeared finally in HEP experiments in the 1980s (Turala, 2005; Heijne, 2008; Ansari *et al.*, 1989) and revolutionized particle detectors in terms of spatial resolution and rate capabilities. They quickly became "the choice" for inner detector particle tracking in HEP experiments and are until today the unchallenged detector concept for this application. A particular issue that grew in importance with increasing collider luminosity and is the main subject of this chapter, is radiation damage introduced by the particle and photon flux traversing the tracking detectors calling for advancements in semiconductor pixel detectors for particles and imaging.

Silicon Detector Properties and Their Usage in HEP

The advantages of silicon detectors compared to other tracking detector concepts, most prominently based on the ionization of gases, are manifold. Without exhausting the technical details we briefly report on some of the benefits offered by silicon detectors, namely their light weight, mechanical stability, and the fact that they can be built self-supporting. Charge transport toward the electrodes is executed by the fast drift (some 10^6 cm/s) of electrons and holes in the conduction valence bands of the semiconductor, resulting in full collection times of below 20 ns in a 300 μm thick device. This fact allows operating the LHC, today's flagship collider, with a particle interaction rate of 40 MHz at very high luminosity. Compared to gases, the ionization density of charged particles passing the solid-state detector is very high, approximately 80 electron-hole pairs per micron for silicon, allowing detection with a good signal-to-noise ratio from thin sensor layers. A typical silicon sensor with a thickness of 300 μm collects about 23.000 e–h pairs when traversed by a highly energetic ionizing particle, a so-called minimum ionizing particle (mip). A major technology step occurred in 1980 with the introduction of the planar detector process (Kemmer, 1980). The possibility to finely segment the charge-collecting electrodes on the sensor employing photolithographic etching techniques allows to produce of electrode patterns with micron precision and any kind of shape, for example, in the form of pixels or strips. The resulting tracking position resolution can reach the micron level, a quality that revolutionized particle tracking and vertexing. The rise of the silicon detector technology was accompanied by the development of the corresponding readout electronics (Heijne, 2021). Both, sensors and readout electronics, were largely boosted by the semiconductor industry, which today allows for combining electronics and sensors into a single monolithic device. The increasing use and application of silicon sensors in HEP are reflected in Fig. 4 showing the surface covered by silicon sensors in the indicated experiment against the first year of operation.

Radiation Damage Challenges for the Long-Term Operation

The ambition to deepen the understanding of particle physics and the advancements in accelerator and detector technologies fostered the construction of ever-higher intensity accelerators with tracking detectors placed closer and closer to the interaction point, thus facing strongly increasing particle interaction rates.

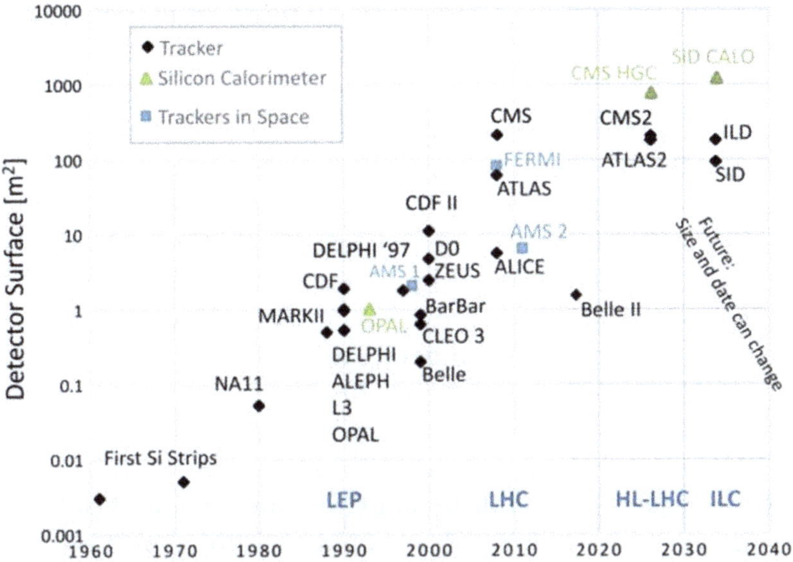

Fig. 4. Evolution of silicon detectors area. In logarithmic scale, the plot displays the increase in area over the last 60 years including as well the detectors approved and considered for the following decades (taken from Hartmann (2017)).

While the higher particle rates and the required increase in spatial resolution posed challenges by themselves, a new issue appeared, namely the aging of detectors induced by the traversing particles. For the inner pixel detector layers at the High Luminosity LHC (HL-LHC), for example, the cumulated particle fluence will reach some 10^{16} highly damaging particles per cm^2 and a dose of about 10 MGy. The resulting radiation damage to the semiconductor base material of the sensors as well as to the surface layers poses severe challenges to the operation of detectors (Moll, 2018). To cope with these constraints, not faced by any other silicon detector application before, the HEP community embarked on a self-contained and profound R&D program. It called for in-depth studies on the microscopic mechanisms leading to the detector damage and on mitigation techniques to handle the effects and guarantee the long-term operation of sensors close to the interaction points.

CERN-Centered R&D Collaborations Focusing on Silicon Detector Developments

History and organization of the research collaborations

In the late 1980s, when the plans for a new generation of proton supercolliders were discussed, the advantages of silicon-based tracking and preshower detectors against their gaseous detector counterparts in terms of position resolution and rate capability were obvious. The instrumentation of the region closest to the interaction point with silicon detectors was therefore a clear need for the LHC experiments, but a full technical feasibility study including the proof or radiation hardness allowing for the anticipated 10 years of operation had still to be achieved. In this light, the RD2 project was established at CERN in 1990[9] and put under the control of the Detector Research & Development Committee (DRDC 1990–1995[10]). The RD2 project was focusing on both, the silicon sensor and the readout electronics development, while later R&D on sensors and on front-end electronics was conducted in separate projects. With the formal approval of the LHC construction in 1994, the R&D needed to bring the experiments to operation was re-evaluated. R&D projects were formed and monitored by a new LHC Committee (LHCC, 1995) replacing the former DRDC. The required systematic and coordinated silicon sensor R&D was covered first by the RD48 project (1995–2000)[11] and then, already in the light of the long-term LHC operation and the LHC luminosity upgrades, by the RD50 project (since 2002).[12] The scope was complemented with two further projects

---

[9] RD2 Collaboration (1990 -1995) R&D proposal, CERN, Geneva, Switzerland. CERN/DRDC/90-27 and DRDC/P3, August 1990.

[10] Detector Research & Development Committee (DRDC) (1990–95) LHC preparations change gear. CERN Courier, April/May 1995 and "The CERN Detector R&D Programme 1990–1994", CERN/DRDC 94-48.

[11] RD48/ROSE — Research and development on silicon for future Experiments, 1995–2000. https://rd48.web.cern.ch/; R&D proposal CERN/LHCC 96-23 P62 and LHC R&D 23, April 1996.

[12] RD50 — Radiation hard semiconductor devices for very high luminosity colliders, since 2002. https://rd50.web.cern.ch/; R&D proposal CERN/LHCC 2002–2003 and LHCC P6, February 2002.

with a more specific focus: RD39 on Cryogenic Tracking Detectors (1993–2017)[13] and RD42 on the Development of Diamond Tracking Detectors (since 1994).[14]

The CERN R&D on silicon particle detectors has thus been carried out inside the community for more than 30 years. The participating institutions were mostly directly involved in the design and construction of one of the LHC particle detectors but also reinforced by solid-state physics, material science, and electronics-oriented institutes. RD2 (1990–1995) comprised between 9 and 14 institutions, RD48 (1996–2000) between 23 and 37 institutes, and the RD50 collaboration grew from its formation in 2002 from 45 member institutes to 64 international member institutes with a total of 420 members in the year 2021. The collaborations have been mostly led by a team of two or three co-spokespersons and were structured in about five research lines, each guided by a research line convener, and covering the different aspects of the research. The R&D is ranging from the study of the microscopic origin of radiation damage and material science to device physics and development of sensor and material characterization tools toward system aspects for full-size silicon detectors. The RD projects are not receiving central funding. Every participating institute has to assure its own resources through its national funding agencies. A small financial contribution to a common fund is asked to every participating institute to centrally organize the bi-annual workshops and provide seed funding to some common activities like silicon wafer procurement, irradiation experiments or the production of new prototypes or test sensors. Publications are authored by the teams that have performed the corresponding research work, only articles summarizing the work of the collaboration or a wider research field are published by the whole collaboration.

[13] RD39 Collaboration: R&D proposal CERN/DRDC 93-53 "Superconducting Microstrip Detectors", 1993 and from 1998 to 2017 CERN/LHCC 98-27 (DRDC P53 Add.1) "Cryogenic Tracking Detectors"; remaining R&D merged into RD50 program in 2017; https://rd39.web.cern.ch/.

[14] RD42 Collaboration (since 1994), CVD Diamond Radiation Detector Development, Proposal: CERN/DRDC 94021, DRDC/P56, May 1994; https://rd42.web.cern.ch/rd42/.

Research strategies

The RD collaborations target the mitigation of radiation damage by three fundamental strategies (see footnotes 11 and 12) that are translated into structured research plans which leave some room for 'blue sky research[15]:

- Material engineering, which stands for the deliberate modification of the detector bulk material. One approach is the defect engineering of silicon, which includes, e.g., the enrichment of the silicon with oxygen, carbon, or other impurities as well as the exploitation of different silicon growth techniques. Another approach is the use of other semiconductor materials than silicon, such as diamond, silicon carbide or gallium nitride.
- Device engineering, which stands for the improvement of sensor structures by, e.g., the modification of the electrode configuration, the thinning of the bulk material or the introduction of additional doping layers to enhance impact ionization and the development of new detector geometries such as 3D sensors or monolithic devices combining sensor and front-end electronics within the same device.
- Variation of detector operational conditions, which investigates the optimum detector operational conditions including operation at cryogenic temperatures or under forward bias.

Naturally, the implementation of these strategies builds on a deep understanding of the physics underlying the detector operation and radiation damage as well as the constraints of system aspects. Basic studies on the radiation-induced microscopic defects, the parameterization of damage effects going along with defect modeling and device simulation, and the tests of segmented sensors and detector systems are therefore a core activity of the research work. Furthermore, the development and operation of dedicated

[15] RD50 Prolongation Request 2018, May 2018, CERN-LHCC-2018-017; LHCC-SR-007.

detector characterization tools and irradiation facilities are essential.

Achievements of the Research Collaborations in Material Science and Beyond

Characterization and parameterization of radiation damage to detectors

High energetic particles traversing a semiconductor detector undergo ionizing and non-ionizing energy losses, IEL and NIEL, respectively. The ionization, manifested by the generation of electron-hole pairs, constitutes the basis for the signal detection and is not producing damage to the semiconductor bulk of the detector. It does however lead to the build-up of interface and oxide charges that can have a detrimental impact. The NIEL results in displacement damage to the semiconductor bulk. Assuming that radiation damage is proportional to the NIEL (the so-called NIEL-hypothesis) allows the scaling of complex radiation fields with various particles and particle energy spectra to an equivalent 1 MeV neutron fluence that is producing the same damage (1 MeV equivalent fluence — neq/cm^2). The establishment of comprehensive NIEL functions was therefore one of the first activities when the LHC targeted RD projects started their work in 1995 (Vasilescu and Lindstroem, 2000). In the following years, it was demonstrated that NIEL scales well for the leakage current, but has, for example, severe shortcomings in describing the change of detector depletion voltage (see next section). The work on damage characterization was throughout the first decade mainly performed on simple *n*-type FZ silicon pad detectors of typically 300 microns thickness, an area in the order of 0.25 cm^2, and realized either (in the beginning) as Schottky diodes or (later) as devices with diffused or implanted electrodes. The detector degradation in terms of increase of leakage current, loss of charge collection efficiency, and change of the full depletion voltage as a function of particle fluence, particle type, and annealing time and temperature was studied in detail (see, e.g., RD2 Collaboration (1996)). Major findings were the type-inversion of the effective doping and, very unexpectedly, the increase of depletion voltage during the long-term storage at room temperature after irradiation, the so-called "reverse annealing" (Wunstorf *et al.*, 1992). To mitigate the latter detrimental effect, the present LHC silicon detectors have to be kept cold even during periods when no data are taken and the detectors are not operated, a requirement that had a major impact on the conception of the cooling systems and detector maintenance planning. In the following decade, a very comprehensive set of macroscopic damage parametrizations was developed (see, e.g., Moll (2018), Kramberger (2011), Li (2009), Lindström *et al.* (2001)) that is often called the "Hamburg model".[16] It allowed for reliable predictions of silicon detectors' performance degradation in the LHC experiments (Dawson, 2021).

Defect and material engineering

Based on promising radiation damage studies on sensors produced from epitaxially grown silicon a wide range of silicon material engineering approaches was studied by the RD48 collaboration from 1996 onwards. Silicon bulk materials originating from different growth techniques like Floating Zone (FZ), epitaxial (EPI), and Czochralski (CZ) silicon were investigated and various possibilities to enrich silicon with impurities like carbon, nitrogen, oxygen, and tin were tested.[17] Silicon with high oxygen content was found to behave differently under charged hadron irradiation and with

[16] The modelling of the radiation damage was a collaborative effort and is often called "Hamburg model," at least for the leakage and depletion voltage, as major contributions were produced in a series of PhD theses at the Hamburg University under the guidance of G. Lindstroem and E. Fretwurst: Renate Wunstorf (1992), Torsten Schulz (1992), Hennig Feick (1997) and Michael Moll (1999). For the charge trapping, the theses of G. Kramberger (Ljubljana, 2001) and O. Krasel (Dortmund, 2004) are very relevant.

[17] RD48 status reports were submitted to the LHCC in 1997 (CERN LHCC 97-39), 1998 (CERN LHCC 98-39) and 1999 (CERN LHCC 2000-009), see also footnote 11.

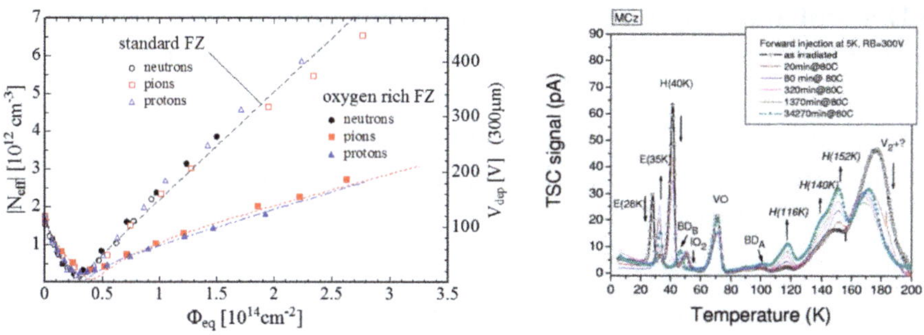

Fig. 5. Dependence of the depletion Voltage for 300 μm thick sensors (left). TSC spectra taken in the course of an isothermal annealing experiment at 80°C. Each peak corresponds to a defect level releasing charge carriers during the heating process of the TSC scan from low to high temperature (right).

Diffusion Oxygenated FZ (DOFZ) a convenient method was found to enrich FZ silicon on wafer level with oxygen. A characteristic result is shown in Fig. 5 (left). The increase of depletion voltage with particle fluence is reduced for the oxygenated devices if the irradiation is performed with protons or pions. However, for neutron irradiations, no difference is observed, which was called "the proton-neutron puzzle" until a deeper understanding was gained (see next paragraph). Oxygen-enriched FZ silicon has been successfully used since 2010 in the pixel detectors of the LHC, i.e., in the detectors closest to the interaction point, exposed to the highest radiation levels and facing mainly charged hadrons. Further successful silicon material modifications are the shift from *n*-type to *p*-type silicon and the carbon enrichment of LGAD gain layers that are discussed in the following.

In the search for radiation harder detectors, other semiconductor materials were evaluated. In particular, diamond proved to be a radiation-hard alternative to silicon for smaller detector applications. The RD42 project "CVD Diamond Radiation Detector Development" (see footnote 14) has successfully brought forward full detector concepts leading to the implementation of diamond-based beam monitors in the LHC experiments. Other semiconductors, in particular SiC and GaN, were studied by the RD48/

RD50 collaborations (2000–2006) (Sellin and Vaitkus, 2006), but found to be not sufficiently radiation hard and not mature enough for larger particle detector applications. Only recently, since about 2018, with the availability of higher quality SiC and GaN wafers, the studies of these materials have been picked up again within the RD50 collaboration.

The microscopic origin of radiation damage — defect characterization and modelling

Radiation damage originates from the interaction of traversing particles that lead to the dislocation of crystal atoms from their lattice sites. The formation of these defects is a temperature and time-dependent process that for part of the defect reactions takes even years at room temperature before stabilization (Moll, 2018). Crystal defects can have electrical properties constituted by energetic levels inside the forbidden bandgap of the semiconductor that impact the device performance in various ways. Generally, this can be described by the so-called Shockley–Read–Hall (SRH) statistics (see, e.g., Sze (2002)) but requires detailed knowledge of the characteristic defect properties, like concentration, energetic position in the bandgap and capture cross-sections for holes and electrons.

Within the HEP community, the emphasis of defect characterization studies is put on the electrical properties of the defects as these directly relate to the sensor performance. Most often the Deep Level Transient Spectroscopy (DLTS) and the Thermally Stimulated Current (TSC) techniques are used. TSC can be employed on highly irradiated devices, a feature that was pioneered and exploited already in the early 1990s in the HEP community (Borchi and Bruzzi, 1994). With the formation of RD48 in 1995 several solid-state physicists joined the study of radiation-induced defects in silicon sensors. This leads to the exploitation of a wide range of defect characterization techniques (see, e.g., Pintilie *et al.* (2016)), the systematic study of sets of identical devices and defect-engineered materials across various characterization techniques and finally to the identification of several defects responsible for the degradation of the different detector properties. Examples are the identification of the mechanism and donor levels leading to the strong differences in space charge generation between proton and neutron irradiated oxygen-rich material (proton-neutron puzzle, see Section 6.2) (Pintilie *et al.*, 2009), the acceptor levels responsible for the reverse annealing (see Fig. 5 (right) and (Moll *et al.*, 1998; Pintilie *et al.*, 2008), the defect responsible for the leakage current increase after gamma irradiation (Pintilie *et al.*, 2009), the identification of the trivacancy defect and its role in leakage current generation and more recently the study of the role of the boron-oxygen defect in the acceptor removal process. Over the last three decades, a very comprehensive set of radiation-induced defect levels with their according capture cross-sections and positions in the bandgap has been built up, that is now serving as an input to numerical device modeling of irradiated silicon sensors (see, e.g., Moll (2018) for more details).

Device engineering

In this section, we describe three outstanding developments of newly engineered devices. While the first spark of innovation often comes from individuals or out of expert discussions, the development of the full sensor concept and the long way toward operation is driven by the community. The progress and success are based on the deep understanding of the physics underlying the device operation and, in these special cases, the radiation damage mechanisms that needed to be mitigated or overcome to bring the devices to high-performance operation in the harsh detector environment of HEP experiments. The RD collaborations are "the place" to foster such innovation and development.

The device engineering approach with the strongest impact by far is the change of the silicon sensor bulk material from *n*-type to *p*-type silicon (see Figs. 6(a) and 6(b)), which comes along with a change of the implant type for the readout electrode from *p*-type to *n*-type and accordingly a change from collecting holes to collecting electrons at the electrode. While present LHC sensors employ *n*-type

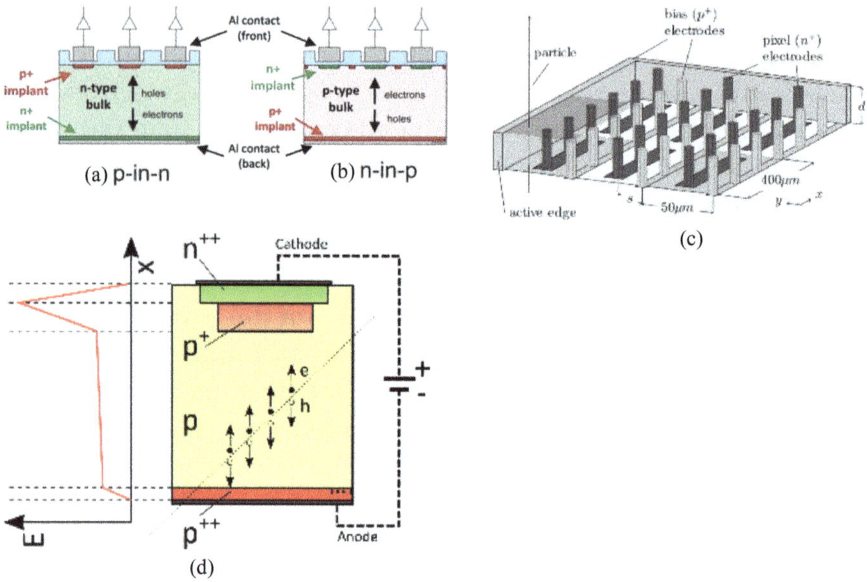

Fig. 6. Schematic representations of the three technology developments described in the text: (a,b) *p*-in-*n* and *n*-in-*p* strip sensors, (c) 3D sensors and (d) LGAD sensors (figures taken from Moll (2018, 2020) and Mathes *et al.* (2008)).

silicon and would not be able to withstand the radiation levels of the LHC upgrade (HL-LHC), the next generation of sensors for the HL-LHC, with a total sensor surface of more than 1000 m^2, is being built from p-type silicon. This paradigm change, triggered by the development within the CERN RD50 collaboration, offers a higher radiation hardness and, for the pixel devices, a lower production cost, but needed two decades of development and characterization in terms of surface isolation techniques, high voltage stability, thickness optimization, and radiation hardness.

3D sensors can be regarded as the most sophisticated device engineering approach. The technology was proposed in 1997 (Parker *et al.*, 1997) and consists of columnar electrodes etched perpendicularly to the wafer surface (see Fig. 6(c)). This geometry decouples the active thickness of the device, i.e., the length penetrated by the ionizing particle, from the charge carrier drift length, i.e., the distance the charges have to travel to the electrodes. The small distance between the columns (about 50–100 μm) translates into small depletion voltage and fast charge collection and is thus an intrinsically more radiation-hard device concept than corresponding planar devices of the same thickness. Less charge gets trapped in heavily irradiated devices due to the shorter drift distance to the electrodes, and since the device can be operated with lower voltages, the power consumption is drastically reduced. Disadvantages of 3D in comparison to planar pixel sensors are the higher pixel capacitance leading to higher noise, a dead region around the columns, and larger production costs. The RD50 collaboration picked up the concept and pushed for technology development in the foundries of CNM, Barcelona, Spain, and FBK, Trento, Italy, being members of the collaboration. The technology matured and several different concepts, like single-column type and double-column type devices, emerged and were tested. The development got a strong push with full industrialization when the ATLAS collaboration evaluated, developed, and tested the 3D sensor concept for their innermost pixel detector layer, the B-layer, and brought the technology from a mature concept toward an operating detector (2007–2014). 3D sensors are to date the radiation hardest silicon sensors and have been chosen for the innermost pixel detector layers of the HL-LHC (see Da Vià *et al.* (2019) for a recent overview of the 3D technology).

Low Gain Avalanche Diodes (LGADs) are the most recent device offspring of the CERN RD50 collaboration. This technology has seen the most aggressive timescale from first device studies to implementation in HEP experiments. In the context of the p-type strip sensor development, it was observed that after high radiation levels, the internal electric field under high voltage operation was strong enough to cause impact ionization that led to a gain in collected charge and, in specific cases, enlarged the radiation hardness (Casse, 2010). This finding triggered investigations on how to exploit and or produce internal charge gain and resulted in the idea to use a low-gain version of the reach-through avalanche photodiode (APD) concept; the LGAD was born (Pellegrini *et al.*, 2014). In LGADs, an additional *p*+ implant, the so-called gain layer, is posed underneath the *n*+ contact of an *n*-in-*p* sensor to achieve in a thin layer a very high electric field (> 250 kV/cm) that leads to charge multiplication by impact ionization (see Fig. 6(d)). It quickly became clear that the concept was providing an excellent precision timing device, which is of interest not only to HEP but also to a wide range of other applications like nuclear physics, hadron therapy or X-ray detection or astrophysics, wherever precision timing in some 10s of picosecond regime is needed. For the HL-LHC ATLAS and CMS experiments (start of operation foreseen in 2027), the LGAD technology was chosen to be implemented in precision timing detectors, which further boosted the development. The number of LGAD producers rose from two in 2015 to about eight in 2021 with a large range of LGAD variants (deep and shallow implant LGAD, Inverted LGAD, etc.) entering the landscape. Improvements in radiation hardness by gain layer engineering (by modifying the depth, width and shape or by adding additional impurities for defect engineering) and improvements of the fill factor of multi-pad devices (by minimizing the inter-pad distance or introducing trenches) matured the technology. The introduction of segmented LGADs that have a homogeneous gain layer and only segment the readout electrode (e.g., AC-LGAD or Inverted LGAD) offer the possibility to have a high-precision device in both, timing and spatial resolution. This device has the potential to offer a 4D tracking capability (see, e.g., Ferrero *et al.* (2021) for details). For further important developments, like monolithic CMOS devices, slim edges, and thin devices the reader is referred to more specific literature given in, e.g., see footnotes 12, 15 and Hartmann (2017).

Research tools

The evaluation of microscopic and macroscopic radiation damage effects in materials and devices requires, like many other research fields, a large set of tools that are partly commercially available, partly require customized adaptations, and often require highly customized developments. The latter are driven in the present case by the community and then shared across all collaboration members. The most prominent example, the transient current technique (TCT), is described in detail in the following. Important further developments are the standardization of irradiation tests, dosimetry, fluence calculation, and annealing procedures, customized read-out electronics as well as damage models, simulation tools, and test-beam and bench-test systems in general. Indispensable elements are also a wide range of irradiation facilities and detector foundries that are closely tied to the collaboration, grant access to processing details, and are open for innovative new developments.

The TCT is a prominent example of a characterization technique that was developed and matured by the CERN R&D collaborations. It is based on the generation of non-equilibrium charge carriers by means of ionizing particles, light pulses or bias pulsing in a sensor under bias and the measurement of the time-resolved signal induced by the moving charge carriers on the electrodes. The technique was exploited already in the late 1960s but gained importance in the 1990s as it opens up a more profound understanding of the signal formation process (Eremin *et al.*, 1996).

V. Eremin *et al.* (1996) analyze the characterization of silicon sensors, which can be fully depleted even if they are rather thick, e.g., 300 microns. Most often subnanosecond laser pulses in the red or infrared are employed to generate the charge carriers, allowing the measurement of a wide range of sensor properties. TCT gives a clear view of the time-resolved detector signal, as identical pulses can be averaged conveniently, and works even if the devices are highly irradiated. Depletion voltage, charge carrier trapping times, charge collection efficiency, and most importantly, with some limitations, the electric field distribution within the bulk of the sensor can be deduced as a function of the applied voltage. The latter is most conveniently obtained with the so-called edge-TCT (extending TCT to 2D resolution, see Kramberger *et al.* (2010)) or the TPA-TCT (extending TCT to 3D resolution, see García *et al.* (2017)). On the technology side, the RD48 and RD50 collaborations developed and produced their own customized lasers. Between 1997 and 2002, a total of 44 lasers were produced at CERN and distributed to collaboration members.[18] Later, the TCT systems were commercialized in

[18]Glaser, Maurice, in private communication. Glaser produced 44 TCT lasers (36x1060nm; 4x980nm; 4x660nm) between 1997 and 2002 for the RD48/RD50 collaborations.

spin-offs and commercial lasers were used. Without TCT, the observation and understanding of the radiation-induced double junction effect in 2002 (Eremin *et al.*, 2002) would not have been possible. TCT is clearly indispensable to develop and benchmark device simulation models that nowadays represent a key element in the development of new devices and radiation damage models.

Conclusions

Over the last 30 years, the CERN RD collaborations charged with the development of radiation hard silicon detectors for HEP applications have brought forward a deep understanding of radiation damage to semiconductor materials and detectors and several new or improved detector concepts. The impact on materials science and device physics deserves special emphasis in its reach beyond the HEP community. The formation of radiation-induced defects in a large range of silicon materials (*n*-type, *p*-type, FZ, CZ, EPI, defect engineered) covering all kinds of particles and a large fluence range has been studied and cast into a comprehensive understanding of the underlying physics, also to the profit of other communities like space or nuclear sciences and radiation monitoring. Models have been established that allow the prediction of defect formation and device operation under various irradiation and temperature conditions in complex radiation fields. Techniques and tools have been developed for in-depth radiation damage characterization but also to measure fundamental semiconductor properties like charge carriers mobility, impact ionization rates or charge carrier generation rates by particles and photons. In device physics, the understanding of signal formation, even in complex multi-electrode systems has been significantly pushed forward in theory, measurement and simulation.

The combination of all these achievements enabled the construction of silicon detector systems that can operate in the very harsh radiation fields of modern HEP experiments with severe challenges as to rate capabilities and constraints in material budget and cost. The development of p-type silicon bulk strip and pixel sensors, which are used for more than 1000 m^2 of sensor area in the High

Luminosity LHC (HL-LHC) experiments, can be regarded as the most relevant outcome. Oxygen-enriched n-type pixel sensors and 3D sensors are already operating in the LHC. For the innermost layers of the HL-LHC pixel detectors also 3D sensors will be used for their supreme radiation hardness and LGADs will be installed in the silicon-based precision timing detectors of the HL-LHC. These are other important examples of innovation in material engineering and detector design that enabled the search for new physics in the hostile operation environment of hadron colliders. These achievements are arising out of several decades of research and knowledge building within a grown community of networking experts from various fields.

The HEP community has concrete plans to build even larger hadron colliders like the Future Circular Collider FCC-hh. The radiation levels will rise again by more than an order of magnitude and the required timing precision will come down to about 10 ps. Today, there are no devices available that could cope with such an environment and therefore it is obvious that the tracking detector research community with its expertise and knowledge base has to be kept alive and has to start now to tackle the challenge, even if operation is anticipated for the second half of the 21st century. Non-existing detectors would be a showstopper and research and development takes, as experience shows, decades.

Selection and Surface Treatment of Materials for Use in the Vacuum of Particle Accelerators*

Cristoforo Benvenuti

In this section, we discuss CERN's role in accelerator technologies and specifically the impact on the development of materials and material techniques.

Vacuum technology has a twofold role in modern particle accelerators that are used both for fundamental physics research and also

---

*Editor's Note: Benvenuti's contribution starts on the next page, with the section "Introduction".

find numerous applications in the industry. To avoid colliding with gas molecules inside the accelerator, the beams of particles in the LHC must travel in a vacuum as empty as interstellar space. In the cryomagnets and the helium distribution line, the vacuum serves a different purpose. Here, it acts as a thermal insulator, to reduce the amount of heat that seeps from the surrounding room-temperature environment into the cryogenics parts which are kept at 1.9 K (–271.3°C).

The first part discusses the role of CERN in the development of non-evaporable getters (NEGs) technology that is successfully used for UHV applications. The need for a distributed pumping system in large particle accelerators and storage rings has been met by the adoption of linear sputter-ion pumps (SIPs) or long NEG strips integrated in the vacuum vessels, thus paving the way for further developments in the field. Benvenuti reflects on the birth of a new technology based on NEG sputtering techniques that were successfully developed by CERN and have been utilized at various synchrotron light sources and ion collider facilities. The aim was to improve the vacuum performance in some conductance-limited vacuum systems in order to minimize the gas bremsstrahlung emission and to reduce the impact of the electron-cloud instability discovered in some machines. The new techniques have resulted in an improved vacuum performance, a reduction of commissioning time for new facilities, and a suitably long lifetime of coating.

Introduction

Modern vacuum technology dates back to the 1950s when R.T. Bayard and D. Alpert proposed a new ionization pressure gauge capable of measuring pressures down to 10^{-11} Torr*, i.e., three orders of magnitude lower than that possible with previous gauges (Bayard and Alpert, 1950). It was thus that their gauge, commonly referred to as the B.A. gauge, marked the beginning of the UHV era. The extended measuring range made it possible to identify the gas desorbed from the vacuum system surfaces as being the main obstacle to obtaining very low pressures. Organic materials were therefore progressively replaced by metal components, leading to an "all metal" design of vacuum systems (Alpert, 1953).

Vacuum is required for particle accelerators to minimize the particle loss due to scattering by the residual gas molecules present in the accelerator chambers. In traditional accelerators, the particles are ejected to hit a target after a few revolutions, and a pressure of the order of 10^{-7} Torr is sufficient. In storage rings the particles are kept circulating for hours, and a much lower pressure, typically 10^{-10}–10^{-11} Torr, is required. It may therefore be contended that the construction of storage rings was made possible by the innovations of D. Alpert.

The Intersecting Storage Rings (ISR) at CERN

The construction of the ISR at CERN was decided in 1964. The ISR consisted of two concentric rings, each about 1 km long, with 8 points of intersection, where protons, of up to 31 GeV in energy and circulating in opposite directions, are brought into frontal collision. The vacuum chambers were made of stainless steel (type 316 LN). The initial evacuation was carried out by turbomolecular pumping stations, while the main pumping was ensured by 350 sputter-ion pumps. In these pumps the residual gas is ionized by a discharge ignited between two electrodes, and the produced ions are buried in the cathode. To maintain the discharge at UHV pressures a high magnetic field is applied to the pump. The pressure was monitored by about 400 B.A. gauges.

In the original ISR vacuum project, two treatments of the vacuum chambers were envisaged, after the standard chemical degreasing.

- Stainless steel contains a few ppm (weight) of hydrogen in solid solution, which is then released at constant rate while pumping (Calder and Lewin, 1967). To decrease the hydrogen content, all the vacuum chambers were outgassed at 900°C in a furnace at 10^{-5} Torr prior to installation on the machine.
- Even the best metal surfaces, when exposed to ambient air, adsorb a few monolayers of water vapor, which is then released

slowly under vacuum while pumping. To speed up the water vapor release and shorten the required pumping time, the vacuum chambers were heated to 150°C during the initial pumping of the ISR. This temperature is high enough to remove entirely the water vapor during a 20 h bakeout cycle.

Thanks to these precautions, the ultimate pressures obtained were in the 10^{-11} Torr range, slightly lower than that specified (Fischer, 1972). The ISR was the first machine of its type, and as such it was not surprising that something unforeseen would occur. Despite initial success, the ISR was plagued by sudden pressure rises produced by the circulating proton beams, which prevented reaching the nominal beam current. This effect was due to the degassing of the vacuum chamber walls bombarded by positive ions produced and accelerated by the circulating protons (Calder *et al.*, 1974). To counteract this effect, the chambers were removed from the machine and their surface individually cleaned by argon ion bombardment (Calder *et al.*, 1977). Furthermore, the temperature of the final bakeout was increased to 300°C, to remove possible hydrocarbon contamination due to exposure to air during the final installation in the machine. In addition to these surface cleaning procedures, the pumping capacity was increased by adding about 500 titanium sublimation pumps. After these improvements, the machine pressure became too low to be measured by the initially installed gauges, which had to be replaced.

The measuring limit of the BA gauge (Fig. 7) is due to the electrons extracted from the ion collector by the X-rays produced by ionizing electrons impinging on the gauge grid. This electron current leaving the collector mimics the positive ion signal which provides the indication of pressure. This is the origin of the measurement limit of 10^{-11} Torr with the B.A. gauge. To remove this limitation, during the 20 years between the B.A. innovation and the operation of the ISR, various gauge designs were proposed in different laboratories, all based on shielding the ion collector from the detrimental X-rays. These new gauges, however, would have required different power supplies, incurring added cost.

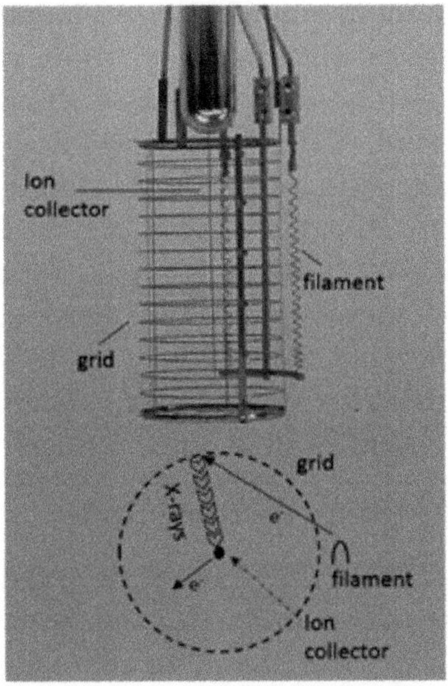

Fig. 7. The Bayard–Alpert gauge. Its functioning is described in the text.

An alternative solution, compatible with the existing power supplies, would have been to reduce the collector diameter. However, Alpert himself had remarked (Alpert, 1958) that there may be a critical size of the collector below which the probability of collecting the ions decreases with the X-ray current, so losing the benefit hoped for. This conclusion was based on the fact that a reduced collector diameter would result in a reduction of the ions collected in their initial flight to the collector, and that, in a perfect radial field, these ions would never be collected.

However, this argument did not consider the perturbation of the radial field by the space charge inside the grid volume resulting from the accumulation of the ions that were not collected. A laboratory investigation showed that by decreasing the collector diameter to 50 microns and properly closing the grid volume to prevent the ions from escaping collection, the B.A. gauge can measure pressures lower than 10^{-11} Torr with an uncertainty of 2×10^{-13} Torr (Benvenuti and Hauer, 1977). The gauges were therefore replaced with this improved version.

The ISR vacuum requirements for the chambers at the intersections in the experimental areas were even more stringent (10^{-13} Torr or better), requiring even better gauges. In this case, a gauge developed by J. Helmer and W.H. Hayward was available (Helmer and Hayward, 1966), making use of an ion collector external to the grid volume and carefully shielded against X-rays. By some modifications, this gauge allowed measuring down to the 10^{-14} Torr range, the lowest pressures measured so far (Benvenuti and M. Hauer, 1980). It was thus that with the ISR the vacuum technology domain was extended from UHV to XHV (Extreme High Vacuum).

The Large Electron Positron Collider (LEP)

The construction of LEP at CERN was approved in 1981. It was to be the largest accelerator ever built. Circular, with a total developed length of about 27 km, it was housed in a tunnel straddling the border between Switzerland and France, near Geneva. LEP was designed to bring electrons and positrons into collision, initially with a center-of-mass energy at the collision points of 91 GeV, and subsequently increased in stages to 209 GeV by the addition of about 290 superconducting accelerating cavities complementing the 56 copper cavities installed initially.

For this type of machine, the vacuum chamber surfaces again play a very crucial role. The electron beams, bent to follow the circular orbit by dipole magnets, produce synchrotron radiation, which then desorbs gas molecules when impinging on the vacuum chamber walls. Fortunately, the radiation results in surface cleaning, the degassing progressively decreases and the beam intensity may be slowly increased until the nominal beam current is reached.

However, the small cross-section of the vacuum chamber limits the flow of the gas molecules to the pumps, thus reducing their pumping efficiency. At LEP, it would have been necessary to install pumps every meter to reach the nominal beam current in a reasonably short time. An elegant remedy, adopted in a similar machine (PETRA at DESY, Hamburg) consists in replacing the appended local pumps by a linear sputter-ion pump making use of the magnetic field of the bending magnets. Unfortunately for LEP, the magnetic field of the dipole magnets at the injection energy was too low to maintain the gas discharge required for pump operation, and this solution could not be adopted. An alternative was therefore considered, based on a NEG strip to be inserted along 23 km of LEP vacuum chambers (Benvenuti, 1983). NEGs are materials able to remove gas molecules by chemical reaction. Doing so, the NEG surface gets progressively covered and its reactivity decreases, finally vanishing at saturation. The NEG reactivity can, however, be restored by heating.

At the time of the LEP project, only one type of NEG was commercially available, in the form of a metal strip coated on both faces with 0.1 mm of a Zr–Al alloy powder. However, it had been used only for industrial applications at much higher pressures (10^{-6}–10^{-8} Torr) and kept at 400°C to avoid surface saturation. The initial activation of this type of NEG requires heating to 750°C.

Laboratory tests were carried out to ascertain the low-pressure behavior of the NEG strip (Benvenuti and Francia, 1988) and of the NEG pump inserted in an LEP chamber (Fig. 8). The results were fully satisfactory, showing that pressures as low as 10^{-12} Torr could be achieved in the LEP chambers with the NEG strip at room temperature and intermittently heated to avoid surface saturation (Benvenuti *et al.*, 1993). Also testing in a real accelerator (PETRA at Desy Hamburg) was conclusive, and finally, the decision was taken to equip the 2000 aluminum LEP chambers,

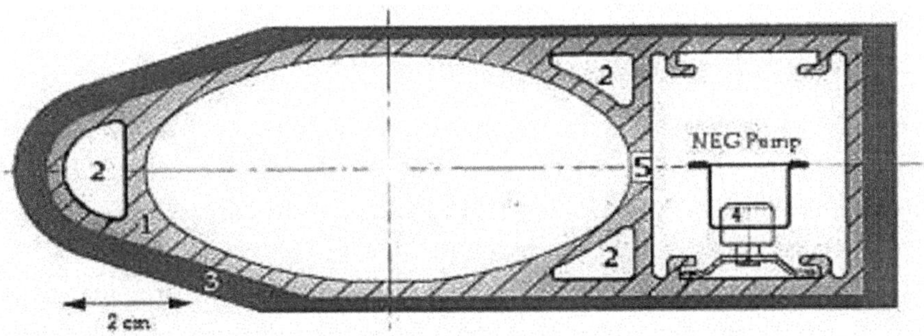

Fig. 8. The LEP vacuum chamber.

most of which 12 m long, with the tested NEG strip. The average pressure obtained after 150°C baking was in the low 10^{-12} Torr range. The NEG pump worked perfectly during the entire 10 years of LEP operation.

Thin-Film Coatings of NEG

Although vital for the realization of LEP, the linear NEG pump presents some inconveniences. It requires a large vacuum chamber to accommodate the insertion of the NEG pump, which must be electrically isolated to allow heating of the strip at 750°C for activation by circulating an electric current of about 100 A. Furthermore, although the NEG pump provides a large pumping speed, degassing of the surface is still present.

The obvious solution to these inconveniences would be to coat the chamber surfaces with a thin getter film (Benvenuti *et al.*, 1999), which would not require an additional space and would transform the vacuum chamber from a gas source into a pump. However, this solution leads to another serious difficulty. The activation of a chamber coated with NEG of the type adopted for LEP would imply heating the whole chamber at the daunting temperature of 750°C, which is higher than any normal construction metal could tolerate. For aluminum chambers, the NEG activation temperature should even be lower than 200°C, so as not to compromise their mechanical stability. A solution was found when using thin-film NEG coatings was envisaged for the last large accelerator built at CERN.

The Large Hadron Collider (LHC)

The LHC is a proton storage machine, which has replaced LEP in the same tunnel, consisting of two rings, where counter-rotating protons of energy up to 7 TeV (Tera electron volts, or 10^{12} eV) are brought into collision at the intersections of the rings.

The LHC is equipped with superconducting bending magnets which, together with the inserted vacuum chambers, are maintained at very low temperatures (<2.3 K). This temperature is low enough to condense all residual gases, therefore in the cold regions of LHC, no pumps are required. However, a large fraction of LHC (about 5 km), and the intersections, are at room temperature, and it was natural to envisage the use of NEG coating. Most of the development work for this application was devoted to finding a coating alloy that could be activated at 200°C or less. The coating was carried out by magnetron sputtering from a central cathode. To avoid the expensive and time-consuming purchase of cathodes of different alloys, the cathodes were produced by twisting together wires made from easily available pure metals. About 20 different alloys were tested in this way, and finally, a titanium-zirconium-vanadium alloy with an activation temperature of about 200°C was identified, and it was this that was adopted for LHC (Benvenuti *et al.*, 2001).

The NEG coatings also provided an initially unexpected benefit — to prevent the multiplication of electrons hitting the vacuum chamber surface (the "electron cloud" problem). This feature has been vital for LHC operation. The NEG coating technique is protected by a CERN patent. Licensed to many companies and institutions, it has subsequently been adopted for accelerators worldwide.

Conclusion

The described evolution of accelerator vacuum technology, from the ISR to LHC, shows opposite trends: the negative role of surface degassing has steadily decreased, and the complex surface treatments needed for the ISR chambers were reduced for LEP and made superfluous for LHC. In contrast, the efficiency of the pumps steadily increased, and finally, the vacuum chamber, traditionally a source of gas, was transformed into a pump by use of the NEG coating.

Superconducting Materials for Accelerators and Detectors

Luca Bottura

Superconductivity has been instrumental to the realization of the magnets for large particle accelerators and detectors. Fast progress led in the mid-1960s to the first large-scale use of superconductors, in the detector solenoid operated as an instrument in the Zero Gradient Synchrotron (ZGS) of Argonne National Laboratory. The following 50 years saw an impressive sequence of superconducting accelerators, from the Tevatron, through HERA and RHIC to the LHC A similar development was followed by superconducting detector magnets where ATLAS and CMS experiments at the LHC are the latest and most impressive realizations. All the above machines and discoveries made using them were only possible thanks to superconducting materials. Truly, we can say that over the last 50 years, superconductivity was the main fuel that advanced the frontier of knowledge in HEP. At the same time, superconductivity is a fascinating state of matter and has significantly influenced other fields of science and society. This chapter reviews the nature and types of superconductors and provides an overview of applications, present and upcoming.

The March of Superconductivity

The most straightforward and intuitive definition of a superconductor is that of a "material that exhibits no electrical resistance," implying that an electric current generated in a fully superconducting loop can flow indefinitely. In fact, being a less intuitive but very fundamental property, a superconductor does not allow magnetic fields to penetrate (also in steady state). This is often referred to as the *Meissner effect* and points to the fact that a superconductor cannot be simply understood as an ideal conductor with zero resistance, but it is rather a specific state of matter.

On a microscopic scale, the charge carriers in a superconductor are bound in a *condensed* state. This is the special state of matter that we refer to. As long as the carriers remain bound, the joined pairs (the so-called *Cooper pairs*) do not interact with lattice defects and vibrations, and can hence move freely in the material. This is the mechanism at the base of the lossless transport of electric current. If the pair is broken on the other hand, e.g., by the energy excitation provided by thermal agitation, the current transport is no longer lossless, and the superconducting state is lost. Because the binding

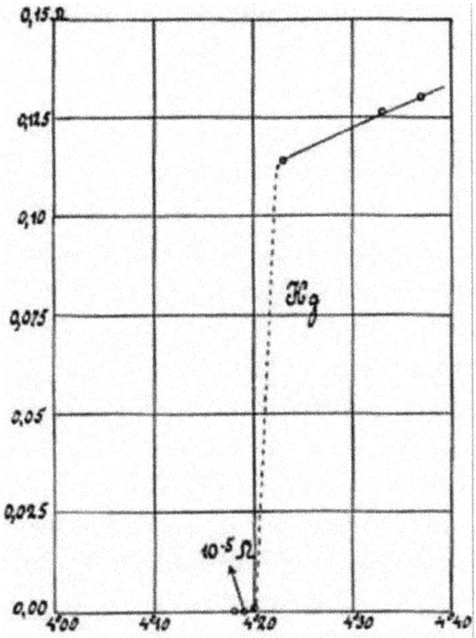

Fig. 9. The first measurement demonstrates the superconducting behavior of mercury. The sample resistance (vertical axis) drops to a value that is below the minimum that can be measured when the temperature is reduced below the liquefaction point of helium 4.2 K (horizontal axis) (Onnes, 1911).

energy of the carrier pair is relatively small, superconductivity is typically achieved only at very low temperatures.

The credit for the discovery of superconductivity goes to Dutch physicist Heike Kamerlingh Onnes. In 1911, Onnes was studying the electrical resistivity of mercury in his laboratory at Leiden University (The Netherlands) as part of the general search toward the understanding of thermo-physical properties of materials in extreme conditions and especially at low temperatures. Mercury could be distilled into a particularly pure metal, whose resistance was expected to drop to zero at the absolute thermodynamic zero. In contrast, Onnes found that the electrical resistance in the mercury completely vanished when he dropped the temperature to below 4.2 K, i.e., −269°C, the temperature of helium liquefaction at ambient pressure. At the time Onnes had the advantage of being the only scientist in the world who could reach such cold temperatures, having achieved the liquefaction of helium in 1908.

Many years elapsed between the discovery of the phenomenon, its understanding, and its applications. In fact, many elements and materials become superconducting when cooled below a certain critical temperature: about half of the elements in nature are superconductors. The finding of the Meissner effect dates to 1933 when physicists Walther Meissner and Robert Ochsenfeld discovered that superconductors "expel" any nearby magnetic fields. This is true only up to modest values of the fields, in the range of few mT to few tens of mT (e.g., 170 mT for pure Nb). This field, indicated with B_{C1}, is called "lower critical" field. As mentioned above, this was a clear indication of the complex nature of the mechanism explaining superconductivity as a change in the state of matter. As discussed later, technical superconductors can sustain much higher fields by allowing flux lines to penetrate, up to a "higher critical" field indicated with B_{C2}.

Theoretical physicists Lev Landau and Vitaly Ginzburg provided in 1950 a macroscopic theory predicting the properties of superconductors while remaining ignorant of what was going on at a microscopic level. Finally, in 1957, physicists John Bardeen, Leon N. Cooper and Robert Schrieffer developed a complete, microscopic theory of superconductivity for the materials known to be superconducting at the time, which are since known as low-temperature superconductors (LTS). Bardeen, Cooper and Schrieffer explained electron pairing, mediated by lattice phonons (sound) to form their theory, known as BCS theory.

LTS, and BCS theory, remained the only known form of superconductivity until 1986, when scientists G. Bednorz and A. Müller, working at the IBM research lab near Zurich, in Switzerland, encountered a particular copper oxide whose resistance dropped to zero at a temperature around 35 K (−238°C). Their results paved the way for a new path of discovery toward what are nowadays called high-temperature superconductors (HTS). A large number of superconducting compounds were rapidly found to be superconducting at temperatures in excess of liquid nitrogen, which is a major change in the cryogenic complexity and power requirements when compared to liquid helium. Though an active field of research, a full understanding and theoretical description of these materials is still missing.

The superconductor with the highest known transition temperature when at ambient pressure are cuprates (mercury, barium, and calcium), at around 133 K. Other materials become superconductors at higher temperatures. The "holy grail" of superconductor research is to find a material that can act as a superconductor at room temperatures. To date, the highest superconducting temperature was reached with extremely pressurized carbonaceous sulfur hydride, which reached superconductivity at 288 K (15°C), but required 267 gigapascals of pressure. This is the pressure expected at the center of a massive planet such as Jupiter, i.e., rather impractical for earth-based applications.

Figure 10 shows the progression in critical temperature of selected materials of different types and classes, either at ambient or high-pressure conditions.

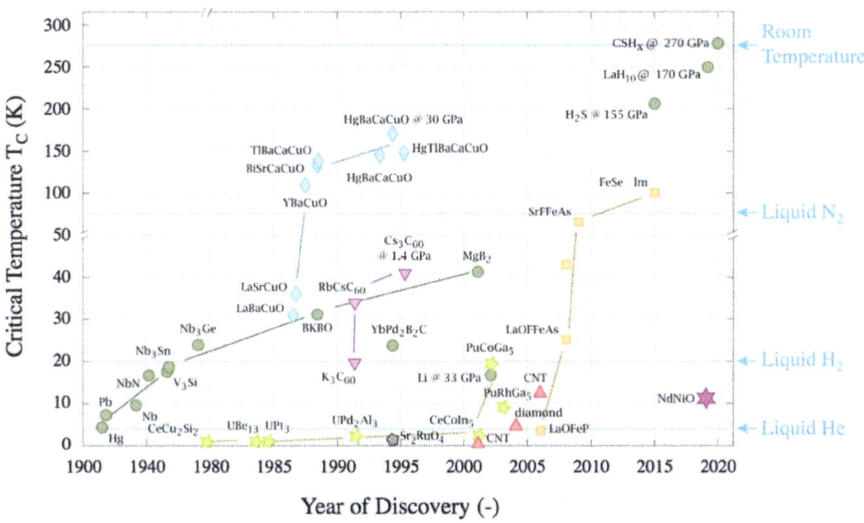

Fig. 10. Timeline of discovery of superconductors, plotted as critical temperature of the superconductor vs. the year of discovery. The plot is a modified version of the material available in PJRay — Own work, CC BY-SA 4.0. https://commons.wikimedia.org/w/index. php?curid=46193149.

Superconductors Types

There are many ways to classify superconductors. Though none of them is a truly rigorous classification, they are useful when referring to specific materials and their characteristics.

Magnetic field exclusion

Two types of superconductors can be defined, depending on whether the material can admit any field in the bulk or not. A superconductor of "Type I" has a single critical field, the lower critical field B_{C1}, below which the field is excluded from the material, and above which superconductivity is lost. A "Type II" superconductor admits flux penetration above the first critical field B_{C1}, still remaining superconducting state up to a second critical field, the higher critical field B_{C2}. The field lines penetrating a Type II superconductor are concentrated in points called vortices.

Underlying theory

A superconductor whose properties can be explained using the BCS theory is said to be "conventional" (typically applying to LTS) or "unconventional" (typically applying to HTS).

Critical temperature

The BCS theory of conventional superconductors predicts a maximum critical temperature of 30 K. The discovery of Bednorz and Müller of a superconducting material at a temperature higher than this upper bound marked a new class of materials, which progressed rapidly up to liquid nitrogen of 77 K. The first class of superconductors is generally named low-temperature superconductors (LTS), referring to materials with a critical temperature below 30 K. In contrast, superconductors with a critical temperature above 30 K are referred to as high-temperature superconductors (HTS). Note that though MgB_2 has a critical temperature close to 40 K, it is generally classified as an LTS material as the explanation of its physics can be understood in terms of a modification of the BCS theory that applies to LTS materials.

Material

Another means to distinguish among superconductors is to classify them by the material class to which they belong. The classes are as follows: elements (e.g., niobium), alloys (e.g., niobium–titanium, idem), compounds (e.g., niobium–tin or magnesium diboride), ceramics (e.g., rare-earth-based REBCO or bismuth-based BSCCO), pnictides (e.g., iron-based superconductors), organic superconductors, sulfides, and other.

Technical superconductors

However important from the point of view of physics, the critical temperature is not the most interesting technical parameter of a superconductor. In fact, the ability to carry current and remain superconducting under an external field is what makes a superconductor *worthy* of technical applications. This is why we rely on two additional characteristic parameters, namely the *critical field* B_C and the *critical current density* $J_C(B,T)$. Table 1 reports a summary of the

Table 1. Material reference and typical values of critical parameters for technical superconductors.

		LTS				HTS		
Material		Nb–Ti	Nb_3Sn	Nb_3Al	MgB_2	YBCO	BSCCO	IBS
Year of discovery		Hulm and Blaugher (1961)	Matthias *et al.* (1954)	Corenzwit (1959)	Nagamatsu *et al.* (2001)	Wu *et al.* (1987)	Maeda *et al.* (1988)	Kamihara *et al.* (2006)
T_c	(K)	9.2	18.2	19.1	39	≈93	95⁽⁵⁾ 108⁽⁶⁾	16⁽⁷⁾ 38⁽⁸⁾ 55⁽⁹⁾
B_c	(T)	14.5	≈30	33	18⁽¹⁾ 36…74⁽²⁾	≈120⁽³⁾ ≈250⁽⁴⁾	≈200	40⁽⁷⁾ 80⁽⁸⁾ 100⁽⁹⁾

Notes: (1) wires and tapes; (2) thin layers; (3) B parallel to c-axis; (4) B in the ab-plane; (5) BSCCO-2212; (6) BSCCO-2223; (7) IBS-11; (8) IBS-122; (9) IBS-1111.

critical temperature T_C and critical field B_C for the technical superconducting materials that have found practical applications over the past 50 years, as well as materials that are expected to come into use in the next few years.

The figures reported in Table 1 highlight an interesting feature of HTS materials, namely their critical field is much higher than that of LTS materials. This is very relevant to magnet applications, where maximum field is one of the main figures of quality. Indeed, the interest for HTS in magnets is both for their potential of operation at higher temperatures than LTS as well as to achieve high operating fields at low temperatures. The same feature appears clearly when plotting in Fig. 12 the *engineering critical current density* J_E of materials listed in Table 1, measured at the temperature of liquid helium, 4.2 K.

The typical benchmark for use in magnets is to achieve a J_E in the range of a few hundreds of A/mm². We see from the values

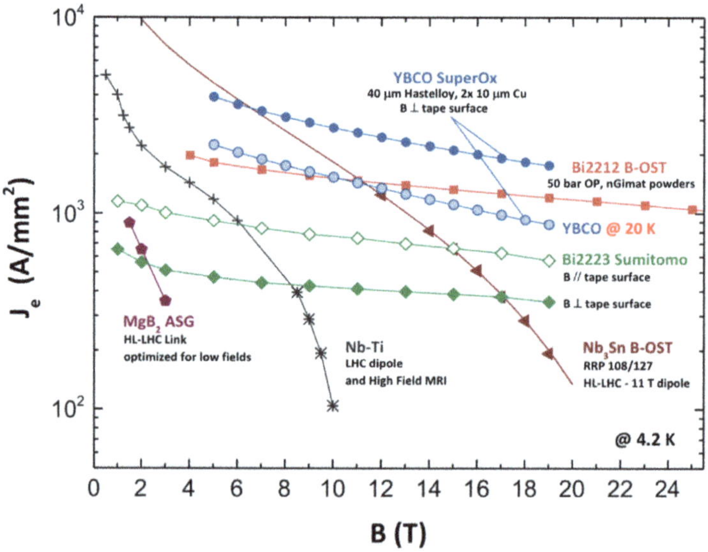

Fig. 12. Engineering critical current density of technical superconductors measured at 4.2 K. JE is referred to the whole cross-section of the superconductor (wire or tape) including all materials as required for its manufacturing and operation (Senatore, 2022).

reported in Fig. 12 that taking a typical J_E of 500 A/mm^2 LTS materials can reach maximum fields of the order of 6...8 T, using Nb–Ti, and 16...18 T with Nb_3Sn. Higher fields can be achieved by cooling below liquid helium temperature, at the cost of higher refrigeration power. This is done at CERN in the LHC, using superfluid helium operating at 1.9 K to cool the km-long strings of magnets. As an alternative, higher fields can be reached by LTS making the coils larger, i.e., accepting to work at reduced engineering current density. This is done, as an example, in ultra-high-field Nuclear Magnetic Resonance magnets built with Nb_3Sn that have reached record fields just short of 24 T. We also see that HTS YBCO and BSCCO cross the performance of LTS at around 12...14 T. Their field dependence is modest because the critical field is so large. These materials offer potential beyond the field reach of LTS, which explains the interest and development of the past years.

History and Characteristics of Technical Superconductors

Nb–Ti — the workhorse

The alloy of Nb–47%Ti (Senatore, 2022) is undoubtedly the most successful practical superconductor and has been used in all superconducting particle accelerators and detectors built to date. Present worldwide production ranges at 600 tons/year, mostly driven by the MRI market. Nb–Ti is a ductile and tough material, easily available in long lengths (piece length up to a few km) in the form of multifilamentary wires where the superconductor is dispersed in a copper matrix of high purity and low electrical resistivity.

Its origins can be traced back to the highly cold-worked Nb wires, used by Yntema in his first superconducting magnet (1955). Nb has the highest T_C of all the elements, 9.5 K, and it was found that heavy deformation by cold drawing could lead to a significant increase in the current carrying capacity. The critical field of Nb, however, remained low. Adding alloying elements was an expedient to increase B_C, leading first to Nb–Zr (the first commercial superconductor by Avco-Everett), and in 1961 to Nb–Ti (first offered commercially by Westinghouse). Fabrication of Nb–Ti became an industrial-type process with the advent of the multi-filamentary conductor, first commercially produced by Imperial Metal Industries Ltd.

In the mid-1960s, Nb–Ti wires were mainly used to build the first large-size magnets such as the bubble chamber magnet at ANL, BNL, or the BEBC at CERN. By that time wire producers had understood the role of heat treatment, inducing precipitation of a normal-conducting phase in the alloy (α-Ti) which provided for efficient pinning centers. It was also around that time that scientists suggested that the superconductor should be subdivided into fine filaments, at most 50–100 μm, to prevent flux-jumps that could lead to a resistive transition — a magnet quench. A further result of the development and construction of Nb–Ti magnets was the realization of the benefits of transposition through twisting, avoiding electromagnetic coupling of the filaments and reducing charge losses, and cabling, making it possible to decrease magnet inductance, reducing voltages, and ease of protection in case of quench.

By the beginning of the 1970s, when the development and construction of the Tevatron started, each supplier had a preferred Nb–Ti alloy composition. Given the sheer size of the production requested for the Tevatron magnets, approximately 90% of the Nb–Ti ever produced, it was only natural to take a step toward standardization. Standard composition materials and procedures were provided under the responsibility of Fermi National Accelerator Laboratory, who led this first industrial large-scale production.

The Tevatron-class Nb–Ti wire had a J_C of 1750 A/mm^2 at 4.2 K and 5 T, a good performance, but short of the perceived potential of the material. A few years later, during the studies driven by the R&D for the Superconducting Super Collider (SSC) it was realized that the secret to high and reproducible J_C was in the high-homogeneity (HiHo) of the composition of the Nb–Ti alloy. This unlocked the way to a 50% increase in J_C, and became rapidly another industry standard.

All the above expedients are used in modern Nb–Ti. Billets are extruded and drawn from initial typical dimensions of 20–30 cm diameter, about 1 m height and 300 kg weight, to single piece lengths of several tens of km of wire. The HiHo alloy, and the combination of cold work, annealing heat treatments and precipitation of segregated phases (α-Ti) in the Nb–Ti filaments, have raised the J_C of industrial material to values of 3000 A/mm^2 at 4.2 K and 5 T. Cooling to 1.9 K shifts this point up to about 8 T.

This industrial product has now reached a high level of quality, also thanks to the demand of the large production (1300 tons) of Nb–47.0%Ti for the LHC. Owing to stable production methods and procedures, and online diagnostics to detect defects, modern Nb–Ti has the guaranteed stable performance that is mandatory for the competitive MRI business: a result of which the pioneers of HEP in the 1960s and 1970s should be justly proud.

Nb3Sn — the next step

The inter-metallic A15 compound Nb_3Sn Matthias *et al.* (1954) is the second LTS material that found its way from material research to large-scale applications. Nb_3Sn is a brittle and fragile compound, and all modern manufacturing routes involve the assembly of the precursor elements Nb and Sn, and a certain amount of Cu into billets that are extruded and/or drawn to the final diameter wire. The Nb_3Sn is then formed by a chemical reaction induced by a heat treatment to temperatures in the range of 600–700°C for durations of few tens to few hundreds of hours. Worldwide production is estimated at 10 tons/year, mostly required for high-field solenoids for NMR and laboratory applications.

The origin of Nb_3Sn as a technical superconductor can be traced back to the work of Kunzler, at Bell Labs. In 1960, Kunzler filed the first patent for a 1.5 T superconducting magnet wound in layers with a Mo_3Re wire, cooled in a cryostat by a bath of liquid He. His 1960 patent on "Superconducting Magnet Configuration" was issued four years later, in April 1964, and can be regarded as the actual start of superconducting magnet technology.

Early in 1961, Kunzler's group reported the discovery of superconductivity in Nb and Sn compound wires up to a field of about 9 T (Kunzler *et al.*, 1961), with an impressive J_C of 1000 A/mm^2 at 9 T and 4.2 K. The wire was made of Nb tubes filled with crushed powder of Nb and Sn, The tube was sealed, compacted, and drawn to long wires. This primitive powder-in-tube (PIT) technique required reaction at high temperature, in the range of 1000–1400°C, to form the superconducting Nb_3Sn phase. Thanks to this wire, Kunzler and his team produced a record field of 6.8 T in a solenoid in 1961, marginally beating similar magnets wound at the same time using Nb–Zr, maxing out at 6.6 T. Nb_3Sn, however, clearly had more potential, and it only took a few years to build the first 10 T solenoid.

Although successful in demonstrating the use of Nb_3Sn in high-field magnets reaching 10 T, as well as a principle demonstration of a 1000 MVA power transmission line at BNL, neither wire fabrication technique was practical. Besides the complication of a relatively high-temperature heat treatment, and the difficulty of dealing with a brittle material, the original PIT had exceedingly large filaments, and tapes have a large aspect ratio by construction. This invariably resulted in huge trapped magnetization and flux jump instabilities. The next step required resolving these issues.

The need for high-temperature heat treatment is the result of the presence of two undesirable compounds ($NbSn_2$ and Nb_6Sn_5) that are stable in the binary system of Nb and Sn below approximately 930°C. In the late 1960s, Tachikawa introduced an alternative concept for the formation of the desirable superconducting phase based on solid-state diffusion. In his original work on V_3Ga (a superconducting intermetallic of the same A15 family), he used small filaments of V surrounded by a Cu matrix alloyed with Ga (Tachikawa, 1971). The presence of Cu destabilizes the Ga-rich compounds (Nagamatsu *et al.*, 2001; Kamihara*et al.*, 2006), and the superconducting phase can be formed at lower temperatures. At the same time, the lower temperature prevents the excessive grain growth that is inevitable in binary systems, thereby increasing the pinning efficiency. This principle is now exploited to fabricate Nb_3Sn wires by the so-called bronze route, one of the leading modern manufacturing techniques used at present to produce most industrial Nb_3Sn.

Bronze route billets are assembled by inserting Nb or Nb–alloy filaments in a bronze matrix surrounded by a Cu can. The assembly can be extruded, and drawn, with excellent homogeneity and yield. The application in NMR and laboratory solenoids was instrumental in motivating the development of bronze-route Nb_3Sn wires with properties suitable for a very high field. During the 1970s and early 1980s it became common to use alloying elements such as Ta or Ti, either in the Nb, or in the bronze, or mixed in the Sn source, to increase B_C and J_C at high field, above 12 T. A further advantage of bronze-route wires is their inherent strength, which can be further improved adding reinforcements, making them capable to withstand the large electromagnetic forces occurring at high field. At the same time, the main limitation of the bronze route is the amount of Sn that can be made available during the heat treatment to form Nb_3Sn. Industrially produced bronze route achieve J_C in the range of 600–1000 A/mm^2 at 12 T and 4.2 K.

To overcome this limitation, alternative routes have been developed based on assembling a large number of Nb or Nb–alloy and Sn or Sn-alloy rods. Because of the presence of an Sn source internal to the assembly they are referred to as internal tin techniques. The alternative path is based on bronze-route wires that can be manufactured using a technique similar to the PIT process devised by Kunzler.

The main advantage of both internal tin and PIT techniques is that there is no hard limit to the amount of Sn that can be assembled in the wire, and it is hence possible to increase the final fraction of useful superconductor. Values in excess of approximately 2750 A/mm^2 at 12 T and 4.2 K were obtained in PIT and recorded 3300 of A/mm^2 at 12 T and 4.2 K in the internal tin Restacked Rod Process (RRP). The main drawback of internal tin and PIT production routes is that the presence of pure, alloyed or inter-metallic Sn in the billet makes an initial extrusion step difficult, if not impossible. The billet size is hence much smaller (typically five times lower mass) than what can be done in the case of bronze route wires, resulting in limitations in length, process scale, and ultimately higher cost.

Nb_3Sn has remained for many years a niche material confined to the high-field portion of solenoids until the efforts in the production of approximately 600 tons of material for the TF and CS coils of the ITER fusion experiment materialized in 2016. Though instrumental to the large scale-up at several worldwide producers, ITER was not very demanding in terms of performances. The ITER specifications were essentially met already in the 1980s by both bronze route and internal tin wires. The record values quoted above are the fruit of the development pursued under the US-DOE Conductor Development Program, started in 1999 as an initiative to make Nb_3Sn suitable for high-field accelerator magnets. In parallel, several initiatives in Europe during the 1990s (Twente University and CERN, INFN, CEA) converged to the EU-financed FP6-CARE, and FP7-EuCARD, now continuing as the High Field Magnet program led by CERN in collaboration with EU laboratories. The main result of this initiative was to develop the PIT route to industrial maturity.

These achievements, and the presence of an industry capable to produce these high-performance wires, have led to the decision to use Nb_3Sn to realize the new quadrupole magnets for the low-beta interaction points 1 and 5 of the LHC, the backbone of the LHC High-Luminosity Upgrade (HL-LHC). This medium size production, a total of approximately 30 tons of high-performance material, is presently ongoing, with completion expected around the end of 2022.

Though Nb_3Sn has achieved the spectacular record performance quoted above, guaranteed values for a large industrial production are much reduced. The HL-LHC critical current density specifications are 2450 A/mm^2 at 12 T and 4.2 K, i.e., nearly 50% below the best result obtained industrially.

Recent studies on the effect of irradiation, causing a large increase of current density at high field, and recent developments on grain refinement, also leading to increased current density, suggest that the full potential of the material is not yet exploited. This is the reason why CERN has launched within the scope of the Future Circular Collider (FCC) study a worldwide initiative to

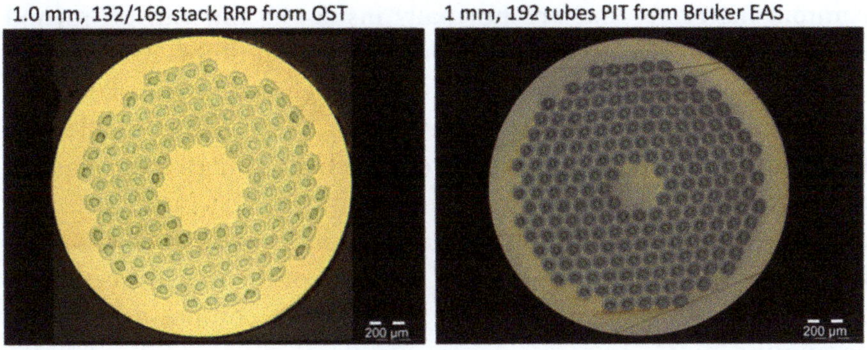

Fig. 13. Cross-section of typical OST-RRP (132/169) and Bruker-PIT (192) Nb_3Sn wires. Wire cross-sections taken at a diameter of 1 mm, as used for the construction of the FRESCA2 dipole (see footnote 7). Similar architecture is used for the construction of the HL-LHC magnets.

push Nb_3Sn performance to its ultimate limits. As discussed in Ballarino and Bottura (2015), the FCC performance targets go beyond the J_C, and include the Residual Resistivity Ratio (RRR), the effective filament diameter, stress and strain tolerance, piece length and, most important, cost.

The FCC conductor R&D can be compared to the R&D on Nb–Ti that was triggered by the SSC and led to the LHC class of wires. It could result in a new class of high-performance Nb_3Sn material suitable not only for accelerator magnets, but also for other large-scale applications such as ultra-high-field NMR and other laboratory solenoids, MRI scanners for medical research, or thermonuclear fusion magnets.

HTS materials — a new age

The 1986 discovery of superconductivity at 35 K in a La–Ba–Cu–O compound by A. Mueller and G. Bednorz (IBM Zurich Labs) ushered applied superconductivity in a new age. The discovery was the result of years of search for superconducting materials that would beat the 1973 record value of Nb_3Ge, 23.3 K. In 1983, Bednorz and Mueller focused on crystalline oxides such as Sr–Ti–O and Li–Ti–O, ceramic materials that are normally insulators but can be turned into conductors by tuning their chemical composition. Both exhibited superconductivity, though below the record value of Nb_3Ge. Attracted by a report that appeared in 1985 that a Ba–La–Cu Perovskite was exhibiting metallic conductivity at high temperature, they focused their research on tuning the Ba/La ratio, and produced several variants of this brittle ceramic compound. The search led to the 1986 measurement of a superconducting transition at 30 K, then increased to 35 K by varying the composition of the compound.

The results could be reproduced rapidly by several groups, and the excitement spread at a high pace as more material variants achieved increasing critical temperatures. By December 1986 a La–Sr–Cu–O had attained a T_C of 45 K, and reached 50 K under pressure in January 1987. By Spring 1987, and especially following the discovery of superconductivity in Y–Ba–Cu–O at 90 K, above liquid N_2 temperature, the news had conquered the media. The exceedingly high critical temperature, by the standard at the time, was at the origin of the name of this new class of materials, the High-Temperature Superconductors, or HTS.

Besides the beneficial effect on spurring scientific research in the field, and the dreams of a fully superconducting tomorrow that made the first page of "Time," the rapid progression created by the new discovery originated expectations that, with hindsight, were by far too optimistic. To gauge the extent of the above statements, following the March 1987 Meeting of the American Physical Society, where the above results were announced, discussions started as to whether the proposed Superconducting Super Collider in the US, or the next step in Thermonuclear Experimental Reactor then in the design phase, should be built using HTS rather than more conventional LTS Nb–Ti or Nb_3Sn.

From our previous discussion, it is evident that the time required from the material discovery to the practical use on a large scale is of the order of tens of years. This is the case also for HTS materials, notably the Rare Earth-based RE–Ba–Cu–O, and the Bi-based Bi–Sr–Ca–Cu–O, whose progress from discovery to applications is coming to fruition only now. Similarly to the development of LTS materials, a flurry of compounds followed the first discovery. The present record for the highest T_C material is held by a Hg–Ba–Ca–Cu–O compound, with a verified value of 134 K (at ambient pressure). Till 2008, all HTS belonged to the class of "Cuprates," but they have been complemented since then by iron-based compounds (iron pnictides).

It was recently discovered that hydrogen-rich composites such as H_2S (hydrogen sulfide) and modifications obtained introducing carbon, CSH (carbonaceous sulfur hydrides) become a metallic conductor under high pressure. H_2S undergoes superconducting transition when cooled below 203 K, or −70°C, under a pressure of 155 GPa. CSH has been reported to exhibit superconductivity up to a temperature of 287.7 K (about 15°C) under a pressure of 267 GPa. This is the highest critical temperature ever announced. The need for extreme pressure frustrates the hope of using these materials as technical superconductors. Still, these results prove the possibility to achieve superconductivity at room temperature.

Back to the more practical Rare Earth RE–Ba–Cu–O (shortened in REBCO) and Bismuth Bi–Sr–Ca–Cu–O (shortened in BSCCO) Cuprates, these are the two HTS materials that seem to bear the largest potential for future use in accelerator magnets, mainly because of the high critical field at low operating temperature (see Table 1) that makes them the ideal candidates to build magnets that achieve fields beyond what is possible with LTS materials.

Both REBCO and BSCCO are ceramic compounds, their production is a complex process that may involve assisted crystal growth and texturing, at high temperatures, and under very well-controlled chemical conditions. One of the reasons for the complex production is that a fundamental limitation to critical current in HTS materials comes from the effect of grain boundaries. Coherence length is very small in HTS, and misaligned grain boundaries are a barrier to the free flow of the super-currents in principle, this means that HTS conductors would require solving the tantalizing task of growing single crystals of km length. Different ways have been found to deal with this limitation in REBCO and BSCCO.

Finally, iron-based chemical compounds, the so-called iron-based superconductors (IBS), were discovered in 2006 and already found early demonstration in coils in 2008. IBS belong to the class of ferropnictides, a vast collection of compositions, where iron arsenides have been singled out for their ease of synthesis, and hence the perspective of a simple manufacturing route toward HTS. IBS have a wide range of critical temperatures, approximately from 25 to 55 K. This is well below the critical temperature range of REBCO and BSCCO, well below liquid nitrogen. The main interest is rather because of the high critical fields, ranging from 40 to 100 T depending on composition and crystal structure. IBS are an active field of research, now mainly devoted to increasing the engineering current density of tapes and wires to the range of technical interest.

MgB2 — another surprise

The most recent of the technical superconductors (Nagamatsu *et al.*, 2001), the intermetallic compound MgB_2 found to be superconducting in 2001, caused much excitement, and for several reasons.

With a critical temperature of 39 K, MgB_2 is superconducting at liquid H_2 temperature (20.3 K), and in principle can operate using helium-free cryogenics. The critical temperature exceeds the upper BCS limit, thus suggesting it may belong to the category of HTS materials. In fact, it was found already soon after its discovery that its behavior resembles rather closely that of a metallic, type II, LTS material. Its properties can be explained in terms of a modification of BCS, considering that two types of electrons, associated with two energy gaps, are bound by phonons to form the superconducting carriers (two-band superconductivity). Besides the high critical temperature, MgB_2 has a large coherence length, and the grain boundaries are transparent to the current flow.

A second reason for excitement was the observation that the critical field in fibers and thin films could reach values as high as 55–75 T, respectively (Eisterer, 2007). These values exceed by far any of the other LTS materials, though up to now it was not possible to reproduce them in wires.

Finally, wire producers were attracted by the ease by which multifilamentary tapes and wires could be manufactured starting from the relatively inexpensive Mg and B precursor elements. Practical manufacturing of MgB_2 wire is presently based on a powder-in-tube (PIT) process, compatible with the hard nature of the basic constituents. Variants of this process exist, depending on whether the MgB_2 is formed at the end of the wire processing from Mg and B precursors (in-situ variant), or rather powders of pre-reacted MgB_2 are sintered in the finished wire (ex-situ variant). In both cases, a heat treatment is required in the range of 600–1000°C, depending on the manufacturing route.

In summary, MgB_2 came with the promise of an inexpensive, high-field superconductor, operating in a temperature range, 10–20 K that can be reached easily using helium-free cryocoolers. Actual results have not yet quite matched these expectations, and the applications of MgB_2 are presently moderate-field MRI, and cables for the transport of large currents (Ballarino and Flükiger, 2017), with ongoing R&D to extend this domain to a wider spectrum of MRI magnets, as well as higher field application. Present performance records for MgB_2 wires are in the range of a current density of 3000 A/mm^2 at 5 T and 4.2 K, with an upper critical field of 15 T.

Applications Beyond Particle Physics

The applications of superconducting technology, present and potential, are numerous. They cover particle physics, but go much beyond, as detailed in the following. The uses that have been considered for superconductors range from small-scale electronics applications, such as superconducting junctions and filters, to Big Science projects such as the LHC and the International Experimental Tokamak Reactor. A non-exhaustive list can be found in Table 2, together with the main advantage of the use of superconductors for the specific application listed. The table is not intended as an exhaustive list, but rather to present the variety of uses by examples. We expand in the following on some of the most relevant

Table 2. A non-exhaustive summary of applications of superconductivity and commentary on their advantages.

Superconducting Device	Application advantage
SQUID Superconducting quantum interference devices (SQUID)	High-resolution (10^{-14} T) magnetometers
Digital circuits (Josephson junctions and rapid single flux quantum)	High switching speed (50 ps), compact circuitry because of low power consumption, absolute flux and voltage references
RF filters	High-quality factor (10^6) filters for mobile phone base stations, as well as military ultra-sensitive/selective receivers
Particle and radiation detectors (transition edge, bolometer, tunnel junction, kinetic inductance, and nanowire single-photon)	High sensitivity detectors (single photon, meV energy gap), high counting speed (GHz), and low jitter (10 ps)
Power transmission cables	High power density and decrease of transmission loss
Fault current limiters	Fast switching and large resistance development (quench), compatible with network frequency
Motors	High power density and reduced losses suitable for compact propulsion trains
Generators	High power density and reduced losses suitable for compact generation systems
Magnetic energy storage	Potential for high storage capacity (MJ to GJ) and high delivered power (MW to GW)
Magnetically levitated transportation	Energy efficient, compact, and high acceleration propulsion thanks to high-field magnets
Magnetic resonance imaging (MRI) magnets	Medium (1.5 T) to high fields (11.7 T) in persistent mode, at low operating costs
Nuclear magnetic resonance (NMR) magnets	Ultra-high fields (28 T) in persistent mode, at low operating costs
Magnets for particle accelerators	Compact high-field (8 T) beam magnets as required for beam guidance over long lengths (30 km)

Superconducting Device	Application advantage
Magnets for particle detectors	Large volume, high-field (4 T) magnets as required for trace bending
Radio-frequency cavities	Very high-quality factor ($5\ 10^{10}$) at high electric field (30 MV/m)
Magnets for magnetically confined fusion (tokamaks, stellarators, helical devices, magnetic traps)	Large volume, high-field (13–20 T) magnets as required for plasma containment, heating and control
Magnets for high-field physics	Steady-state ultra-high fields (32 T) at low operating costs

applications, concentrating on large-scale applications in the fields of magnets and power applications, starting from the magnets for accelerators and detectors. The reader will get a taste of the potential and variety of usage of superconductors, but also on the difficulty of converting ideas into technologies and, eventually, successful applications.

Accelerator and detector magnets

Superconducting magnets have found critical applications as technology enablers in large instruments, such as particle accelerators and colliders. The history of good companionship of superconductivity and particle physics is dating back to the early 1960s, when superconducting solenoids were used in particle detectors of various dimensions and sizes, including the BEBC detector magnet at CERN that started data taking in 1973. This impressive realization, a 3.5 T solenoid with a stored energy of 800 MJ, remained a reference till it was surpassed only in the 2000s by the CMS solenoid, producing a comparable field of 4 T, but storing a spectacular energy of 2.7 GJ. Superconductivity is necessary for modern particle detectors to produce the field required to bend the product of particle interactions over the large volume of the detector, still at a reasonable operating cost.

In the case of accelerators, the use of superconducting magnets traces its beginning at the end of the 1970s, with the development of the low-beta quadrupoles for the ISR at CERN, and of the superconducting ring magnets of an "energy doubler" at Fermi National Laboratory. This machine, named Tevatron, came online in 1983 and was the first large-scale use of Nb–Ti for accelerators. It was followed by HERA (1991), RHIC (2000), and eventually the LHC (2008).

The LHC, operating at CERN, is built in its near entirety with superconducting dipoles, quadrupoles, and other multipole magnets, as well as large superconducting particle detector magnets. Superconductors offer in this case the only possibility to build compact magnets, generating high field (the LHC dipoles run at about 8 T) and accelerating particle beams at spectacular energies (the protons in an LHC beam reach approximately 7 TeV). This would not be possible without superconducting magnets, e.g., by using classical electromagnets.

Magnetic Resonance Imaging (MRI)

MRI is the dominating application of superconductivity. In an MRI machine, the superconductor is used to generate the homogeneous solenoid field that acts as a reference for the proton resonance excited and measured by the RF antennas. MRI solenoids have a large bore, of the order of 1 m, as needed to accommodate the patient. The field generated in the majority of MRI systems in the world ranges from 1.5 T (70% of the systems in operation worldwide) to 3 T (20% of the systems in operation worldwide). More recently, 7 T machines have been clinically approved and are coming into operation (less than 100 units worldwide in 2022). The highest field achieved in a full-body MRI is 11.7 T, in the recently commissioned ISEULT magnet. The vast majority of these machines, across the whole field range, are built with Nb–Ti.

Generating the required field (several T), in the large bore (1 m), and high homogeneity (level of ppm or better) is only possible thanks to the high current density of superconductors, and the lack of resistive losses that allows for operation in persistent mode.

Advances in cryogenic engineering have reduced the burden of cooling to liquid helium temperature to a periodic refill which can come at months' intervals. Newer machines will most likely make use of HTS to further ease this matter, making helium-free operation possible, and reducing maintenance to a minimum. The present MRI market has been estimated to generate an 8 BUSD business (2022) and is the true driver of the superconductor industry.

Nuclear Magnetic Resonance (NMR)

NMR is the second industrial large-scale application of superconductivity. Ultra-high fields, of high spatial homogeneity (level of 10^{-9} required), and stable in time (less than 1 ppm per year) are made possible by LTS, and more recently HTS. Fields in the range of 30 T have been achieved in solenoids with relatively small bores (few centimeters). High fields are beneficial to NMR because they increase the resolution of the nuclear resonance spectra, and have resulted in major advances in understanding of the physical, chemical, and biological properties of matter. As a figure of reference, the present NMR market is estimated to be well in excess of 1 BUSD per year (2022).

High magnetic field science

A wealth of physical phenomena and properties are affected by magnetic fields. Scientists use powerful magnets to explore semiconductors, superconductors, crystals, and other organic and inorganic materials to reveal their nature and structure, and develop new technologies. Field strength is instrumental to discovery in this field, and superconductors, both LTS and HTS, have been used to make these fields both possible and affordable. While the record fields for research purposes are still detained by resistive electromagnets, the toll to pay in terms of power consumption is considerable. A magnetic field laboratory has typical consumptions in the range of tens of MW to achieve fields of 30–40 T. Superconducting solenoid magnets built with LTS are commonly used to produce the background field for resistive inserts, thus limiting the power demand of a given facility. It is to be noted that the low field grade of a research solenoid is usually the most massive part of the magnet. More recently all-superconducting solenoids built with a combination of LTS and HTS have reached steady state fields in excess of 30 T and are run as user facility magnets. A worldwide effort is ongoing to extend the range of steady-state all-superconducting solenoids to 40 T and beyond.

Radio frequency cavities

The ultra-low electrical resistivity of a superconducting material allows an RF resonator to obtain an extremely high-quality factor, Q. As an example, it is possible to obtain a quality factor up to $Q = 5\ 10^{10}$ in a 700 MHz niobium resonant cavity at 1.8 K generating an electric field of 20 to 30 MV/m. Such a very high Q resonator stores energy with very low loss and very narrow bandwidth. These properties can be exploited for a variety of applications, including the construction of high-performance accelerating structures. Though a high Q implies no loss of RF power, hence in principle less costly operation, power efficiency is not the main reason for the use of superconducting materials in radiofrequency. The main motivators are in reality the following: the possibility to use superconducting RF in continuous mode (CW operation), which would not be possible using copper resonators; a low beam impedance associated with a large aperture, as the beam passage of a superconducting resonator does not need to be limited in size to concentrate the electric field, which is required on the contrary in copper resonators; excellent coupling of the RF power to the beam, especially important as the generation of RF power is energy intensive and costly, and any power loss in the coupling to the beam reflects negatively on the overall efficiency and cost of the chain.

Thermonuclear fusion magnets

Conventional thermonuclear fusion experiments (e.g., JET, JT-60, ASDEX, NTSX-U or MAST) make use of copper electromagnets. The limited current density and resistive loss set an upper bound to the field that they can achieve in the range of 1–3 T. At the same time, plasma containment, heating and control of future fusion machines call for an increase in the field produced by the toroidal and solenoid magnets. Most important, the fusion power grows dramatically with the field, e.g., with the fourth power of the toroidal field in a tokamak. This is why superconducting magnets have found application in thermonuclear experiments already built (e.g., T-15, Tore-Supra, K-STAR, SST), in the commissioning phase (e.g., JT-60SA), construction (e.g., ITER, SPARC) or planned (e.g., DEMO, ARC). The use of HTS for fusion magnets may yield a quantum step for a new generation of tokamaks.

Power applications

The potential power applications of superconductivity can be broadly separated into two classes. The first class includes all applications where a superconductor replaces a normal conductor such as copper or aluminum: power transmission cables, motors and generators, or transformers. The second class includes applications where superconductivity is the enabling technology, making the device possible in a way that would have little or at most limited capability if conventional materials were used. Examples of this class are superconducting magnetic energy storage systems (SMES) or fault-current limiters (FCLs) and fault-current controllers (FCCs). Although power applications have been considered among the first uses of technical superconductors such as Nb–Ti, it is the discovery of HTS materials that has sparked renewed interest in industrial and commercial applications in this field. The main reason is that HTS can operate at temperatures close to that of liquid nitrogen, with a considerable simplification of the cryogenic system, a reduction of the refrigeration cost by over an order of magnitude and an increase in the acceptance of a new technology, which remains a big hurdle in a field of application where established technologies are already at hand and widely distributed. While many of the applications described in the following have been demonstrated successfully in real operating situations, a significant market penetration is yet to come.

One of the characteristic applications are *power transmission cables* — besides having lower impedance and losses, superconductors can carry more current than conventional conductors of the same cross-section. This is true even when considering the additional thermal shield and vacuum duct that encloses the cold cable. This characteristic allows superconducting cables to carry several times as much power in a given envelope as a conventional cable. For this reason, superconducting cables are of special interest when considering retrofitting existing trenches in areas with high population density (e.g., city centers) requiring an increase of transmitted power, or in locations where bottlenecks need a sizeable increase in power. Besides, large DC currents flow without loss in superconducting cables. This makes low-voltage/large-current DC power transmission based on superconducting cables an interesting alternative to the high-voltage/low-current AC lines optimized to reduce resistive loss in conventional conductors.

A second potential field of applications is *motors and generators.* Superconducting electric machines, motors and generators, have one or more superconducting elements. Their advantage when compared to normal conducting machines is that they have greater efficiency, because of the absence of resistive loss, and their volume can be substantially smaller for the same power. The latter is in fact the parameter of utmost interest, i.e., the power density, which in superconducting machines is due to the ability to generate high magnetic fields that cannot be reached in a conventional machine. Superconducting motors, for example, are expected to reach power densities in the range of 10 kW/kg, an order of magnitude higher than conventional motors. This can be achieved at comparable or higher torque density of tens of Nm/kg. The advantage is that the same propulsion system can fit in smaller volume, or conversely higher power and torque can be achieved using the same volume as a conventional system. One example where this is particularly relevant is wind generators. In this case, the weight of the generator, which is located at the top of the tower, is an important design parameter for the tower and foundations, thus influencing the cost. With the increasing trend toward larger turbines (more than 150 m in diameter), higher towers (above 100 m) and increased power (10 MW), an increase in power density is a definite asset to increase the generated power while decreasing the weight of the power train.

Propulsion

Superconductors are also considered in the field of mobility. The most advanced example is the construction of the Chuo Shinkansen superconducting Maglev Line, a project of Central Japan Railway under construction between Tokyo and Nagoya, with plans for extension to Osaka. The train will employ the superconducting maglev technology. The levitating force is generated between superconducting magnets on the trains, built with Nb–Ti, and coils on the track. The distance of about 500 km between Tokyo and Nagoya will be covered in less than one hour.

The recent focus on aviation sustainability, and in particular, the combination of fuel efficiency and noise reduction, is providing opportunities for an airborne electric power train. HTS machines, with their superior power density and efficiency, are considered as a viable path toward an all-electric aircraft. Indeed, once coupled with the use of liquid hydrogen as a fuel, this would yield an integrated power and cooling system that could produce the required energy transition toward decarbonized transportation. Major players in the airplane industry are considering this possibility, e.g., studying and prototyping systems that aim at medium-sized powertrains, of the order of 500 W, with power densities in excess of 30 kW/kg and an efficiency of more than 97%.

Some Final Considerations

Superconductors are fascinating materials, being one of the few macroscopic and visible examples of quantum theory. They have revolutionized fields of research, medicine and society. Still, after more than a century since their discovery, their mysteries are not yet completely unveiled, and their potential is not exhausted. Besides the present and sustained need for medical and scientific applications that are possible only thanks to superconductors, fields related to the ongoing energy transition could see a widespread application of superconductors in the near future. But, most important, better knowledge and understanding of materials and technology could lead the way into emerging and new applications, as may be the case for the upcoming quantum revolution.

References

Adeva, B. *et al.* (1990). The construction of the L3 experiment. *Nuclear Instruments and Methods in Physics Research Section A* 289, 35–102.

Alpert, D. (1953). *Journal of Applied Physics* 24, 860.

Alpert, D. (1958). *Encyclopaedia of Physics*, 12, 624.

Amsler, C. (1998). Proton-antiproton annihilation and meson spectroscopy with the crystal barrel. *Reviews of Modern Physics* 70, 1293.

Annenkov, A., Auffray, E. and Borisevich, A. *et al.* (1999). Suppression of the radiation damage in lead tungstate scintillation crystal. *Nuclear Instruments and Methods in Physics Research Section A* 426, 486–490.

Ansari, R. *et al.* (1 July 1989). The silicon detectors in the UA2 experiment. *NIMA* 279, (1–2), pp. 388–395 https://doi.org/10.1016/0168-9002(89)91111-X.

Ballarino A., and Bottura, L. (2015). Targets for R&D on Nb3Sn Conductor for High Energy Physics. *IEEE Transactions on Applied Superconductivity* 25(3), 1–6. doi: 10.1109/TASC.2015.2390149.

Ballarino, A. and Flükiger, R. (2017). Status of MgB2 wire and cable applications in Europe. *Journal of Physics: Conference Series* 871, 012098.

Bayard, R.T. and Alpert, D. (1950). *Review of Scientific Instruments* 21, 571.

Bell, P. R. (1948). The use of anthrance as a scintillation counter. *Physical Review* 73, 1405–1406.

Benvenuti, C. (1983). *Nuclear Instruments and Methods* 205, 391.

Benvenuti, C. and Francia, F. (1988). *Journal of Vacuum Science & Technology A* 6, 2528.

Benvenuti, C. and Hauer, M. (1977). *Nuclear Instruments and Methods* 140, 453.

Benvenuti, C. and Hauer, M. (1980). *Le Vide* 201(Supp), 199.

Benvenuti, C., Bojon, J. P., Chiggiato, P., Lösch, G. (1993). *Vacuum* 44, 509.

Benvenuti, C., Cazeneuve, J. M., Chiggiato, P. and Cicoira, F., Escudeiro Santana, A., Johanek, V., Ruzinov, V. and Fraxedas, J. (1999). *Vacuum* 53, 219.

Benvenuti, C., Chiggiato, P., Costa Pinto, P., Escudeiro Santana, A., Hedley, T., Mongelluzzo, A., Ruzinov, V. and Wevers, I. (2001). *Vacuum* 60, 57.

Borchi, E. and Bruzzi, M. (1994). Radiation damage in silicon detectors. *Rivista Del Nuovo Cimento* 17(11).

Broser, V. I. and Kallmann, H. (1947). Uber die Anregung von Leuchtstoffen durch schnelle Korpuskularteilchen I. *Z Naturforschg* 2A, 439–440.

Calder, R. S. and Lewin, G. (1967). *British Journal of Applied Physics* 18, 1459.

Calder, R. S., Fischer, E., Gröbner, O. and Jones, E. (1974). In *Proceedings of the 9th Inernational Conference on High Energy Accelerators,* Stanford.

Calder, R. S., Grillot, A., Le Normand, F. and Mathewson, A. (1977). In *7th International Vacuum Congress,* Vienna.

Casse, G. (2010). Enhanced efficiency of segmented silicon detectors of different thicknesses after proton irradiations up to 10^{16} neq cm^2. *NIMA* 624, 401. https://doi.org/10.1016/j.nima.2010.02.134.

Collins, G. B. and Hoyt, R. C. (1948). Detection of beta-rays by scintillations. *Physical Review* 73, 1259–1260.

Coltman, J. W. and Marshall, F. H. (1947). Some characteristics of the photo-multiplier radiation detector. *Physical Review* 72, 528.

Corenzwit, E. (1959). Superconductivity of Nb3Al. *Journal of Physics and Chemistry of Solids* 9(1), 93.

Da Vià, C., Dalla Betta, G.-F. and Parker, S. (2019). *Radiation Sensors with Three-Dimensional Electrodes* (CRC Press). ISBN: 9780429621598, 0429621590.

Dawson, I. (2021). Radiation effects in the LHC experiments, Impact on detector performance and operation. CERN Yellow Reports: Monographs CERN-2021-001. http://dx.doi.org/10.23731/CYRM-2021-001.

Eisterer, M. (2007). Magnetic Properties and Critical Currents of MgB2. *Superconductor Science & Technology* 20. 10.1088/0953-2048/20/12/R01.

Eremin, V., Strokan, N., Verbitskaya, E., Li, Z. (1996). Development of transient current and charge techniques for the measurement of effective net concentration of ionized charges (N-eff) in the space charge region of p-n junction detectors. *Nuclear Instruments and Methods in Physics Research Section A* 372, 388. https://doi.org/10.1016/0168-9002(95)01295-8.

Eremin, V., Verbitskaya, E., Li, Z. (2002). The origin of double peak electric field distribution irradiated silicon detectors", *Nuclear Instruments and Methods in Physics Research Section A* 476, 556. https://doi.org/10.1016/S0168-9002(01)01642-4.

Fasoli, M. *et al.* (2011). Band-gap engineering for removing shallow traps in rare-earth Lu3Al5O12 garnet scintillators using Ga3+ doping. *Physical Review B* 84, 081102(R). https://the10ps-challenge.org.

Fischer, E. (1972). *Journal of Vacuum Science & Technology A* 9, 1208.

Ferrero, M., Arcidiacono, R., Mandurrino, M., Sola, V., Cartiglia, N. (2021). *An Introduction to Ultra-Fast Silicon Detectors* (Boca Raton: CRC Press).

García, M. F., Sánchez, J. G., Echeverría, R. J., Moll, M., Montero, R., Moya, D., Pinto, R. P., Vila, I. (2017). On the determination of the substrate effective doping concentration of irradiated HV-CMOS sensors using an edge-TCT technique based on the two-photon-absorption process. *Journal of Instrumentation* 12(1), C01038.

Hartmann, F. (2017). Evolution of silicon sensor technology in particle physics. *Modern Physics*, https://doi.org/10.1007/978-3-319-64436-3

Heijne, E. (2008). 1980, a revolution in silicon detectors, from energy spectrometer to radiation imager: Some technical and historical details. *NIMA* 591, 6–13. https://doi.org/10.1016/j.nima.2008.03.015

Heijne, E. (2021). History and future of radiation imaging with single quantum processing pixel detectors. *Radiation Measurements* 140, 106436, https://doi.org/10.1016/j.radmeas.2020.106436.

Helmer, J. and Hayward, W. H. (1966). *Review of Scientific Instruments* 37, 1652.

Hofstadter, R. (1949). The detection of gamma-rays with thallium-activated sodium iodide crystals. *Physical Review* 75, 796–810.

Hulm, J. K. and Blaugher, R. D. (1961). Superconducting solid solution alloys of the transition elements. *Physical Review* 123(5), 1569.

Kallmann, H. (1949). Quantitative measurements with scintillation counters. *Physical Review* 75, 623–626.

Kallmann, H. (1950). Scintillation counting with solutions. *Physical Review* 78, 621–622.

Kallmann, H. and Furst, M. (1950). Fluorescence of solutions bombarded with high energy radiation (energy transport in liquids). *Physical Review* 79, 857–870.

Kallmann, H. and Furst, M. (1951a). Fluorescence of solutions bombarded with high energy radiation (energy transport in liquids). Part II. *Physical Review* 81, 853–864.

Kallmann, H. and Furst, M. (1951b). High energy induced fluorescence in organic liquid solutions (energy transport in liquids). Part III. *Physical Review* 85, 816–825.

Kamihara, Y. *et al.* (2006). Iron-based layered superconductor: LaOFeP. *Journal of the American Chemical Society* 128, 10012.

Kemmer, J. (1980). Improvement of detector fabrication by the planar process. *Nuclear Instruments and Methods A* 169, 499.

Kramberger, G. (2011). Solid state detectors for high radiation environments. *Landolt-Börnstein New Series I/21B2*. https://materials.springer.com/bp/docs/978-3-642-14142-3.

Kramberger, G., Cindro, V., Mandić, I., Mikuž, M., Milovanović, M., Zavrtanik, M., Žagar, K. (2010). Investigation of irradiated silicon detectors by edge-TCT. *IEEE Transactions on Nuclear Science* 57 (4), 2294–2302.

Kubota, Y. *et al.* (1991). The CleoII detector. CLNS 91/1122.

Kunzler, J. E., Buehler, E., Hsu, F. S. L., Wernick, J. H. (1961). Superconductivity in Nb3Sn at high current density in a magnetic field of 88 kgauss. *Physical Review Letters* 6(3), 89. https://doi.org/10.1103/PhysRevLett.6.89

Lecoq, P. (September 2017). Pushing the limits in time-of-flight PET imaging. *IEEE Transactions on Radiation and Plasma Medical Sciences,* 1(6), 473–485.

Lecoq, P. *et al.* (2020). Roadmap towards the 10ps time-of-flight PET challenge. *Physics in Medicine & Biology* 65, 21RM01.

Lecoq, P., Li, P. J. and Rostaing, B. (1991) BGO radiation damage effects: Optical absorption, thermoluminescence and thermoconductivity. *Nuclear Instruments and Methods in Physics Research Section A* 300, 240–258.

Li, Z. (2009). Radiation damage effects in SI materials and detectors and rad-hard Si detectors for SLHC. *JINST* 4, P03011. https://doi.org/10.1088/1748-0221/4/03/P03011.

Lindström, G. *et al.* (2001). Radiation hard silicon detectors — Developments by the RD48 (ROSE) collaboration. *Nuclear Instruments and Methods in Physics Research, Section A* 466(2), 308–326. https://doi.org/10.1016/S0168-9002(01)00560-5.

Maeda, H., Tanaka, Y., Fukutomi, M., Asano, T. (1988). A new high-Tc oxide Superconductor without a rare earth element. *Japan Journal of Applied Physics* 27(2), L361-L364.

Marshall, F. H. and Coltman. J. W. (1947). The photo-multiplier radiation detector. *Physical Review* 72, 528.

Mathes, M. *et al.* (December 2008) Test beam characterization of 3-D silicon pixel detectors. *IEEE Transactions on Nuclear Science* 55(6), 3731–3735.

Matthias, B.T. *et al.* (1954). Superconductivity of Nb3Sn. *Physical Review* 95(6), 1435.

Moll, M. (2020). Acceptor removal — Displacement damage effects involving the shallow acceptor doping of p-type silicon devices. In *Proceedings of Science (VERTEX2019)*, p. 12. https://pos.sissa.it/373/027/.

Moll, M. (August 2018). Displacement damage in silicon detectors for high energy physics. *IEEE Transactions on Nuclear Science* 65(8). https://doi.org/10.1109/TNS.2018.2819506.

Moll, M., Feick, H., Fretwurst, E. and Lindstroem, G. (1998). Correlation between a deep hole trap and the reverse annealing effect in neutron-irradiated silicon detectors. *Nuclear Instruments and Methods in Physics Research A* 409, 194–197.

Moon, R. J. (1948). Inorganic crystals for the detection of high energy particles and quanta. *Physical Review* 73, 1210.

Nagamatsu, J., Nakagawa, N., Muranaka, T., Zenitani, Y. and Akimitsu, J. (1 March 2001). Superconductivity at 39 K in magnesium diboride. *Nature* 410(6824), 63. doi:10.1038/35065039. PMID 11242039.

Novotny, R., Riess, R., Hingmann, R. *et al.* (1987). Detection of hard photons with BaF2 scintillators. *Nuclear Instruments and Methods in Physics Research Section A* 262, 340–346.

Onnes, H. K. (1911). Further experiments with liquid Helium gas on the electrical resistance of pure metals, etc. VI. On the sudden change in the rate at which the resistance or mercury disappears. Communications from the Physics Laboratory of the University of Leiden, 124, p. 21.

Oreglia, M. *et al.* (1982). Study of the reaction psi → gamma gamma J/psi. *Physical Review D* 25, 2295–2277.

Parker, S. I., Kenney, C. J. and Segal, J. (1997). 3D — A proposed new architecture for solid-state radiation detectors. *Nuclear Instruments and Methods in Physics Research, Section A* 395(3), 328–343.

Pellegrini, G. *et al.* (November 2014). Technology developments and first measurements of Low Gain Avalanche Detectors (LGAD) for high energy physics applications. *NIMA* 765, 12–16.

Pintilie, I. *et al.* (2016). Experimental techniques for defect characterization of highly irradiated materials and structures. *Proceedings of Science* 1–12.

Pintilie, I., Fretwurst, E. and Lindström, G. (2008). Cluster related hole traps with enhanced-field-emission — The source for long term annealing in hadron irradiated Si diodes. *Applied Physics Letters* 92, 024101.

Pintilie, I., Lindstroem, G., Junkes, A. and Fretwurst, E. (2009). Radiation-induced point- and cluster-related defects with strong impact on damage properties of silicon detectors. *Nuclear Instruments and Methods in Physics Research A* 611, 52–68.

Ray, R. E. (1994). The KTeV pure CsI calorimeter. In *Proceedings of the 5th International Conference on Calorimetry in High Energy Physics*, World Scientific, New Jersey, pp. 110–114.

RD2 Collaboration (1990–95). (August 1990). *R&D Proposal* (Geneva, Switzerland: CERN), CERN/DRDC/90-27 and DRDC/P3.

RD2 Collaboration (1996). Radiation induced bulk damage in silicon detectors. *NIM A* 383, 144–154.

Reynolds, G. T. (1950). *Nucleonics* 6, 488–489.

Schorr, M. G. and Torney, F. L. (1950). Solid non-crystalline scintillation phosphors. *Physical Review* 80, 474–475.

Sellin, P. J. and Vaitkus, J. (February 2006). New materials for radiation hard semiconductor dectectors. *NIMA* 557, 479. https://doi.org/10.1016/j.nima.2005.10.128.

Senatore, C. (February 2022). *Private Communication* (University of Geneva).

Snell, A. H. (1960). Chairman, subcommittee on instruments and techniques, USA National Academy of Sciences. In *Proceedings Conference Semiconductor Nuclear Particle Detectors*, 28–30 September 1960, Asheville NC, USA, NAS-NRC Pub 871.

Swank, R. K. (1954). *Nucleonics* 12, 4–22.

Sze, S. M. (2002). *Semiconductor Devices, Physics and Technology* (2nd ed.) (Hoboken, NJ, USA: Wiley) see chapter 3 for the Shockley-Read-Hall statistics.

Tachikawa, K. (1971). Studies on superconducting V3Ga tapes. In proceedings of the International Cryogenics Engineering Conference, Berlin, 25–27 May 1970 (Guildford: Iliffe Science and Technology Publications), 339.

Turala, M. (2005). Silicon tracking detectors — Historical overview. *NIMA* 541, 1–14. https://doi.org/10.1016/j.nima.2005.01.032.

Vasilescu, A. and Lindstroem, G. (2000). Displacement Damage in Silicon. https://rd50.web.cern.ch/niel/.

Weber, M. J. and Monchamp, R. R. (1973). *Journal of Applied Physics* 44, 5496.

Wu, M. K. *et al.* (1987). Superconductivity at 93-K in a new mixed-phase Y-Ba-Cu-O compound system at ambient pressure. *Physical Review Letters*, 58(9), 908.

Wunstorf, R., Benkert, M., Claussen, N., Croitoru, N., Fretwurst, E., Lindström, G. and Schulz, T. (1 May 1992). Results on radiation hardness of silicon detectors up to neutron fluences of 1015 n/cm^2. *NIMA* 315(1–3), 149–155. https://doi.org/10.1016/0168-9002(92)90696-2.

Yuan, L. C. L. (1961). Application of solid-state devices for high-energy particle detection, pp. 177–181 in report UCRL-9343. In *1st International Conference on Instrumentation for High-energy Physics*, 12–14 September 1960, Berkeley, CA, USA, New York, NY: Interscience, p. 338.

Chapter 6.3

Will Small Be Beautiful? Making Policies for Our Nanotech Future*

W. Patrick McCray

"It was as if nanotechnology had gone through a phase transition; what had once been perceived as blue sky research ... was now being seen as the key technology of the 21st century."[1]

---

*This chapter is a reproduction of W. Patrick McCray. "Will Small Be Beautiful? Making Policies for Our Nanotech Future." *History and Technology* 21, 2 (2005): 177–203.

[1]W. Patrick McCray is an assistant Professor in the History Department at the University of California at Santa Barbara. He is the author, most recently, of *Giant Telescopes: Astronomical Ambition and the Promise of Technology* (Cambridge, MA: Harvard University Press, 2004). He is currently engaged in a pilot research project that examines the recent history of nanotechnology with especial emphasis on science policy, community formation, public perceptions, and the collection of oral histories. The research for this article was supported by a grant from the California NanoSystems Institute and National Science Foundation Grant SES 0304727 (as part of a Nanoscale Interdisciplinary Research Team project). Several references cited in this article are available in on-line format only. Where this is the case, a URL link is provided and a hard copy of the document is in the author's working files. He wishes to acknowledge the assistance and suggestions of Nicole Archambeau, Larry Badash, Justin Bengry, Evelyn Hu, John Krige, Kristen Shedd, and the anonymous reviewers. "Interview with Mihail C. Roco," http://www.nano.gov/html/interviews/MRoco.htm.

On October 24, 2003, California politicians, academics, and civic leaders attended the groundbreaking ceremony for the California NanoSystems Institute in Santa Barbara. This $55 million high-tech building, with its ultra-filtered clean rooms and modular laboratories, represented only a fraction of the burgeoning national investment in nanotechnology.[2] Since 2000, when President William J. Clinton announced a new national initiative to foster the tools of the "next industrial revolution," dozens of universities and corporations have initiated programs where scientists and engineers research phenomena and technologies at the nanoscale.[3]

Following proclamations by Clinton and other leaders, government and commercial investment into nanotechnology soared. In 2004, federal agencies in the United States dispensed almost a billion dollars a year to eager researchers (Fig. 1 shows the federal funding profile for nanotechnology research).[4] Meanwhile, forecasts, not from intemperate prognosticators but sober-minded science managers, predicted that the international market for nano-goods would be $1 trillion by 2015. The production of these goods, supporters said, would require a new high-tech (and highly paid) workforce of some two million people, potentially leading to a major restructuring of the global workforce.[5]

To its advocates, nanotechnology is a "transcendent realm," a regime where research in the physical and biotechnological sciences may converge with information technologies and the cognitive sciences.[6] While definitions of what nanotechnology is vary — indeed,

[2] A second facility is being built at the University of California at Los Angeles. Almost all publications, regardless of audience, tend to use "nanoscience" and "nanotechnology" interchangeably with a strong preference for the latter and presumably more all-encompassing term. Despite the obvious historiographical conundrum this poses for the historian of technology, I have adopted this style except where I am explicitly referring to the actual scientific research on nanoscale phenomena and techniques.

[3] The White House (January 21, 2000).

[4] Roco (2003), Office of Science and Technology Policy (2004).

[5] Roco (2001).

[6] Roco (2000).

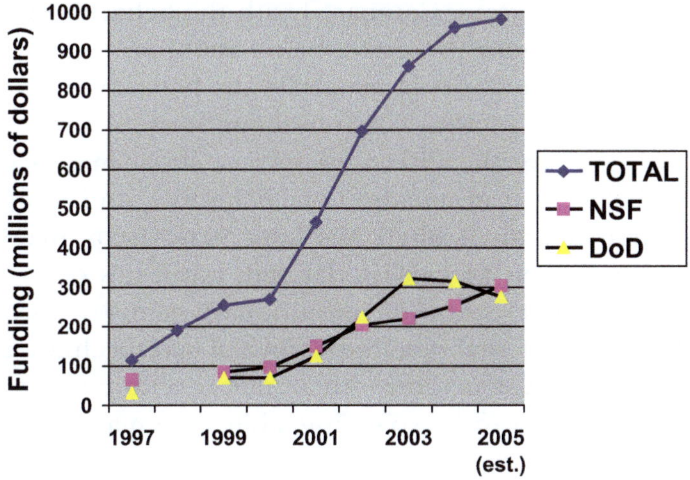

Fig. 1. Funding profile for nanotechnology research in the United States, 1997–2005. Total funding is shown along with monies allocated specifically to the National Science Foundation and the Department of Defense. 2005 figures are projected; breakdown by agency not available for 1998.

Source: Roco (2004).

one of its rhetorical strengths would seem to be its interpretive flexibility — its basic feature is "smallness."[7] According to official government definitions in the United States, nanotechnology involves three key components; research and technology development at the length scale of 1–100 nanometers; creating structures and devices that have novel properties and applications because of their small to intermediate size; and the ability "to control or manipulate on the atomic scale."[8]

Nanotechnology, in other words, is engineering done at the molecular scale and represents the ability to deliberately shape new

[7] A nanometer (the prefix is the Greek word for "dwarf") is one-billionth of a meter and an oft-quoted unit of measurement cited in nanotech literature, the width of a human hair, is some 10,000 nanometers wide.

[8] Definition adapted from www.nano.gov, the official website for the National Nanotechnology Initiative.

structures, atom by atom. Compared with the behavior of individual atoms or molecules (the latter being about one nanometer in size) or the properties of bulk materials, the behavior of structures in the size range of a few to a hundred nanometers is quite different. In this regime, the effects of surface area, interfaces, and quantum behavior can create new behaviors and properties. Because the nanoscale is where the basic structure of materials is established and their properties defined, advocates of nanotechnology predict the creation of revolutionary nanostructures from the "bottom up." If realized, the ability of nanotechnologists to shape materials and structures with atomic precision would be the culmination of humans' expanding ability to manipulate their material world.

Humans, with varying degrees of intent, have been nanotechnologists for millennia. Ancient artisans consciously manipulated metal, glass, and ceramic materials to create certain properties. Consider the Lycurgus Cup, a 4th-century glass vessel in the collection of the British Museum. The object's unusual optical properties are caused by a haphazard dispersion of nanometer-sized particles of a gold–silver alloy in a glass matrix.[9] Nanotechnology's potential, its advocates claim, rests in its promise for *precise control.* In their view, it offers the ability to organize atoms and molecules into usable novel materials, structures, or systems with physical, chemical, and biological properties that are either known or predictable via sophisticated computer modeling.

The marshaling of Congressional support for the National Nanotechnology Initiative (NNI) occurred against the background of post-Cold War (and, later, post-September 11) science and technology policies. Proponents of nanotechnology offered seductive visions of a tomorrow that appealed to citizens, scientists, and policy makers, visions which this chapter examines. As one Nobel laureate said in a government brochure designed to make nanotechnology visible to the public, "Nanotechnology has given us the tools … to play with the ultimate toy box of nature ….

[9] Barber and Freestone (1990).

The possibilities to create new things appear endless."[10] In testimony to Congress and in books written for the public as well as science and policy communities, technological visionaries and science administrators alike predicted that nanotechnology could "change the nature of almost every human-made object in the next century" and, in doing so, not only alter our "comprehension of nature and life ... [but] even influence societal and international relations."[11] These predictions, in turn, helped spur lawmakers to allocate billions of dollars for research and commercial development. Despite arguments from skeptics attempting to re-direct policy and commercial research and development efforts, government leaders eventually adopted and ratified these utopian predictions in national science policy when they approved the National Nanotechnology Initiative in 2000.

The sheer dollar amount, the rapidity of the funding increases, and the accompanying flood of books, websites, and pop culture references to what one author called the "nanocosm" are sufficient to command attention from a wide array of disciplines in the academy.[12] While scholars have made initial efforts to analyze the economic, social, ethical, and legal challenges posed by the predicted "next wave" of technological innovation, historians have paid comparatively little attention to nanotechnology.[13] While this lacuna is understandable as nanotechnology's potential benefits still reside largely in research labs (the major commercial products exploiting nanoscale research thus far are stain-resistant pants and improved cosmetics), the relative dearth of historiography concerning

---

[10] Horst Stormer quoted in Amato (1999), p. 1.

[11] First quote from John Armstrong, formerly chief scientist at IBM, quoted on p. viii of Roco *et al.* (1999); second quote from Roco (2004), p. 6.

[12] Atkinson (2003).

[13] See, for example, the papers collected in Baird *et al.* (2003). The issue of patents is presented in Meyer (2001) while a view of potential risks is articulated in Swiss-Re (2004) Risk perception Series. See also Crow and Sarewitz (2001) and Fogelberg and Glimell (2003). An example of the few articles with a focus on the history of nanotechnology include Johnson (2003) which addresses policy issues and Mody (2003) which discusses instrumentation.

nanotechnology with which one can engage makes the historian's task more difficult. It is too early to assess the economic or societal effects of nanotechnology; it is not premature, however, to examine the roots of and rationale for the United States' massive investment in the nanocosm. This chapter is therefore to be read as an exploratory reconnaissance of the funding and policy terrain of nanotechnology in the last decade and is written with the understanding that future documentation will add much more depth and detail to the story presented here.

What is required to create a successful new science initiative? Previous studies of policy-making for science have identified several variables for victory. These include the role of entrepreneurs who lobby for policies that support their goals as well as perceived national needs. Coordinated activity among agencies, interest groups, and researchers as well as budgetary and institutional contexts are other factors.[14] Howard McCurdy, in his discussion of how Americans imagined space exploration, rightly notes that public policy also has an important subjective dimension whose roots are located in imagination and visions of the future.[15] Why was the NNI successful in obtaining too much federal support and what was its "subjective dimension"?

This chapter addresses the area where, in the late 1990s, utopian imaginings and policy rationales met fiscal realities. National science policy is frequently analyzed through, if not indeed seen as commensurate with, the politics and annual changes of science budgets.[16] This chapter goes beyond the conflation of policies with budgets to explore how and why such monies were allocated. In describing how advocates for the NNI gradually orchestrated its funding, this paper offers a historical perspective on the promotion of a major national initiative to support nanotechnology in the

---

[14] See, for example, Kingdon (1995) as well as papers in Sabatier (1999). See also Korsmo and Sfraga (2003) and Cook-Deegan (1991) for examples of how specific research programs were advocated.

[15] McCurdy (1997).

[16] Sarewitz (2003).

United States. In looking at the rationales offered by lawmakers and science managers, I pay special attention to the depiction of future nano-utopias in support of policy goals.

Historians will recognize the marshaling of techno-utopian visions as a theme common to the history of many technologies. From household appliances to nuclear power and the Internet, the introduction of new technologies has often been accompanied, if not presaged, by wildly optimistic claims and unbridled enthusiasm.[17] More often than not, enthusiasts and supporters of new and "revolutionary" technologies have seen their predictions tempered by fiscal, social, or regulatory realities. The belief that the next industrial revolution will come from nanotechnology existed in conjunction with other more prosaic rationales, some relics from the Cold War, for funding the NNI. Besides considering how advocates promoted the NNI and how its funding reflects broader changes in national science and technology policy, it is important to ask why utopian techno-visions continue to flourish given their predilection to ultimately disappoint.

Before the Flood

Historians of science and technology have long recognized the existence of "creation stories," myth-like narratives for research communities that trace the development of a particular idea or invention back to a singularity — lone inventor or small teams who create a revolutionary breakthrough. For geneticists, there is Watson, Crick, and the double helix; for electrical engineers, Shockley, Brattain and Bardeen's invention of the transistor looms large.

Nanotechnologists' creation story can be traced to a precise point in space and time — the evening of December 29, 1959, when Caltech physicist Richard P. Feynman gave an after-dinner speech in Pasadena to members of the American Physical Society. Feynman, one of the most original physicists of the 20th century, gave his

[17] See essays in Corn (1986) as well as the more recent collection by Sturken and Ball-Rokeach (2004).

address, whimsically entitled "There's Plenty of Room at the Bottom," at a time when the mass production of microelectronics was just beginning and computers still occupied entire rooms.[18]

Feynman described the potential benefits of "manipulating and controlling things on a small scale" which he believed could be accomplished while still obeying the basic laws of physics. Scientists, he predicted, would build microscopic machines that could manipulate matter at a very small scale while engineering done at the "bottom" would provide plenty of attractive research opportunities. Human control of matter at the atomic scale was, according to the Caltech physicist, a "development which I think cannot be avoided" and he backed up his assertion by offering a thousand dollars "to the first guy who can take the information on the page of a book and put it on an area 1/25,000 smaller in linear scale in such manner that it can be read by an electron microscope."[19] In his talk, Feynman himself drew upon utopian visions proffered by contemporary fiction writers which itself suggested, ironically, how the scientific and visionary aspects of nanotechnology were interconnected from its very beginning.[20]

Feynman himself never did any research that could be construed as nanoscience.[21] Indeed, the word "nanotechnology" did not appear until 1974 and then in an obscure proceedings of a Japanese industrial conference.[22] Nevertheless, Feynman's address became a crucial touchstone for future advocates of nanotechnology. In both

---

[18] Feynman (February 1960).

[19] Ibid., p. 26.

[20] Feynman's 1959 talk appropriated ideas by science fiction author Robert Heinlein in his 1942 novella *Waldo* and, as Colin Milburn notes, nanotech's real origin can be argued to rest not with Feynman but in science fiction: Milburn (2002).

[21] A search of the World Wide Web in 2004 for the most popular web pages dealing with Feynman returned as many references to his 1959 paper as they did his Nobel Prize. Also worth noting is the fact that the Foresight Institute, a non-profit organization devoted to promoting nanotechnology, annually awards a prize named after Feynman.

[22] Taniguchi (1974). Taniguchi's term, used in a much different fashion from today, described precision machining of devices with tolerances less than a micrometer.

popular treatments as well as policy documents, quotations from it are rampant while Feynman's authorship serves as a rhetorical anchor to scientific authority. This critical linkage was strengthened by the fact that Feynman later shared the Nobel Prize for his research on theoretical quantum electrodynamics. According to what some have appropriately called the "standard story" of nanotechnology, Feynman presented the crucial initial vision of the innovative nano-research that scientists could do, the achievement of which only required the proper tools.[23]

These tools — namely new and improved electron microscopes — were incrementally developed by scientists and engineers during the 1960s and 1970s. This led to the next milestone in the "standard story" which helped set the stage for the eventual promotion of a national nanotechnology initiative. In 1986, Gerd Binnig and Heinrich Rohrer, researchers at an IBM research laboratory in Zurich, received the Nobel Prize for their design of the scanning tunneling microscope. With a properly-functioning instrument (the achievement of which is still something of a black art), researchers now had the ability to "see" individual atoms as their probe scanned a sample's surface and rendered an image of its topology on a computer screen.[24]

Now able to image atoms, the next milestone in realizing Feynman's now-mythic vision was the ability to move them. In a 1990 paper published in *Nature*, two researchers reported that they had manipulated individual xenon atoms on the surface of a single crystal of nickel cooled to almost absolute zero with "atomic precision."[25] In their demonstration of the ability to "fabricate rudimentary structures of our own design," Eigler and Schweizer honored their corporate sponsor by using 35 xenon atoms to spell "IBM."

---

[23] Baird and Shew (2003).

[24] Mody (2003). As Mody notes, scientists in the United States gradually adopted the STM in a variety of settings including corporate and national research labs as well as university departments. In 1986, the same year that Binnig and Rohrer won the Nobel Prize, Digital Instruments was founded in Santa Barbara, California and became the first company to market their version of an STM, the "Nanoscope."

[25] Eigler and Schweize (1990).

A major article in *The New York Times* followed in which a reporter described how he laboriously moved about a few atoms with Eigler's coaching.[26] Engineering at the nanoscale, as the *Times* described it, was not only potentially groundbreaking but entertaining — as Eigler wrote in his 1989 lab notebook entry, "I am really having fun!"

Stunts like this made international headlines and provided the capstone to Feynman's roadmap — in principle, researchers could now organize matter as well as see at the atomic level. However, doing so was still a challenging enterprise that required a good deal of expensive equipment, tacit knowledge, and specialized skills.[27] While a new community of electron microscopists coalesced around the new tools and techniques in the late 1980s and early 1990s, much of this happened out of sight from the larger scientific community and the general public. Another element, one that Feynman did not account for, was needed: a messenger to boldly proselytize for nanotechnology.

In the 1980s, K. Eric Drexler emerged as the most visible and vocal proponent of nanotechnology and its transforming potential. Born in 1955, Drexler attended MIT and, while working at that school's Space Systems Laboratory, wrote a 1981 paper that outlined his early vision for nanotechnology.[28] He argued that the ability to design protein molecules could lead to the manufacturing of molecular machines which, in turn, could make "second-generation machines" and the "construction of devices and materials to complex atomic specifications."[29] Drexler went on to teach the first university course on nanotechnology, published an equation-laden textbook on the subject, and was a Co-Founder of the Foresight Institute, a non-profit group devoted to promoting the benefits of nanotechnology.

The first full articulation of Drexler's nano-utopia was his 1986 book *Engines of Creation*. Written for the general public, Drexler's book considered the potential benefits and risks arising

[26] Siebert (1996).
[27] Mody (2003).
[28] Drexler (1981).
[29] Ibid., p. 5278.

from the wholesale production of nanomachines.[30] At the heart of his vision was what Drexler called "assemblers," hypothetical autonomous machines, protein-sized or smaller, that could produce virtually any material or object. Such ideas were incorporated by science fiction writers who imagined a future in which a person had access to a "utility fog" from which they could call forth the material necessities of life. While skeptical, one review of Drexler's book described it as a "hopeful forecast, remarkable for an unembarrassed faith in progress through technology."[31]

A full treatment of Drexler's activities and the attention they received is beyond the scope of this chapter. It will suffice to say that a 1991 article in *Science* called Drexler "the apostle of nanotechnology" and noted that he was already a lightning rod for praise as well as vitriolic scorn from fellow scientists.[32] Criticisms came from both science journalists and other researchers, with the detractors claiming that Drexler's vision of self-assembling nanomachines not only drew attention (and possibly funding) from actual lab-based endeavors but also would violate basic laws of conservation.[33] Referring to Drexler's missionary zeal and the fervor he inspired among his supporters, one critic said, "You might as well call it nanoreligion."[34]

However, for the sake of promoting research and development of nanotechnology to the top of the national agenda, Drexler and other nano-advocates performed an important function. In promoting radical futurist scenarios, which his critics placed in the realm of fiction rather than science, Drexler and his colleagues brought nanotechnology to the attention of policy makers. As such, Drexler and company may be considered the vanguard of a wave that eventually brought potent government support for more conventional and staid research programs devoted to nanotechnology. This was not long in coming.

---

[30] Drexler (1986).

[31] Monmaney (1986).

[32] Amato (1991).

[33] Stix (1996).

[34] Amato (1991), pp. 1311.

In June 1992, Senator Albert Gore, recently returned from the United Nations Earth Summit in Rio de Janiero, convened a congressional hearing devoted to "New Technologies for a Sustainable World."[35] Drexler was one of the witnesses invited to appear before Gore's subcommittee. He testified that "molecule-by-molecule control," which nanotechnology suggested, could become "the basis of a manufacturing technology that is cleaner and more efficient than anything we know today." Nanotechnology, Drexler argued, certainly met "the criteria for an environmentally critical technology."[36] Gore, for his part, displayed great enthusiasm for Drexler's ideas and his knowledgeable reference to Feynman's 1959 speech also confirmed that nanotechnology's creation myth still possessed great currency.[37]

Like many predictions of technological utopias, Drexler's congressional testimony says more about contemporary circumstances than it makes accurate statements about the future. Drexler's pronouncements resonated with concerns held by American politicians and the public about the environment, sustainable development, and American economic competitiveness. It depicts a future ameliorated and achieved through the adoption of a "technological fix," one where nanotechnology would raise "the material standard of living worldwide, while decreasing resource consumption and environmental impact."[38]

Drexler's appearance marked the most visible discussion yet on Capitol Hill of nanotechnology as a potentially valuable area for federal investment. Topics that politicians care most about — health, the economy, and national security — all looked to benefit from nanotechnology. While yet a minor blip, nanotechnology was now on the Clinton administration's radar screen.

[35] Science Committee on Commerce, Science, and Transportation (1992).

[36] Ibid., p. 21.

[37] Ibid., p. 30.

[38] Corn (1986) and Drexler, *New Technologies for a Sustainable World*, p. 22.

Planting the Flag

The utopian nano-futures portrayed by Drexler and others garnered a great deal of attention in the early 1990s. And the claims could be quite extravagant. In 1995, a nano-enthusiast claimed that nanotechnology could even produce immortality.[39] Consequently, many scientists and science writers were skeptical if not openly derisive and hostile toward such visionary claims. In his review of the book *Nano! Remaking the World Atom by Atom*, chemist David E. H. Jones, for example, concluded that until scientific questions surrounding the most optimistic claims of nano-enthusiasts were answered, nanotechnology itself would "remain just another exhibit in the freak-show that is the boundless optimism school of technical forecasting."[40] Perhaps more seriously, other critics noted that advocates who promised great benefits from nanotechnology and drew on Feynman's reputation for their scientific authority were squandering the political and social capital accumulated by real scientists who were doing actual laboratory work.[41]

Scientists were clearly expending a great deal of effort to establish distinct boundaries between what they saw as legitimate research and nano-visions that seemed to have more in common with science fiction than laboratory realities.[42] In order to gain access to the resources — i.e., funding from the federal government and industry — they believed it necessary to pursue a full-scale research program in nanotechnology, a compelling vision for nanotechnology needed to be articulated by people not on the fringe of the research establishment.

Mihail C. Roco was a key person who took the stage in the mid-1990s determined to strip nanotechnology of its exaggerated claims

---

[39] Du Charme (1995).

[40] Jones (1995) which was a review of Regis (1995).

[41] Stix (1996).

[42] The permeable boundary between science and science fiction is considered in Milburn (2002). Moreover, the efforts of scientists to delineate between "real" science and the long-term visions put forth by people by Drexler stands as a classic example of boundary work; see, for example, Gieryn (1999).

and to promote an ambitious, realistic nano-research program as a national priority. Born in 1947 in Bucharest, Roco was a mechanical engineer who came to the United States in 1980, where he carried out research at a number of institutions before joining the National Science Foundation (NSF) in 1990. "Captivated by the unity and coherence encountered in nature," Roco wanted to likewise promote the "coherence of science and technology."[43] Soon after joining the NSF, he helped create a new, modestly-funded program that investigated the synthesis of nano-sized particles.[44] Gradually, Roco began to lay the foundation for a much more ambitious inter-agency program that would support a whole range of research in nanotechnology.[45]

Roco launched his entrepreneurial plans at a promising time. Following the end of the Cold War, there was much debate among scientists, policy makers, and government agencies about how to restructure US science policy. There was general agreement that that US science policy needed revision yet achieving consensus on how best to do this was difficult. National defense against a monolithic Communist threat no longer served as sufficient justification for funding basic science. Just as significantly, the end of the Cold War signaled to policy makers that the next arena of conflict would be in the global marketplace in the form of increased economic competition from European and Asian countries.[46] This was an especially salient fact to US lawmakers who read reports of how Japanese and European companies were taking the basic research done in American laboratories to the

[43] Roco (2004), p. 10.

[44] This was funded at about $3 million in 1991, a modest amount by NSF standards.

[45] A note on the funding process: The President's budget is typically submitted in January each year. In the US, new actions require a two-step congressional process — authorization and appropriation, two interdependent yet distinct activities. The actual allocation of money requires the prior passage of an authorization statute and the authorization process is done by a pair of committees, one in the House and one in the Senate.

[46] This was a point made in a number of studies including a report by the Council on Competitiveness called (1994).

global marketplace and profiting handsomely in the process. In 1991, Japan's Ministry of International Trade and Industry pledged some $225 million for nanotechnology, with research aimed at benefiting that country's electronics industries.[47] News like this fed existing fears among corporate and government leaders that the US was falling behind in a potentially key economic area and spurred more calls for a revamped national science policy in the wake of the Cold War.

Articulating an appropriate policy to deal with this competitive threat from abroad became a priority for lawmakers and policy analysts. More than half a dozen major reports appeared that suggested how to re-structure US science policy in response to the new global environment.[48] Solutions included improving America's publicly-funded research infrastructure and investing public monies in technologies whose benefits were only on the horizon. While not named as such, advocates soon touted nanotechnology as one of these undersupported fields with great potential. Many of these suggested changes were embraced in the first official review of US science policy in the post-Cold War era. *Science in the National Interest,* released in August 1994, described basic research and technological development not as the keys to national security but as elements that "underpin the nation's economy." Gore stated, in words reminiscent of Vannevar Bush's half-century-old vision, "Technology is the engine of economic growth; science fuels technology's engine."[49] US science policy, the report said, should enhance connections between fundamental research and national goals and one of these goals was, of course, long-term economic growth.

Compared with the structured plans put forth by the Japanese, the US effort in nanotechnology was underfunded and disorganized. For example, since 1978, the NSF had funded the National

[47] Crawford (1991).

[48] Boesman (1997).

[49] Office of the Press Secretary (August 3, 1994).

Nanofabrication Facility at Cornell University.[50] The $40 million facility served as a place where researchers from around the country could come to do research and use equipment their home institutions lacked. The National Institute of Standards and Technology also began an initiative in nano-research. Given these disparate initiatives, which *in toto* were spread over twelve different agencies, Roco and others claimed that the US lagged behind other countries when it came to a major national investment in nanotechnology.[51]

As a first step in coordinating the people, infrastructure, and funding for nanotechnology research in the US, Roco, along with Richard Siegel (Rensselaer Polytechnic Institute) and Evelyn Hu (University of California at Santa Barbara), organized a workshop in Arlington, Virginia. This 1997 meeting was held under the auspices of the World Technology Evaluation Center (WTEC), formerly known as the Japanese Technology Evaluation Center. Charged with examining foreign research in selected technology areas and operated out of Loyola College in Baltimore in cooperation with the NSF, WTEC provided information for policy makers concerned about American industrial competitiveness.

The WTEC meeting marked the beginning of an effort to ascertain research trends in nanotechnology worldwide.[52] Roco claimed that nanotechnology was at "a level of development comparable to that of computer/information technology in the 1950s" yet was poised — with sufficient funding — to become a key emerging technology for the 21st century.[53] At the meeting, attendees described

---

[50] This was originally called the National Research for Submicron Structures; the name was changed in 1988. Six years later, the NSF funded the National Nanofabrication Users Network which expanded the Cornell model to four other schools:

[51] Roco (2004), p. 2.

[52] A report from the May 1997 workshop was later published; Siegel *et al.* (1998). According one person involved in the process, some scientists were uncomfortable with the fact that Roco appeared to have already decided that the US presence in nanotechnology was not sufficient. Evelyn Hu, January 4, 2005 personal communication with the author.

[53] Ibid., p. 1.

the current level of US spending on nanotechnology — in fiscal year 1997, this was $115 million with more than half of the money coming from the NSF — and emphasized the need for interagency collaboration. The report Roco and his colleagues prepared also noted that governments in Japan and Europe were already funding comprehensive programs.[54] However, compared to the far-reaching and visionary claims by Drexler and other nano-utopians, the soberly written WTEC report served to clarify the current state of nano-research rather than promote nano-utopian visions.

After the May 1997 meeting, Roco helped form an Interagency Working Group on Nanoscience, Engineering, and Technology (IWGN) which he chaired. The IWGN answered to the cabinet-level National Science and Technology Council established by President Clinton in 1993. Throughout 1998, the IWGN met and worked out a vision for what ultimately became the National Nanotechnology Initiative. Like all major government initiatives, putting the pieces in place took a great deal of time and coalition building.

While Roco and his colleagues were formulating their proposal for the NNI, the US government reviewed its investment in scientific research yet again. In February 1997, soon after the second Clinton administration began, House Speaker Newton L. Gingrich asked the House Committee on Science to develop a long-range science and technology policy. The committee's report, released in September 1998, reaffirmed that, with the end of the Cold War, the principal contributions of science and technology would not be for national defense but economic competitiveness.[55] As Gingrich framed it, "the competitions we are engaged in now are less military and largely economic."[56]

The 1998 report also articulated aspects of America's new "mega-era" of science policy which advocates of the National

[54] A follow-up report addressed the specific question of where the US stood in relation to Japan and Europe; Siegel *et al.* (1999).

[55] House of Representatives Committee on Science, 105th Congress (1998).

[56] February 12, 1997 letter from Gingrich to House Committee on Science Chairman F. James Sensenbrenner, Jr.

Nanotechnology Initiative soon embraced.[57] Grants to individual scientists and engineers were the "primary channel by which the government stimulates knowledge-driven basic research," while the research done at universities and national labs must be continually "applied to the development of new products or processes," so as to further strengthen linkages between university labs and commercial ventures formed after the passage of the Bayh–Dole Act in 1980.[58] Just as significantly for the success of the NNI, the report acknowledged that private industry was unlikely to fund areas like nanotechnology that had no foreseeable practical application; this role would therefore devolve to the federal government.

As members of Congress reexamined national science policy, several prominent people joined the campaign for a national nanotechnology effort. At this point, as the push for the NNI's funding intensified, the first noticeable strains of nano-utopian visions crept into the rhetoric of scientists and administrators. In April 1998, Neal Lane, President Clinton's science advisor, appeared before the House Basic Research Subcommittee. Lane, whose specialty was theoretical atomic and molecular physics, made a statement that was to be widely quoted: "If I were asked for an area of science and engineering that will most likely produce the breakthroughs of tomorrow, I would point to nanoscale science and engineering."[59] Noting how NSF support had enabled nanotechnology to transform "from the realm of science fiction to science fact," Lane went on to describe ideas that drew as much from the former as the latter: nano-sized medical probes, new classes of materials, and — straight from Drexler's writings — "nanomanufactured objects that could

---

[57] According to the 1998 report, the pre-World War Two period was the first "mega-era" of science policy while the US support of science and technology for national defense during the Cold War typified the second "mega-era."

[58] The Bayh–Dole Act encouraged the utilization of inventions produced under federal funding and promoted the participation of universities in the commercialization of ideas derived from basic research.

[59] Lane's April 22, 1998 testimony can be found on the web site for the House Committee on Science: http://www.house.gov/science/.

change their properties automatically or repair themselves." As Lane put it, the "possibilities of nanotechnology are endless."

On March 11, 1999, Roco formally presented plans for the National Nanotechnology Initiative to Neal Lane and Thomas Kalil, Clinton's Deputy Assistant for Technology and Economic Policy. During their 2-hour meeting, Roco described an initiative for fiscal year 2001 costing some $500 million. At the end of the meeting, nanotechnology won out over competing initiatives and now had two well-placed advocates in the White House.

According to Roco, the next task was to educate both the public as well as Congress as to what nanotechnology was and what it could offer. A glossy brochure entitled *Nanotechnology: Shaping the World Atom by Atom* made the case to the public and drew upon nano-utopian rhetoric that was rapidly becoming *de rigueur* — the vision of Feynman's 1959 talk, the progress made in manipulating atoms via electron microscopy, the potential to fabricate nanoscale electronic devices, the now-famous IBM logo writ in xenon atoms, and the vast commercial and societal benefits that investment in the NNI could open up.[60]

Congress still required persuading, however. On June 22, 1999, the House of Representatives' Committee on Science met to review the federal government's support of nanotechnology and to discuss its potential economic implications. Chairman Nick Smith opened the meeting by noting that the United States did not dominate nanotechnology. "How much effort," he asked, "should the Federal Government be putting into tax-payer-funded research in this area?"[61]

Several witnesses answered Chairman Smith; all drew in some capacity upon the utopian visions of nanotechnology's potential that had been gestating for over a decade. Of those who spoke that afternoon, chemist Richard E. Smalley of Rice University was perhaps the most persuasive. Smalley was a Co-Discoverer of carbon-60

---

[60] Amato (1999). What is unknown is how effective this campaign was or how many people it reached.

[61] House of Representatives Committee on Science (1999), pp. 1–2.

molecules, better known as "buckyballs," for which he had shared the 1996 Nobel Prize. Buckyballs were closely related to carbon nanotubes that Japanese scientists had discovered in 1991, and advocates of nanotechnology referred to these new structures as clear examples of how matter could be deliberately controlled at the atomic level.

"There is a growing sense in the scientific and technical community," Smalley told Congress, "that we are about to enter a golden new era."[62] Citing his own experience battling cancer, Smalley connected research at the nanoscale with an array of societal benefits that could result, including medical breakthroughs. The Nobel laureate shrewdly pointed out that research in nanotechnology was "intrinsically small science." Because research in nanoscience was not centralized in one or two national labs, funding the NNI would combat the trend toward "Big Science," offering a way to fund individual researchers. Funding "small science" (in both senses) would, Smalley argued, help attract students into science careers just as *Sputnik's* launch in 1957 had inspired his generation.

The NNI, Smalley avowed, would also support broader national policy goals. "There's this immediacy between the ivory tower pure scientist and the technologist" in nano-research, Smalley said, and the NNI would simultaneously strengthen both while creating valuable linkages between basic research and commercial applications.[63] Nanotechnology, Smalley concluded, presented a "tremendously promising new future." What was needed was someone bold enough to "put a flag in the ground and say: 'Nanotechnology, this is where we are going to go …'"[64]

In urging Congress to fund a major national initiative for nanotechnology, advocates of the NNI deployed an array of effective arguments, some new and some already familiar: the specter of economic competition with Europe and Japan; the opportunity to

---

[62] Ibid., p. 7.
[63] Ibid., p. 21.
[64] Ibid., p. 12.

recruit a new generation of scientists and engineers; and support for traditional individual, peer-reviewed research rather than giving a handout to a few giant national labs. Persuasive utopian images and frontier metaphors were added to the rhetorical brew. One witness that afternoon, for example, compared nanotechnology with Kennedy's declaration to put a man on the moon.[65] What the US needed to do, advocates insisted, was to be bold once again and plant a flag on a new frontier.

The Floodgates Open

Richard Smalley and the others who testified in June 1999 were largely pushing against an open door as they made their case for a major national investment in nanotechnology. Reflecting new ideas in circulation about national science policy, questions from congressional representatives were more praising of nanotechnology than they were probing. Such enthusiasm is understandable as the House committee that heard Smalley and others was convened to gather information, not appropriate taxpayers' money. Congressional questioning would become more intense once real dollars were at stake.

Roco and other NNI advocates focused their efforts next on federal agencies such as the NSF, NASA, and the Department of Defense which would likely administer funding for a major new initiative. By November 1999, the Office of Management and Budget and the White House were studying plans for the NNI. That same month, the Presidential Council of Advisors in Science and Technology (PCAST) issued its formal review of the proposed NNI.[66] More circumspect in its espousal of the now-familiar utopian predictions for nanotechnology, PCAST nevertheless identified nanotechnology as a potential path to a "new industrial revolution," one that might lead to unforeseen advances in "materials and

[65] Ibid., p. 10.

[66] Pres. George H. W. Bush established PCAST in 1990. It is a non-governmental advisory group formed without Congressional approval. Therefore, each President must renew the group's existence. President Clinton established his PCAST in November 1993 by executive order.

devices of all sorts … medicine and health, biotechnology and agriculture, as well as national security."[67] In an increasingly-familiar comparison, nanotechnology was likened to the state of solid-state physics research in the 1950s but, unlike the lead America enjoyed in transistors and microchips when their commercialization began, the United States lacked "an early lead in nanotechnology." PCAST argued that the US could not afford to be in second place: "The country that leads in discovery and implementation of nanotechnology will have great advantage in the economic and military scene for many decades to come." The NNI, therefore, represented an opportunity to swiftly create an infrastructure for nano-research and perhaps avoid the precarious state the US found itself in during the 1980s when fears of Japanese technological supremacy ran rampant. A subsequent letter from the PCAST to President Clinton offered firm and clear support of the NNI.[68]

With these endorsements in hand, President Clinton traveled to Pasadena, California where he addressed a standing-room-only crowd at Caltech's Beckman Auditorium. On January 21, 2000, Clinton articulated the need to strengthen America's investment in science and technology. After paying homage to Caltech and Feynman (as well as an image suspended above him of the Western Hemisphere constructed with individual gold atoms), Clinton announced the inclusion of a $2.8 billion "21st Century Research Fund" in his 2001 budget, which was an element in a larger strategy to double the NSF's budget over the next five years. A "top priority" was a major increase in the NSF's funding for nanotechnology — over 124% or $227 million in the next year alone — which, in conjunction with funding other agencies, would nearly double the federal investment. Clinton rhapsodized

> "Just imagine, materials with 10 times the strength of steel and only a fraction of the weight; shrinking all the information at the Library of Congress into a device the size of a sugar cube; detecting

[67] This and subsequent quotes from the PCAST review; PCAST's review available at: http://www.ostp.gov/PCAST/pcastnano2.html.

[68] December 14, 1999: http://www.ostp.gov/PCAST/pcastnano.html.

cancerous tumors that are only a few cells in size. Some of these research goals will take 20 or more years to achieve. But that is why — precisely why ... there is such a critical role for the federal government."[69]

A major Presidential address now incorporated aspects of the utopian views suggested by Drexler and adopted by science managers. Returning to Washington, DC on Air Force One, Neal Lane and NSF Director Rita Colwell reiterated the promise of nanotechnology for reporters. According to Lane, "nanotechnology may not be something that everybody talks about every day, but they will, because it's really the whole next generation in manufacturing.[70] Not surprisingly, Colwell and Lane praised the speech and predicted wide bipartisan support of the President's plan. In fact, just a few months before Clinton's speech, an editorial in *The Washington Post* by former House Speaker Gingrich claimed nanotechnology would "have as big an impact on our lives as transistors and chips did in the past 40 years" and called for a substantial increase in federal support for science.[71]

Despite the utopian gloss advocates could put on the importance of nanotechnology to national needs, the annual appropriations process grounded the American nanotechnology initiative to more prosaic political and budgetary struggles. Initial congressional response to the NNI was positive. At a Senate roundtable discussion in April 2000, Senator Evan Bayh, whose father, 20 years older, cosponsored the Bayh–Dole Act that helped re-shape the relationship between universities and the federal government, called nanotechnology "extremely important to future rates of innovation" and the overall economy in the US.[72] He and several of his colleagues heard presentations from Richard Smalley as well as Donald Eigler of IBM. Improvising on his earlier testimony, Smalley noted how basic sciences like physics and chemistry were at the core of

[69]William J. Clinton; January 21, 2000 speech. Text available at: http://pr.caltech.edu/events/presidential_speech/pspeechtxt.html.
[70]Lane quote from Jones (2000).
[71]Jones (1999).
[72]Leath (2000b).

nanotechnology. A vote for the NNI was, in effect, a vote for basic science research and education.[73]

As spring arrived in Washington, it became clear that the science advocates would not secure funding for the NNI without a struggle. Concerns over the NNI centered around several issues. Some Republican lawmakers warned that Roco's multi-agency plan would be hard to manage and result in a duplication of research. The NNI lacked, one congressman complained, "a clear strategy to ensure coordination within the government."[74] Meanwhile, a non-partisan report from the Congressional Research Service noted that, as "nanotechnology was still in its infancy," large-scale practical applications were at least a decade away.[75] Finally, those lobbying Congress to double the entire NSF budget opposed any funding for the NNI that came from research programs already established.

In May 2000, the Senate subcommittee which controlled appropriations for both NASA and the NSF met. Senator Christopher Bond, the panel's chair, expressed doubts not about the NNI's value *per se* but about how the NSF would manage the rapid infusion of funding.[76] The Missouri Republican told NSF officials to "count me skeptical" about multi-agency initiatives like the NNI and asked the agency to clarify how it would allocate the requested increase in funds for nano-research. Bond's minority counterpart, Senator Barbara Mikulski of Maryland, expressed her own worries about allocating monies to the NNI without any kind of long-term plan. As she told Lane and Colwell, nanotechnology seems "like our little secret ... we need visibility."

A few weeks before the Senate hearing, nanotechnology did indeed receive greater visibility, although not the kind its advocates wanted. In April 2000, the technology magazine *Wired* published an article by Bill Joy called "Why the Future Doesn't

---

[73] Leath (2000a).
[74] Southwick (2000).
[75] Davey (2000).
[76] Leath (2000c).

Need Us."[77] Joy, one of the Co-Founders of Sun Microsystems, could not be dismissed as a neo-Luddite. In warning about future technological dystopias, Joy singled out nanotechnology — especially the Drexlerian visions of autonomous and self-replicating nano-assemblers — as a potential threat to humanity. Joy's article became especially infamous for highlighting the "grey goo" problem, the idea that uncontrolled self-replication by nanomachines might destroy entire ecosystems and threaten what it meant to be human. In contrast to the utopian views of Drexler or the more attainable goals Roco and other NNI supporters proposed, Joy, writing as a "skeptical nanoist," sounded the tocsin for a decidedly dystopian vision of a nano-future and urged a moratorium on further research.[78]

Nano-advocates recognized the potential danger opponents like Joy posed. Just as utopian visions can generate public support and shape favorable policy, dystopian ones can derail them. At the May 2000 Senate hearing, Senator Bond remarked on the "hysteria and fear" already surrounding genetically-modified foods and it was easy for nano-advocates to imagine a similar public reaction to nanotechnology.[79] There is no direct evidence that Joy's concerns influenced the congressional debate of the NNI in the summer of 2000.[80] Furthermore, Joy did not represent the dominant view of industry. According to one participant in the establishment of the NNI, a "critical factor" in the selling of the NNI in the summer of 2000 was

---

[77] Joy (2000).

[78] With apologies to Robert Boyle and his 1661 book *The Sceptical Chymist.*

[79] For example, the popular science fiction show *Star Trek: Voyager* featured a main character who could assimilate other life forms by injecting them with "nanoprobes."

[80] Joy's article did encourage policy makers to stress the need to better understand (i.e. fund studies of) the environmental and societal risks of the nano-frontier. The article was also seen as a sufficient threat so as to warrant a response from John S. Brown, a researcher at Xerox PARC at a September 2000 conference the NSF convened to address the ethical and societal issues of nanotechnology; see Brown (2001). Glimell (2003) discusses Joy's article and Roco's reaction — What Glimell calls "doing the 'Roco'-motion — to it."

the vocal support industry representatives, especially those in the semiconductor business, gave on the Hill.[81]

Over the summer of 2000, these lobbying efforts paid off as a swell of bipartisan support for the NNI swept away concerns, fiscal, dystopian, or otherwise. Maryland's Senator Mikulski, renowned in science circles as a staunch advocate of NASA, jumped on the utopian bandwagon. In a speech at NASA headquarters in June 2000, Mikulski told her audience that "we are on the verge of a new revolution — the nanotechnology revolution" and urged her colleagues to support the NNI, one of the "least noted and most important documents of the Clinton administration."[82] Across the senate aisle, Republican Senator Trent Lott praised nanotechnology in a letter to his colleagues on the Senate Appropriations Committee. Lott hailed the NNI not so much for its utopian potential but because the national research infrastructure it would help create might be based in regions like Lott's home state of Mississippi, which traditionally received few federal research dollars.[83]

At the end of the summer, Roco's committee formally unveiled its proposal for the NNI, now honed by months of fine-tuning. *The National Nanotechnology Initiative: The Initiative and its Implementation Plan* requested \$495 million — \$150 million of which the NSF would dispense — to expand research at the nanoscale.[84] At the heart of the interagency plan was an increase of some \$90 million for basic research to fund individual scientists and small teams. This was a major increase, up about 50% from the previous fiscal year to \$177 million. While defined as "research leading to new

---

[81] Evelyn Hu, January 4, 2005 personal communication with the author. More work clearly needs to be done to understand the complexities of the support industry provided to the NNI.

[82] Quote from p. 272 of Roco and Bainbridge (2001). Mikulski's support of science, and NASA in particular, is detailed in Munson (1993) which provides an insightful look at the appropriations process.

[83] Roco and Bainbridge (2001), p. 272.

[84] National Science and Technology Council (2000), this was submitted by the Subcommittee on Nanoscale Science, Engineering, and Technology (NSET), chaired by Roco, which replaced the earlier IWGN.

fundamental understanding and discoveries," the proposal clearly tied this research to the "capacity to create new affordable products with dramatically improved performance."[85]

The report also identified several "Grand Challenges," broad areas of research targeted for interagency funding.[86] These included creating new nanostructured materials, nanoelectronics, and bio-nanosensors and linked basic science and engineering to ambitious, long-term economic goals. Congress directed the remaining money to construct a national infrastructure with "centers of excellence" where nano-researcher would work with the stated goal of producing new discoveries to be "rapidly commercialized by industry."[87]

After last-minute political wrangling, Congress approved budgets for the NNI's various agencies in the fall of 2000. Some $465 million of federal money was directed to fund the NNI, representing a 72% increase in the national investment in nanotechnology compared to the previous year.[88] Several states followed suit. California, with 12% of the US population and the world's fifth-largest economy, has frequently been on the leading edge of new technological developments. In late 2000, the then Governor, Gray Davis approved a $100 million state initiative to foster nanotechnology in California: "It's my hope," Davis explained, "to replicate Silicon Valley."[89]

The overall funding profile for the national investment in nanotechnology is shown in Fig. 1. Three important facts stand out. In the span of little more than half a decade,

---

[85] National Science and Technology Council (2000), p. 42.

[86] The "Grand Challenges" areas were to receive a total of some $133 million, $62 million more than the previous year. Some of this was to administered by the NSF but the Department of Defense was slated to be the single largest recipient of "Grand Challenge" funds.

[87] Some $28 million (about 6% of the NNI funding) was set also aside for workforce training and studying societal and ethical implications. *National Nanotechnology Initiative,* 30.

[88] Of this, the NSF received $150 million; the Department of Defense $125 million.

[89] Markoff (2000).

nanotechnology advocates like Roco and two different administrations had increased national funding for nanoscale science and engineering by some 850%. Second, the NSF and the Department of Defense control the largest amounts of NNI funding and the relative amounts of money going to civilian versus defense-related research in nanotechnology reflect changing national priorities after the September 11 attacks. Finally, national enthusiasm for nanotechnology, at least in terms of budget allocations, only increased after George W. Bush was elected President in 2000. The new Republican administration generously supported nanotechnology, acting in part on the belief that the nascent field would contribute significantly to national security.[90] On December 3, 2003, Bush signed the "21st Century Nanotechnology Research and Development Act" which authorized $3.7 billion more for nano-research over the next four years.[91] If one accepts this premise that budgets reflect policy, any doubts about nanotechnology's place as the first "boom science" of the 21st century or immediate concerns about environmental or societal consequences were swept away by the flood of money that began pouring into universities and national labs in 2001.[92]

Nano-Boom or Gloom?

It is too soon to judge the effects the NNI may have on society, industry, or the scientific community. Indeed, one might even hesitate to call nanotechnology a full-fledged "technology" as so very few of the

---

[90] The effects of the terrorist attacks of September 11, 2001 on funding for nanotechnology is complex and cannot be more than noted here. An August 3, 2003 letter from John H. Marburger, Pres. Bush's science advisor noted that nanotechnology was "essential to achieving the President's top three priorities: winning the war on terrorism, securing the homeland, and strengthening the economy;" included in National Science and Technology Council. (2003).

[91] $982 million was requested for fiscal year 2005 alone.

[92] Sarewitz (2003) notes the relation, for better or worse, between budgets and policy in the US.

devices advocates envisioned have materialized. Furthermore, nanotechnology has yet to infiltrate public awareness fully, as a recent poll indicated.[93] However, we can try to understand the reasons and rationales that prompted lawmakers to direct vast sums of taxpayer dollars to support a realm of technoscience that retains strong links to fiction as much as facts.

The NNI emerged at a salient point in US history. In the late 1980s and throughout the 1990s, economic competitiveness replaced the twilight military struggle of the Cold War, and science advocates could no longer claim national defense as the prime rationale for funding basic research. Lawmakers were still struggling to adapt national policies to the new global environment and to the implications of the increasing commercialization of science. Advocates of the NNI proposed it at a propitious time — between the end of the Cold War and a renewed preoccupation with national security after September 11, 2001 — when lawmakers were trying to reshape national science policy.

One recent analysis correctly situates the NNI in the period of "post-academic" science.[94] This regime is characterized by an emphasis on the utility of science and the enlistment of academic research as a "wealth-creating technoscientific motor for the whole economy," views clearly expressed in the documents and testimony supporting the NNI.[95] While the end of the Cold War is certainly relevant, the changing nature of research funded by the federal government and conducted at universities is even more significant. Since the passage of the Bayh–Dole Act in 1980, the "triple helix" of relations between the academy, industry, and government has been significantly altered and strengthened.[96] The borders between science and technology, as the NNI implementation plan shows, have

---

[93] Fewer than 30% of people polled had heard of "nanotechnology" and less than 20% could offer a reasonable definition of it. The Royal Society (2004), p. 11.

[94] Johnson (2003), while the term post-academic science is itself from Ziman (2000).

[95] Ziman (2000), p. 73.

[96] The literature on this phenomenon is vast; a classic paper on the topic is Etzkowitz (1989).

blurred while the commercialization of academic science has become a key driver for its funding.

The favorable timing of the NNI only partly accounts for its success. It cannot explain fully the justifications deployed in support of a major new initiative that would benefit science, technology, and industry. As we have seen, nano-advocates presented an array of heterogeneous arguments to Congress and the public. Despite the new period of "post-academic science," many of these justifications were framed in familiar ways. They included the appeal to national competitiveness, introduced in the 1980s in the face of Japan's economic success, and the need for America to "plant a flag," as did the Apollo astronauts, on the cutting edge of a new technology. While American scientists claimed two-thirds of the 1996 Nobel Prize for their discovery of buckyballs, such honors did not ensure that the US would reap commercial benefits from them. That Japan and Europe had or were planning major government programs for nanotechnology allowed advocates of the NNI to argue that it was imperative that the US do likewise.

Congressional witnesses predicted that nanotechnology would draw American youth into science, a rationalization for science funding that dates to the *Sputnik* era at least. Chemist Richard Smalley reached back even further in time in his testimony, claiming that nano-research as "small science" was a way for the individual scientist to do more personal research commensurate with the days of sealing wax and string. And, after September 11, national security, the major rationale for so much Cold War-era research and development, emerged once again as a major justification for funding nanotechnology.[97]

The NNI also offered lawmakers and science managers an opportunity to revitalize areas of research they saw as underfunded during the 1990s. Clinton's announcement in 2000 of major federal support for nanotechnology, for example, was part of a larger

[97] For example, for fiscal year 2003 the Bush Administration added another "Grand Challenge" to the NNI which would focus on homeland security via detection and protection against chemical, biological, explosive, and radiological (CBER) threats.

strategy to significantly increase the NSF's budget and an opportunity for physics, chemistry, and engineering to regain lost ground due to stagnant budgets during the 1990s. The NNI, one editorial noted, "may prove to be one of the most brilliant coups in the marketing of basic research" since Nixon announced the 'War on Cancer.'[98] However, while the public could understand the necessity of research on cancer, the NNI relied on strategic and utopian marketing to enable its success.

Indeed, one can question whether the NNI is really even *science* policy per se. After leaving the NSF, Neal Lane, one of the advocates for the NNI inside the White House, admitted that the NNI was actually much more a *technology* policy and those who advocated the applications (as opposed to the fundamental knowledge) enabled by nano-research had won the day.[99] The NNI funded the basic research that advocates like Smalley called for, but much of this was directed toward specific commercial applications and national goals.[100]

Finally, one must acknowledge the energy that entrepreneurial science managers like Roco invested in crafting a policy and plan for the NNI. As Roco explained, the NNI was "prepared with the same rigor as a science project between 1997 and 2000: we developed a long-term vision for research and development."[101] These efforts succeeded in winning over lawmakers initially skeptical of a complex multi-agency effort as well as the NSF's ability to manage a major infusion of funds.

One can compare the success the NNI had with other major research initiatives. Opponents of the ill-fated Superconducting Supercollider, for example, criticized it as a mega-project that would benefit a small and elite research community as well as a

---

[98] Editors (2001).

[99] Baird and Shew (2003), p. 10.

[100] The focus on applications is clear in Congressional debates over science budgets. As Jones (2004) notes, discussion of the fiscal 2005 budget referring to nanotechnology specifically asks the NSF to increase research support for practical applications such as improving semiconductor devices.

[101] Roco (2004), p. 3.

single state (Texas).[102] In comparison, the NNI's funding was not earmarked for a single project, discipline, or region. Many parts of the country would presumably benefit from Congressional support for the NNI and the research monies would benefit a large number of researchers in both the life and physical sciences. In this sense, the organization and funding of the NNI was a decentralized form of federally-supported Big Science similar to the Human Genome Project. Advocates for both claimed the programs would foster pluralism and decentralization in science as well as local investments and research initiatives.[103] The establishment of national policies for nanotechnology also resembles the rush to fund research in high-temperature superconductivity in the late 1980s. Descriptions of levitating trains, medical breakthroughs, and robust economic competitiveness all spurred federal and commercial investment in fundamental research conducted with an eye toward future applications from the outset.[104]

Overshadowing all of these rationales are two meta-reasons, both familiar and suspect to historians of technology, which also help account for the success of the NNI. While technological determinism may be out of fashion with scholars, nano-advocates regularly made arguments employing various shades of it. A typical line of reasoning invoked Moore's Law (which states that the computing power on a microchip doubles roughly every 18 months) and described the emergence of economic crises once engineers reached the physical limits of what was possible with microtechnologies.[105] Indeed, industry advocates used the seeming end of the technological improvements described by Moore's Law to push for the NNI.[106] This depiction of technological inevitability arose from nothing more than what experts claimed was the "natural" development of computing technology. Advocates presented nanotechnology as one

---

[102] Riordan (2000).

[103] Heilbron and Kevles offer a comparison between the SSC and the Human Genome Project in (1998); also Cook-Deegan (1991).

[104] Nowotny and Felt (1997).

[105] Roco *et al.* (1999), x.

[106] Evelyn Hu, January 4, 2005 personal communication with the author.

way to avert these limitations.[107] Indeed, according to some enthusiasts, revolutionary developments at the nanoscale were themselves inevitable, yet another phase in human mastery over nature.[108]

Even more powerful as a motivating force were predictions of the utopian benefits nanotechnology could bring to American society and the economy. Visions of tomorrow that Roco, Smalley, and others successfully presented — which themselves tapped into Richard Feynman's mythic vision of nanotechnology's potential and the futuristic, sometimes outlandish, ideas put forth by people like Drexler — attracted and excited lawmakers. With the end of the Cold War and lawmakers' continued interest in economic competitiveness, nanotechnology offered to reshape, literally, how our world was constructed. As David Nye has pointed out, American narratives of technological utopias come in a variety of categories: *natural* where technologies are a natural outgrowth of society; *ameliorative* in which new machines improve everyday life; and *transformative* in which technologies reshape social reality.[109] Nanotechnology appears as a hybrid example as advocates drew upon elements from all three narratives in their lobbying efforts.

While the most extreme scenarios of a future made better through nanotechnology originated with people like Drexler, who operated at the vanguard (and sometimes the fringe) of nanotechnology, they were embraced and repackaged by mainstream science managers and renowned researchers. Mihail Roco and his colleagues described nanotechnology with an exuberance that matched the prognostications of Drexler and his allies.[110] Roco promoted the coming convergence of nanotechnology with biotechnology, information technology, and cognitive science ("Nano-Bio-Info-Cogno" in policy jargon) which could produce a "golden age ... an epochal turning point in human history."

---

[107] Roco (2004), p. 6.

[108] For example, an article by Drexler's wife and partner at the Foresight Institute draws a direct line from the Greek concept of the atom to the future benefits of nanotechnology, suggesting the inevitability of this path; Peterson (1995).

[109] Nye (2004), p. 171.

[110] Roco (2002), Roco and Bainbridge (2002).

As expressed, "If the Cognitive Scientists can think it, the Nano people can build it, the Bio people can implement it, and the IT people can monitor and control it."[111] Fully espousing an understanding of history predicated on an unshakable belief in technological progress, the funding of nanotechnology put humanity "at the threshold of a New Renaissance of science and technology."[112] Whether this convergence is indeed real or whether it represents another phase in the marketing of nanotechnology is yet to be seen.[113]

Visions of a future enabled and ennobled by nanotechnology share many common themes with scenarios presented by advocates of nuclear power, space exploration, and genetics technology. For example, supporters of nuclear power claimed that it would leave practically no part of society unimproved. Funded by the federal government and viewed with fear and fascination by the public, advocates presented the power of the nucleus as the solution to a host of economic, social, and national security problems. By eliminating hunger, unemployment, environmental degradation, and the drudgery of housework, nuclear power promised "dazzling possibilities of life, liberty and the pursuit of happiness."[114] Outer space, meanwhile, offered Americans new frontiers to explore as well as endless research opportunities for scientists.[115] Decades later, advocates of genetic engineering assured the possibility of disease prevention and therapy along with an improved understanding of what it meant to be human. Supporters of nanotechnology, deliberately or not, conflated these disparate utopian aspects. Nanotechnology could "offer the opportunity to understand life processes ... cure and prevent disease, heal injured bodies, and protect society" all

---

[111] Quotes from Roco and Bainbridge (2002), p. 285 and p. 289.

[112] Roco and Bainbridge (2002), p. 281.

[113] According to Evelyn Hu, Roco's emphasis on nano's convergence with others areas of research may be part of a "deliberate repackaging" of the NNI to ensure that interest and funding remain strong. Evelyn Hu, January 4, 2005 personal communication with the author.

[114] Del Sesto (1986), p. 61.

[115] McCurdy (1997).

while providing more "endless frontiers" for exploitation and exploration by scientists and entrepreneurs.[116]

Dystopian warnings, put forth initially by Joy and others, did not abate after the NNI was funded. If anything, the surge of new money directed toward nano-research and development fueled anxieties.[117] Researchers spoke out publicly on the possible toxicity of newly designed nanoparticles and the insurance industry, still engaged in asbestos-related lawsuits, began to estimate litigation costs stemming from nanotechnologies.[118] Politicians were also aware of a potential backlash against the initiative they had approved. At an April 2003 hearing on nanotechnology's societal effects, Vicki L. Colvin, Director of the Center for Biological and Environmental Nanotechnology at Rice University, warned Congress that nanotechnology needed to avoid the "wow-to-yuck" arc that genetically modified organisms had taken.[119] At the other end of the spectrum, Eric Drexler, arguably the most well-known and visionary advocate of nanotechnology's transformative potential, criticized the NNI for its timidity and business-as-usual funding strategy. Leaders of the NNI, Drexler charged, had neglected the original "Feynman vision" of nanomachines because "public concern regarding its dangers might interfere with research funding."[120] Despite the ample and growing funding directed toward nanotechnology, the debate over the trajectory its research and development would take remained robust and unsettled.

In the 1920s, visionaries like Henry Dreyfuss and Norman Bel Geddes believed that utopia could literally be designed and society thereby perfected.[121] Advocates of nanotechnology expressed similar views, namely that the future could be crafted atom by atom.

---

[116] Committee for the Review of the National Nanotechnology Initiative (2003).

[117] The 2003 publication of Michael Crichton's bestselling potboiler *Prey*, the plot of which centers around hostile swarms of nanomachines, undoubtedly tarnished the limited public perception of nanotechnology as well.

[118] Monastersky (2004), Swiss-Re (2004).

[119] House of Representatives Committee on Science (2003), p. 49.

[120] Drexler (2004).

[121] Seagal (1985).

Whether future generations will look at predictions about nanotechnology's ability to reshape the world with the same bemusement that the phrase "energy too cheap to meter" draws remains to be seen. Nanotechnology remains as much a futuristic vision as an active area of real-life research and development. With the inevitable disappointments that almost always accompany predictions of technological utopias, one may wonder why they remain so attractive to scientists, lawmakers, and the public. As the history of nanotechnology continues to unfold, perhaps historians will address this question again and arrive at a more nuanced and specific answer. Until that time, a simpler explanation remains — visions of technological utopias, despite their dubious claims to success and the darker visions that accompany them, still sell and compel.

References

Amato, I. (1991). The Apostle of Nanotechnology. *Science* 254(6036), 1310–1311.

Amato, I. (1999). *Nanotechnology: Shaping the World Atom by Atom* (Washington, DC: National Science and Technology Council).

Atkinson, W. I. (2003). *Nanocosm: Nanotechnology and the Big Changes Coming from the Inconceivably Small* (New York City: Amacom Publishing).

Baird, D., *et al.*, (eds.). (2003). *Discovering the Nanoscale: A Reader of Workshop Manuscripts; from an International Conference at Darmstadt Technical University, October 9–12, 2003.* (Darmstadt: Darmstadt Technical University).

Baird, D. and Shew, A. (2003). Probing the history of scanning tunneling microscopy. In Davis Baird, *et al.*, (eds.), *Discovering the Nanoscale: A Reader of Workshop Manuscripts; from an International Conference at Darmstadt Technical University, October 9–12, 2003.* (Darmstadt: Darmstadt Technical University), pp. 3–19.

Barber, D. J. and Freestone, I. C. (1990). An investigation of the origin of the color of the Lycurgus Cup by analytical transmission electron microscopy. *Archaeometry* 32(1), 33–45.

Boesman, W. C. (coordinator). (1997). *Analysis of Ten Selected Science and Technology Policy Studies.* (Washington, DC: Congressional Research Service).

Brown, J. S. (2001). Don't count society out: A response to Bill Joy. In M. C. Roco and W. S. Bainbridge (eds.), *Societal Implications of Nanoscience and Nanotechnology.* (Boston: Kluwer Academic Publishers).

Committee for the Review of the National Nanotechnology Initiative. (2003). *Small Wonders, Endless Frontiers: A Review of the National Nanotechnology Initiative.* (Washington, DC: Committee for the Review of the National Nanotechnology Initiative, Division on Engineering and Physical Sciences, National Research Council).

Cook-Deegan, R. M. (1991). The human genome project: The formation of federal policies in the United States, 1986–1990. In K. Hanna (ed.), *Biomedical Politics* (Washington, DC: National Academy Press), pp. 99–167.

Corn, J., (ed.). (1986). *Imagining Tomorrow: History, Technology, and the American Future* (Cambridge: MIT Press).

Corn, J. J. (1986). Epilogue. In J. J. Corn (ed.), *Imagining Tomorrow: History, Technology, and the American Future* (Cambridge: The MIT Press), pp. 219–229.

Council on Competitiveness. (1994). *Science in the National Interest* (Washington, DC: National Science and Technology Council).

Crawford, R. (1991). Japan starts a big push toward the small scale. *Science* 254(5036), 1304–1305.

Crow, M. M. and Sarewitz, D. (2001). Nanotechnology and societal transformation. In A. H. Teich, *et al.* (eds.), *AAAS Science and Technology Policy Yearbook* (Washington, DC: AAAS), pp. 89–101.

Davey, M. E. (2000). *Manipulating Molecules: The National Nanotechnology Initiative.* (CRS Report # RS20589).

Del Sesto, S. L. (1986). Wasn't the future of nuclear engineering wonderful? In J. J. Corn (ed.), *Imagining Tomorrow: History, Technology, and the American Future* (Cambridge: The MIT Press), pp. 58–76, p. 61.

Drexler, K. E. (1981). Molecular engineering: An approach to the development of general capabilities for molecular manipulation. *Proceedings of the National Academy of Science* 78(9), 5275–5278.

Drexler, K. E. (1986). *Engines of Creation: The Coming Era of Nanotechnology* (New York: Anchor Books).

Drexler, K. E. (2004). Nanotechnology: From Feynman to funding. *Bulletin of Science, Technology, and Society* 24(1), 21–27.

Du Charme, W. M. (1995). *Becoming Immortal: Nanotechnology, You, and the Demise of Death* (Evergreen, CO: Blue Creek Ventures).

Editors. (2001). Megabucks for nanotech. *Scientific American* 285(3), 8.

Eigler, D. M. and Schweizer, E. K. (1990). Positioning single atoms with a scanning tunneling microscope. *Nature* 344(6266), 524–526.

Etzkowitz, H. (1989). Entrepreneurial science in the academy: A case of transformation of norms. *Social Problems* 36(1), 14–29.

Feynman, R. P. (February 1960). There's plenty of room at the bottom. *Engineering and Science*, 22–26.

Fogelberg, H. and Glimell, H. (2003). *Bringing Visibility to the Invisible: Towards a Social Understanding of Nanotechnology.* (Goteborgs: Goteborgs Universitet). http://www.sts.gu.se/publications/STS_report_6.pdf.

Gieryn, T. (1999). *Cultural Boundaries of Science: Credibility on the Line* (Chicago: The University of Chicago Press).

Gingrich, February 12, 1997 letter to House Committee on Science Chairman F. James Sensenbrenner, Jr. http://www.house.gov/science/science_policy_study.htm.

Glimell, H. (2003). Grand visions and lilliput politics: Staging the exploration of the 'endless frontier'." In D. Baird, *et al.* (eds.), *Discovering the Nanoscale: A Reader of Workshop Manuscripts; from an International Conference at Darmstadt Technical University, October 9–12, 2003.* (Darmstadt: Darmstadt Technical University), pp. 147–164.

Heilbron, J. L. and Kevles, D. (1988). Finding a policy for mapping and sequencing the human genome: Lessons from the history of particle physics. *Minerva* 26(3), 299–314.

House of Representatives Committee on Science (1998). *Unlocking our Future: Toward a New National Science Policy* (Washington, DC: US Government Printing Office). http://www.house.gov/science/science_policy_report.htm.

House of Representatives Committee on Science (1999). *Nanotechnology: The State of Nano-Science and its Prospects for the Next Decade* (Washington, DC: US Government Printing Office).

House of Representatives Committee on Science (2003). *The Societal Implications of Nanotechnology: Hearing before the Committee on Science* (Washington, DC: US Government Printing Office).

Johnson, A. (2003). The end of pure science? Science policy from Bayh-Dole to the national nanotechnology initiative. In D. Baird, *et al.* (eds.), *Discovering the Nanoscale: A Reader of Workshop Manuscripts; from an International Conference at Darmstadt Technical University, October 9–12, 2003* (Darmstadt: Darmstadt Technical University), pp. 20–33.

Jones, D. E. H. (1995). Technical boundless optimism. *Nature* 374(6525), 835–837.

Jones, R. M. (1999). Former house speaker Gingrich on doubling research funding. *FYI: The AIP Bulletin of Science Policy News*, #154. http://www.aip.org/fyi/1999/fyi99.154.htm.

Jones, R. M. (2000). Aboard air force one: Press briefing by Neal Lane and Rita Colwell. *FYI: The AIP Bulletin of Science Policy News*, #10. http://www.aip.org/fyi/2000/fyi00.010.htm.

Jones, R. M. (2004). House appropriations bill recommends cut in FY 2005 NSF funding. *FYI: The AIP Bulletin of Science Policy News* #99. http://www.aip.org/fyi/2004/099.html

Joy, W. N. (2000). Why the future doesn't need Us. *Wired* 8(4), 238–262.

Kingdon, J. W. (1995). *Agendas, Alternatives, and Public Policies* (New York: Addison, Wesley, Longman).

Korsmo, F. L. and Sfraga, M. P. (2003). From interwar to cold war: Selling field science in the United States, 1920s through 1950s. *Earth Sciences History* 22(1), 55–78.

Leath, A. (2000a). Senate science and technology caucus holds briefing on nanotechnology. *FYI: The AIP Bulletin of Science Policy News*, #44. http://www.aip.org/fyi/2000/fyi00.044.htm.

Leath, A. (2000b). Administration and congress see promise in nanotechnology. *FYI: The AIP Bulletin of Science Policy News* # 46. http://www.aip.org/fyi/2000/fyi00.046.htm.

Leath, A. (2000c). NSF hearing: Good news, bad news. *FYI: The AIP Bulletin of Science Policy News* #51. http://www.aip.org/fyi/2000/fyi00.051.htm.

Markoff, J. (2000). California sets up 3 centers for basic scientific research. *The New York Times*, December 20, A30.

McCurdy, H. E. (1997). *Space and the American Imagination* (Washington, DC: Smithsonian Institution Press).

Meyer, M. (2001). Patent citation analysis in a novel field of technology: An exploration of nano-science and nano-technology. *Scientometrics* 51(1), 163–183.

Milburn, C. (2002). Nanotechnology in an age of posthuman engineering: Science fiction as science. *Configurations* 10(2), 261–295.

Mody, C. (2003). From the topografiner to the STM and the AFM: What the history of probe microscopy might tell us about nanoscience. In D. Baird, *et al.* (eds.), *Discovering the Nanoscale: A Reader of Workshop Manuscripts; from an International Conference at Darmstadt Technical University, October 9–12, 2003* (Darmstadt: Darmstadt Technical University), 59–77.

Monastersky, R. (2004). The dark side of small: As nanotechnology takes off, researchers scramble to assess its risks. *The Chronicle of Higher Education*, September 10, A12.

Monmaney, T. (1986). Nanomachines to our rescue. *The New York Times*, August 10, BR 8.

Munson, R. (1993). *The Cardinals of Capitol Hill: The Men and Women who Control Government Spending* (New York: Grove Press).

National Science and Technology Council (2000). *National Nanotechnology Initiative: The Initiative and Its Implementation Plan* (Washington DC: Subcommittee on Nanoscale Science, Engineering and Technology).

National Science and Technology Council (2003). *National Nanotechnology Initiative: Research and Development Supporting the Next Industrial Revolution, Supplement to the President's FY 2004 Budget* (Washington, DC: Subcommittee on Nanoscale Science, Engineering, and Technology).

Nowotny, H. and Felt, U. (1997). *After the Breakthrough: The Emergence of High Temperature Superconductivity as a Research Field* (Cambridge: Cambridge University Press).

Nye, D. E. (2004). Technological prediction: A promethean problem. In M. Sturken, D. Thomas and S. Ball-Rokeach (eds.), *Technological Visions: The Hopes and Fears that Shape New Technologies* (Philadelphia: Temple University Press), pp. 159–176.

Office of Science and Technology Policy (2004). *National Nanotechnology Initiative: Research and Development Funding in the President's 2005 Budget* (Washington, DC: Office of Science and Technology Policy). http://www.ostp.gov/html/budget/2005/FY05NNI1-pager.pdf.

Office of the Press Secretary (August 3, 1994). *White House Releases National Science Policy Report* (Washington, DC: The White House).

Peterson, C. L. (1995). Nanotechnology: Evolution of the concept. In M. Krummenacker and J. Lewis (eds.), *Prospects in Nanotechnology: Toward Molecular Manufacturing* (New York: Wiley), pp. 173–186.

Regis, E. (1995). *Nano! Remaking the World Atom by Atom* (Boston: Little, Brown).

Riordan, M. (2000). The demise of the superconducting super collider. *Physics in Perspective* 2(4), 411–425.

Roco, M. C. (2000). Towards a US national nanotechnology initiative. *Journal of Nanoparticle Research* 1(5–6), 435–438.

Roco, M. C. (2001). International strategy for nanotechnology research and development. *Journal of Nanoparticle Research* 3(5–6), 353–360.

Roco, M. C. (2002). Coherence and divergence of megatrends in science and engineering. *Journal of Nanoparticle Research* 4, 9–19

Roco, M. C. (2003). Broader societal issues of nanotechnology. *Journal of Nanoparticle Research* 5(3–4) 181–189.

Roco, M. C. (2004). The US national nanotechnology initiative after 3 years (2001–2003). *Journal of Nanoparticle Research* 6, 1–10.

Roco, M. C. and Bainbridge, W. S. (2002). Converging technologies for improving human performance: Integrating from the nanoscale. *Journal of Nanoparticle Research* 4, 281–295.

Roco, M. C., Williams, S. and Alivisatos, P. (eds.). (1999). *Nanotechnology Research Directions: IWGN Workshop Report* (Baltimore, MD: World Technology Evaluation Center).

Roco, M. C. and Bainbridge, W. S. (eds.). (2001). *Societal Implications of Nanoscience and Nanotechnology* (Boston: Kluwer Academic Publishers).

Sabatier, P. A. (ed.). (1999). *Theories of the Policy Process* (Boulder, CO: Westview Press).

Sarewitz, D. (2003). *Does Science Policy Exist, and If So, Does it Matter: Some Observations on the U.S. R&D Budget* (Discussion Paper for Earth Science Institute; Science, Technology, and Global Development Seminar).

Science Committee on Commerce, Science, and Transportation (Subcommittee on Science, Technology, and Space), United States Senate (1992). *New Technologies for a Sustainable World* (Washington, DC: Subcommittee on Science, Technology, and Space of the Committee on Commerce, Science, and Transportation, United States Senate, 102nd Congress, Second Session).

Seagal, H. P. (1985). *Technological Utopianism in American Culture* (Chicago: The University of Chicago Press).

Siebert, C. (1996). The next frontier: Invisible. *The New York Times Magazine*, September 29, pp. 137–139.

Siegel, R. W., Hu, E. and Roco, M. C. (1999). *Nanostructure Science and Technology: A Worldwide Study.* (Baltimore, MD: World Technology Evaluation Center).

Siegel, R. W., Hu, E. and Roco, M. C. (eds.). (1998). *WTEC Workshop Report on R&D Status and Trends in Nanoparticles, Nanostructured Materials, and Nanodevices in the United States* (Baltimore, MD: International Technology Research Institute).

Southwick, R. (2000). Nanotechnology, the study of minute matter, becomes a big priority in the budget. *The Chronicle of Higher Education*, A38.

Stix, G. (1996). Waiting for breakthroughs. *Scientific American* 274(4), 94–99.

Sturken, M., Thomas, D. and Ball-Rokeach, S. J. (eds.). (2004). *Technological Visions: The Hopes and Fears that Shape New Technologies* (Philadelphia: Temple University Press).

Swiss-Re (2004). *Nanotechnology: Small Matter, Many Unknowns* (Swiss Reinsurance Company).

Taniguchi, N. (1974). On the basic concept of 'nano-technology'. In *Proceedings of the International Conference of Production Engineering*, Vol. 2. (Tokyo: Japan Society of Precision Engineering.

The Royal Society (2004). *Nanoscience and Nanotechnologies: Opportunities and Uncertainties* (London: The Royal Society).

White House: Office of the Press Secretary (January 21, 2000). National Nanotechnology initiative: Leading to the next industrial revolution. This document is available at: www.whitehouse.gov/news/releases/2003/12/20031203-7.html.

Ziman, J. (2000). *Real Science: What It is and What it Means* (New York: Cambridge University Press).

Index

G

Q

R

S

T

www.ingramcontent.com/pod-product-compliance
Ingram Content Group UK Ltd.
Pitfield, Milton Keynes, MK11 3LW, UK
UKHW020356150926
8624IPUK00004B/183

9 789811 284335